RAND McNALLY

# THE ROAD ATLAS

LARGE SCALE

Photo Credits: ©Laurie Borman/Rand McNally ii (all); ©Brett Gover/Rand McNally iii (all); ©Hilary Solan/Rand McNally iv (all); ©Erin Vorhies/Rand McNally 137 (all); ©Nathalie Strassheim/Rand McNally 138 (tl, tr, mr, br); Courtesy of the Bretton Woods Canopy Tour 138 (bl); ©Getty Images 7



Library of Congress Catalog Number: 92-060588

For licensing information and copyright permissions, contact us at licensing@randmcnally.com

If you have a comment, suggestion, or even a compliment, please visit us at go.randmcnally.com/contact or e-mail us at consumeraffairs@randmcnally.com.

or write to:
Rand McNally Consumer Affairs
P.O. Box 7600
Chicago, Illinois 60680-9915

Published in U.S.A.
Printed in China

1 2 3 LE 11 10



## CONTENTS


### Travel Information

*New!* **Best of the Road**® .......... 2-6
Our editors have mapped out five terrific road trips. Each trip features the best attractions, shops, and places to eat on the road.

*Updated!* **Numbers to Know** .......... 7
Toll-free phone numbers and websites for hotel chains and cell phone emergency numbers.

**Mileage Chart** .......... inside front cover
Driving distances between 77 North American cities.

**Mileage and Driving Times Map** .......... inside back cover
Distances and driving times between hundreds of North American cities and national parks.

### Maps

**Map Legend** .......... 7
**United States Overview Map** .......... 8-9
**States and Cities** .......... 10-235

### Index .......... 236-264

## Quick Map References

**Alabama**
Northern .......... 10-11
Southern .......... 12-13
**Alaska** .......... 14-15
**Arizona**
Northern .......... 16-17
Southern .......... 18-19
Cities .......... 20-21
**Arkansas**
Western .......... 22-23
Eastern .......... 24-25
**California**
Northern .......... 26-27
San Francisco .......... 28
East-Central .......... 29
Northern Cities .......... 30-31
Southern Cities .......... 32-33
West-Central .......... 34-35
Southern .......... 36-37
Los Angeles .......... 38-41
**Colorado**
Western .......... 42-43
Eastern .......... 44-45
Cities .......... 46-47
**Connecticut** .......... 48-49
**Delaware** .......... 50-51
**Florida**
Cities .......... 51
Northern .......... 52-53
Southern .......... 54-55
Panhandle .......... 54
Cities .......... 56-57
**Georgia**
Northern .......... 58-59
Southern .......... 60-61
Atlanta .......... 62
**Hawaii** .......... 63
Honolulu .......... 62
**Idaho** .......... 64-65
**Illinois**
Northern .......... 66-67
Southern .......... 68-69
Chicago .......... 70-71
Cities .......... 72-73
**Indiana**
Cities .......... 72-73
Northern .......... 74-75
Southern .......... 76-77
**Iowa**
Western .......... 78-79
Eastern .......... 80-81
**Kansas**
Western .......... 82-83
Eastern .......... 84-85

**Kentucky**
Western .......... 86-87
Eastern .......... 88-89
**Louisiana** .......... 90-91
**Maine** .......... 92-93
**Maryland**
Western .......... 94
Baltimore .......... 95
Eastern .......... 96-97
**Massachusetts**
Western .......... 98
Boston .......... 99
Eastern .......... 100-101
**Michigan**
Northern .......... 102-103
Southern .......... 104-105
Cities .......... 106
Detroit .......... 107
**Minnesota**
Northern .......... 108-109
Southern .......... 110-111
Minneapolis/St. Paul .......... 112-113
**Mississippi** .......... 114-115
**Missouri**
Western .......... 116-117
Eastern .......... 118-119
St. Louis .......... 120
Cities .......... 121
**Montana**
Western .......... 122-123
Eastern .......... 124-125
**Nebraska**
Western .......... 126-127
Eastern .......... 128-129
**Nevada** .......... 130-131
Cities .......... 133
**New Hampshire** .......... 132-133
**New Jersey**
Northern .......... 134-135
Southern .......... 136-137
**New Mexico** .......... 138-139
**New York**
Southern .......... 140-141
Northwestern .......... 142-143
Northeastern .......... 144-145
New York City .......... 146-149
**North Carolina**
Western .......... 150-151
Eastern .......... 152-153
Cities .......... 154-155
**North Dakota** .......... 156-157

**Ohio**
Northwestern .......... 158-159
Northeastern .......... 160-161
Southwestern .......... 162-163
Southeastern .......... 164-165
**Oklahoma**
Western .......... 166-167
Eastern .......... 168-169
**Oregon**
Western .......... 170-171
Eastern .......... 172-173
**Pennsylvania**
Northwestern .......... 174-175
Southwestern .......... 176-177
Northeastern .......... 178-179
Southeastern .......... 180-181
Philadelphia .......... 182
Pittsburgh .......... 183
**Rhode Island** .......... 184-185
**South Carolina** .......... 186-187
**South Dakota** .......... 188-189
**Tennessee**
Western .......... 190-191
Eastern .......... 192-193
Cities .......... 195
**Texas**
Houston .......... 194-195
Cities .......... 195
Dallas/Fort Worth .......... 196-197
Southeastern .......... 198-199
Southwestern .......... 200-201
Northeastern .......... 202-203
Southeastern .......... 204-205
**Utah**
Western .......... 206-207
Eastern .......... 208-209
**Vermont** .......... 210-211
**Virginia**
Western .......... 212-213
Eastern .......... 214-215
Cities .......... 216-217
**Washington**
Western .......... 218-219
Eastern .......... 220-221
Cities .......... 222-223
**Washington, DC** .......... 224-225
**West Virginia** .......... 226-227
**Wisconsin**
Northern .......... 228-229
Southern .......... 230-231
Cities .......... 232-233
**Wyoming** .......... 234-235

**National Park Maps**
Acadia National Park .......... 93
Arches National Park .......... 208
Black Hills Region
(Wind Cave National Park) .......... 188
Bryce Canyon National Park .......... 208
Canyonlands National Park .......... 208
Capitol Reef National Park .......... 208
Colonial National
Historical Park .......... 216
Crater Lake National Park .......... 172
Denali National Park
& Preserve .......... 15
Gettysburg National
Military Park .......... 177
Grand Canyon National Park .......... 20
Great Smoky Mountains
National Park .......... 155
Isle Royale National Park .......... 102
Joshua Tree National Park .......... 34
Mammoth Cave National Park .......... 89
Mesa Verde National Park .......... 42
Mount Rainier National Park .......... 223
Petrified Forest National Park .......... 21
Rocky Mountain National Park .......... 46
Waterton-Glacier
International Peace Park .......... 123
Yellowstone and
Grand Teton National Parks .......... 234
Yosemite National Park .......... 27
Zion National Park .......... 207

# Best of the Road®

**10TH ANNIVERSARY**

***Each year*** **our editors drive five new road trips to share with you those special things we call *Best of the Road*®.**

EDITOR'S PICKS

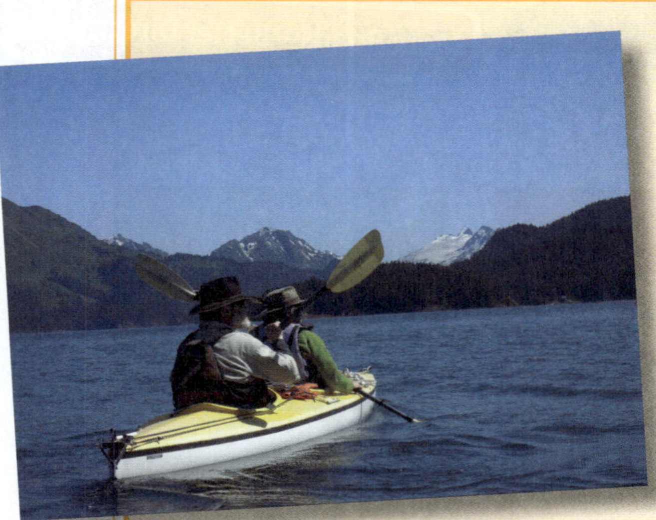

**A Seaside Adventure (Homer)**

On this kayak ecotour, gather sea vegetables that taste surprisingly like garden veggies to toss in a seafood soup on the beach. Lunch is followed by a tour around uninhabited islands in Kachemak Bay. Paddlers may spy sea otters, eagles, a variety of gulls, oystercatchers, and other wildlife on this day-long adventure. *Little Tutka Bay, (907) 235-6672, www.seasideadventure.com*

**Mill Bay Coffee and Pastries (Kodiak)**

Sip rich, fresh, air-roasted coffee and munch on gourmet pastries at this Kodiak restaurant. French chef/owner Joël Chenet prepares elegant desserts from scratch. Daily lunch specials include locally caught fish. *3833 Rezanof Dr. East, (907) 486-4411, www.millbaycoffee.com*

## WILD, WARM ALASKA

### Anchorage to Valdez, Alaska

Magnificent scenery and wildlife abound on this loop route. For road-trippers, there are places to explore that cruise ship passengers will never see. Miles of marine highway ferry system connect islands and mainland routes, transporting cars as well as passengers. Miles of two-lane highways connect villages and cities on roads once traversed mainly by dog sled in winter and reduced to thick mud in summer. The drive ranges from Kodiak Island, up the Kenai Peninsula to Anchorage, then into the interior as far as Fairbanks. Nudge down into Valdez before returning to Anchorage.

**Best known:** Alaska Native Heritage Center in Anchorage; bear-watching at Katmai National Park's Brooks Falls; Mt. McKinley in Denali National Park; Thompson Glacier near Valdez.

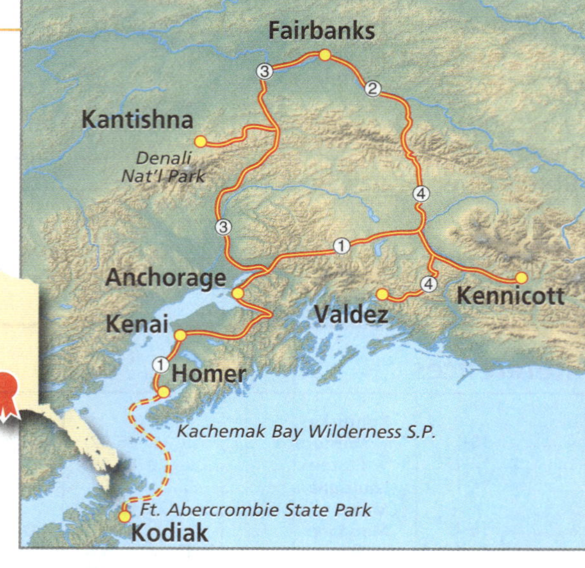


**St. Elias Alpine Guides — Root Glacier Hike (Kennicott)**

A safe, guided hike onto the Root Glacier in Wrangell-St. Elias National Park lets even greenhorns tackle tramping on the ice with crampons. Half-day hikes involve a two-mile easy-to-moderate walk to the glacier. *Main St., (907) 345-9048, www.steliasguides.com*

**Camp Denali (Denali Park)**

Camp in harmony with the wilderness in rustic cabins, enjoy gourmet meals, and explore the tundra on guided naturalist hikes. Three- and four-day stays include all meals, services, and transportation from the park entrance. *(907) 683-2290, www.campdenali.com*

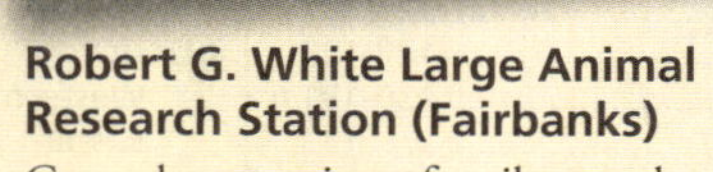

**Robert G. White Large Animal Research Station (Fairbanks)**

Get a close-up view of caribou and prehistoric-looking musk oxen on a student-led tour of this facility. The gift shop sells handsome qiviut goods, made from the soft underwool of musk oxen, that are warm enough for Fairbanks winters. Proceeds benefit the research station. *2220 Yankovich Rd., (907) 474-7945, www.uaf.edu/lars*

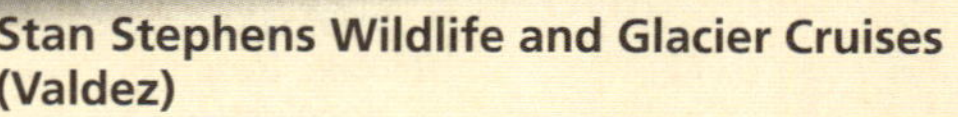

**Stan Stephens Wildlife and Glacier Cruises (Valdez)**

Watch whales breach and dive, see puffins take flight, spot bald eagles perched on an iceberg. Small excursion boats maneuver through iceberg-littered waters near the Columbia and Meares Glaciers. *Small Boat Harbor, 112 N. Harbor Dr., (907) 835-4731, www.stanstephenscruises.com*

**MORE GREAT STOPS**

- **Alaska Rivers Company — Kenai River Float Trip**
  Mile 50 South, Sterling Hwy.
  Cooper Landing, AK
  (907) 595-1226
  *www.alaskariverscompany.com*
- **Denali National Park Sled Dog Demonstration**
  Park Headquarters
  Denali National Park, AK
  (907) 683-2294
  *www.nps.gov/denal*
- **Big Daddy's BarB-Q**
  107 Wickersham St.
  Fairbanks, AK
  (907) 452-2501
  *www.bigdaddysbarb-q.com*
- **Museum of the North**
  907 Yukon Dr.
  Fairbanks, AK
  (907) 474-7505
  *www.uaf.edu/museum/*
- **Kennicott Glacier Lodge**
  Main St.
  Kennicott, AK
  (907) 258-2350
  *www.kennicottlodge.com*

### The B.B. King Museum and Delta Interpretive Center (Indianola)

The Mississippi Delta is filled with blues museums, but this one ranks at the top. Well-designed multi-media exhibits tell the story of King's remarkable life and times. Highlights include King's beloved guitar, Lucille; his home recording studio; and, adjoining the museum, an old brick cotton gin where King worked in the 1940s. *400 Second St., (662) 887-9539, www.bbkingmuseum.org*

### The Viking Tour (Greenwood)

This two-hour tour begins with a walk through the plant where Viking makes the sleek ranges that have won a devoted following around the world. Other stops include Viking's corporate headquarters, the elegant Viking-owned Alluvian Hotel and Spa, and the Viking Cooking School, which draws cooking aficionados from all over the United States. *111 Front St., (662) 455-1200, www.vikingrange.com*

## Following Old Man River

### Natchez to Tunica, Mississippi

William Faulkner, a Mississippi native, once wrote "The past isn't dead. It isn't even past." Along this trip, the past does indeed seem to occupy the present. Layers of history lie as deep as the Delta soil. Travelers can explore ancient Indian mounds, an early American wilderness trail, antebellum homes, Civil War sites, the realm of King Cotton, and the birthplace of the blues. But history isn't the only thing that makes this trip special. There's great scenery, from eerie cypress swamps to majestic forests to endless pancake-flat expanses of cotton fields. There are countless places to enjoy wonderful down-home Southern food. And there's the graciousness and hospitality of the people, who seem genuinely glad you're there.

**Best known:** Antebellum homes in Natchez and Vicksburg; Vicksburg National Military Park; the Natchez Trace Parkway; the Blues Highway; casino hotels near Tunica.

### Tunica RiverPark (Robinsonville)

The centerpiece of this 130-acre park is the Mississippi River Museum, which features interactive exhibits exploring the natural and human history of the river. An observation deck provides great views of the river itself, and the *Tunica Queen* riverboat offers sightseeing cruises. Visitors can also hike a two-mile trail through wetland forest. *1 RiverPark Dr., (662) 357-0050, www.tunicariverpark.com*

### McCartys Pottery (Merigold)

The attractive, understated pottery of Lee and Pup McCarty is famous throughout Mississippi. Here in tiny Merigold it's displayed in a former mule barn set in a lush, shady garden. There are functional items like bowls and vases as well as whimsical pieces such as rabbits, raccoons, and frogs. *101 Saint Mary St., (662) 748-2293, www.mccartyspottery.com*

### Natchez in Historic Photographs (Natchez)

Displayed in the First Presbyterian Church's Stratton Chapel, this collection brings Natchez's history to life. Most of the photos were taken between 1870 and 1913 by Henry C. Norman. He captured evocative images of the city's people, its homes and gardens, its downtown streets, and its lively riverfront. *First Presbyterian Church / Stratton Chapel, 405 State St., (601) 442-2581,*

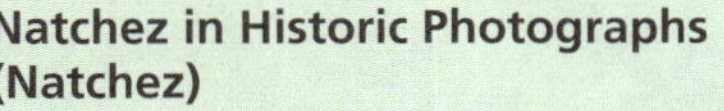

*www.fpcnatchez.org/natchezinhistoricalphotographs.php*

### The Tomato Place (Vicksburg)

Part old-fashioned produce stand, part café, and part craft gallery, this colorful, quirky place sits along US 61 just south of Vicksburg. It offers locally grown fruit and vegetables, homemade jams and preserves, all-fruit smoothies, and fresh-squeezed lemonade. Diners can enjoy tasty po' boys, BLTs, sweet potato fries, and peach cobbler. *3229 Hwy. 61 S, (601) 661-0040, www.thetomatoplace.com*

### MORE GREAT STOPS

- **Natchez City Cemetery**
  Cemetery Road
  Natchez, MS
  (601) 445-5051
  *www.natchezcemetery.com*
- **Saint Catherine Creek National Wildlife Refuge**
  21 Pintail Lane
  Natchez, MS
  (601) 442-6696
  *www.fws.gov/saintcatherinecreek/*
- **Mount Locust Inn**
  2680 Natchez Trace Pkwy.
  Tupelo, MS
  (800) 305-7417
  *www.nps.gov/natr*
- **Grand Gulf Military Monument**
  12006 Grand Gulf Rd.
  Port Gibson, MS
  (601) 437-5911
  *www.grandgulfpark.state.ms.us*
- **The Mississippi Gift Company**
  300 Howard St.
  Greenwood, MS
  (662) 455-6961
  (800) 467-7763
  *www.themississippigiftcompany.com*

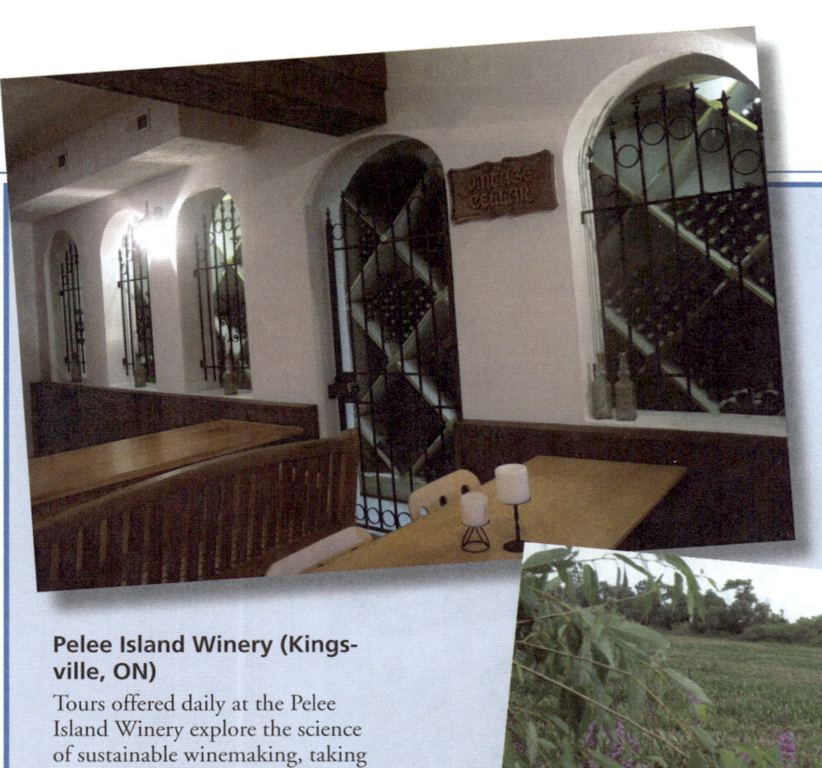

### Ottawa National Wildlife Refuge Complex (Oak Harbor, OH)

The Ottawa National Wildlife Refuge Complex—though teeming with waterfowl, shorebirds, and songbirds—is more than just a favorite spot for birders. Only 15 miles from Toledo, it's an accessible getaway for photographers, students, and outdoor enthusiasts. The complex, which is home to three wildlife refuges, covers nearly 10,000 acres. *14000 W. OH 2, (419) 898-0014, www.fws.gov/midwest/ottawa/*

### Pelee Island Winery (Kingsville, ON)

Tours offered daily at the Pelee Island Winery explore the science of sustainable winemaking, taking visitors through every step of the process. For a nominal tour fee, adults get a tasting of both reds and whites in the company cellar as well as a keepsake glass. *455 Seacliff Dr., (519) 733-6551, (800) 597-3533, www.peleeisland.com*

### Chelsea Gallery (Chelsea, MI)

Stop in for a cup of coffee and take in works from about 70 artists at this large gallery on Chelsea's historic Main Street. Featured artists live and create their art within a 90-mile radius. The gallery sells media including sculpture, stained glass, oil paintings, and jewelry at affordable prices. *115 S. Main St., (734) 475-1008, http://chelsea-gallery.com/*

## Around Lake Erie

### Cleveland, Ohio, to Windsor, Ontario

Lake Erie unifies places large and small, American and Canadian, well-known and undiscovered. The area around this Great Lake is as teeming with wildlife as it is urban adventure. This trip starts in downtown Cleveland, and the drive west is dotted with historic lakeside towns that look remarkably like they did a century ago. After a stop in Toledo, the route heads north into Michigan. Ann Arbor and Detroit are the big names in the region, but the smaller towns and villages are not to be missed. Crossing over into Canada, Windsor and its surrounding area offer a mix of cosmopolitan and country. The trip ends with a scenic ferry ride back to the States, an opportunity to reflect on the roads traveled and the ones ahead.

**Best known:** The Rock and Roll Hall of Fame and Museum in Cleveland; Cedar Point Amusement Park in Sandusky; The Henry Ford in Detroit; Caesars Windsor in Windsor.

### Berlin Fruit Box Company (Berlin Heights, OH)

Since 1858, descendants of Samuel Patterson have been following his designs and instructions at this small factory. On the company tour, logs are transformed into artist-signed, handcrafted baskets of all shapes and sizes. *51 Mechanic St., (419) 588-2081, (888) 905-1858, www.samuelpattersonbaskets.com/*

### Boardwalk Trail at Point Pelee National Park (Leamington, ON)

Just a few kilometers from the southernmost point of Canada, the most popular spot in Point Pelee National Park, lies a marshland recreation area that can be missed easily if you aren't looking for it. Tucked behind a row of trees off one of the park's main thoroughfares are a boardwalk trail, a boat dock, and observation tower. *1118 Point Pelee Dr., (519) 322-5700, www.pc.gc.ca/pn-np/on/pelee/index_E.asp*

### Zingerman's Delicatessen (Ann Arbor, MI)

A trip to Zingerman's is a far cry from a typical stop by the corner deli. The outgoing and informative staff, the expansive cheese and deli counters, and the oversized sandwiches made with fresh ingredients are just a few reasons that make Zingerman's Delicatessen an Ann Arbor institution and a must-visit for foodies. College students and locals gather on the large patio to share tables and samples of their corned beef, side salads, and pastries. *422 Detroit St., (734) 663-3354, www.zingermansdeli.com/*

### MORE GREAT STOPS

- **Toft Dairy**
  3717 Venice Rd.
  Sandusky, OH
  (419) 625-4376
  (800) 521-4606
  *www.toftdairy.com*
- **Liberty Street Robot Supply & Repair / 826 Michigan**
  115 E. Liberty St.
  Ann Arbor, MI
  (734) 761-3463
  *www.826michigan.org*
- **Arab American National Museum**
  13624 Michigan Ave.
  Dearborn, MI
  (313) 582-2266
  *www.arabamericanmuseum.org*
- **John Freeman Walls Historic Site & Underground Railroad Museum**
  859 Puce Rd.
  Lakeshore, ON
  (519) 727-6555
  *www.undergroundrailroadmuseum.com*
- **Canadian Club Brand Center**
  2072 Riverside Dr. E
  Windsor, ON
  (519) 973-9503
  *www.canadianclubwhisky.com*

**Westbrook Wine Farm (O'Neals, CA)**

Wine is an experience at Westbrook Wine Farm, where proprietors Ray and Tammy Krause blend winemaking passion with scientific principle. Ray explains his consideration of everything from gravity irrigation and sunlight angles to the best cork for the bottle, then pours samples of Fait Accompli and Gambono for guests in the wine cellar. *49610 House Ranch Rd., (559) 868-3499, www.westbrookwinefarm.com*

**Yosemite Mountain Sugar Pine Railroad (Fish Camp, CA)**

The vintage Shay steam engine number 10 chugga-chuggas through the Sierra National Forest on the old Madera Sugar Pine Lumber Company railroad. Before or after the ride, a stop at the gold panning station is a must: Visitors learn to pan for real gold with Mike the Prospector. *56001 Yosemite Hwy. 41, (559) 683-7273, www.yosemitesteamtrains.com*

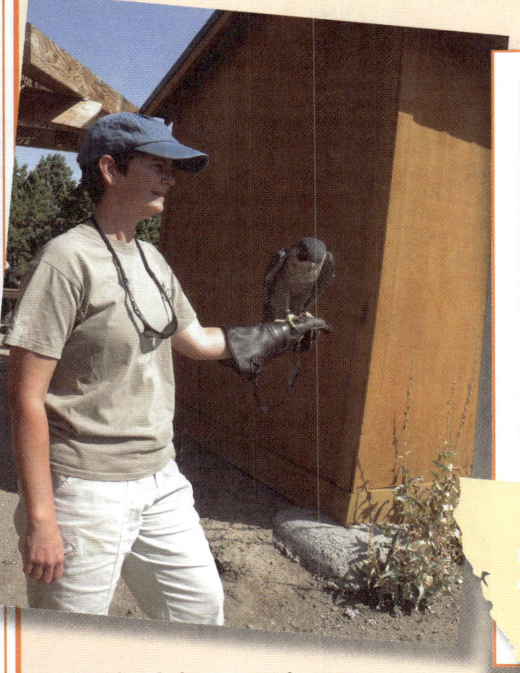

**Animal Ark (Reno, NV)**

Rescued animals thrive outdoors at the Animal Ark Sanctuary in spacious, natural habitats that allow visitors an up-close experience. Don't miss Effie the Arctic Fox posing for a photo-op, and Yogi the Bear devouring a colorful yard treat of fruit and "buffalo popsicles," frozen blocks of raw buffalo meat. *1265 Deerlodge Rd., (775) 970-3111, www.animalark.org*

## EARTH TONES

### Reno, Nevada to Yosemite, California

This drive through the obtuse angle of Nevada and California shows off all of earth's dramatic tones. The earnest brown of the high desert in Reno, Virginia City, and Carson City plays host to memories of mining and railroads. The route then sinks into a deep blue and green panorama at Lake Tahoe, circling from Incline Village to the north, along the west side of the lake through Homewood and Emerald Bay, and around to South Lake Tahoe. Meandering south through the Sierra Nevada, the journey continues along the Tioga Pass* and on to the towering white and gray bluffs in Yosemite National Park. The trip culminates further south in Madera County, decorated with the red, soaring sequoias of the Sierra National Forest.

**Best known:** Casinos in Reno and Stateline; Nevada State Capitol in Carson City; Squaw Valley (site of 1960 Winter Olympics); Yosemite National Park.

**Note: The Tioga Pass is closed in winter.*

**TNT Stagelines (Virginia City, NV)**

It's the wildest, bumpiest, most exhilarating ride of your life, and it's found on a 19th-century-style stagecoach ride. Gary and Nancy Teel operate the authentic thrill ride on dusty, bumpy roads that promise to send travelers airborne in a wagon traveling up to 22 mph. *F Street at the railroad crossing, (775) 721-1496, www.tntstagelines.com*

**Tahoe Maritime Museum (Homewood, CA)**

The Tahoe Maritime Museum is as nautical outside—modeled after old boathouses—as it is inside. Historic boats such as CrisCraft and GarWood models and a collection of outboard motors are exhibit highlights. Little skippers can test their skills at the boat simulator and craftsmanship areas. *5205 West Lake Blvd., (530) 525-9253, www.tahoemaritimemuseum.org*

**MORE GREAT STOPS**

- **Locals BBQ & Grill**
  3849 S. Carson St.
  Carson City, NV
  (775) 841-9999
  *www.LBBQ.com*
- **Thunderbird Lodge**
  5000 Hwy. 28
  Incline Village, NV
  (775) 832-8750
  *http://thunderbirdlodge.org*
- **Camp Richardson Historic Resort**
  1900 Jameson Beach Rd.
  South Lake Tahoe, CA
  (530) 541-1801
  *www.camprichardson.com*
- **Tahoe Trout Farm**
  1023 Blue Lake Ave.
  South Lake Tahoe, CA
  (530) 541-1491
- **The Nature Center at Happy Isles**
  Yosemite National Park
  (209) 372-0200
  *www.nps.gov/yose*

EDITOR'S PICKS

**Via Lactea Farm (Brookfield)**

Andy and Jenny Tapper have developed their farm into a local source of chèvre, feta, yogurt, and chickens for restaurants and individual customers. Between March and December, visitors can stop by the farmhouse store for goat dairy products, fresh eggs, maple syrup, and soap. The Tappers also give tours (call ahead). *366 Stoneham Rd., (603) 522-3626, www.vialacteafarm.com*

**Byrne & Carlson Chocolatier Confectioner (Portsmouth)**

When in Portsmouth, do as its residents do: stop by the Byrne & Carlson chocolate boutique to see what kinds of handmade crème confections grace the truffle case that week. Across from that case you'll find chocolate bars inlaid with crystallized pansies, almonds, orange slices, and molded-chocolate seashells. Other shop delights: tasting bars featuring cocoa flavors from Madagascar, Venezuela, and the Caribbean island of Guanaja, and one variety even sprinkled with sea salt. *121 State St., (888) 559-9778, www.byrneandcarlson.com*

## Seaside to Summit

### Manchester to Bretton Woods, New Hampshire

New Hampshire packs many kinds of vacation adventures into one small state. You need not drive very far to transition from sea spray to wooded lakesides to alpine vistas. This drive begins in Manchester, then slides seaward to soak up Portsmouth. Only an hour's drive northwest brings lakes Winnepesaukee and Squam, edged with pines and beloved by generations of boaters. And not very far north of there, the White Mountains sit astride the state. Swing up around the westerly edge of the range to find Bretton Woods in the shadow of Mount Washington, then wind gradually back south, past the lakes and the farms, until nearly at sea level again.

**Best known:** The Flume Gorge; Conway Scenic Railway and outlet shopping in North Conway; Kancamagus Highway; the cog railroad to Mount Washington's summit.

**Highland Center (Bretton Woods)**

A small shop offers critical gear, workshops teach compass navigation and other hiking basics, the cafeteria serves meals 3x/day, and the lodging rooms are comfortable and reasonably priced. Adventurers can try out gear from the L.L. Bean room lined with boots, underlayers, snowshoes, sleeping bags, tents, you name it. *Rte. 302, (603) 278-4453 (HIKE), www.outdoors.org/lodging/lodges/highland/index.cfm*

**Bretton Woods Canopy Tour (Bretton Woods)**

The storied Mount Washington Resort offers a canopy tour: a series of nine ziplines, two rope bridges, and three rappels through the hemlocks at the Bretton Woods ski resort. "Zippers" take the tour in groups of eight, guided by two experts. The longest zip stretches 830 feet, long enough to allow speeds of around 30 miles per hour. The entire adventure lasts for three-and-a-half hours. Reservations and a steep fee are required. *Omni Mount Washington Resort, Rte. 302, (603) 278-4947, (603) 278-8989, www.MountWashingtonResort.com*

**Hampshire Pewter (Wolfeboro)**

Visitors watch longtime craftspeople pour liquid metal into the handmade molds, turn out racks of Christmas ornaments, or finish pieces using handcrafted tools. Many of the company's products can be purchased in the shop. Candlesticks, sconces, and tableware are legion, while the ornament choices include a series of New Hampshire's covered bridges. *43 Mill St., (603) 569-4944, (800) 639-7704, www.hampshirepewter.com*

### MORE GREAT STOPS

- **Ash Street Inn**
  118 Ash St.
  Manchester, NH
  (603) 668-9908
  *www.AshStreetInn.com*
- **Castle in the Clouds Conservation Area**
  Ossipee Park Rd.
  Moultonborough, NH
  (603) 253-3301
  *www.lrct.org/hiker-patch.html*
- **Squam Lakes Natural Science Center**
  23 Science Center Rd.
  Holderness, NH
  (603) 968-7194
  *www.nhnature.org*
- **Thompson House Eatery**
  193 Main St.
  Jackson, NH
  (603) 383-9341
  *www.thompsonhouseeatery.com*
- **Weather Discovery Center**
  2779 White Mountain Hwy. (NH 16)
  North Conway, NH
  (603) 356-2137
  *www.mountwashington.org*

# NUMBERS TO KNOW

## HOTEL RESOURCES

**Adam's Mark Hotels & Resorts**
(800) 444-2326
www.adamsmark.com

**America's Best Inns & Suites**
(800) 237-8466
www.americasbestinns.com

**AmericInn**
(800) 396-5007
www.americinn.com

**Baymont Inn & Suites**
(877) 229-6668
www.baymontinn.com

**Best Western**
(800) 780-7234
www.bestwestern.com

**Budget Host**
(800) 283-4678
www.budgethost.com

**Clarion Hotels**
(877) 424-6423
www.clarionhotel.com

**Coast Hotels & Resorts**
(800) 716-6199
www.coasthotels.com

**Comfort Inn**
(877) 424-6423
www.comfortinn.com

**Comfort Suites**
(877) 424-6423
www.comfortsuites.com

**Courtyard by Marriott**
(888) 236-2427
www.courtyard.com

**Crowne Plaza Hotel & Resorts**
(877) 227-6963
www.crowneplaza.com

**Days Inn**
(800) 329-7466
www.daysinn.com

**Delta Hotels & Resorts**
(888) 890-3222
www.deltahotels.com

**Doubletree Hotels, Guest Suites, Resorts & Clubs**
(800) 222-8733
www.doubletree.com

**Drury Hotels**
(800) 378-7946
www.druryhotels.com

**Econo Lodge**
(877) 424-6423
www.econolodge.com

**Embassy Suites Hotels**
(800) 362-2779
www.embassysuites.com

**Extended Stay Hotels**
(800) 804-3724
www.extstay.com

**Fairfield Inn by Marriott**
(800) 228-2800
www.fairfieldinn.com

**Fairmont Hotels & Resorts**
(800) 257-7544
www.fairmont.com

**Four Points by Sheraton**
(800) 368-7764
www.fourpoints.com

**Four Seasons Hotels & Resorts**
(800) 819-5053
www.fourseasons.com

**Hampton Inn**
(800) 426-7866
www.hamptoninn.com

**Hilton Hotels**
(800) 445-8667
www.hilton.com

**Holiday Inn Hotels & Resorts**
(888) 465-4329
www.holidayinn.com

**Homewood Suites**
(800) 225-5466
www.homewood-suites.com

**Howard Johnson**
(800) 446-4656
www.hojo.com

**Hyatt Hotels & Resorts**
(888) 591-1234
www.hyatt.com

**InterContinental Hotels & Resorts**
(888) 424-6835
www.intercontinental.com

**Jameson Inns**
(800) 526-3766
www.jamesoninns.com

**Knights Inn**
(800) 843-5644
www.knightsinn.com

**La Quinta Inns & Suites**
(800) 753-3757
www.lq.com

**Le Méridien Hotels & Resorts**
(800) 543-4300
www.lemeridien.com

**Loews Hotels**
(866) 563-9792
www.loewshotels.com

**MainStay Suites**
(877) 424-6423
www.mainstaysuites.com

**Marriott International**
(888) 236-2427
www.marriott.com

**Microtel Inns & Suites**
(800) 771-7171
www.microtelinn.com

**Motel 6**
(800) 466-8356
www.motel6.com

**Omni Hotels**
(888) 444-6664 (U.S. only)
(402) 952-6664 (outside U.S.)
www.omnihotels.com

**Park Inn**
(888) 201-1801
www.parkinn.com

**Preferred Hotels & Resorts**
(800) 323-7500
www.preferredhotels.com

**Quality Inn & Suites**
(877) 424-6423
www.qualityinn.com

**Radisson Hotels & Resorts**
(888) 201-1718
www.radisson.com

**Ramada Worldwide**
(800) 272-6232
www.ramada.com

**Red Lion Hotels**
(800) 733-5466
www.redlion.com

**Red Roof Inn**
(800) 733-7663
www.redroof.com

**Renaissance Hotels & Resorts by Marriott**
(800) 468-3571
www.renaissancehotels.com

**Residence Inn by Marriott**
(800) 331-3131
www.residenceinn.com

**The Ritz-Carlton**
(800) 542-8680
www.ritzcarlton.com

**Rodeway Inn**
(877) 424-6423
www.rodewayinn.com

**Sheraton Hotels & Resorts**
(800) 325-3535
www.sheraton.com

**Sleep Inn**
(877) 424-6423
www.sleepinn.com

**Super 8**
(800) 800-8000
www.super8.com

**Travelodge Hotels**
(800) 578-7878
www.travelodge.com

**Westin Hotels & Resorts**
(800) 937-8461
www.westin.com

**Wyndham Hotels & Resorts**
(877) 999-3223
www.wyndham.com

*To find a bed-and-breakfast at your destination, log on to www.bedandbreakfast.com.®*
*NOTE: All toll-free reservation numbers are for the U.S. and Canada unless otherwise noted. These numbers were accurate at press time, but are subject to change. Find more listings or book a hotel online at randmcnally.com.*

## CELL PHONE EMERGENCY NUMBERS

**Alabama** .......... *47
**Alaska** .......... 911
**Arizona** .......... 911
**Arkansas** .......... 911
**California** .......... 911
**Colorado** .......... 911; *277
**Connecticut** .......... 911
**Delaware** .......... 911
**District of Columbia** .......... 911
**Florida** .......... 911; *347
**Georgia** .......... 911; *477
**Hawaii** .......... 911
**Idaho** .......... *477
**Illinois** .......... 911
**Indiana** .......... 911
**Iowa** .......... 911; *55
**Kansas** .......... 911; *47
**Kentucky** .......... 911; (800) 222-5555 (in KY)
**Louisiana** .......... 911; *577 (road emergencies)
**Maine** .......... 911
**Maryland** .......... 911; *77
**Massachusetts** .......... 911
**Michigan** .......... 911
**Minnesota** .......... 911
**Mississippi** .......... 911; *47
**Missouri** .......... *55
**Montana** .......... 911
**Nebraska** .......... *55
**Nevada** .......... *647
**New Hampshire** .......... *77
**New Jersey** .......... 911, *77
**New Mexico** .......... 911
**New York** .......... 911
**North Carolina** .......... 911; *47
**North Dakota** .......... 911; *2121
**Ohio** .......... 911
**Oklahoma** .......... 911
**Oregon** .......... 911
**Pennsylvania** .......... 911
**Rhode Island** .......... 911
**South Carolina** .......... 911; *47
**South Dakota** .......... 911
**Tennessee** .......... 911; *847
**Texas** .......... 911
**Utah** .......... 911; *11
**Vermont** .......... 911
**Virginia** .......... 911
**Washington** .......... 911
**West Virginia** .......... 911; *77
**Wisconsin** .......... 911
**Wyoming** .......... 911

# Map Legend

## Roads and related symbols

Limited-access, multilane highway—free; toll
New road (under construction as of press time)
Other multilane highway
Principal highway
Other through highway
Other road (conditions vary — local inquiry suggested)
Unpaved road (conditions vary — local inquiry suggested)
Ramp; one way route
Toll car ferry (unless otherwise indicated on map)
Tunnel; mountain pass
Railroad; Intracoastal Waterway
Interstate highway; Interstate highway business route
U.S. highway; U.S. highway business route
Trans-Canada highway; Autoroute
Mexican highway or Central American highway
State/provincial highway; secondary state/provincial, or county highway
Great River Road; Great Circle Route
Lewis & Clark Highway; Lincoln Highway; Route 66
Scenic route; Best of the Road™ route
Service area; toll booth or fee booth
Interchanges and exit numbers
For most states, the mileage between interchanges may be determined by subtracting one number from the other.
Highway distances (segments of one mile or less not shown):
Cumulative miles (red): the distance between arrows
Cumulative kilometers (blue) the distance between arrows
Intermediate miles (black): the distance between intersections & places
Comparative distance
1 mile = 1.609 kilometers 1 kilometer = 0.621 mile

## Cities & towns size of type on map indicates relative population

National capital; state or provincial capital
County seat or independent city
City, town, or recognized place—incorporated; unicorporated
Urbanized area
Separate cities within metropolitan area

## Parks, recreation areas, & other points of interest

U.S. or Canadian national park
U.S. national recreation area or U.S. Fish & Wildlife Service location
Other U.S. National Park Service location, or state/provincial park system location
National forest, national grassland, or city park; wilderness area
State park system—with campsites; without campsites
Campsite; wayside or roadside park
Point of interest, historic site or monument
Airport
Building
Foot trail
Golf course or country club; ski area
Hospital or medical center
Native American tribal lands
Tourist information center; port of entry
Military or governmental installation; military airport
Rest area—with toilets; without toilets

## Physical features

Mountain peak; highest point in state/province
Lake; intermittent lake; dry lake
River; intermittent river
Desert; glacier
Swamp or mangrove swamp
Continental divide

## Other symbols

Area shown in greater detail on inset map
52 Inset map page indicator (if not on same page)
Pg.65 Map continuation indicator
COOK County or parish boundary and name
State or provincial boundary
National boundary
Time zone boundary
33°N 95°W Latitude; longitude

## Map abbreviations

Listed below are some of the commonly used abbreviations on our maps. For a complete list of abbreviations that appear on the maps, go to **www.randmcnally.com/ABBR.**

| | | | |
|---|---|---|---|
| Bfld. | battlefield | N.P. | National Park |
| Cr. | creek | N.R.A. | National Recreation Area |
| I. | island | N.W.R. | National Wildlife Refuge |
| Int'l | international | S.H.S. | State Historic Site |
| L. | lake | S.N.A. | State Natural Area |
| N.H.P. | National Historic Park | S.P. | State Park |
| N.H.S. | National Historic Site | S.R.A. | State Recreation Area |
| N.M. | National Monument | W.M.A. | Wildlife Management Area |

Population figures used in this atlas are from the latest available census or are Census Bureau or Rand McNally estimates.

Map legend Pg. 7

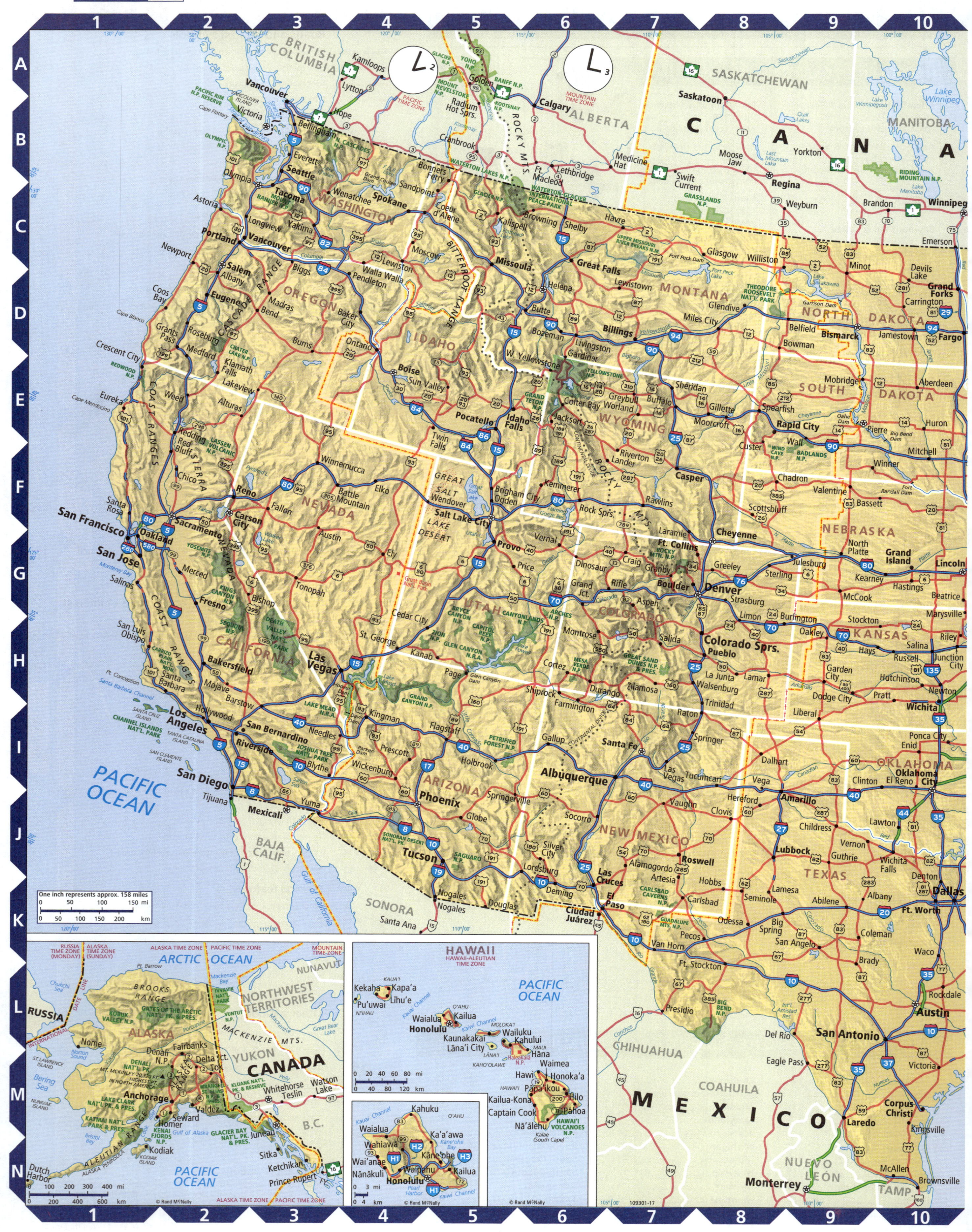


## Selected places of interest

- Acadia National Park, D-20
- Arches National Park, G-6
- Badlands National Park, F-9
- Big Bend National Park, L-8
- Biscayne National Park, N-18
- Bryce Canyon National Park, H-5
- Canyonlands National Park, H-6
- Capitol Reef National Park, H-5
- Carlsbad Caverns National Park, K-7
- Channel Islands National Park, I-1
- Congaree National Park, J-17
- Crater Lake National Park, D-2
- Cuyahoga Valley National Park, F-16
- Death Valley National Park, H-3
- Denali National Park, M-1
- Dry Tortugas National Park, N-17
- Everglades National Park, N-17
- Glacier Bay National Park, M-2
- Glen Canyon National Recreation Area, H-5
- Grand Canyon National Park, H-4
- Grand Teton National Park, E-6
- Great Sand Dunes Nat'l Park & Preserve, H-7
- Great Smoky Mountains National Park, I-15
- Guadalupe Mountains National Park, K-7

**Land area:** 3,537,438 sq. mi.

**Population:** 304,059,724

**Largest city:** New York, 8,363,710

- Haleakalā National Park, L-6
- Hawai'i Volcanoes National Park, M-6
- Hot Springs National Park, J-12
- Isle Royale National Park, D-13
- Kings Canyon National Park, G-2
- Lake Mead National Recreation Area, H-4
- Lassen Volcanic National Park, E-2
- Mammoth Cave National Park, I-14
- Mesa Verde National Park, H-6
- Mount Rainier National Park, C-3
- North Cascades National Park, B-4
- Olympic National Park, B-3
- Petrified Forest National Park, I-5
- Redwood National Park, E-1
- Rocky Mountain National Park, G-7
- Sequoia National Park, H-2
- Shenandoah National Park, G-17
- Theodore Roosevelt National Park, D-8
- Voyageurs National Park, D-12
- Waterton-Glacier International Peace Park, C-5
- Wind Cave National Park, F-8
- Yellowstone National Park, E-6
- Yosemite National Park, G-2
- Zion National Park, H-5

Index of cities Pg. 236 | Map legend Pg. 7

Mileages between cities

| | Andalusia | Anniston | Auburn | Birmingham | Chattanooga, TN | Columbus, GA | Dothan | Florence | Gadsden | Grove Hill | Huntsville | Meridian, MS | Mobile | Montgomery | Selma | Tuscaloosa |
|---|---|---|---|---|---|---|---|---|---|---|---|---|---|---|---|---|
| Atlanta, GA | 252 | 90 | 108 | 146 | 117 | 106 | 206 | 263 | 119 | 294 | 181 | 289 | 328 | 160 | 210 | 201 |
| Birmingham | 181 | 64 | 109 | | 146 | 141 | 196 | 118 | 61 | 155 | 102 | 146 | 258 | 90 | 87 | 58 |



**Alabama state facts**

Nickname: The Heart of Dixie
Capital: Montgomery, J-8

Population: 4,661,900 (rank: 23rd)
Largest city: Birmingham, 228,798, F-7

Land area: 50,744 sq. mi. (rank: 28th)
Highest point: Cheaha Mountain, 2,407 ft., F-9

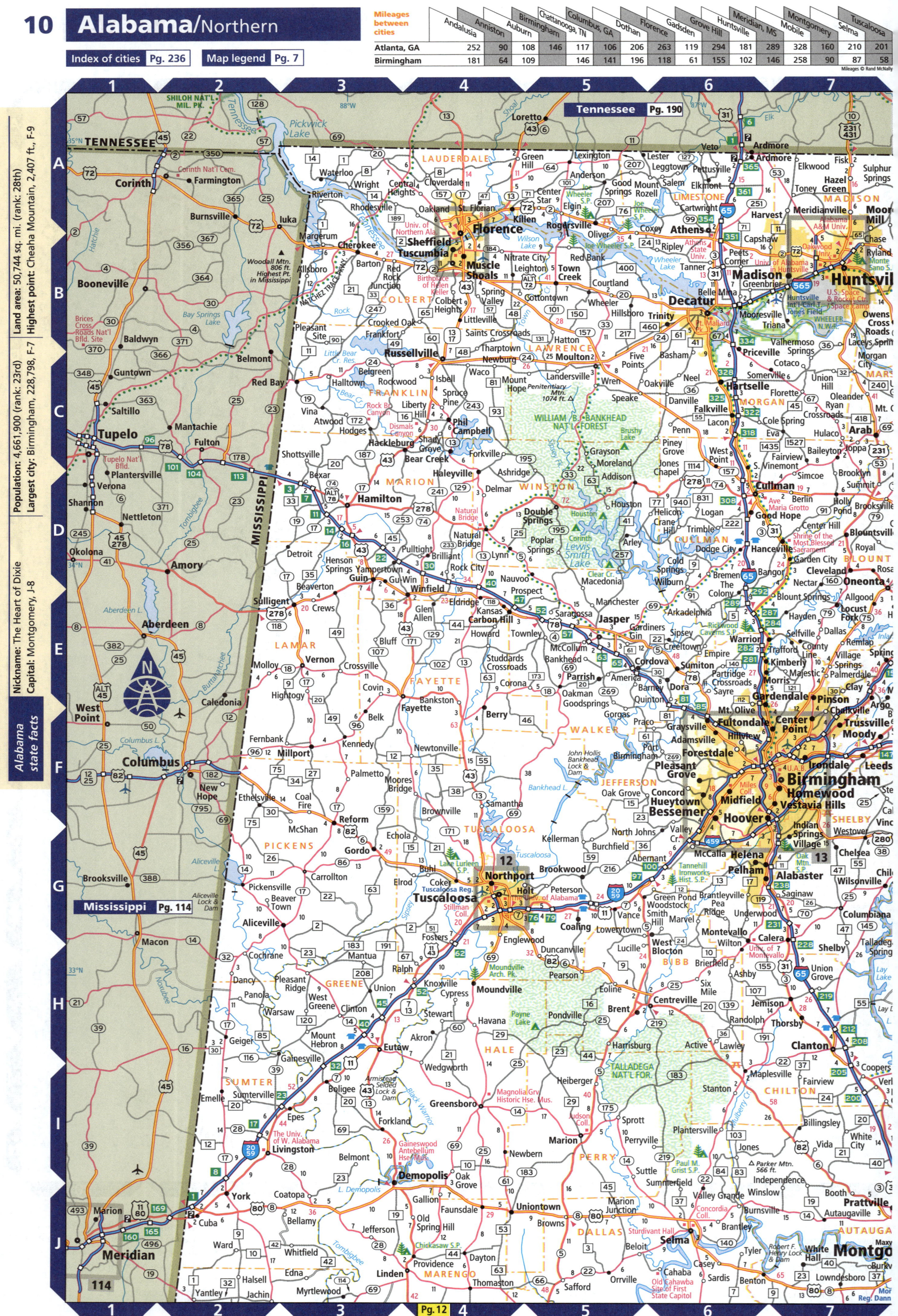

Mileages between cities

| | Andalusia | Anniston | Auburn | Birmingham | Chattanooga, TN | Columbus, GA | Dothan | Florence | Gadsden | Grove Hill | Huntsville | Meridian, MS | Mobile | Montgomery | Selma | Tuscaloosa |
|---|---|---|---|---|---|---|---|---|---|---|---|---|---|---|---|---|
| Chattanooga, TN | 322 | 119 | 221 | 146 | | 219 | 319 | 166 | 89 | 300 | 102 | 291 | 399 | 232 | 228 | 203 |
| Dothan | 74 | 207 | 118 | 196 | 319 | 99 | | 311 | 252 | 169 | 294 | 253 | 196 | 103 | 148 | 210 |

Mileages © Rand McNally

continued p. 12

10 66 miles; 20 215 miles; 59 241 miles; 65 367 miles — Total mileages through Alabama

More mileages at go.randmcnally.com/MC

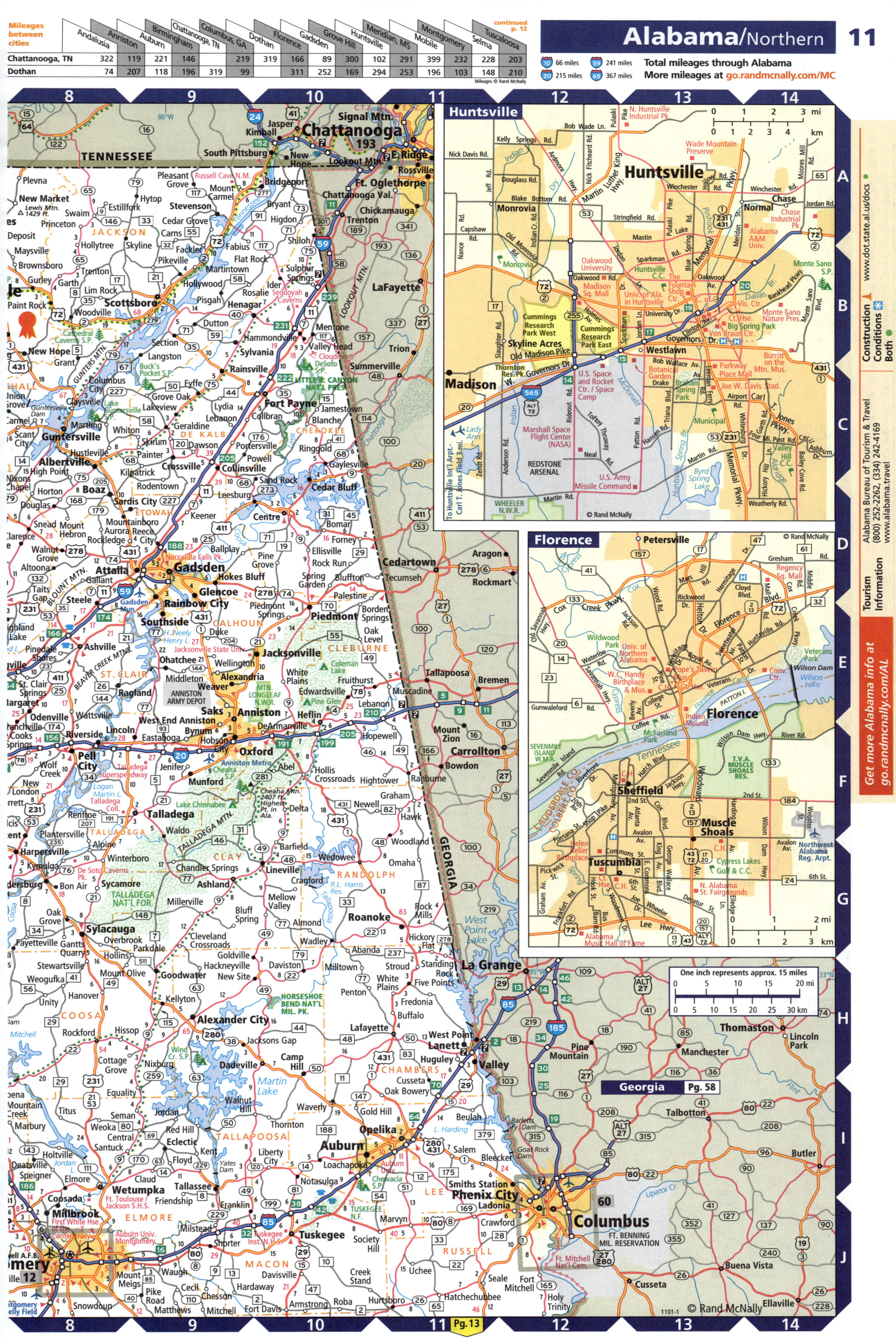

Index of cities Pg. 236 | Map legend Pg. 7

**Mileages between cities**

| | Andalusia | Anniston | Auburn | Birmingham | Chattanooga, TN | Columbus, GA | Dothan | Florence | Gadsden | Grove Hill | Huntsville | Meridian, MS | Mobile | Montgomery | Selma | Tuscaloosa |
|---|---|---|---|---|---|---|---|---|---|---|---|---|---|---|---|---|
| Huntsville | 279 | 104 | 210 | 102 | 102 | 243 | 294 | 64 | 72 | 254 | | 244 | 356 | 189 | 188 | 155 |
| Mobile | 123 | 280 | 222 | 258 | 399 | 256 | 196 | 376 | 313 | 82 | 356 | 133 | | 168 | 159 | 203 |



**Alabama state facts**

Nickname: The Heart of Dixie
Capital: Montgomery, J-8
Population: 4,661,900 (rank: 23rd)
Largest city: Birmingham, 228,798, F-7
Land area: 50,744 sq. mi. (rank: 28th)
Highest point: Cheaha Mountain, 2,407 ft., F-9

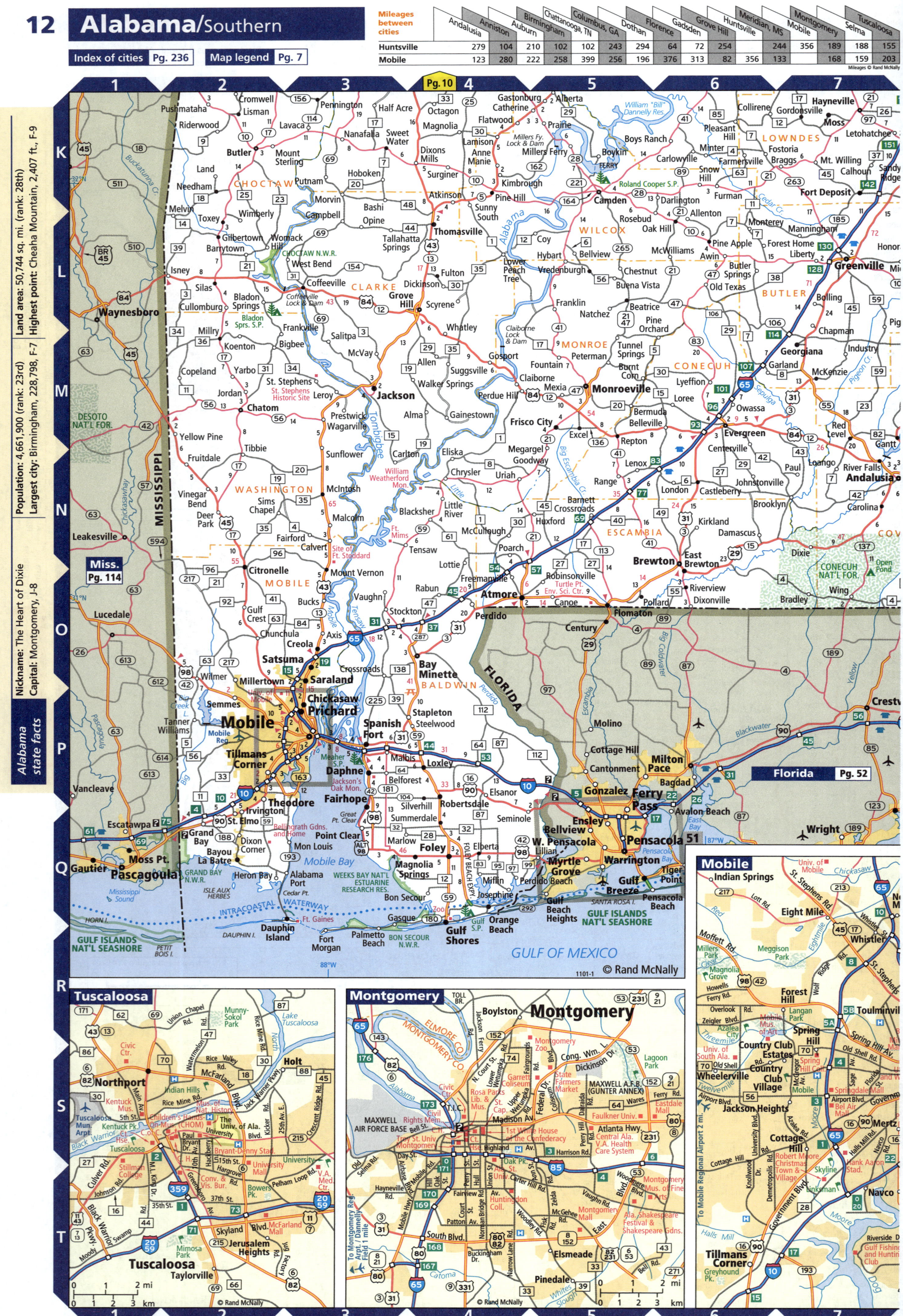

**Mileages between cities**

| | Andalusia | Anniston | Auburn | Birmingham | Chattanooga, TN | Columbus, GA | Dothan | Florence | Gadsden | Grove Hill | Huntsville | Meridian, MS | Mobile | Montgomery | Selma | Tuscaloosa |
|---|---|---|---|---|---|---|---|---|---|---|---|---|---|---|---|---|
| **Montgomery** | 91 | 110 | 54 | 90 | 232 | 87 | 103 | 205 | 148 | 134 | 189 | 153 | 168 | | 50 | 104 |
| **Tuscaloosa** | 194 | 118 | 159 | 58 | 203 | 192 | 210 | 123 | 118 | 121 | 155 | 93 | 203 | 104 | 75 | |



10 66 miles
20 215 miles
59 241 miles
65 367 miles

**Total mileages through Alabama**
**More mileages at go.randmcnally.com/MC**

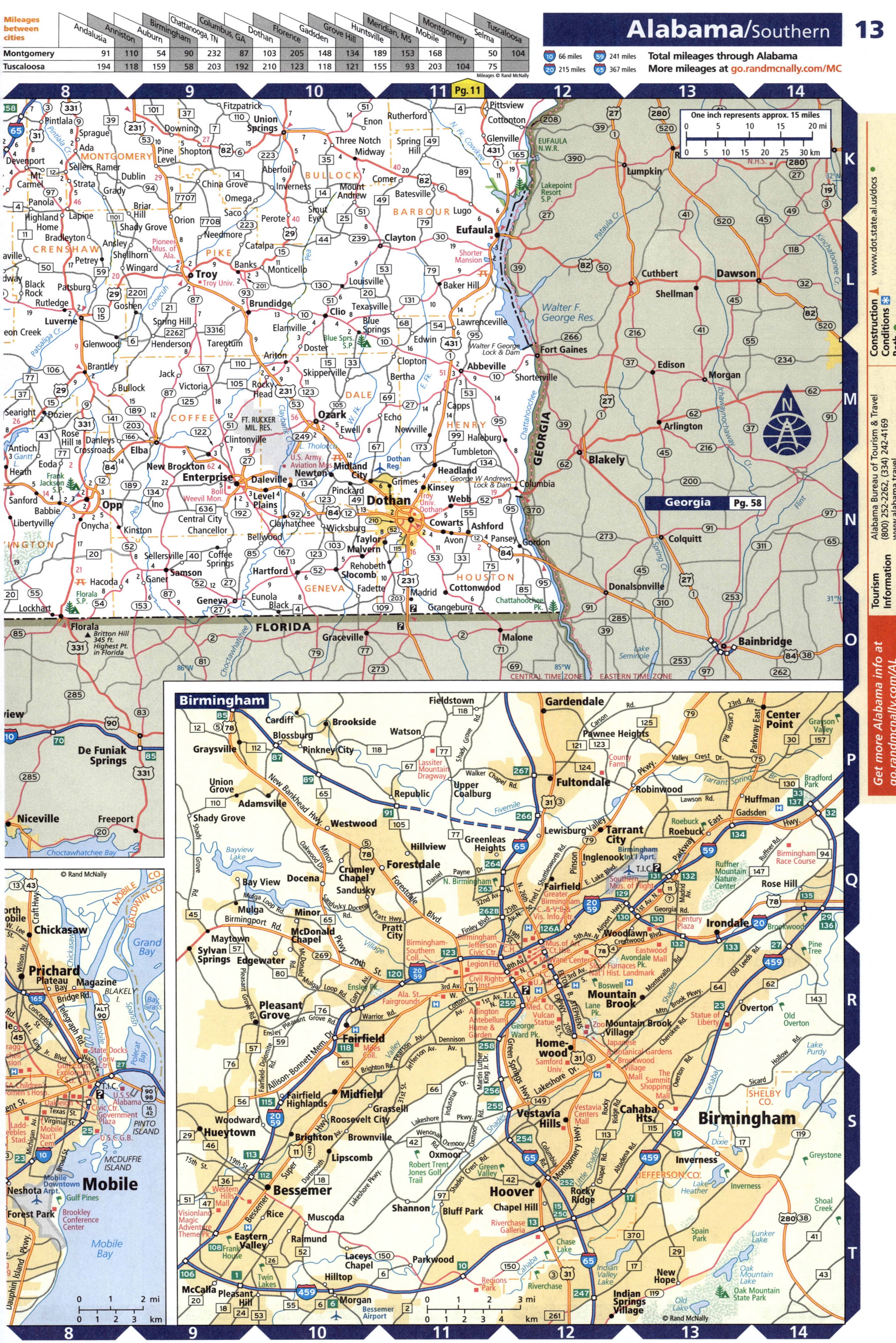

Index of cities Pg. 236 | Map legend Pg. 7

**Mileages between cities**

| | Anchorage | Denali N.P. | Fairbanks | Haines | Homer | Prince Rupert, BC | Tok | Valdez |
|---|---|---|---|---|---|---|---|---|
| Anchorage | | 236 | 358 | 756 | 221 | 1557 | 317 | 297 |
| Fairbanks | 358 | 122 | | 640 | 578 | 1441 | 202 | 362 |
| Haines | 756 | 762 | 640 | | 975 | 919 | 438 | 691 |
| Homer | 221 | 457 | 578 | 975 | | 1776 | 537 | 277 |



**Alaska state facts**

Nickname: The Last Frontier
Capital: Juneau, H-12
Population: 686,293 (rank: 47th)
Largest city: Anchorage, 279,671, G-8
Land area: 571,951 sq. mi. (rank: 1st)
Highest point: Mt. McKinley, 20,320 ft., F-8

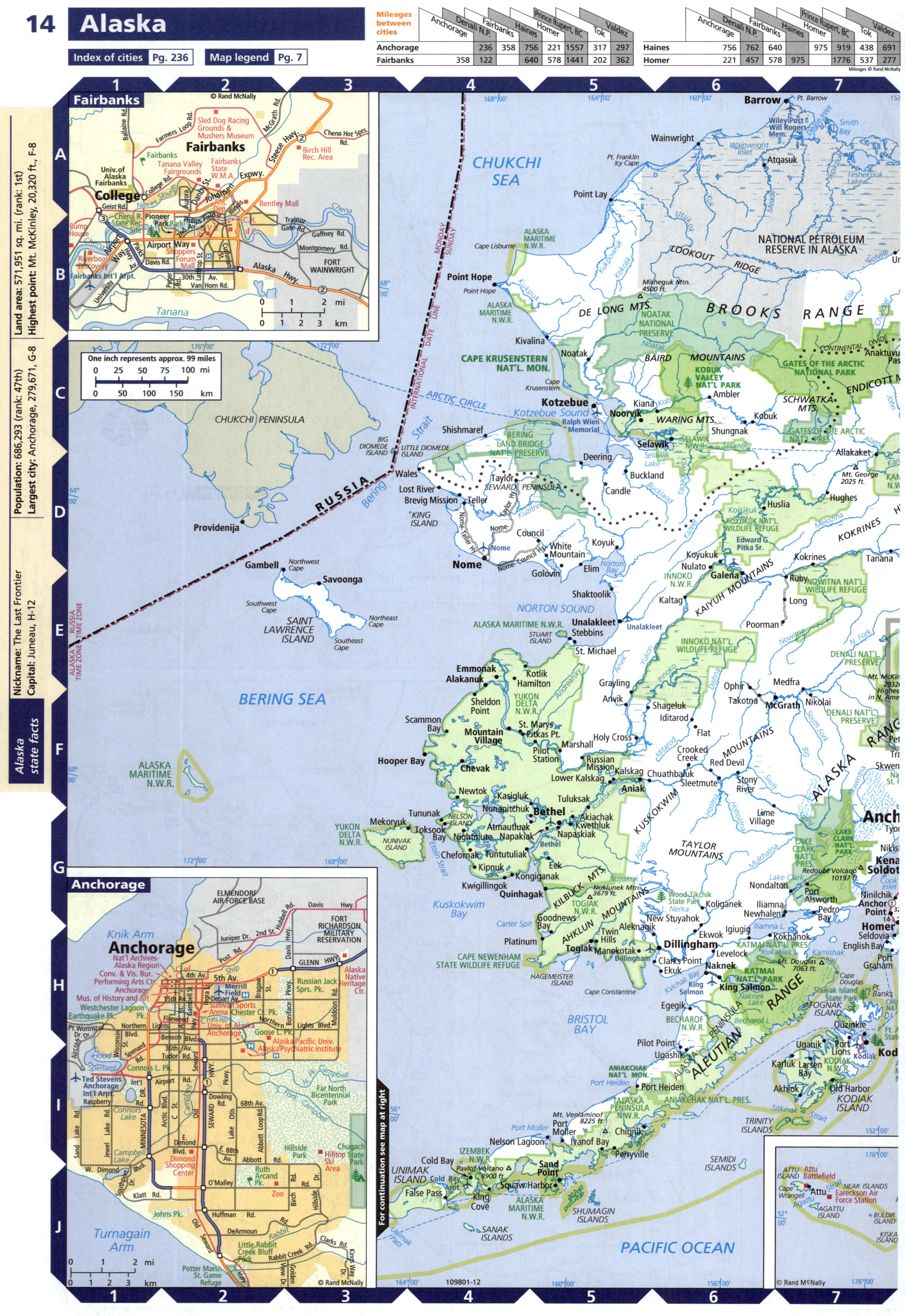

**Mileages between cities**

| | Anchorage | Denali N.P. | Fairbanks | Haines | Homer | Prince Rupert, BC | Tok | Valdez |
|---|---|---|---|---|---|---|---|---|
| Kenai | 157 | 393 | 514 | 911 | 83 | 1713 | 473 | 213 |
| Seward | 126 | 362 | 483 | 880 | 168 | 1682 | 442 | 182 |
| Tok | 317 | 324 | 202 | 438 | 537 | 1240 | | 252 |
| Valdez | 297 | 346 | 362 | 691 | 277 | 1493 | 252 | |



① 408 miles ② 202 miles ③ 325 miles

**Total mileages through Alaska**
**More mileages at go.randmcnally.com/MC**

Index of cities Pg. 236 | Map legend Pg. 7

**Mileages between cities**

| | Casa Grande | Chinle | Eagar | Flagstaff | Gallup, NM | Grand Canyon | Holbrook | Kingman | Lake Havasu City | Las Vegas, NV | Lordsburg, NM | Nogales | Page | Phoenix | Tucson | Yuma |
|---|---|---|---|---|---|---|---|---|---|---|---|---|---|---|---|---|
| Flagstaff | 191 | 213 | 176 | | 185 | 79 | 90 | 146 | 204 | 250 | 374 | 321 | 133 | 139 | 255 | 318 |
| Holbrook | 220 | 123 | 86 | 90 | 94 | 167 | | 237 | 295 | 340 | 264 | 304 | 214 | 230 | 238 | 409 |



*Arizona state facts*

**Nickname:** The Grand Canyon State
**Capital:** Phoenix; J-7
**Population:** 6,500,180 (rank: 14th)
**Largest city:** Phoenix, 1,567,924, J-7
**Land area:** 113,635 sq. mi. (rank: 6th)
**Highest point:** Humphreys Peak, 12,633 ft., E-8

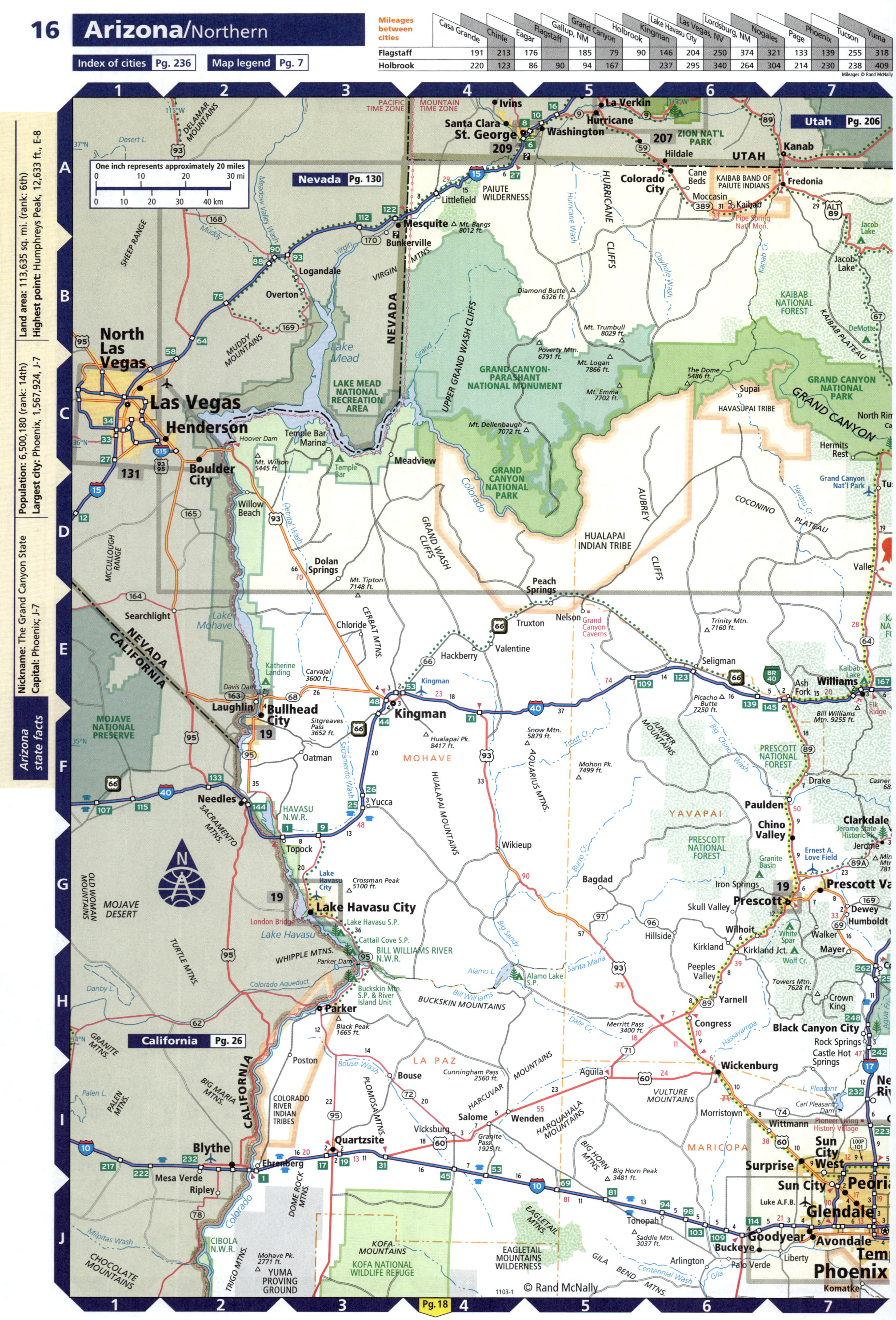

**Mileages between cities**

continued p. 18

| | Casa Grande | Chinle | Eagar | Flagstaff | Gallup, NM | Grand Canyon | Holbrook | Kingman | Lake Havasu City | Las Vegas, NV | Lordsburg, NM | Nogales | Page | Phoenix | Tucson | Yuma |
|---|---|---|---|---|---|---|---|---|---|---|---|---|---|---|---|---|
| Las Vegas, NV | 336 | 463 | 427 | 250 | 435 | 275 | 340 | 104 | 152 | | 558 | 467 | 271 | 285 | 401 | 292 |
| Page | 324 | 204 | 301 | 133 | 255 | 137 | 214 | 281 | 340 | 271 | 499 | 455 | | 275 | 390 | 453 |



Total mileages through Arizona

| Interstate | Miles |
|---|---|
| 8 | 178 miles |
| 10 | 392 miles |
| 17 | 146 miles |
| 40 | 359 miles |

More mileages at go.randmcnally.com/MC

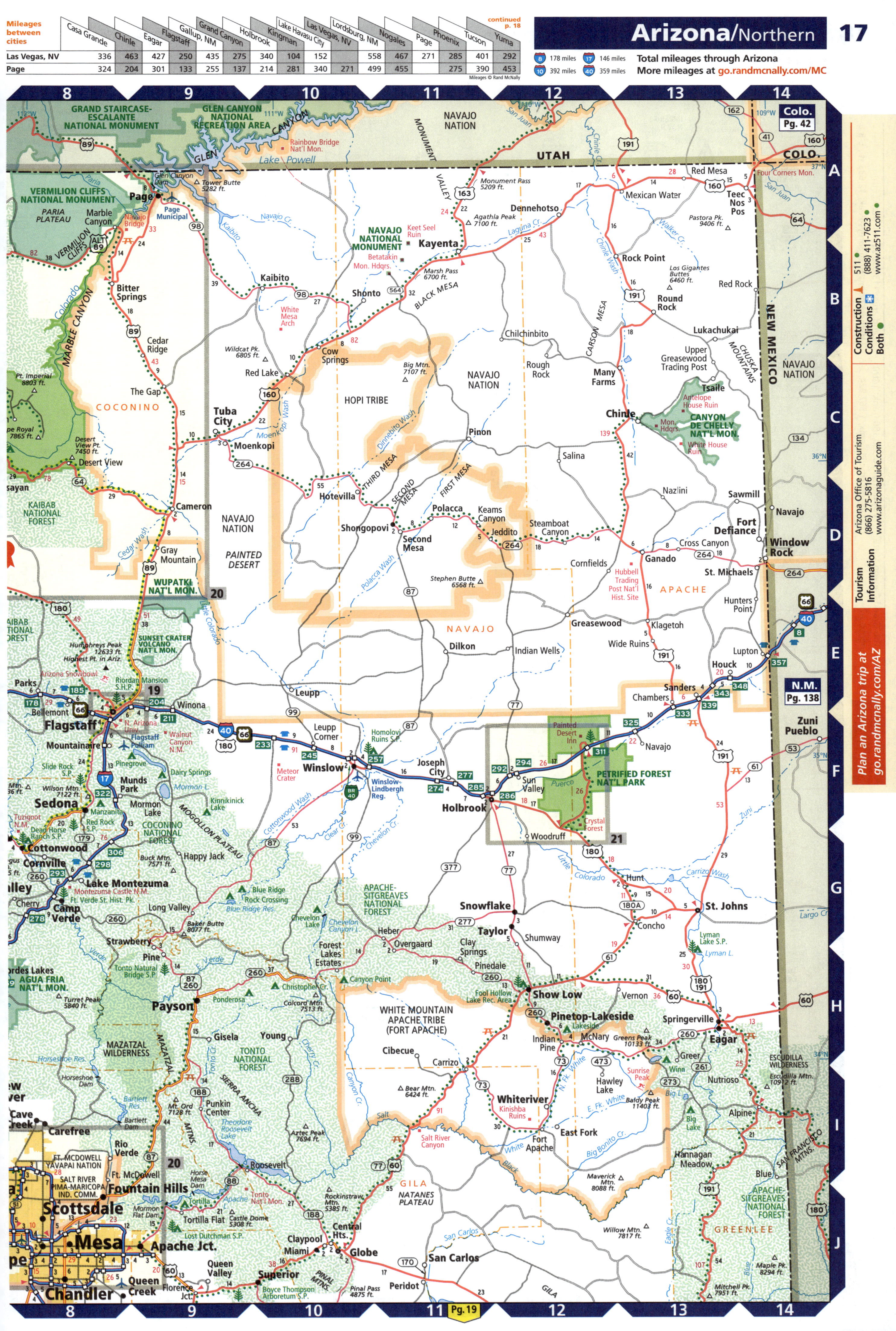

Index of cities Pg. 236 | Map legend Pg. 7

Mileages between cities

| | Casa Grande | Chinle | Eagar | Flagstaff | Gallup, NM | Grand Canyon | Holbrook | Kingman | Lake Havasu City | Las Vegas, NV | Lordsburg, NM | Nogales | Page | Phoenix | Tucson | Yuma |
|---|---|---|---|---|---|---|---|---|---|---|---|---|---|---|---|---|
| Phoenix | 48 | 353 | 226 | 139 | 324 | 218 | 230 | 182 | 198 | 285 | 268 | 179 | 275 | | 116 | 181 |
| Prescott | 148 | 306 | 270 | 93 | 278 | 126 | 184 | 148 | 206 | 251 | 368 | 278 | 227 | 97 | 213 | 214 |



**Arizona state facts**

Nickname: The Grand Canyon State
Capital: Phoenix; J-7
Population: 6,500,180 (rank: 14th)
Largest city: Phoenix, 1,567,924, J-7
Land area: 113,635 sq. mi. (rank: 6th)
Highest point: Humphreys Peak, 12,633 ft. E-8

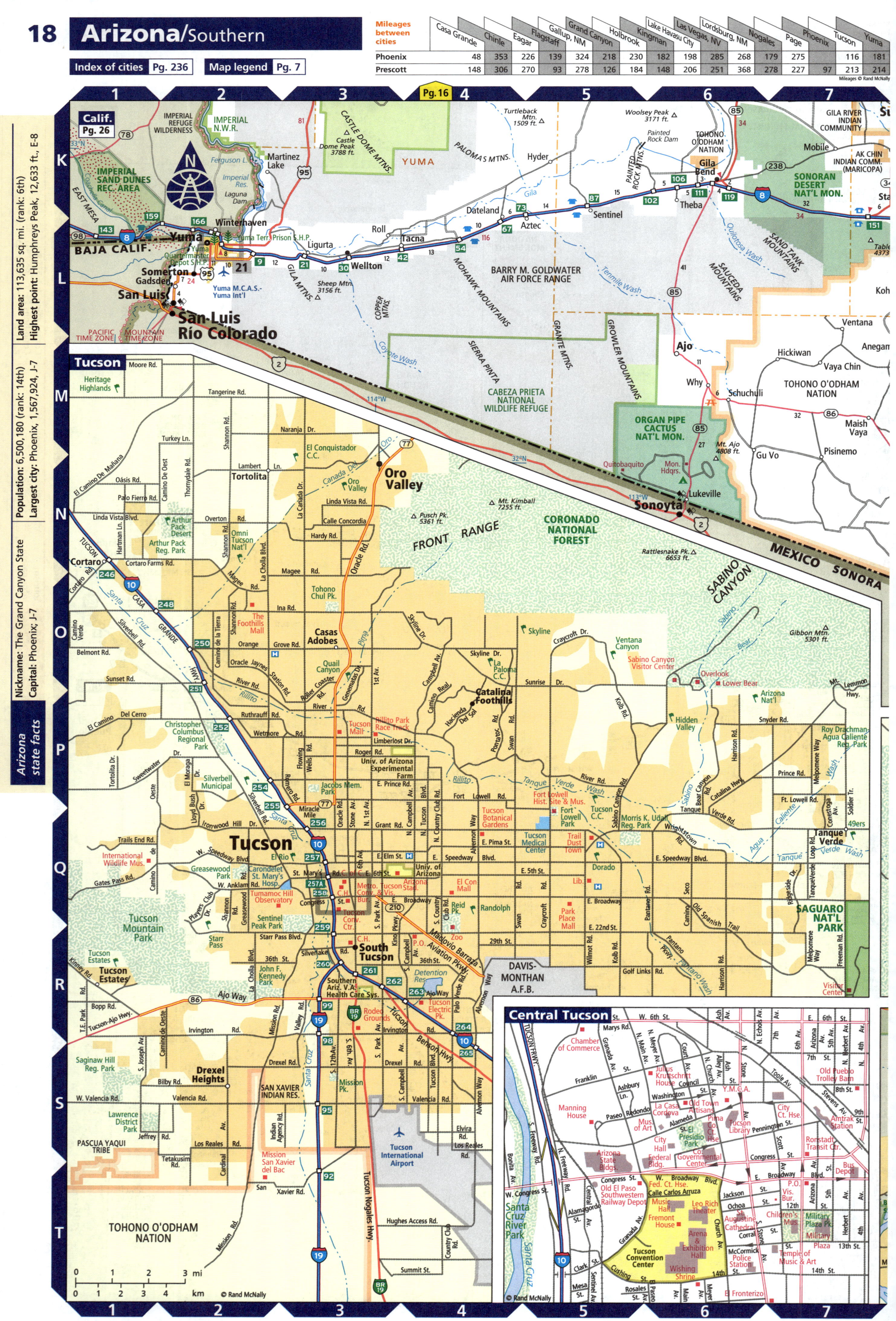

Mileages between cities

| | Casa Grande | Chinle | Eagar | Flagstaff | Gallup, NM | Grand Canyon | Holbrook | Kingman | Lake Havasu City | Las Vegas, NV | Lordsburg, NM | Nogales | Page | Phoenix | Tucson | Yuma |
|---|---|---|---|---|---|---|---|---|---|---|---|---|---|---|---|---|
| Tucson | 66 | 361 | 238 | 255 | 333 | 334 | 238 | 297 | 314 | 401 | 156 | 66 | 390 | 116 | | 236 |
| Yuma | 172 | 532 | 399 | 318 | 502 | 397 | 409 | 213 | 155 | 292 | 392 | 301 | 453 | 181 | 236 | |

Mileages © Rand McNally

Total mileages through Arizona: 8 178 miles; 10 392 miles; 17 146 miles; 40 359 miles

More mileages at go.randmcnally.com/MC

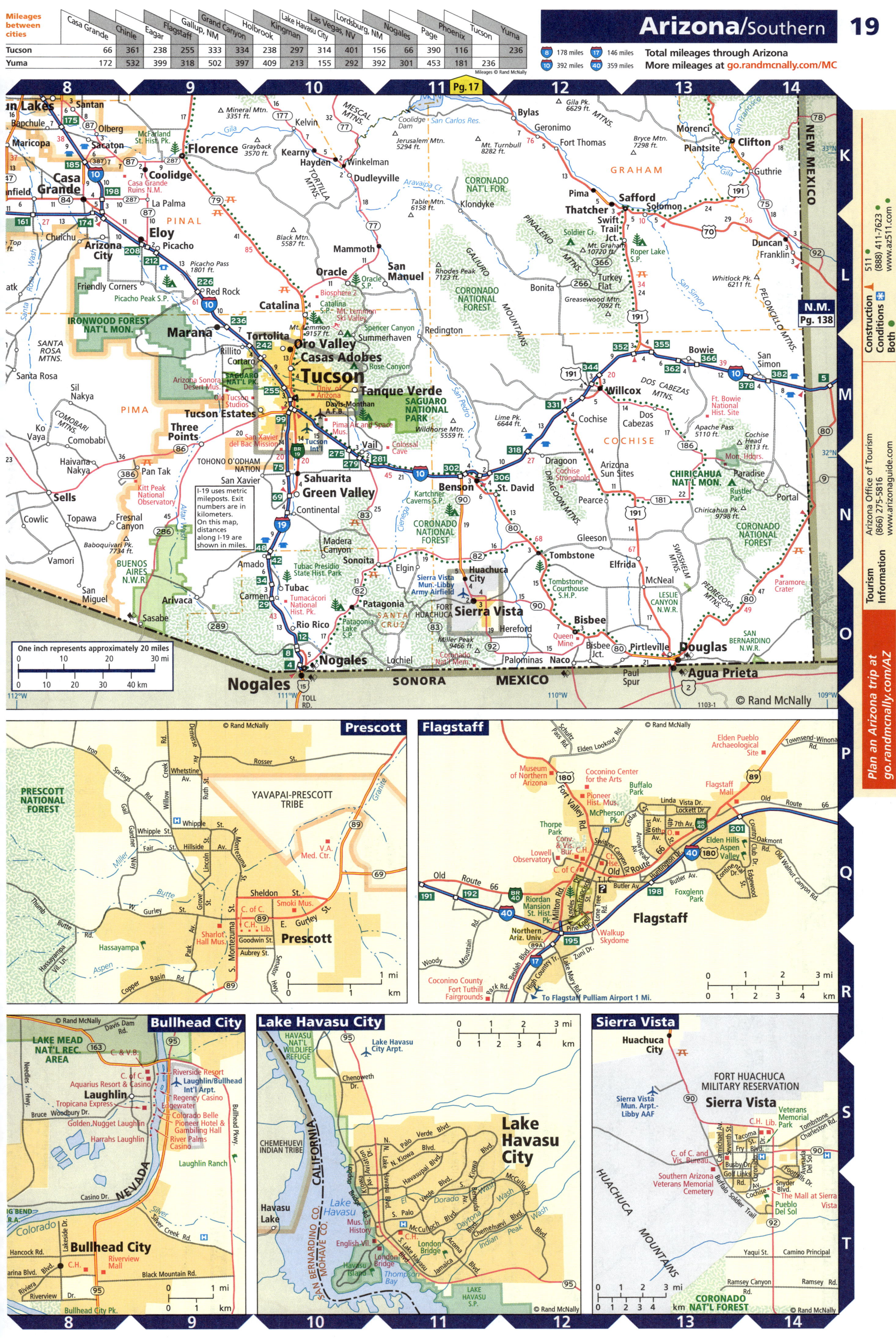

## City sights to see

- Arizona Historical Society Sanguinetti House Museum, Yuma, L-6
- Arizona Science Center, Phoenix, M-3
- Arizona State Capitol, Phoenix, M-1
- Heard Museum, Phoenix, L-2
- Mesa Southwest Museum, Mesa, J-7
- Painted Desert Inn Museum, Petrified Forest National Park, L-10

- Phoenix Art Museum, Phoenix, L-2
- Taliesin West, Scottsdale, H-7
- Tusayan Ruin and Museum, Grand Canyon National Park, D-9
- Yavapai Observation Station, Grand Canyon National Park, D-8
- Yuma Territorial Prison State Historic Park, Yuma, L-6

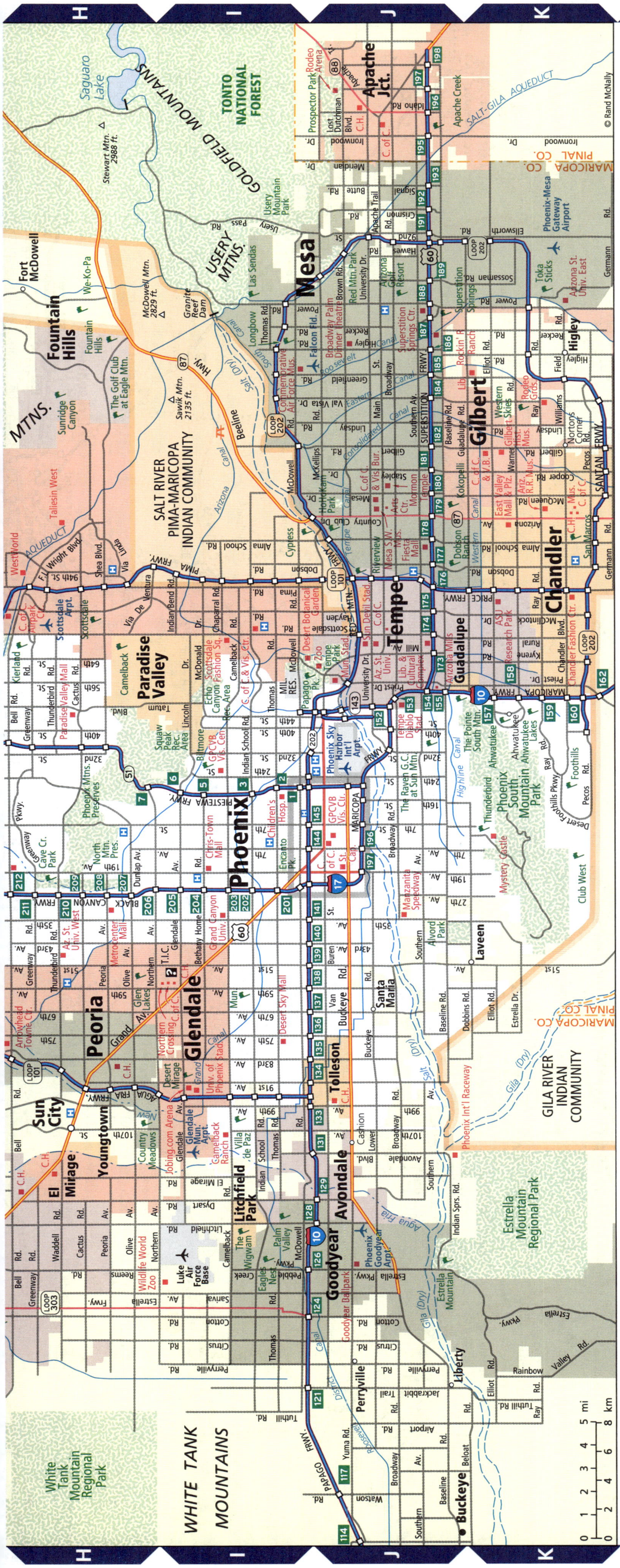

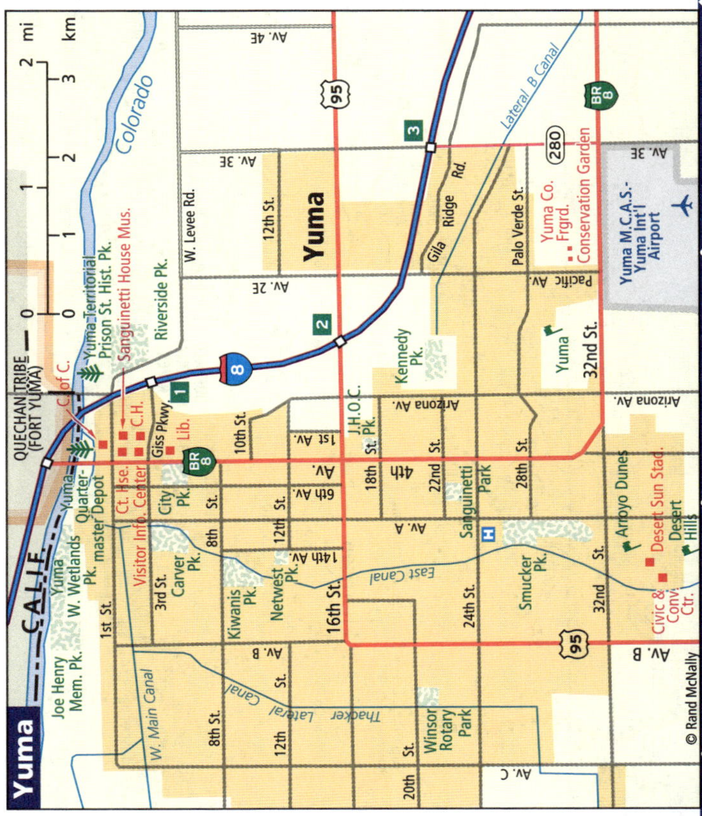

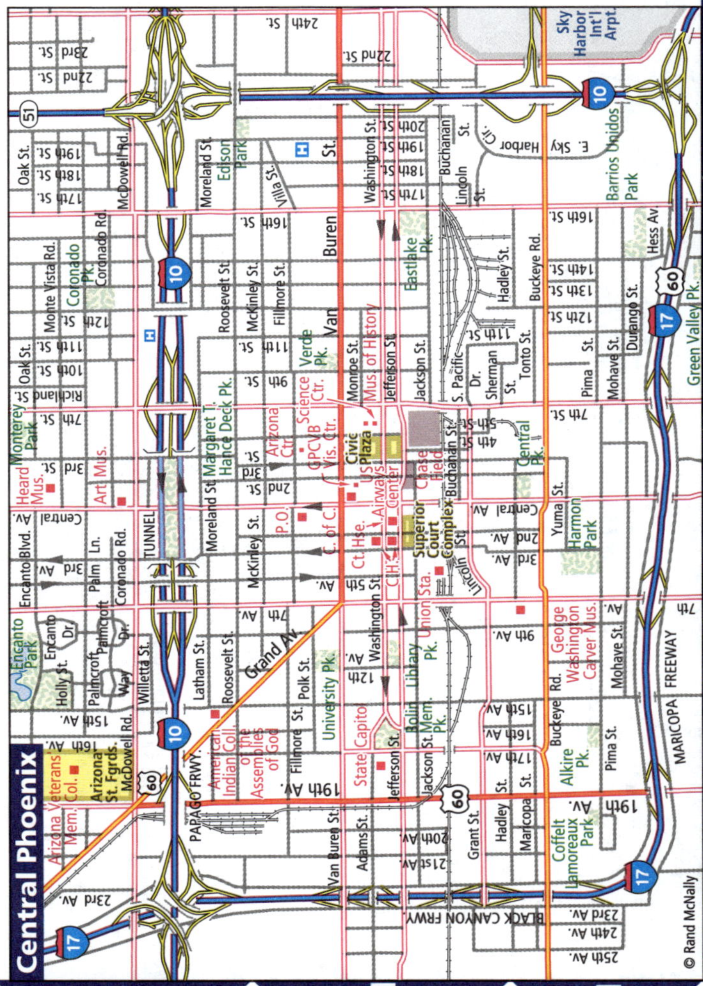

Index of cities Pg. 237 | Map legend Pg. 7

Arkansas state facts

Nickname: The Natural State
Capital: Little Rock, G-7
Population: 2,855,390 (rank: 32nd)
Largest city: Little Rock, 189,515, G-7
Land area: 52,068 sq. mi. (rank: 27th)
Highest point: Magazine Mtn., 2753 ft., F-3

Mileages between cities

| | Batesville | Branson, MO | DeQueen | El Dorado | Fayetteville | Fort Smith | Greenville, MS | Hot Springs | Jonesboro | Little Rock | Memphis, TN | Mountain Home | Pine Bluff | Rogers | Russellville | Texarkana |
|---|---|---|---|---|---|---|---|---|---|---|---|---|---|---|---|---|
| El Dorado | 209 | 287 | 141 | | 304 | 227 | 109 | 121 | 245 | 118 | 250 | 268 | 91 | 325 | 190 | 88 |
| Fayetteville | 251 | 98 | 184 | 304 | | 58 | 335 | 184 | 250 | 188 | 318 | 123 | 231 | 24 | 115 | 236 |



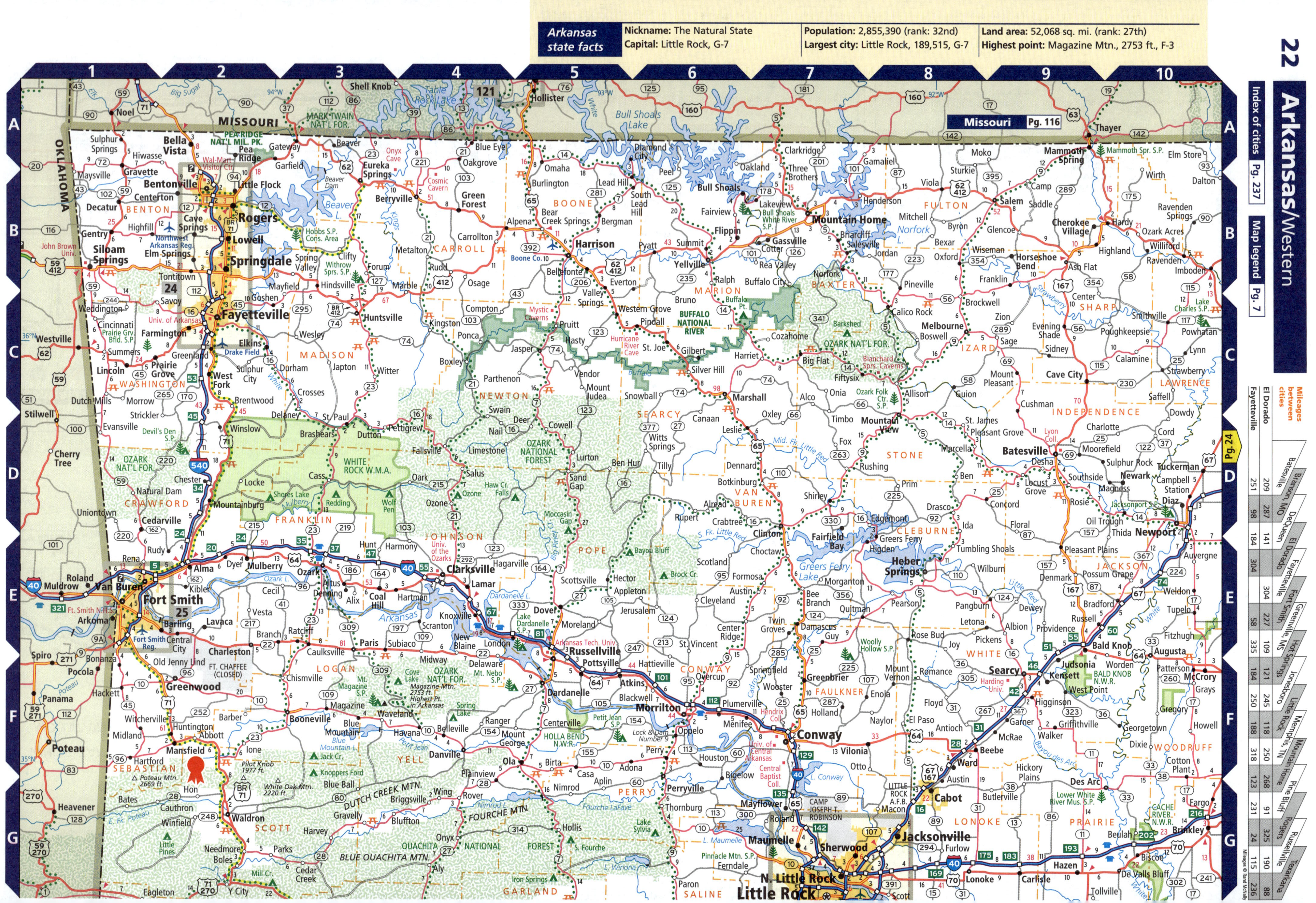

Mileages between cities

| | Batesville | Branson, MO | DeQueen | El Dorado | Fayetteville | Fort Smith | Greenville, MS | Hot Springs | Jonesboro | Little Rock | Memphis, TN | Mountain Home | Pine Bluff | Rogers | Russellville | Texarkana |
|---|---|---|---|---|---|---|---|---|---|---|---|---|---|---|---|---|
| Fort Smith | 219 | 158 | 130 | 227 | 58 | | 304 | 130 | 261 | 158 | 286 | 187 | 199 | 81 | 84 | 182 |
| Jonesboro | 68 | 203 | 272 | 245 | 250 | 261 | 219 | 182 | | 130 | 70 | 126 | 171 | 253 | 173 | 270 |

continued p. 24



Total mileages through Arkansas

| | | | |
|---|---|---|---|
| 30 | 143 miles | 55 | 72 miles |
| 40 | 284 miles | 65 | 309 miles |

More mileages at go.randmcnally.com/MC

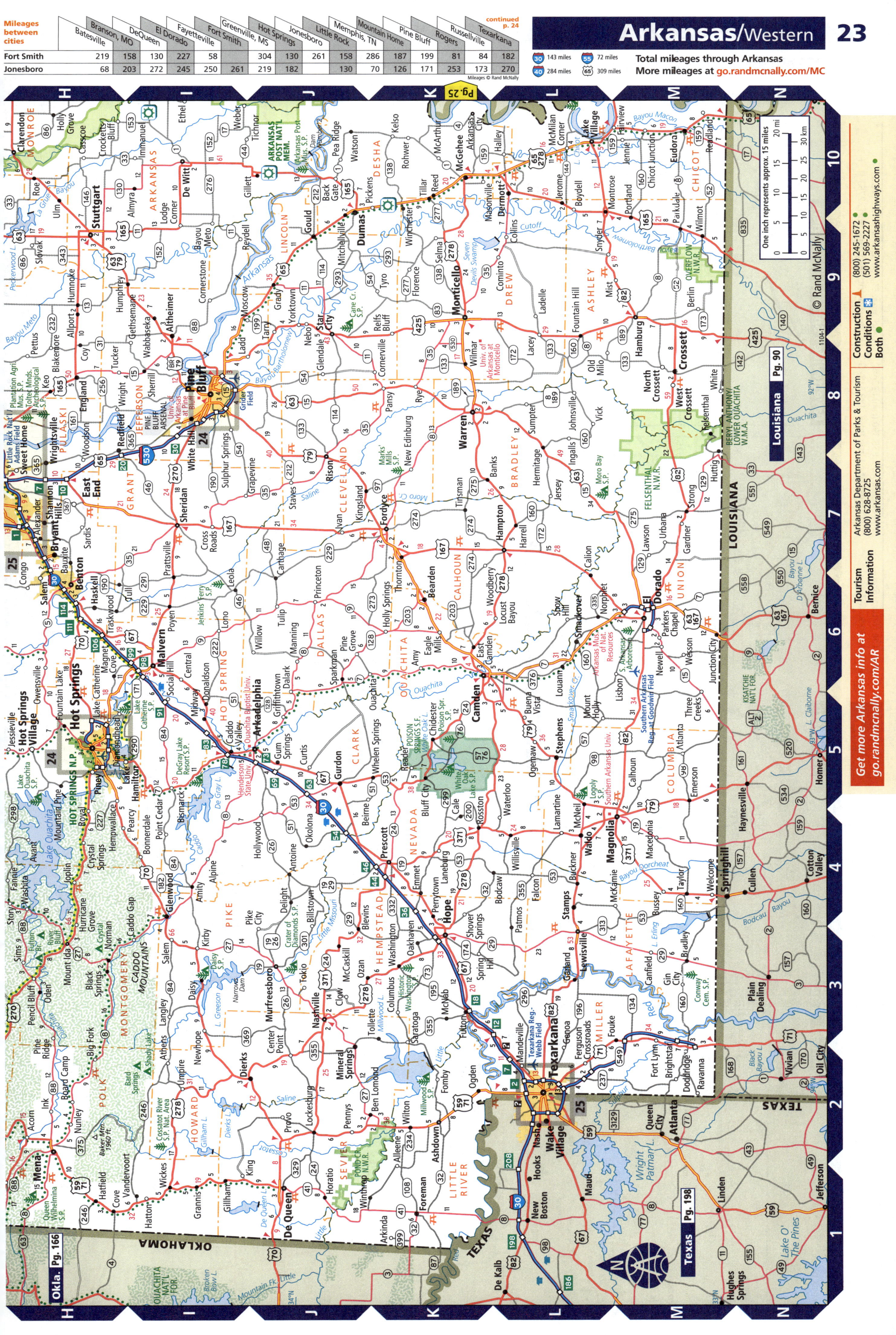

Index of cities Pg. 237 | Map legend Pg. 7

| Mileages between cities | Batesville | Branson, MO | DeQueen | El Dorado | Fayetteville | Fort Smith | Greenville, MS | Hot Springs | Jonesboro | Little Rock | Memphis, TN | Mountain Home | Pine Bluff | Rogers | Russellville | Texarkana |
|---|---|---|---|---|---|---|---|---|---|---|---|---|---|---|---|---|
| **Little Rock** | 94 | 172 | 143 | 118 | 188 | 158 | 147 | 54 | 130 | | 137 | 151 | 43 | 208 | 74 | 142 |
| **Memphis, TN** | 119 | 274 | 278 | 250 | 318 | 286 | 152 | 188 | 70 | 137 | | 195 | 152 | 339 | 204 | 276 |



### Arkansas state facts

**Nickname:** The Natural State
**Capital:** Little Rock, G-7
**Population:** 2,855,390 (rank: 32nd)
**Largest city:** Little Rock, 189,515, G-7
**Land area:** 52,068 sq. mi. (rank: 27th)
**Highest point:** Magazine Mtn., 2753 ft., F-3

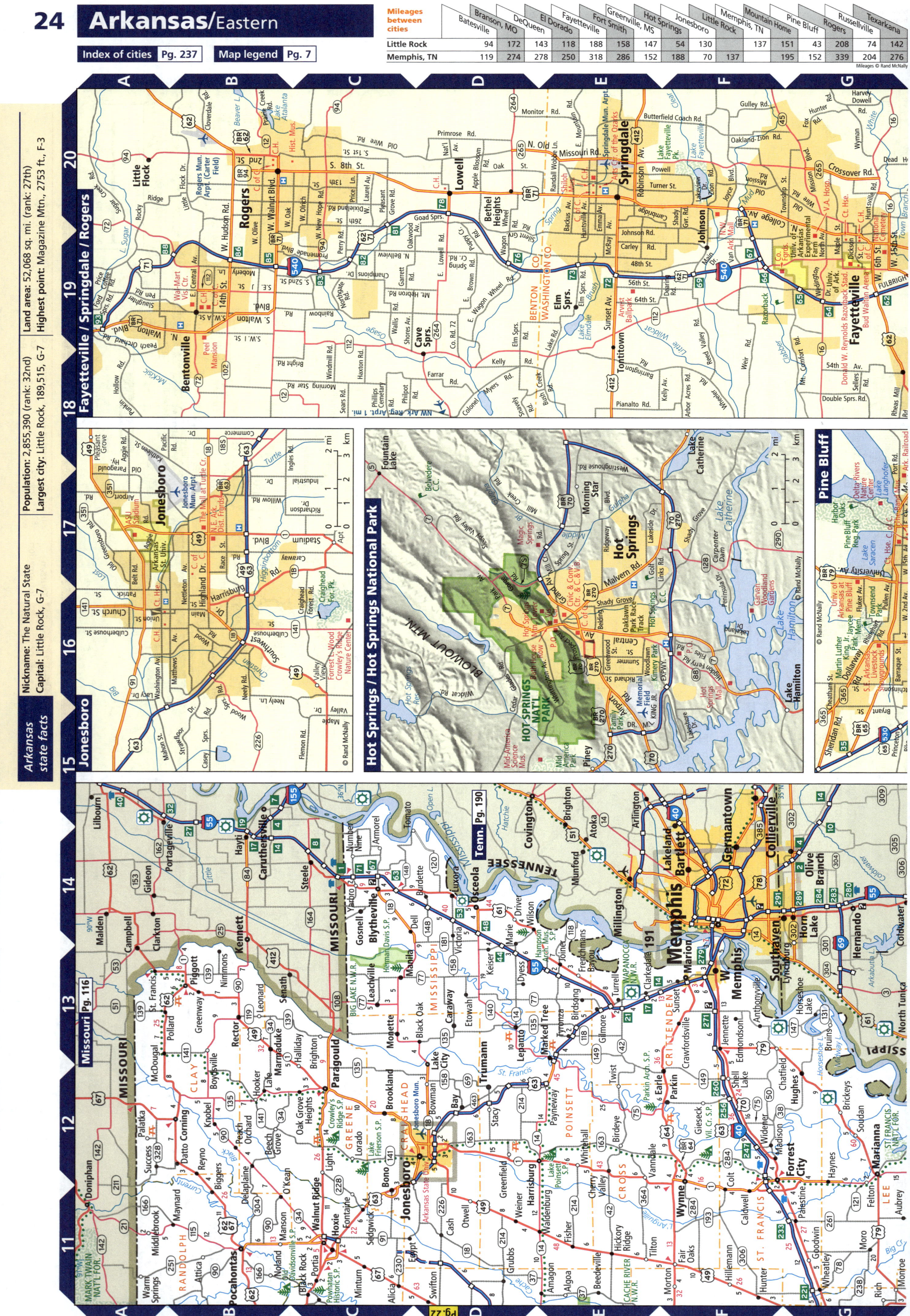

**Mileages between cities**

| | Batesville | Branson, MO | DeQueen | El Dorado | Fayetteville | Fort Smith | Greenville, MS | Hot Springs | Jonesboro | Little Rock | Memphis, TN | Mountain Home | Pine Bluff | Rogers | Russellville | Texarkana |
|---|---|---|---|---|---|---|---|---|---|---|---|---|---|---|---|---|
| Mountain Home | 78 | 83 | 287 | 268 | 123 | 187 | 298 | 198 | 126 | 151 | 195 | | 194 | 126 | 125 | 287 |
| Texarkana | 234 | 306 | 54 | 88 | 236 | 182 | 198 | 110 | 270 | 142 | 276 | 287 | 152 | 258 | 209 | |

Mileages © Rand McNally

30: 143 miles · 40: 284 miles · 55: 72 miles · 65: 309 miles

Total mileages through Arkansas
More mileages at go.randmcnally.com/MC

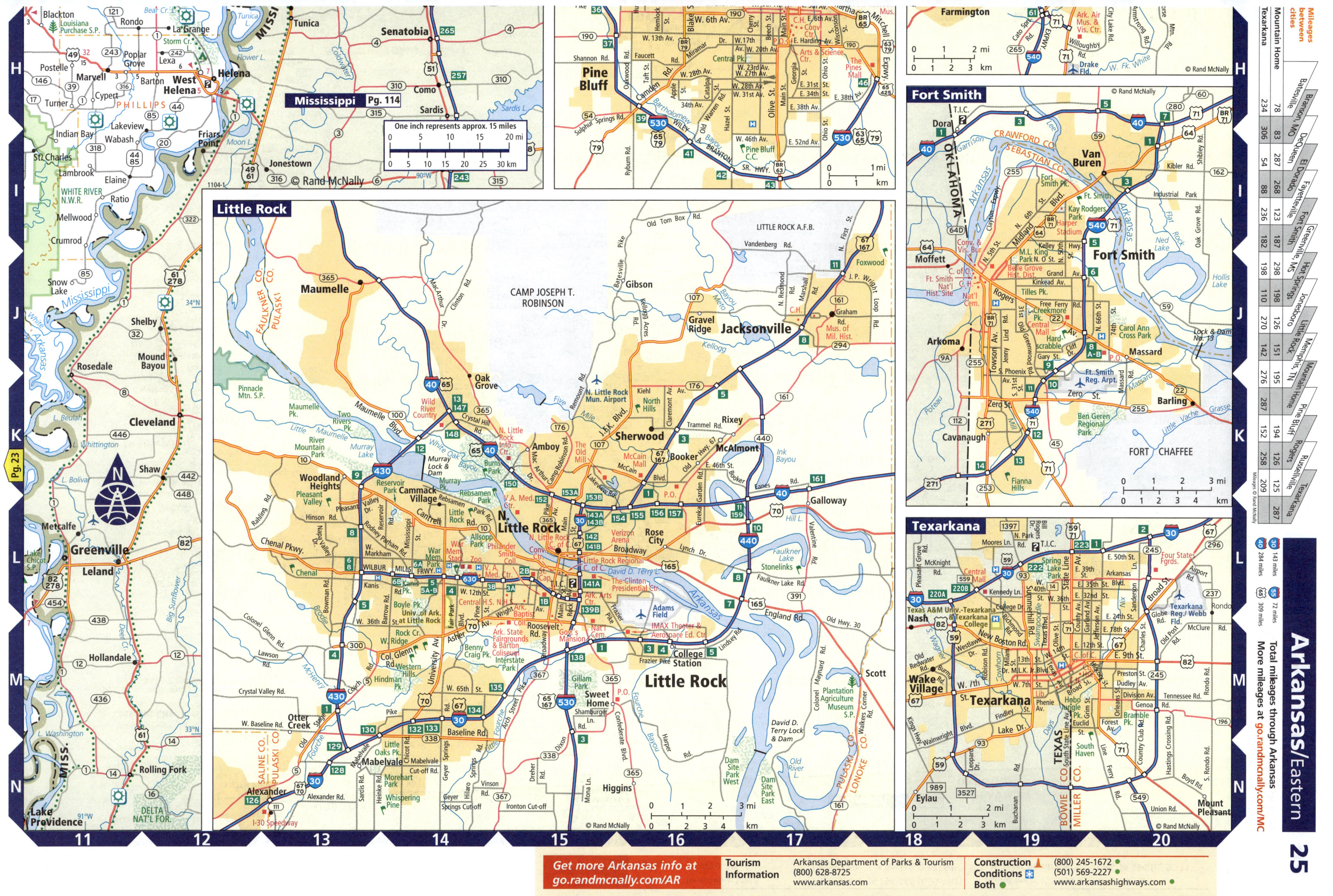

**California state facts**

Nickname: The Golden State
Capital: Sacramento, NK-7
Population: 36,756,666 (rank: 1st)
Largest city: Los Angeles, 3,833,995, SJ-11
Land area: 155,959 sq. mi. (rank: 3rd)
Highest point: Mt. Whitney, 14,494 ft., SC-11



Index of cities Pg. 237 | Map legend Pg. 7

Mileages between cities

| | Bishop | Crescent City | Los Angeles | Oroville | Redding | Sacramento | San Francisco | San Jose | Santa Rosa | S. Lake Tahoe | Stockton | Susanville | Ukiah | Vallejo | Yosemite N.P. | Yreka |
|---|---|---|---|---|---|---|---|---|---|---|---|---|---|---|---|---|
| Alturas | 371 | 280 | 648 | 225 | 144 | 302 | 357 | 385 | 365 | 228 | 349 | 103 | 330 | 329 | 392 | 176 |
| Bishop | | 614 | 265 | 326 | 400 | 269 | 295 | 290 | 364 | 176 | 224 | 286 | 418 | 328 | 138 | 454 |



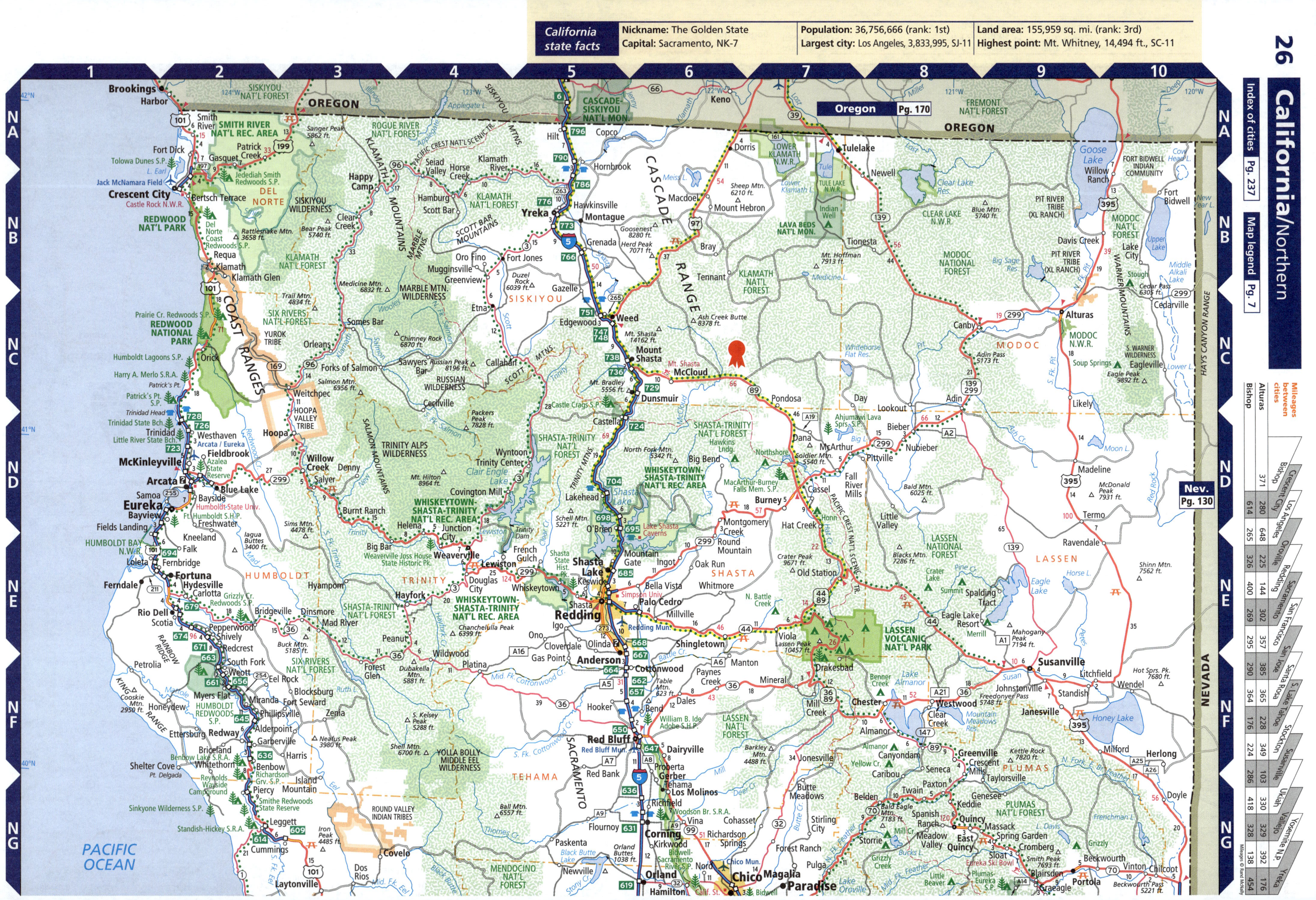

**Mileages between cities**

| | Bishop | Crescent City | Los Angeles | Oroville | Redding | Sacramento | San Francisco | San Jose | Santa Rosa | S. Lake Tahoe | Stockton | Susanville | Ukiah | Vallejo | Yosemite N.P. | Yreka |
|---|---|---|---|---|---|---|---|---|---|---|---|---|---|---|---|---|
| **Eureka** | 546 | 81 | 644 | 222 | 146 | 289 | 272 | 315 | 217 | 392 | 325 | 259 | 158 | 262 | 454 | 198 |
| **Redding** | 400 | 208 | 544 | 94 | | 161 | 216 | 244 | 198 | 264 | 209 | 112 | 188 | 187 | 332 | 98 |

Mileages © Rand McNally

continued p. 28

5 797 miles 101 791 miles 80 199 miles

**Total mileages through California**
**More mileages at go.randmcnally.com/MC**

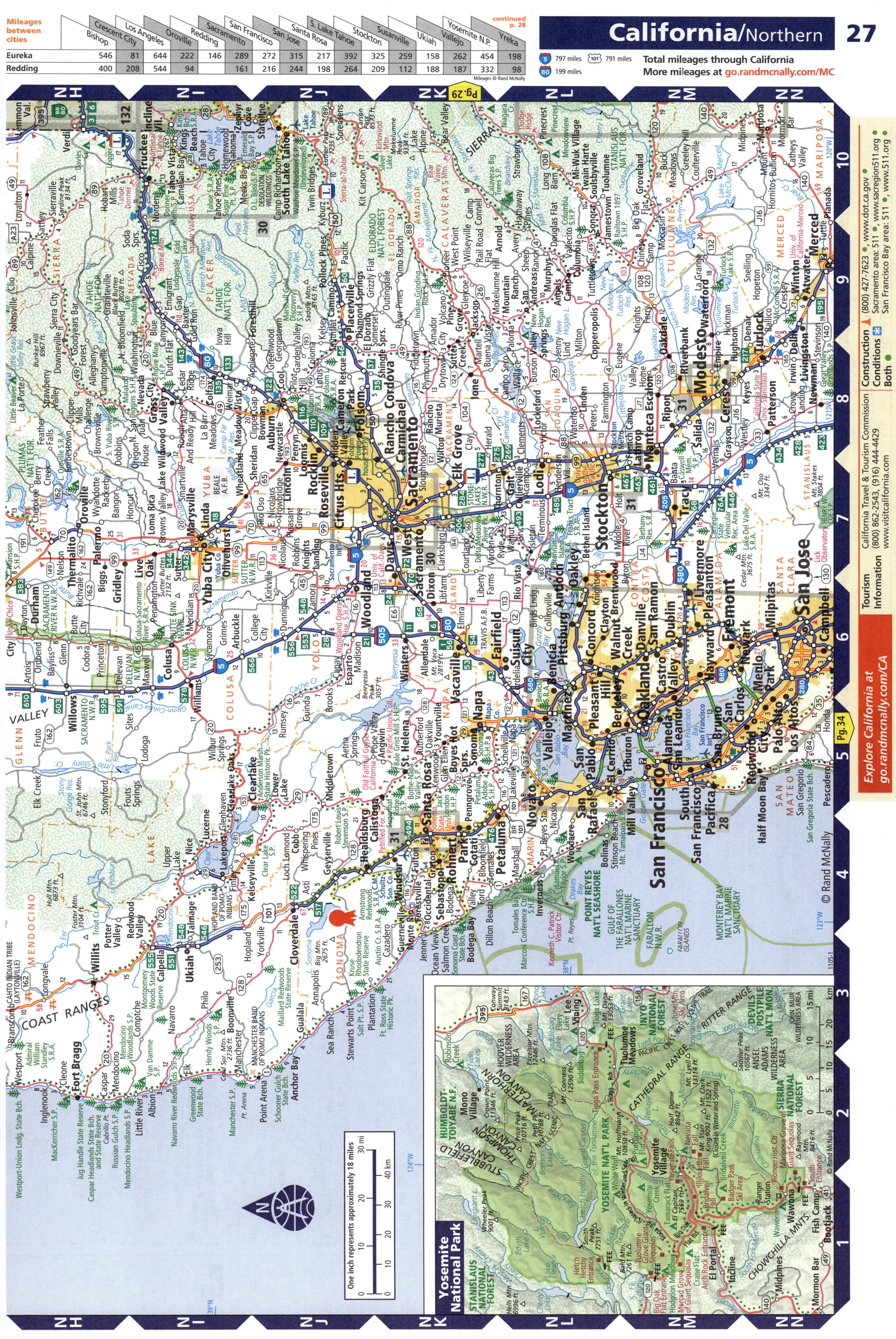


*Explore California at go.randmcnally.com/CA*

| | |
|---|---|
| **Tourism Information** | California Travel & Tourism Commission (800) 862-2543, (916) 444-4429 www.visitcalifornia.com |
| **Construction** ▲ **Conditions** ❄ **Both** ● | (800) 427-7623 ●, www.dot.ca.gov ● Sacramento area: 511 ●, www.sacregion511.org ● San Francisco Bay area: 511 ●, www.511.org ● |

Index of cities Pg. 237
Map legend Pg. 7

**Mileages between cities**

| | Bishop | Crescent City | Los Angeles | Oroville | Redding | Sacramento | San Francisco | San Jose | Santa Rosa | S. Lake Tahoe | Stockton | Susanville | Ukiah | Vallejo | Yosemite N.P. | Yreka |
|---|---|---|---|---|---|---|---|---|---|---|---|---|---|---|---|---|
| **Sacramento** | 269 | 372 | 383 | 68 | 161 | | 87 | 115 | 95 | 100 | 47 | 217 | 145 | 58 | 160 | 257 |
| **San Francisco** | 295 | 355 | 380 | 150 | 216 | 87 | | 45 | 55 | 187 | 82 | 303 | 115 | 30 | 189 | 312 |



*California state facts*

| | | |
|---|---|---|
| **Nickname:** The Golden State | **Population:** 36,756,666 (rank: 1st) | **Land area:** 155,959 sq. mi. (rank: 3rd) |
| **Capital:** Sacramento, NK-7 | **Largest city:** Los Angeles, 3,833,995, SJ-11 | **Highest point:** Mt. Whitney, 14,494 ft., SC-11 |

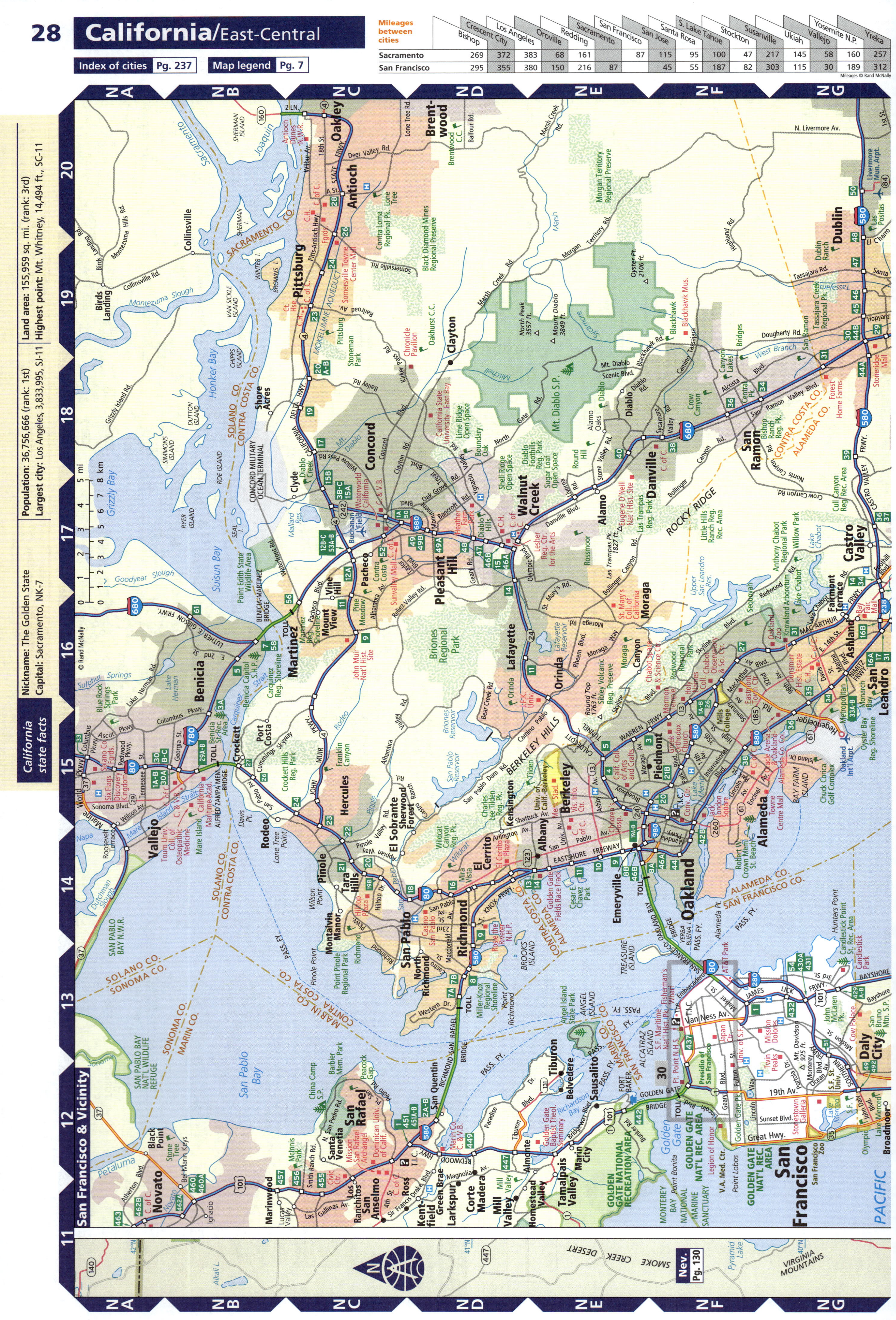

Mileages between cities

| | Bishop | Crescent City | Los Angeles | Oroville | Redding | Sacramento | San Francisco | San Jose | Santa Rosa | S. Lake Tahoe | Stockton | Susanville | Ukiah | Vallejo | Yosemite N.P. | Yreka |
|---|---|---|---|---|---|---|---|---|---|---|---|---|---|---|---|---|
| San Jose | 290 | 396 | 340 | 178 | 244 | 115 | 45 | | 96 | 215 | 74 | 330 | 156 | 64 | 182 | 340 |
| S. Lake Tahoe | 176 | 472 | 445 | 157 | 264 | 100 | 187 | 215 | 195 | | 147 | 143 | 248 | 159 | 189 | 311 |



5 797 miles · 101 791 miles · 80 199 miles

Total mileages through California
More mileages at go.randmcnally.com/MC

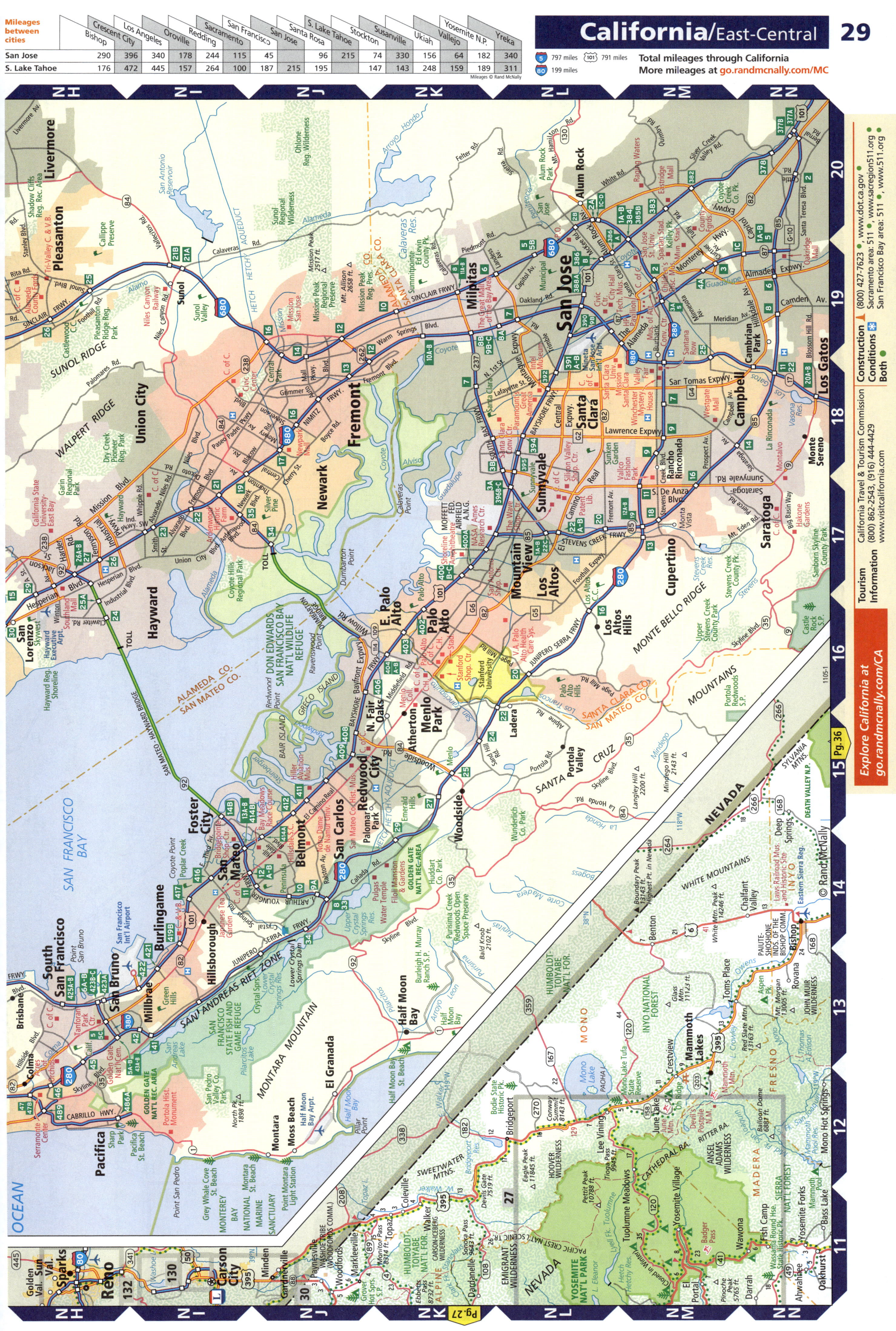


Tourism Information: California Travel & Tourism Commission (800) 862-2543, (916) 444-4429 www.visitcalifornia.com

Construction / Conditions / Both: (800) 427-7623 • www.dot.ca.gov • Sacramento area: 511 • www.sacregion511.org • San Francisco Bay area: 511 • www.511.org

Explore California at go.randmcnally.com/CA

**City sights to see**

- AT&T Park, San Francisco, E-10
- California State Capitol, Sacramento, I-6
- California State Railroad Museum, Sacramento, H-6
- Chinatown, San Francisco, C-8
- Coit Memorial Tower, San Francisco, B-8
- Crocker Art Museum, Sacramento, I-5
- Fisherman's Wharf, San Francisco, A-7

- Ghirardelli Square, San Francisco, B-7
- Golden Gate Bridge, San Francisco, A-2
- Monterey Bay Aquarium, Monterey, M-1
- National Steinbeck Center, Salinas, K-5
- Pier 39, San Francisco, A-8
- San Francisco Cable Car Museum, San Francisco, C-8
- Squaw Valley U.S.A., Olympic Valley, F-8

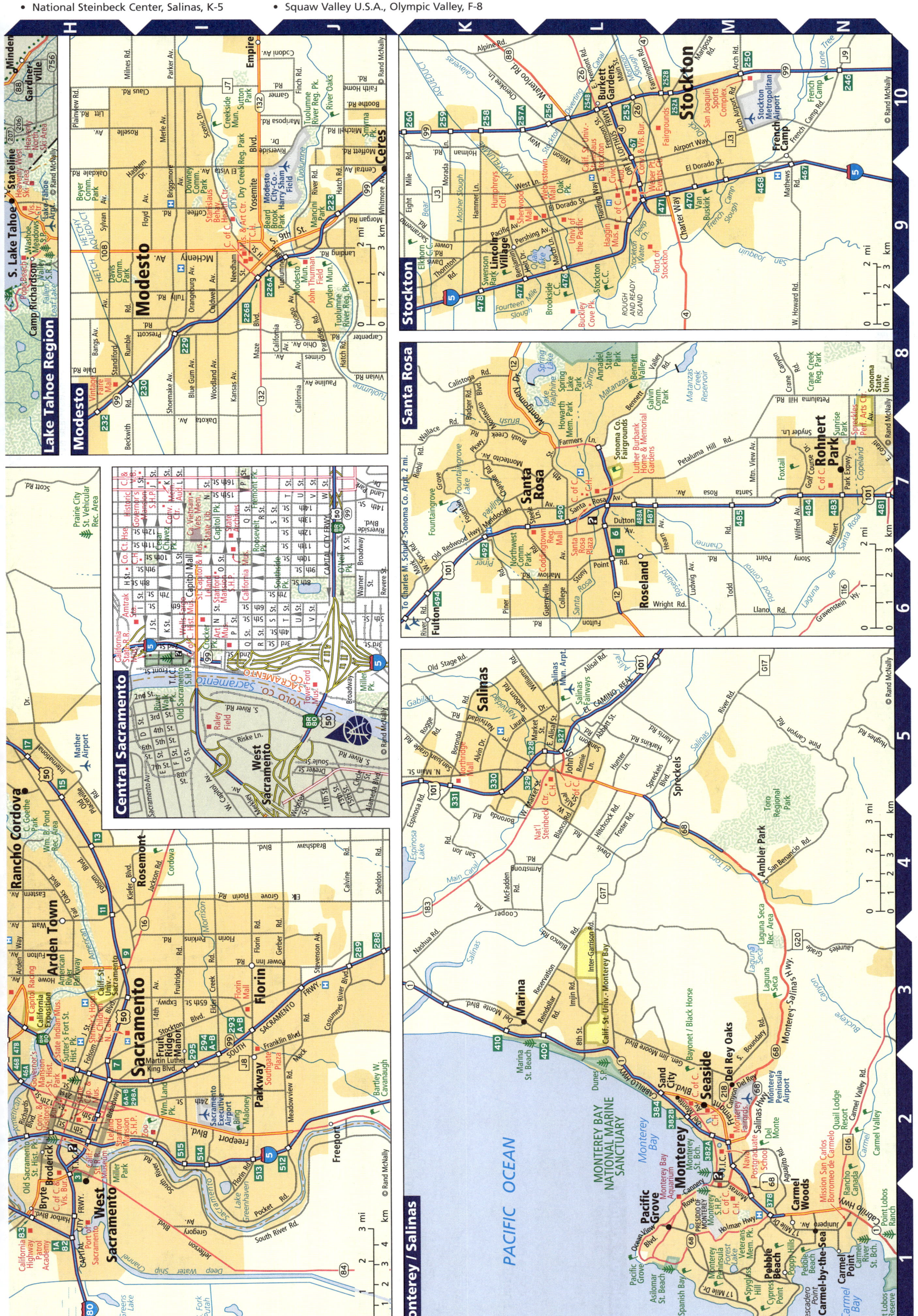

## City sights to see

- Balboa Park, San Diego, K-10
- Birch Aquarium at Scripps Institute, San Diego, G-1
- Cabrillo National Monument, San Diego, K-1
- Channel Islands National Park Visitor Center & Headquarters, Ventura, B-8
- Gaslamp Quarter Historic District, San Diego, M-9

- LEGOLAND California, Carlsbad, J-8
- The Living Desert Nature Preserve, Palm Desert, G-10
- Museum of Contemporary Art, San Diego, L-8
- Palm Springs Desert Museum, Palm Springs, E-7
- San Diego Zoo, San Diego, J-3
- SeaWorld, San Diego, I-2
- Stearns Wharf, Santa Barbara, B-5

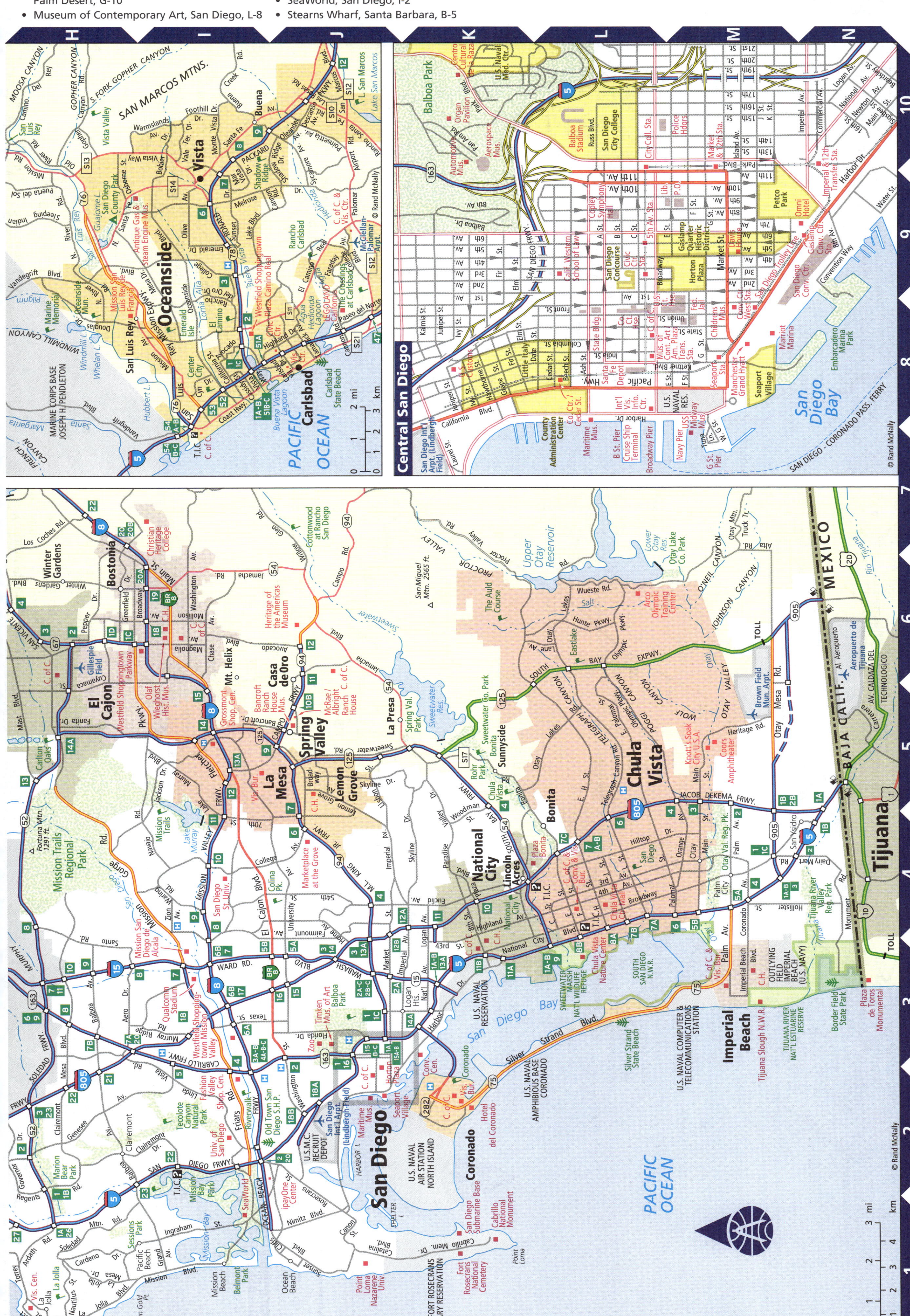

Index of cities Pg. 237

Map legend Pg. 7

**Mileages between cities**

| | Bakersfield | Barstow | El Centro | Fresno | Las Vegas, NV | Los Angeles | Monterey | Needles | Palm Springs | Riverside | San Bernardino | San Diego | San Francisco | San Luis Obispo | Santa Barbara | Sequoia N.P. |
|---|---|---|---|---|---|---|---|---|---|---|---|---|---|---|---|---|
| **Bakersfield** | | 129 | 322 | 109 | 286 | 112 | 222 | 272 | 216 | 166 | 166 | 232 | 284 | 130 | 147 | 122 |
| **Fresno** | 109 | 239 | 429 | | 395 | 218 | 150 | 381 | 323 | 271 | 273 | 339 | 183 | 130 | 254 | 77 |

Mileages © Rand McNally

***California state facts***

**Nickname:** The Golden State
**Capital:** Sacramento, NK-7

**Population:** 36,756,666 (rank: 1st)
**Largest city:** Los Angeles, 3,833,995, SJ-11

**Land area:** 155,959 sq. mi. (rank: 3rd)
**Highest point:** Mt. Whitney, 14,494 ft., SC-11

Total mileages through California

| | |
|---|---|
| I-5 | 797 miles |
| I-10 | 243 miles |
| I-15 | 287 miles |
| I-40 | 155 miles |

More mileages at go.randmcnally.com/MC

**Mileages between cities**

| | Bakersfield | Barstow | El Centro | Fresno | Las Vegas, NV | Los Angeles | Monterey | Needles | Palm Springs | Riverside | San Bernardino | San Diego | San Francisco | San Luis Obispo | Santa Barbara | Sequoia N.P. |
|---|---|---|---|---|---|---|---|---|---|---|---|---|---|---|---|---|
| Las Vegas, NV | 286 | 156 | 312 | 395 | | 270 | 507 | 110 | 278 | 234 | 225 | 331 | 569 | 415 | 358 | 410 |
| Los Angeles | 112 | 114 | 212 | 218 | 270 | | 319 | 256 | 107 | 54 | 60 | 120 | 380 | 189 | 94 | 232 |

continued p. 36



Pg. 37

## Bakersfield

## Sequoia & Kings Canyon National Parks



**Tourism Information**
California Travel & Tourism Commission
(800) 862-2543, (916) 444-4429
www.visitcalifornia.com

**Construction** (800) 427-7623 • www.dot.ca.gov
**Conditions** San Diego area: 511 • www.511sd.com
**Both**

Get more California info at go.randmcnally.com/CA

Index of cities Pg. 237 | Map legend Pg. 7

Mileages between cities

| | Bakersfield | Barstow | El Centro | Fresno | Las Vegas, NV | Los Angeles | Monterey | Needles | Palm Springs | Riverside | San Bernardino | San Diego | San Francisco | San Luis Obispo | Santa Barbara | Sequoia N.P. |
|---|---|---|---|---|---|---|---|---|---|---|---|---|---|---|---|---|
| Monterey | 222 | 350 | 530 | 150 | 507 | 319 | | 494 | 424 | 372 | 373 | 439 | 112 | 142 | 237 | 226 |
| Palm Springs | 216 | 123 | 108 | 323 | 278 | 107 | 424 | 188 | | 52 | 54 | 139 | 486 | 296 | 201 | 338 |

Mileages © Rand McNally

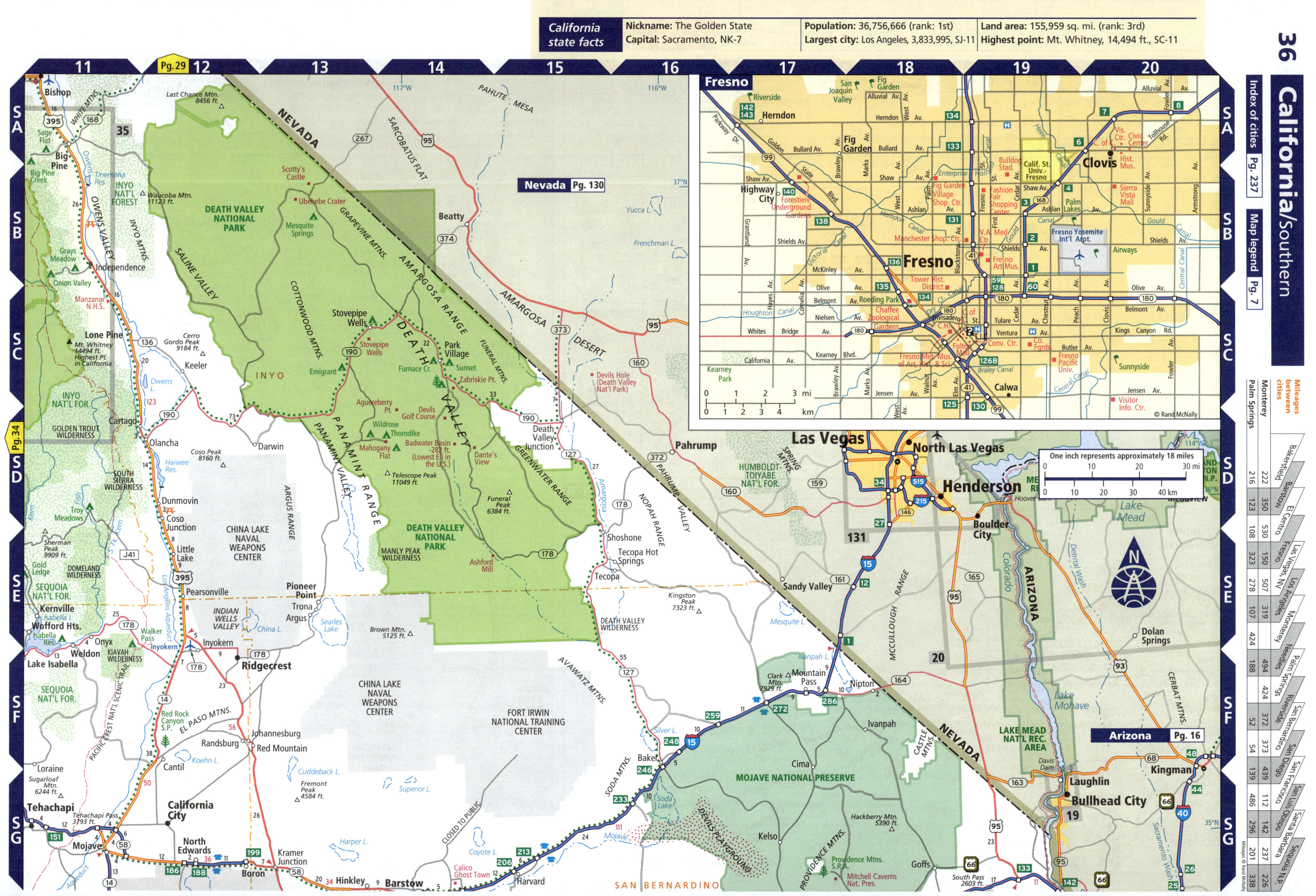

Mileages between cities

| | Bakersfield | Barstow | El Centro | Fresno | Las Vegas, NV | Los Angeles | Monterey | Needles | Palm Springs | Riverside | San Bernardino | San Diego | San Francisco | San Luis Obispo | Santa Barbara | Sequoia N.P. |
|---|---|---|---|---|---|---|---|---|---|---|---|---|---|---|---|---|
| San Diego | 232 | 176 | 113 | 339 | 331 | 120 | 439 | 317 | 139 | 97 | 106 | | 501 | 313 | 214 | 352 |
| Santa Barbara | 147 | 203 | 306 | 254 | 358 | 94 | 237 | 345 | 201 | 148 | 150 | 214 | 325 | 94 | | 268 |

Mileages © Rand McNally

5: 797 miles · 15: 287 miles · 10: 243 miles · 40: 155 miles

Total mileages through California
More mileages at go.randmcnally.com/MC

Get more California info at go.randmcnally.com/CA

Tourism Information: California Travel & Tourism Commission (800) 862-2543, (916) 444-4429 www.visitcalifornia.com

Construction / Conditions / Both: (800) 427-7623, www.dot.ca.gov; San Diego area: 511, www.511sd.com

City sights to see (pages 38-41)

- Aquarium of the Pacific, Long Beach, I-8
- Disneyland, Anaheim, I-11
- Dodger Stadium, Los Angeles, I-1
- El Pueblo de Los Angeles, Los Angeles, K-2
- Exposition Park, Los Angeles, F-7
- Getty Center, Los Angeles, E-4
- Grand Central Market, Los Angeles, L-2
- Hollywood Bowl, Los Angeles, D-6

- Huntington Library, San Marino, D-9
- Japanese American National Museum, Los Angeles, K-3
- Knott's Berry Farm, Buena Park, H-10
- Los Angeles Co. Art Mus., Los Angeles, E-6
- Los Angeles Maritime Museum, Los Angeles, J-7
- Los Angeles Zoo and Botanical Gardens, Los Angeles, D-7

list continued p.40

**City sights to see (pages 38-41)**

- Mission San Juan Capistrano, San Juan Capistrano, M-14
- Mount Wilson Observatory, Mt. Wilson, C-9
- Old Pasadena, Pasadena, D-8
- Oldest Winery in Calif., Rancho Cucamonga, D-15
- The Queen Mary, Long Beach, J-8
- Richard M. Nixon Library & Birthplace, Yorba Linda, H-12

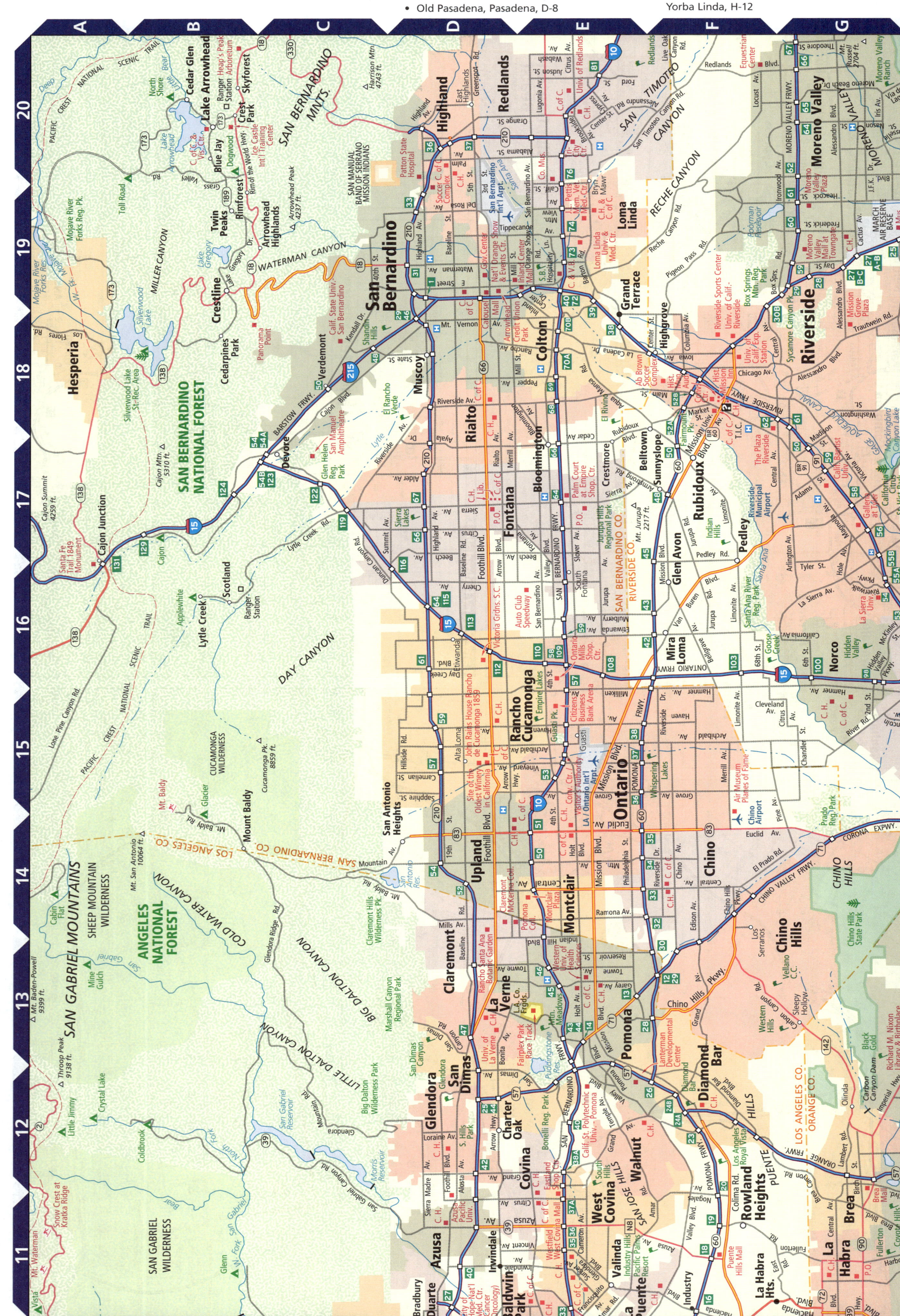

- Rose Bowl, Pasadena, D-8
- Santa Monica Pier, Santa Monica, F-4
- Universal City, Los Angeles, D-6
- Venice Boardwalk, Los Angeles, F-4
- Walt Disney Concert Hall, Los Angeles, K-2
- Warner Bros. Studio, Burbank, D-6
- Will Rogers State Historic Park, Pacific Palisades, E-4

Index of cities Pg. 238 | Map legend Pg. 7

**Mileages between cities**

| | Alamosa | Aspen | Burlington | Colorado Springs | Craig | Denver | Durango | Estes Park | Fort Collins | Grand Junction | Gunnison | Lamar | Leadville | Pueblo | Sterling | Trinidad |
|---|---|---|---|---|---|---|---|---|---|---|---|---|---|---|---|---|
| Burlington | 311 | 363 | | 151 | 363 | 166 | 460 | 222 | 220 | 408 | 324 | 108 | 265 | 189 | 142 | 230 |
| Colorado Springs | 163 | 155 | 151 | | 264 | 69 | 313 | 133 | 133 | 309 | 166 | 158 | 121 | 42 | 194 | 128 |



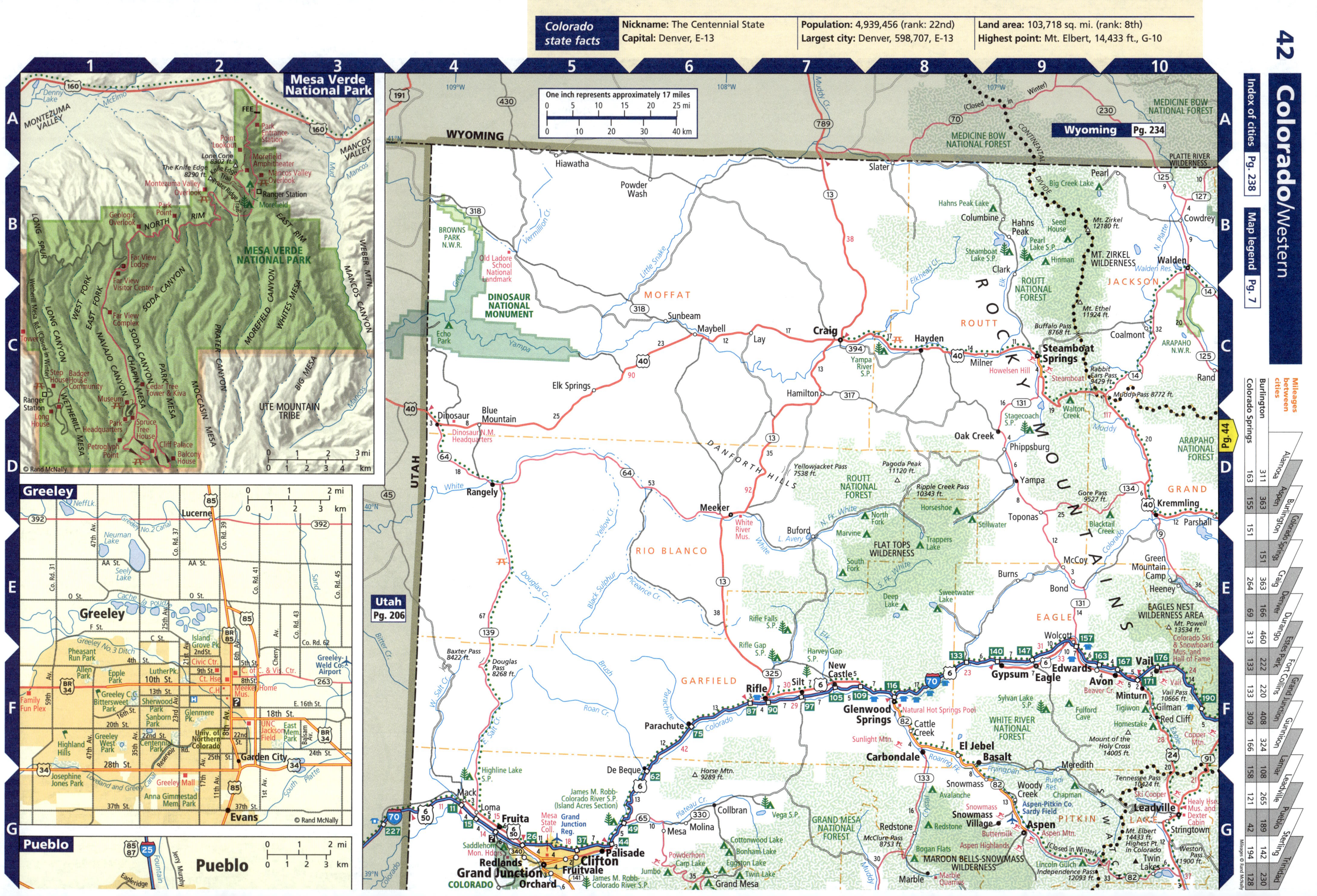

continued p. 44

Mileages between cities

| | Alamosa | Aspen | Burlington | Colorado Springs | Craig | Denver | Durango | Estes Park | Fort Collins | Grand Junction | Gunnison | Lamar | Leadville | Pueblo | Sterling | Trinidad |
|---|---|---|---|---|---|---|---|---|---|---|---|---|---|---|---|---|
| Denver | 234 | 197 | 166 | 69 | 197 | | 336 | 64 | 63 | 243 | 200 | 208 | 99 | 112 | 125 | 198 |
| Durango | 149 | 246 | 460 | 313 | 312 | 336 | | 402 | 396 | 168 | 142 | 351 | 253 | 269 | 458 | 258 |



25 300 miles 76 185 miles
70 451 miles 50 467 miles

Total mileages through Colorado
More mileages at go.randmcnally.com/MC

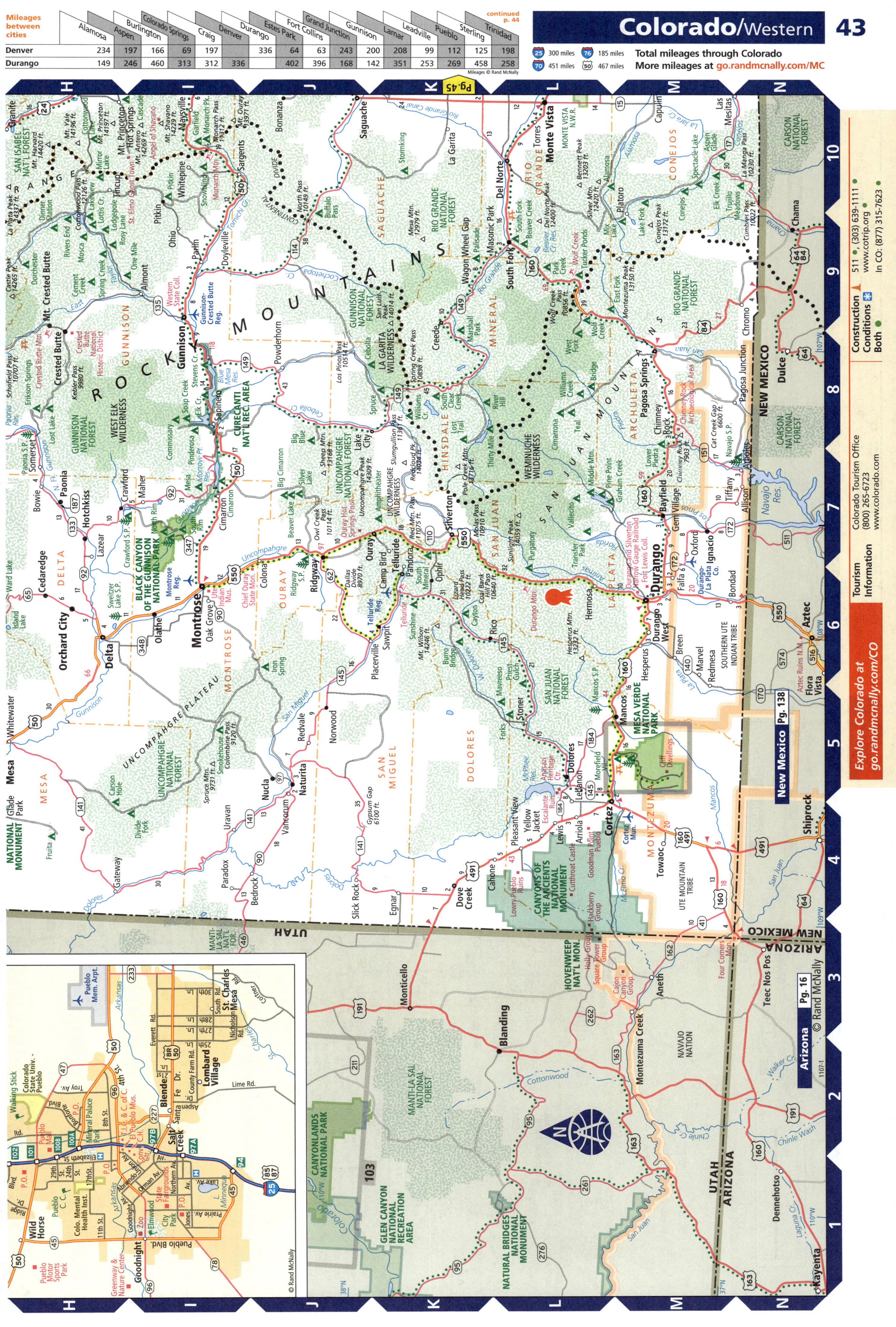

Index of cities Pg. 238 | Map legend Pg. 7

Mileages between cities

| | Alamosa | Aspen | Burlington | Colorado Springs | Craig | Denver | Durango | Estes Park | Fort Collins | Grand Junction | Gunnison | Lamar | Leadville | Pueblo | Sterling | Trinidad |
|---|---|---|---|---|---|---|---|---|---|---|---|---|---|---|---|---|
| Fort Collins | 296 | 258 | 220 | 133 | 201 | 63 | 396 | 42 | | 303 | 260 | 261 | 160 | 175 | 102 | 261 |
| Grand Junction | 247 | 128 | 408 | 309 | 151 | 243 | 168 | 258 | 303 | | 126 | 448 | 174 | 287 | 364 | 370 |



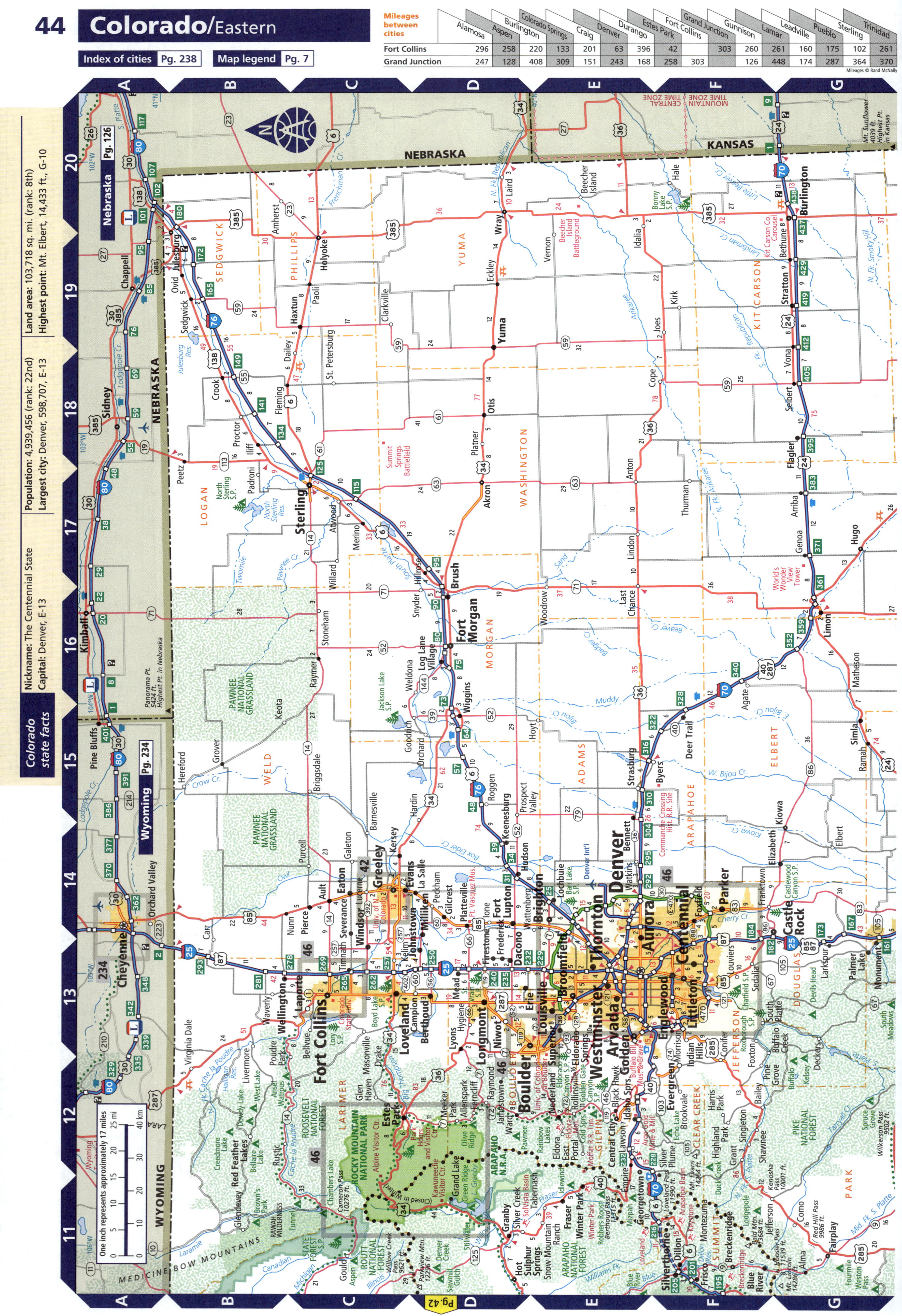

**Mileages between cities**

| | Alamosa | Aspen | Burlington | Colorado Springs | Craig | Denver | Durango | Estes Park | Fort Collins | Grand Junction | Gunnison | Lamar | Leadville | Pueblo | Sterling | Trinidad |
|---|---|---|---|---|---|---|---|---|---|---|---|---|---|---|---|---|
| Leadville | 135 | 58 | 265 | 121 | 145 | 99 | 253 | 143 | 160 | 174 | 102 | 276 | | 154 | 222 | 204 |
| Trinidad | 109 | 232 | 230 | 128 | 392 | 198 | 258 | 262 | 261 | 370 | 209 | 136 | 204 | 85 | 322 | |

Mileages © Rand McNally

25 300 miles · 76 185 miles · 70 451 miles · 50 467 miles

**Total mileages through Colorado**
**More mileages at go.randmcnally.com/MC**

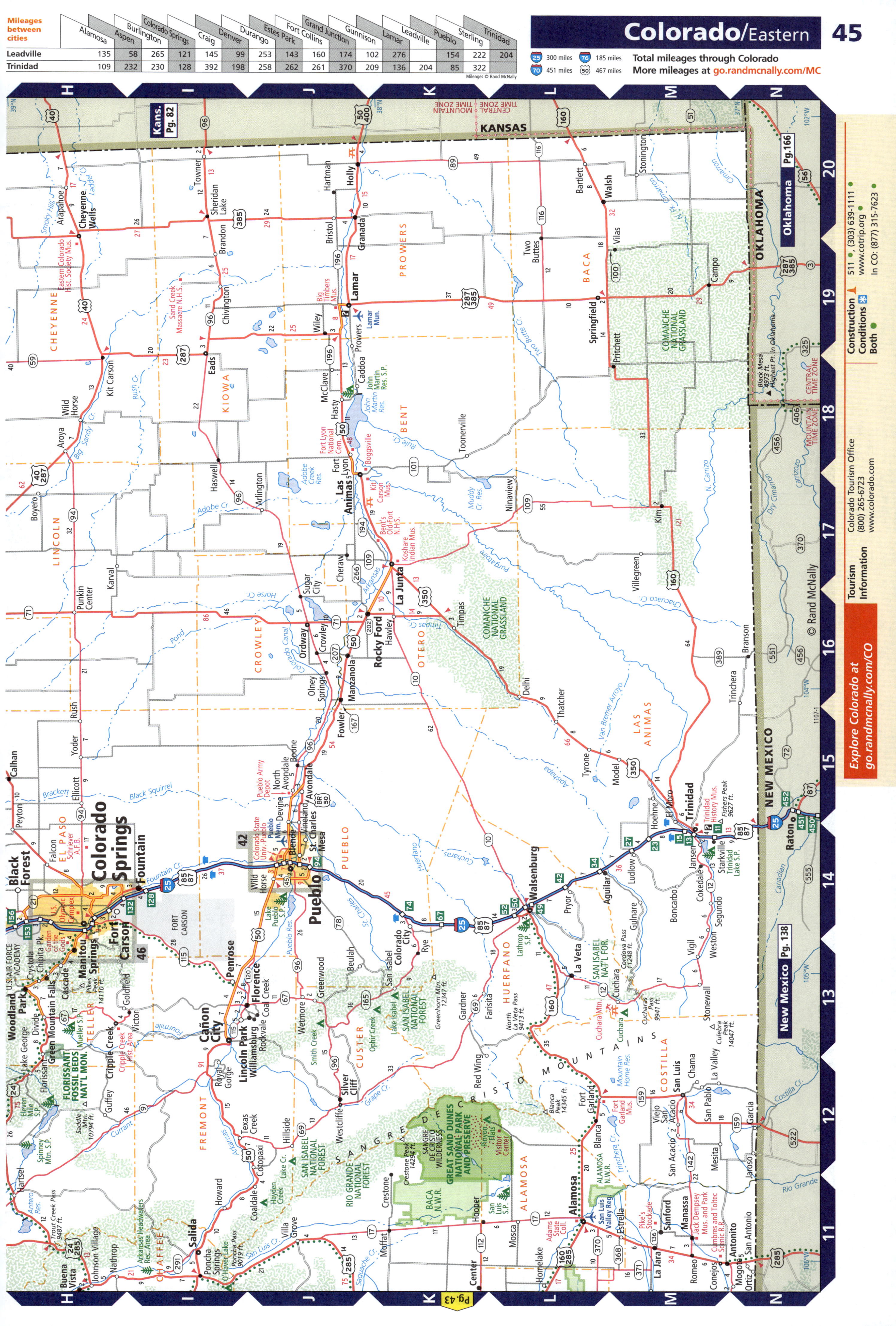

**City sights to see**

- Black American West Museum & Heritage Center, Denver, L-3
- Cave of the Winds, Colorado Springs, G-1
- Colorado History Museum, Denver, M-2
- Colorado State Capitol, Denver, M-2
- Denver Art Museum, Denver, M-2
- Denver Museum of Nature & Science, Denver, L-4
- Garden of the Gods, Colorado Springs, G-1

- National Center for Atmospheric Research, Boulder, D-4
- Old Town National Historic District, Fort Collins, B-9
- ProRodeo Hall of Fame, Colorado Springs, G-2
- Red Rocks Amphitheatre, Morrison, J-4
- U.S. Airforce Academy, Colorado Springs, F-1
- United States Mint, Denver, M-2
- World Figure Skating Hall of Fame, Colorado Springs, I-2

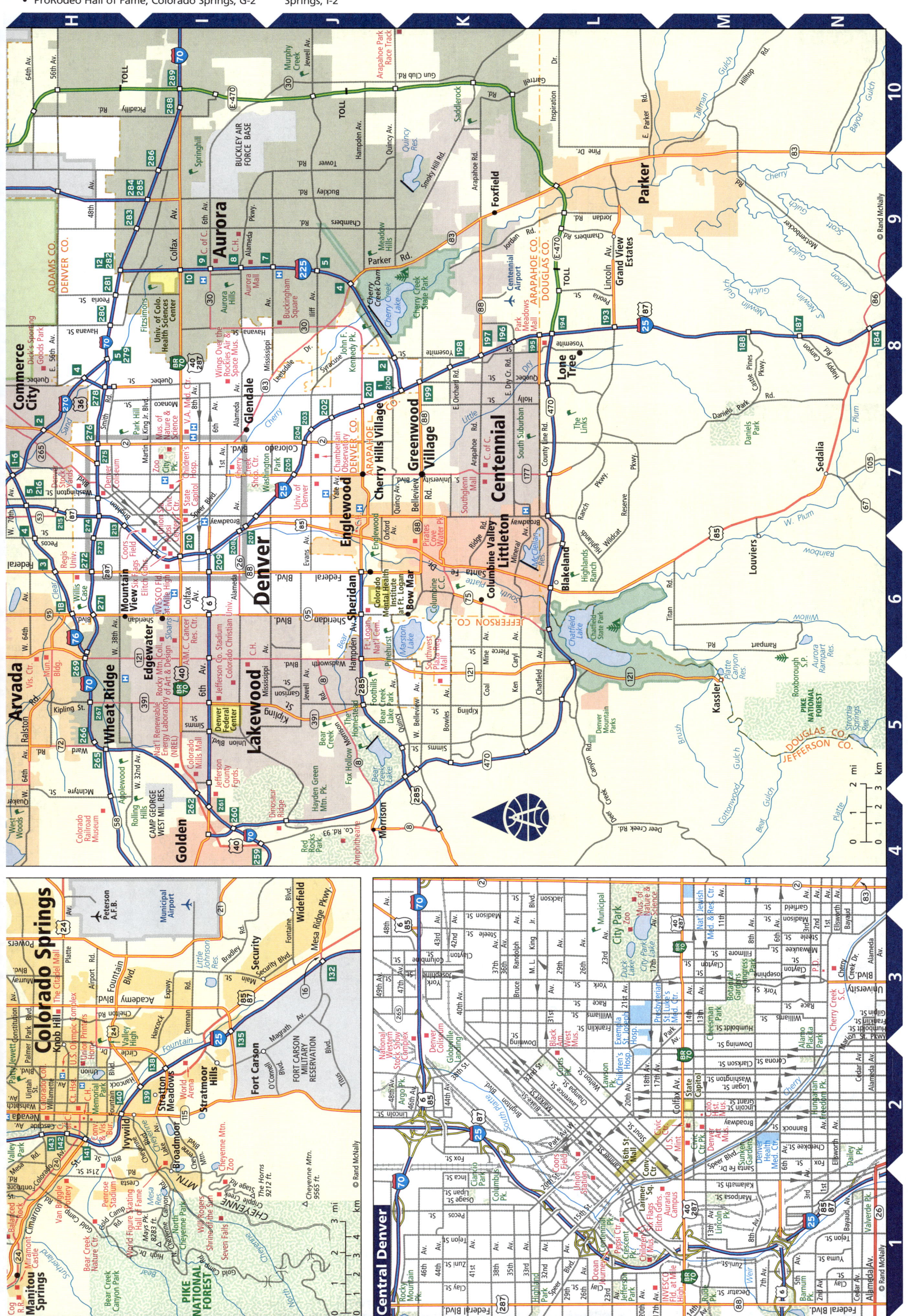

# Connecticut

Index of cities Pg. 239 | Map legend Pg. 7

Mileages between cities

| | Bridgeport | Hartford | New Haven | New London | New York, NY | Putnam | Torrington | Waterbury |
|---|---|---|---|---|---|---|---|---|
| Bridgeport | | 55 | 18 | 64 | 54 | 107 | 50 | 30 |
| Danbury | 29 | 57 | 35 | 81 | 62 | 104 | 47 | 27 |
| Hartford | 55 | | 38 | 45 | 108 | 47 | 26 | 30 |
| New Haven | 18 | 38 | | 46 | 72 | 89 | 43 | 22 |

Mileages © Rand McNally

Connecticut state facts

Nickname: The Constitution State
Capital: Hartford, C-9
Population: 3,501,252 (rank: 29th)
Largest city: Bridgeport, 136,405, H-5
Land area: 4,845 sq. mi. (rank: 48th)
Highest point: Mt. Frissell, 2,380 ft., A-4

New York Pg. 140

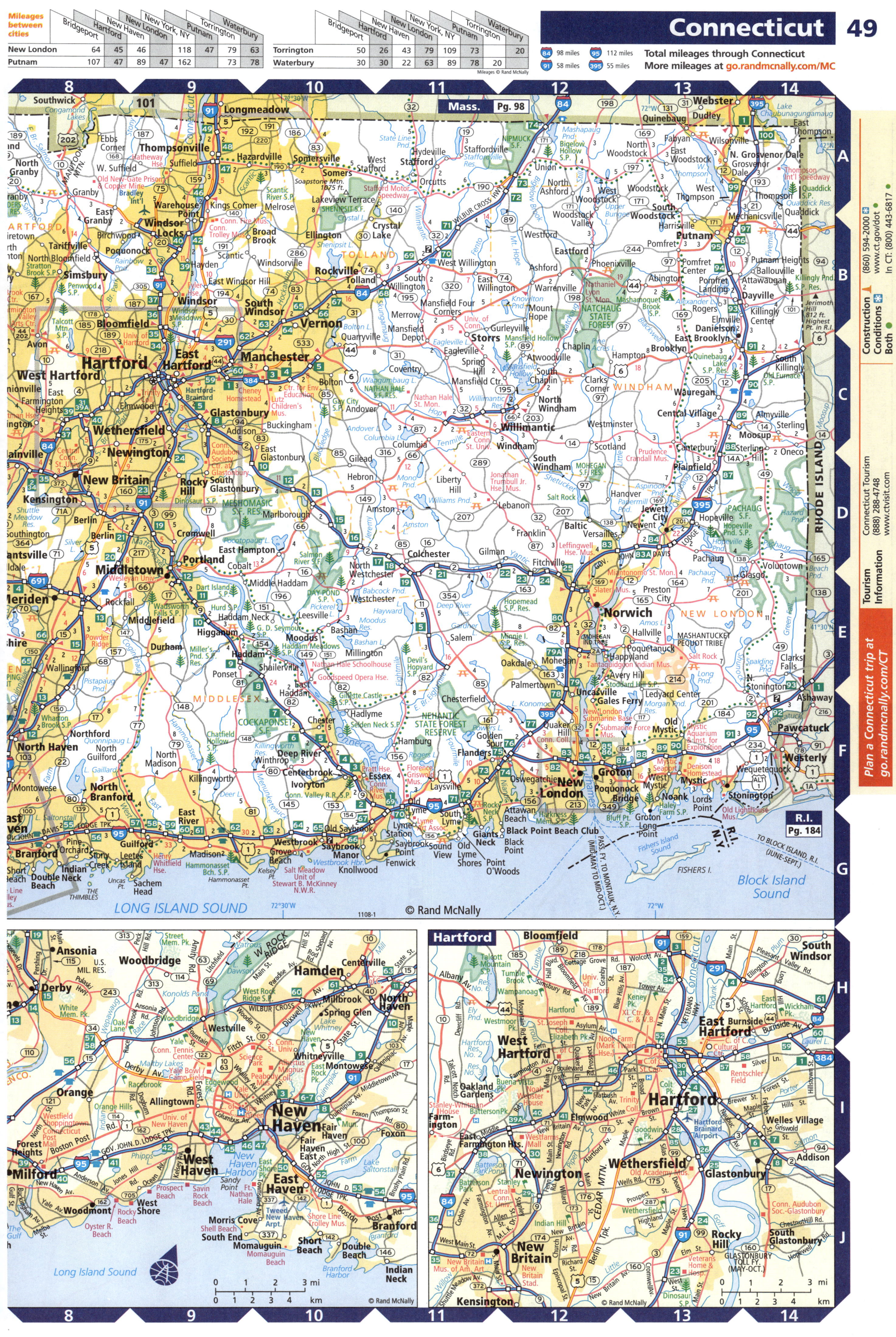

Mileages between cities

| | Bridgeport | Hartford | New Haven | New London | New York, NY | Putnam | Torrington | Waterbury |
|---|---|---|---|---|---|---|---|---|
| New London | 64 | 45 | 46 | | 118 | 47 | 79 | 63 |
| Putnam | 107 | 47 | 89 | 47 | 162 | | 73 | 78 |
| Torrington | 50 | 26 | 43 | 79 | 109 | 73 | | 20 |
| Waterbury | 30 | 30 | 22 | 63 | 89 | 78 | 20 | |

Mileages © Rand McNally

Total mileages through Connecticut
84: 98 miles; 95: 112 miles; 91: 58 miles; 395: 55 miles
More mileages at go.randmcnally.com/MC

Index of cities Pg. 239 | Map legend Pg. 7

**Mileages between cities**

| | Dover | Georgetown | Lewes | Milford | Philadelphia, PA | Salisbury, MD | Selbyville | Wilmington |
|---|---|---|---|---|---|---|---|---|
| Dover | | 36 | 40 | 20 | 80 | 56 | 55 | 50 |
| Georgetown | 36 | | 15 | 16 | 114 | 27 | 20 | 85 |
| Lewes | 40 | 15 | | 21 | 119 | 43 | 29 | 90 |
| Middletown | 26 | 62 | 66 | 46 | 56 | 84 | 81 | 27 |

Mileages between cities

| | Dover | Georgetown | Lewes | Milford | Philadelphia, PA | Salisbury, MD | Selbyville | Wilmington |
|---|---|---|---|---|---|---|---|---|
| Millville, NJ | 94 | 124 | 128 | 108 | 45 | 147 | 143 | 53 |
| Newark | 46 | 80 | 85 | 64 | 43 | 102 | 99 | 14 |
| Selbyville | 55 | 20 | 29 | 36 | 133 | 24 | | 104 |
| Wilmington | 50 | 85 | 90 | 70 | 29 | 107 | 104 | |

Mileages © Rand McNally

95 23 miles · 1 104 miles · 13 108 miles

Total mileages through Delaware
More mileages at go.randmcnally.com/MC

Get more Delaware info at go.randmcnally.com/DE

Tourism Information: Delaware Tourism Office, (866) 284-7483, www.visitdelaware.com

Construction / Conditions / Both: (302) 760-2080 · www.deldot.gov · In DE: (800) 652-5600

Index of cities Pg. 239 | Map legend Pg. 7

**Mileages between cities**

| | Daytona Beach | Fort Myers | Fort Pierce | Gainesville | Jacksonville | Key West | Miami | Orlando | Panama City | Pensacola | St. Petersburg | Sarasota | Tallahassee | Tampa | Titusville | W. Palm Beach |
|---|---|---|---|---|---|---|---|---|---|---|---|---|---|---|---|---|
| Fort Myers | 225 | | 128 | 254 | 312 | 279 | 152 | 171 | 497 | 589 | 117 | 80 | 397 | 130 | 209 | 124 |
| Jacksonville | 92 | 312 | 227 | 72 | | 507 | 349 | 141 | 264 | 355 | 222 | 253 | 164 | 198 | 136 | 284 |



**Florida state facts**

Nickname: The Sunshine State
Capital: Tallahassee, B-2
Population: 18,328,340 (rank: 4th)
Largest city: Jacksonville, 807,815, C-9
Land area: 53,927 sq. mi. (rank: 26th)
Highest point: Britton Hill, 345 ft., Q-5

Mileages between cities

| | Daytona Beach | Fort Myers | Fort Pierce | Gainesville | Jacksonville | Key West | Miami | Orlando | Panama City | Pensacola | St. Petersburg | Sarasota | Tallahassee | Tampa | Titusville | W. Palm Beach |
|---|---|---|---|---|---|---|---|---|---|---|---|---|---|---|---|---|
| Key West | 414 | 279 | 284 | 483 | 507 | | 162 | 387 | 727 | 821 | 390 | 352 | 627 | 402 | 371 | 231 |
| Miami | 256 | 152 | 123 | 336 | 349 | 162 | | 229 | 579 | 663 | 262 | 225 | 479 | 255 | 213 | 68 |

Mileages © Rand McNally

continued p. 54

Total mileages through Florida
4: 132 miles · 10: 362 miles · 75: 471 miles · 95: 382 miles
More mileages at go.randmcnally.com/MC

Index of cities Pg. 239 · Map legend Pg. 7

Mileages between cities

| | Daytona Beach | Fort Myers | Fort Pierce | Gainesville | Jacksonville | Key West | Miami | Orlando | Panama City | Pensacola | St. Petersburg | Sarasota | Tallahassee | Tampa | Titusville | W. Palm Beach |
|---|---|---|---|---|---|---|---|---|---|---|---|---|---|---|---|---|
| Orlando | 54 | 171 | 110 | 114 | 141 | 387 | 229 | | 357 | 451 | 106 | 132 | 257 | 84 | 39 | 159 |
| Pensacola | 442 | 589 | 549 | 338 | 355 | 821 | 663 | 451 | 102 | | 458 | 511 | 193 | 459 | 487 | 594 |

Mileages © Rand McNally

Florida state facts

Nickname: The Sunshine State
Capital: Tallahassee, B-2

Population: 18,328,340 (rank: 4th)
Largest city: Jacksonville, 807,815, C-9

Land area: 53,927 sq. mi. (rank: 26th)
Highest point: Britton Hill, 345 ft., Q-5

| Mileages between cities | Daytona Beach | Fort Myers | Fort Pierce | Gainesville | Jacksonville | Key West | Miami | Orlando | Panama City | Pensacola | St. Petersburg | Sarasota | Tallahassee | Tampa | Titusville | W. Palm Beach |
|---|---|---|---|---|---|---|---|---|---|---|---|---|---|---|---|---|
| Tallahassee | 253 | 397 | 364 | 148 | 164 | 627 | 479 | 257 | 96 | 193 | 257 | 328 | | 273 | 295 | 413 |
| Tampa | 137 | 130 | 151 | 127 | 198 | 402 | 255 | 84 | 373 | 459 | 23 | 60 | 273 | | 124 | 202 |

Mileages © Rand McNally

Total mileages through Florida: I-4 132 miles; I-10 362 miles; I-75 471 miles; I-95 382 miles

More mileages at go.randmcnally.com/MC

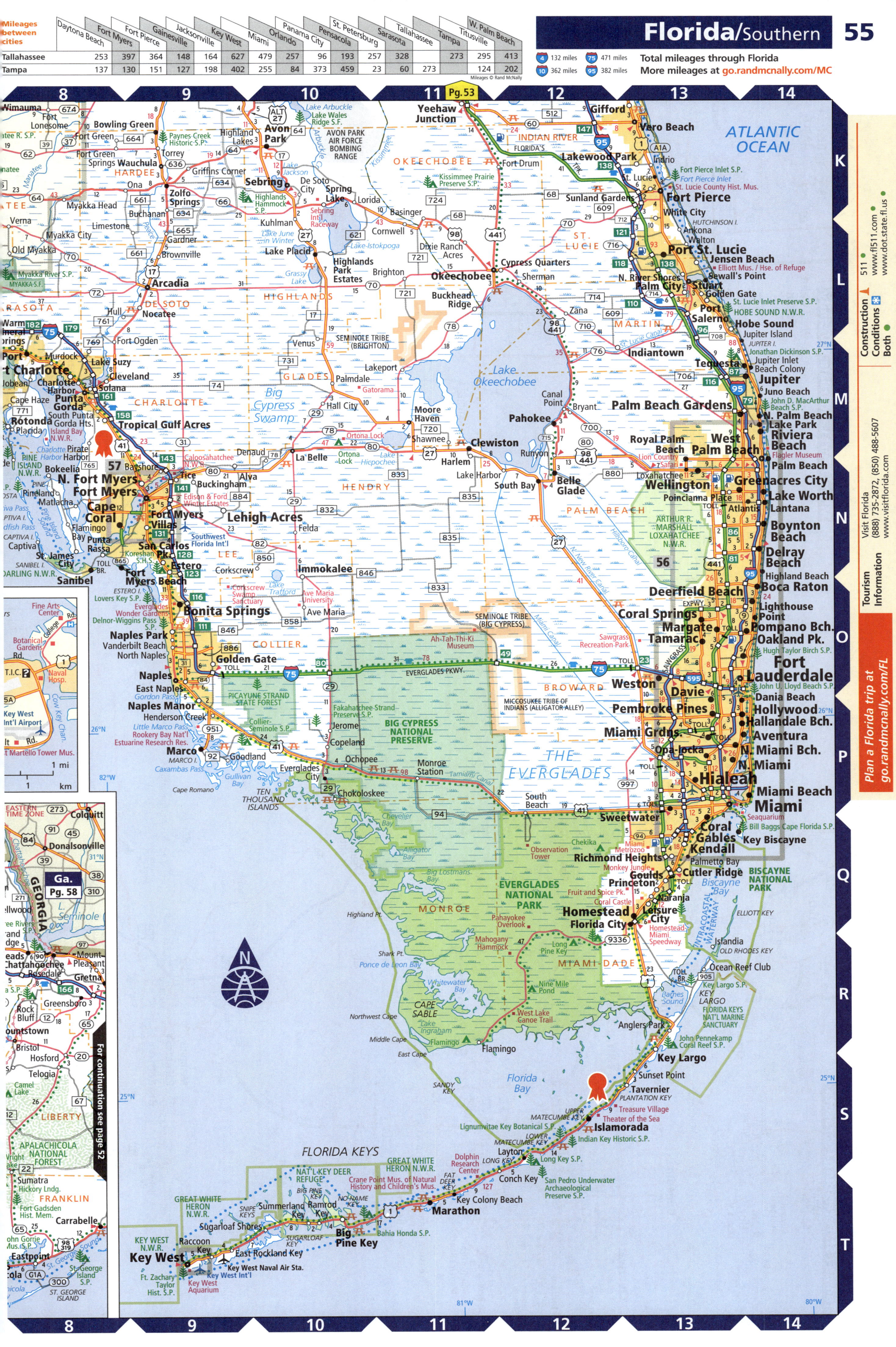

## City sights to see

- Art Deco National Historic District, Miami Beach, L-9
- Busch Gardens, Tampa, B-4
- Goodyear Blimp Base, Pompano Beach, G-9
- Hugh Taylor Birch State Park, Fort Lauderdale, H-9
- Marie Selby Botanical Gardens, Sarasota, H-3
- Miami Seaquarium, Miami, M-9

- Norton Mus. of Art, West Palm Beach, B-10
- Ringling Museum of Art / Ringling Center for the Cultural Arts, Sarasota, G-3
- Salvador Dali Museum, St. Petersburg, D-2
- St. Petersburg Mus. of Hist., St. Petersburg, D-2
- Thomas A. Edison & Henry Ford Winter Homes, Fort Myers, M-1
- Vizcaya Museum and Gardens, Miami, M-8

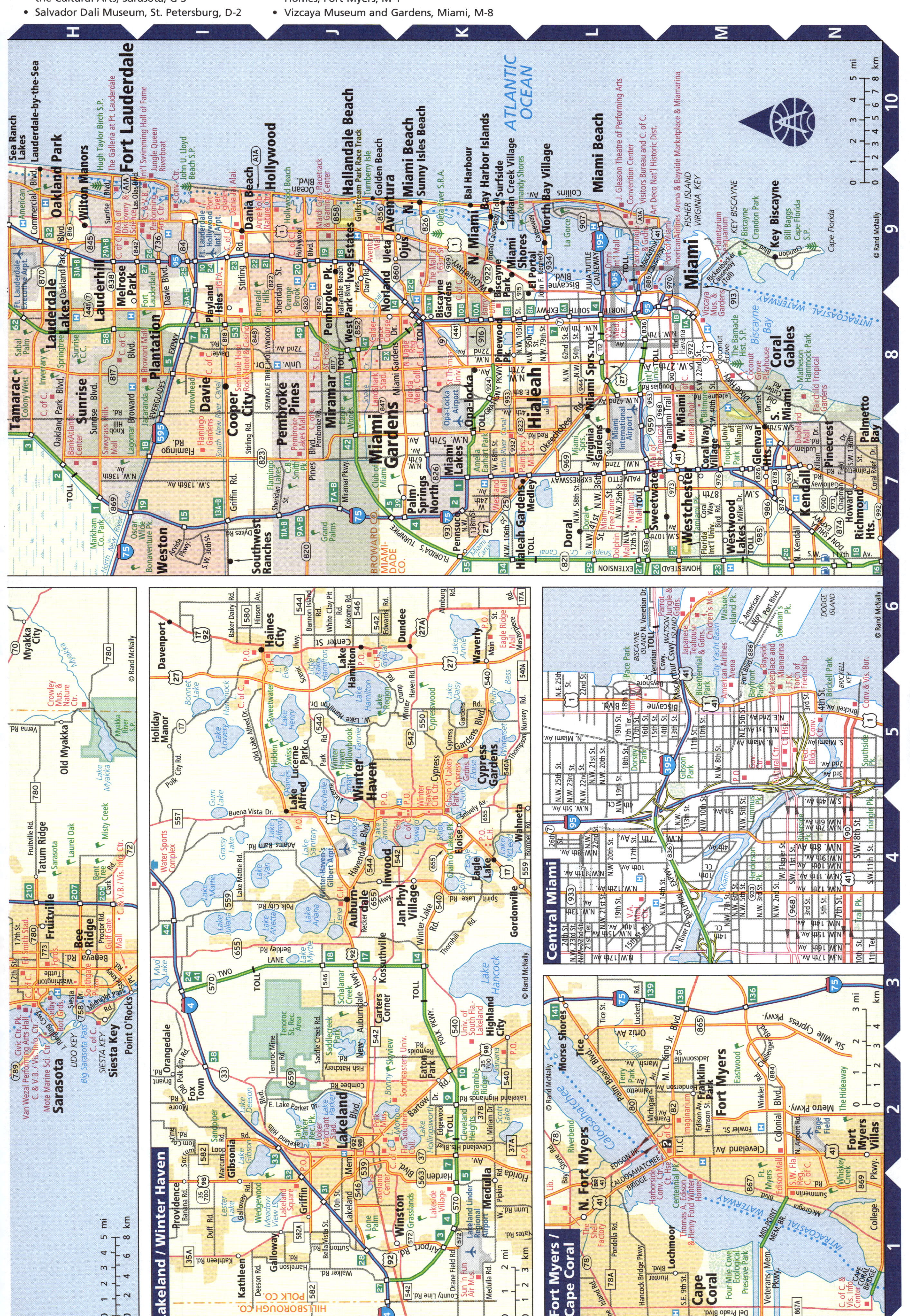

Index of cities Pg. 240 Map legend Pg. 7

Mileages between cities

| | Albany | Athens | Atlanta | Augusta | Bainbridge | Brunswick | Chattanooga, TN | Columbus | Gainesville | Jacksonville, FL | Macon | Rome | Savannah | Toccoa | Valdosta | Vidalia |
|---|---|---|---|---|---|---|---|---|---|---|---|---|---|---|---|---|
| Atlanta | 182 | 69 | | 148 | 240 | 275 | 117 | 106 | 54 | 346 | 82 | 70 | 247 | 94 | 228 | 172 |
| Augusta | 211 | 98 | 148 | | 268 | 193 | 265 | 249 | 140 | 254 | 123 | 217 | 134 | 132 | 217 | 99 |

Mileages © Rand McNally

Georgia state facts

Nickname: The Peach State
Capital: Atlanta, E-4
Population: 9,685,744 (rank: 9th)
Largest city: Atlanta, 537,958, E-4
Land area: 57,906 sq. mi. (rank: 21st)
Highest point: Brasstown Bald, 4,784 ft., B-6

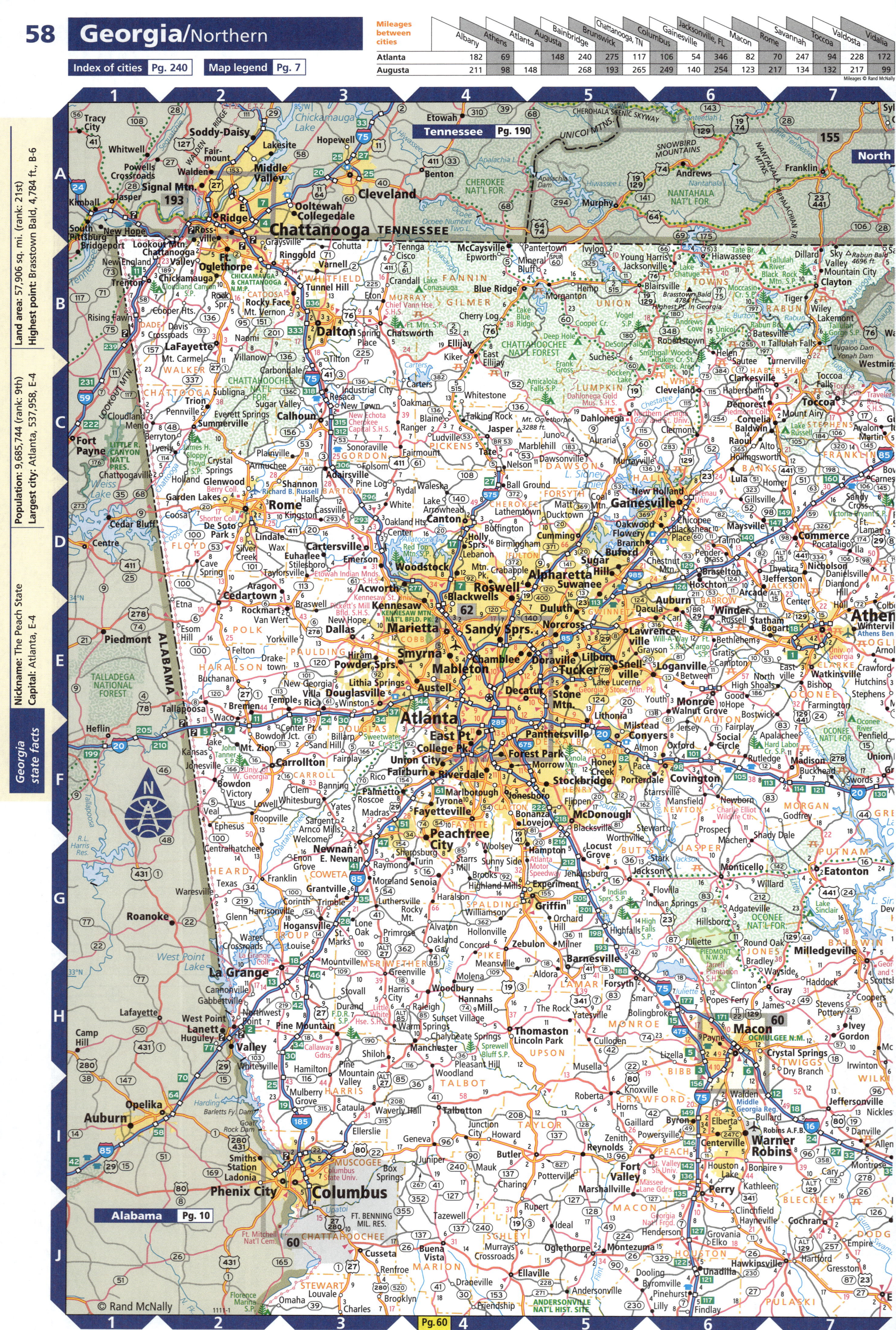

Mileages between cities

| | Albany | Athens | Atlanta | Augusta | Bainbridge | Brunswick | Chattanooga, TN | Columbus | Gainesville | Jacksonville, FL | Macon | Rome | Savannah | Toccoa | Valdosta | Vidalia |
|---|---|---|---|---|---|---|---|---|---|---|---|---|---|---|---|---|
| Chattanooga, TN | 300 | 172 | 117 | 265 | 348 | 397 | | 219 | 121 | 465 | 201 | 71 | 364 | 155 | 346 | 289 |
| Columbus | 85 | 171 | 106 | 249 | 128 | 258 | 219 | | 161 | 292 | 98 | 144 | 249 | 201 | 173 | 175 |

Mileages © Rand McNally

continued p. 60

Total mileages through Georgia: 20 203 miles; 75 355 miles; 85 180 miles; 95 112 miles

More mileages at go.randmcnally.com/MC

Index of cities Pg. 240 | Map legend Pg. 7

**Mileages between cities**

| | Albany | Athens | Atlanta | Augusta | Bainbridge | Brunswick | Chattanooga, TN | Columbus | Gainesville | Jacksonville, FL | Macon | Rome | Savannah | Toccoa | Valdosta | Vidalia |
|---|---|---|---|---|---|---|---|---|---|---|---|---|---|---|---|---|
| Jacksonville, FL | 198 | 310 | 346 | 254 | 204 | 66 | 465 | 292 | 396 | | 270 | 416 | 135 | 375 | 121 | 164 |
| Macon | 106 | 91 | 82 | 123 | 163 | 193 | 201 | 98 | 132 | 270 | | 152 | 165 | 143 | 152 | 90 |



**Georgia state facts**

Nickname: The Peach State
Capital: Atlanta, E-4
Population: 9,685,744 (rank: 9th)
Largest city: Atlanta, 537,958, E-4
Land area: 57,906 sq. mi. (rank: 21st)
Highest point: Brasstown Bald, 4,784 ft., B-6

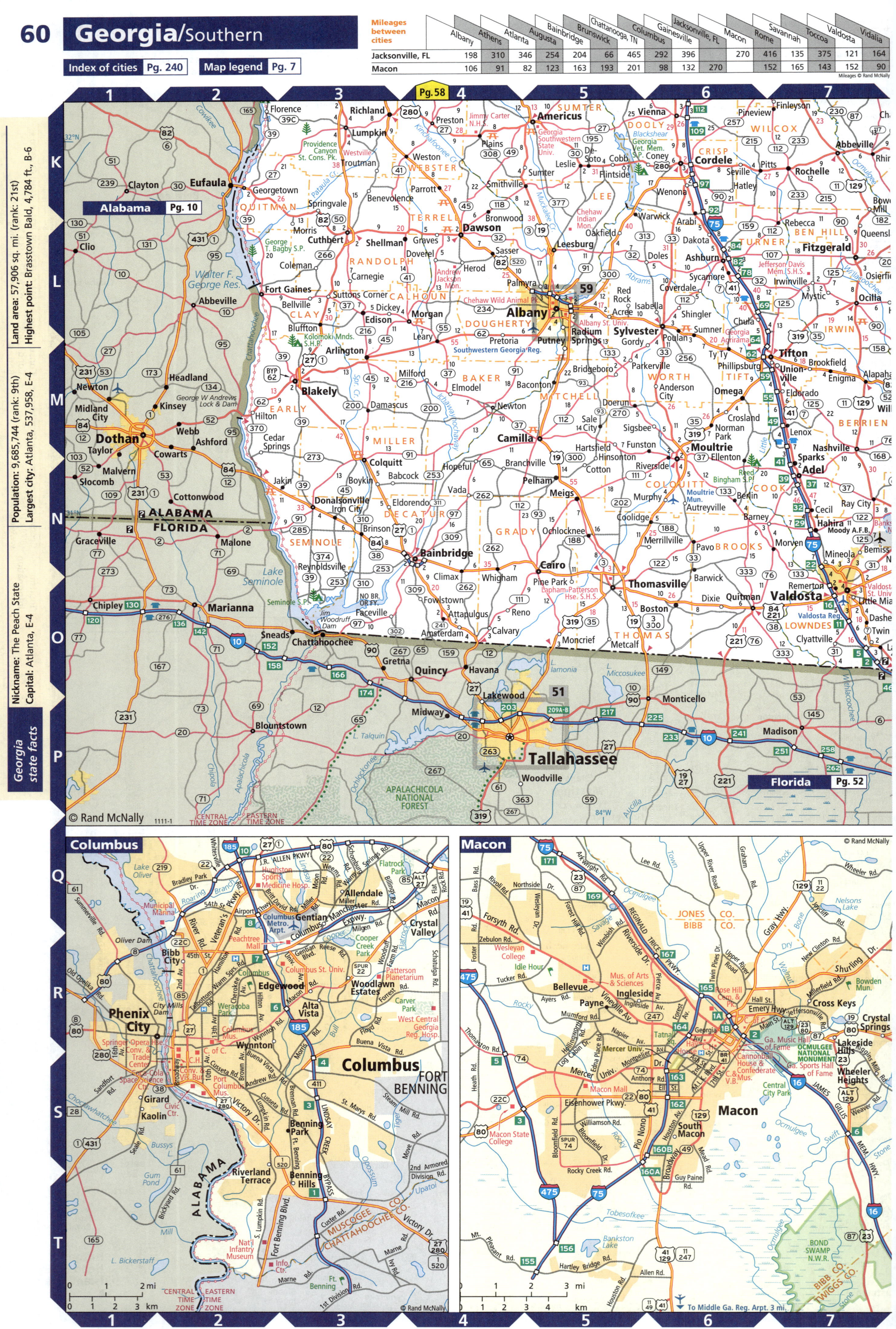

Mileages between cities

| | Albany | Athens | Atlanta | Augusta | Bainbridge | Brunswick | Chattanooga, TN | Columbus | Gainesville | Jacksonville, FL | Macon | Rome | Savannah | Toccoa | Valdosta | Vidalia |
|---|---|---|---|---|---|---|---|---|---|---|---|---|---|---|---|---|
| Savannah | 226 | 222 | 247 | 134 | 249 | 77 | 364 | 249 | 297 | 135 | 165 | 317 | | 255 | 167 | 90 |
| Valdosta | 79 | 243 | 228 | 217 | 83 | 120 | 346 | 173 | 278 | 121 | 152 | 298 | 167 | 317 | | 118 |

Mileages © Rand McNally

20 203 miles · 75 355 miles · 85 180 miles · 95 112 miles

Total mileages through Georgia
More mileages at go.randmcnally.com/MC

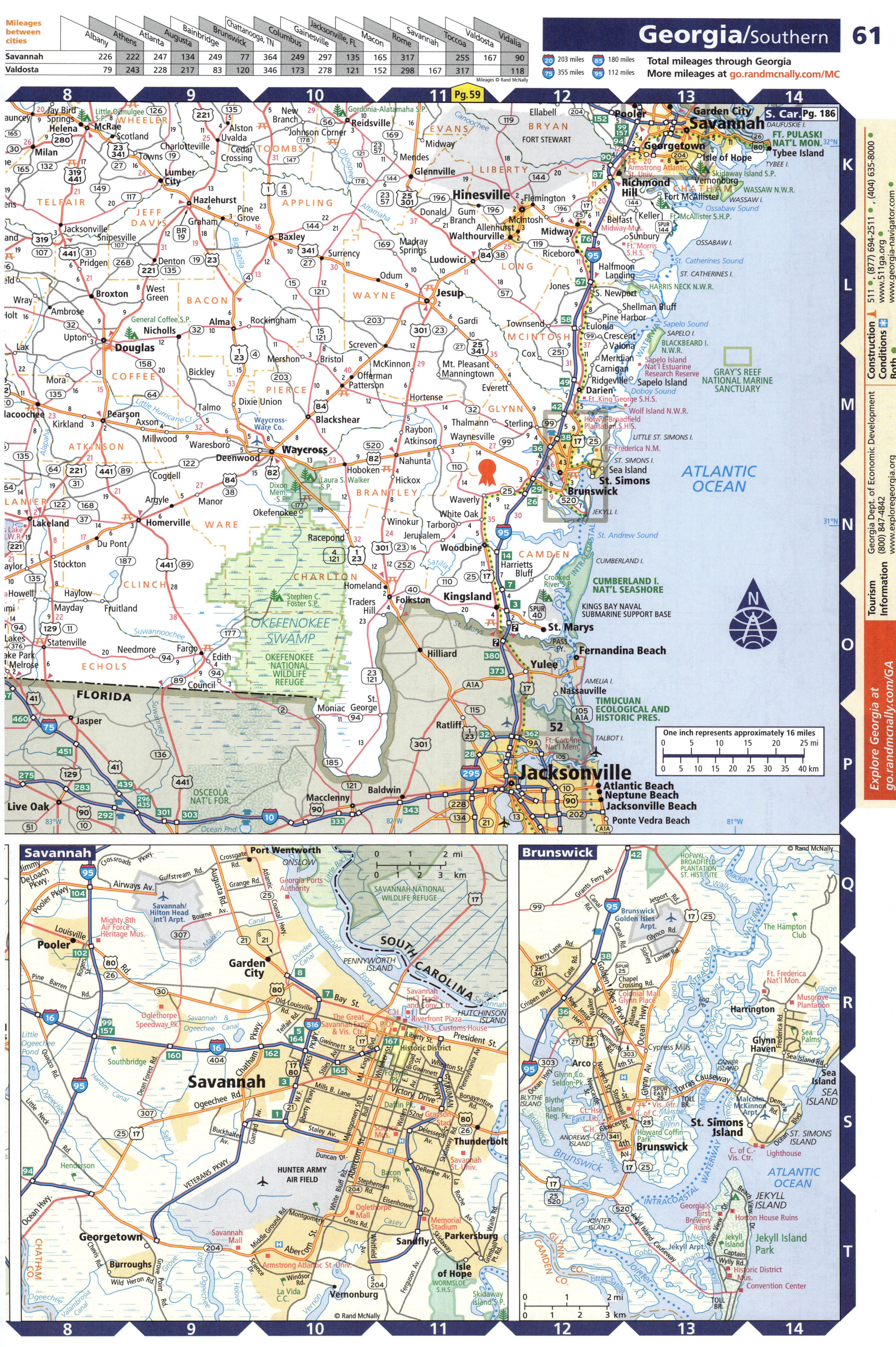

Index of cities Pg. 240 | Map legend Pg. 7

Hawaii state facts

Nickname: The Aloha State
Capital: Honolulu, N-4
Population: 1,288,198 (rank: 42nd)
Largest city: Honolulu, 371,657, N-4
Land area: 113,635 sq. mi. (rank: 6th)
Highest point: Mauna Kea, 13,796 ft., M-9

Mileages between cities

| | Hilo | Honolulu | Kahului | Kailua | Kailua Kona | Kapa'a | Lahaina | Wahiawa |
|---|---|---|---|---|---|---|---|---|
| Hilo | | 225* | 127* | 237* | 74 | 337* | 149* | 236* |
| Honolulu | 225* | | 108* | 11 | 177* | 116* | 130* | 20 |

| | Hilo | Honolulu | Kahului | Kailua | Kailua Kona | Kapa'a | Lahaina | Wahiawa |
|---|---|---|---|---|---|---|---|---|
| Kahului | 127* | 108* | | 22* | 93* | 214* | 22 | 119* |
| Kailua Kona | 74 | 177* | 93* | 188* | | 283* | 116* | 188* |

*via plane

Mileages © Rand McNally

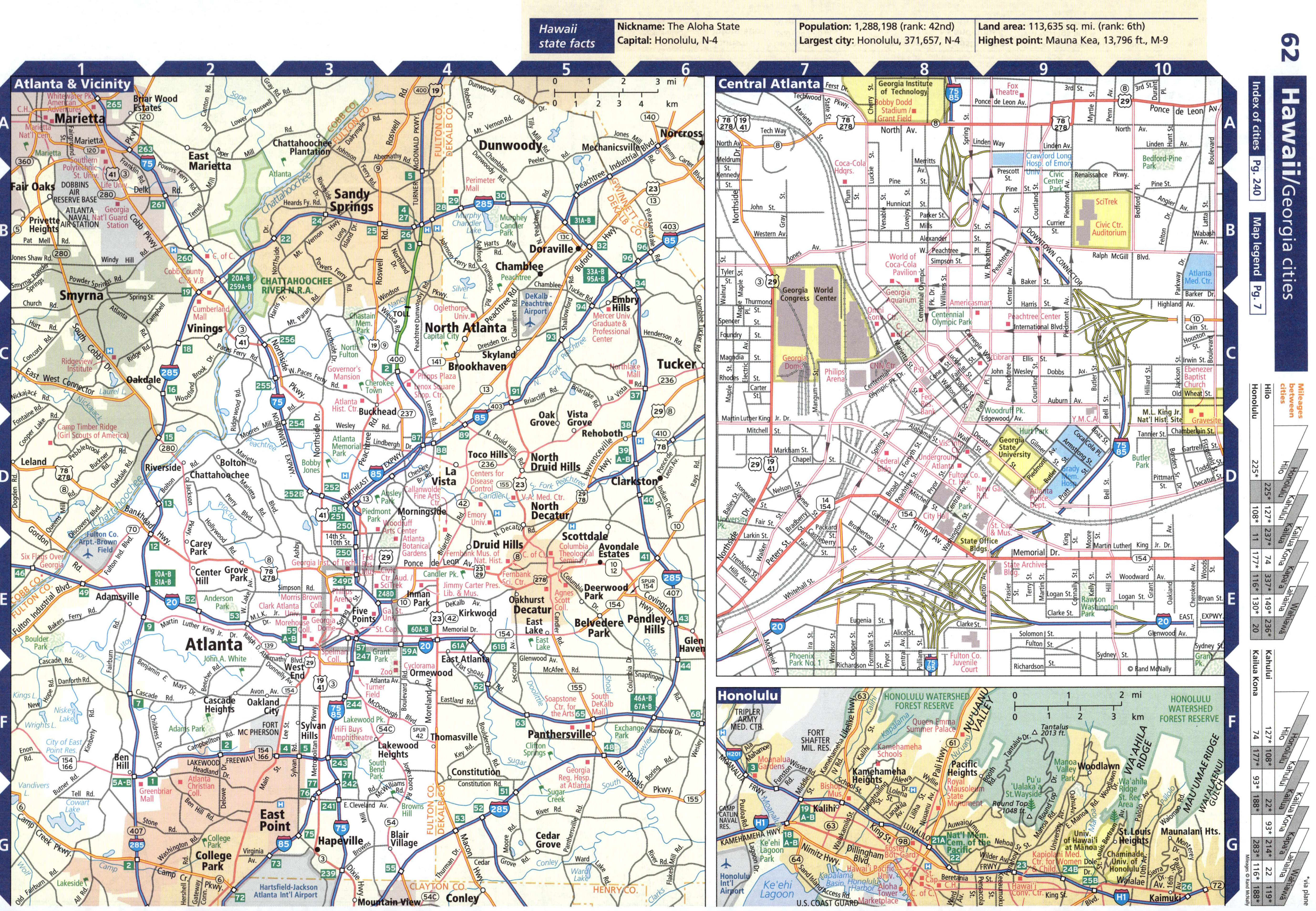

| | Hilo | Honolulu | Kahului | Kailua | Kailua Kona | Kapa'a | Lahaina | Wahiawā |
|---|---|---|---|---|---|---|---|---|
| Kapa'a | 337* | 116* | 214* | 128* | 283* | | 236* | 128* |
| Kaunakakai | 177* | 68* | 55* | 79* | 144* | 174* | 77* | 79* |

| | Hilo | Honolulu | Kahului | Kailua | Kailua Kona | Kapa'a | Lahaina | Wahiawā |
|---|---|---|---|---|---|---|---|---|
| Lahaina | 149* | 130* | 22 | 43* | 116* | 236* | | 141* |
| Wahiawā | 236* | 20 | 119* | 26 | 188* | 128* | 141* | |

Index of cities Pg. 241 | Map legend Pg. 7

**Idaho state facts**

Nickname: The Gem State
Capital: Boise, K-2
Population: 1,523,816 (rank: 39th)
Largest city: Boise, 205,314, K-2
Land area: 82,747 sq. mi. (rank: 11th)
Highest point: Borah Peak, 12,662 ft., J-5

**Mileages between cities**

| | Boise | Coeur d'Alene | Lewiston | Missoula, MT | Mountain Home | Pocatello | Salmon | Twin Falls |
|---|---|---|---|---|---|---|---|---|
| Boise | | 383 | 268 | 367 | 44 | 234 | 247 | 128 |
| Bonners Ferry | 459 | 76 | 191 | 212 | 504 | 573 | 351 | 589 |
| Coeur d'Alene | 383 | | 115 | 166 | 428 | 525 | 303 | 513 |
| Idaho Falls | 279 | 478 | 526 | 312 | 237 | 49 | 160 | 159 |

Mileages © Rand McNally

**Mileages between cities**

| | Boise | Coeur d'Alene | Lewiston | Missoula, MT | Mountain Home | Pocatello | Salmon | Twin Falls |
|---|---|---|---|---|---|---|---|---|
| Lewiston | 268 | 115 | | 214 | 313 | 504 | 332 | 398 |
| Pocatello | 234 | 525 | 504 | 361 | 191 | | 209 | 114 |
| Salmon | 247 | 303 | 332 | 138 | 287 | 209 | | 247 |
| Twin Falls | 128 | 513 | 398 | 384 | 85 | 114 | 247 | |

Mileages © Rand McNally

15 196 miles · 86 63 miles · 84 276 miles · 90 74 miles

Total mileages through Idaho
More mileages at go.randmcnally.com/MC

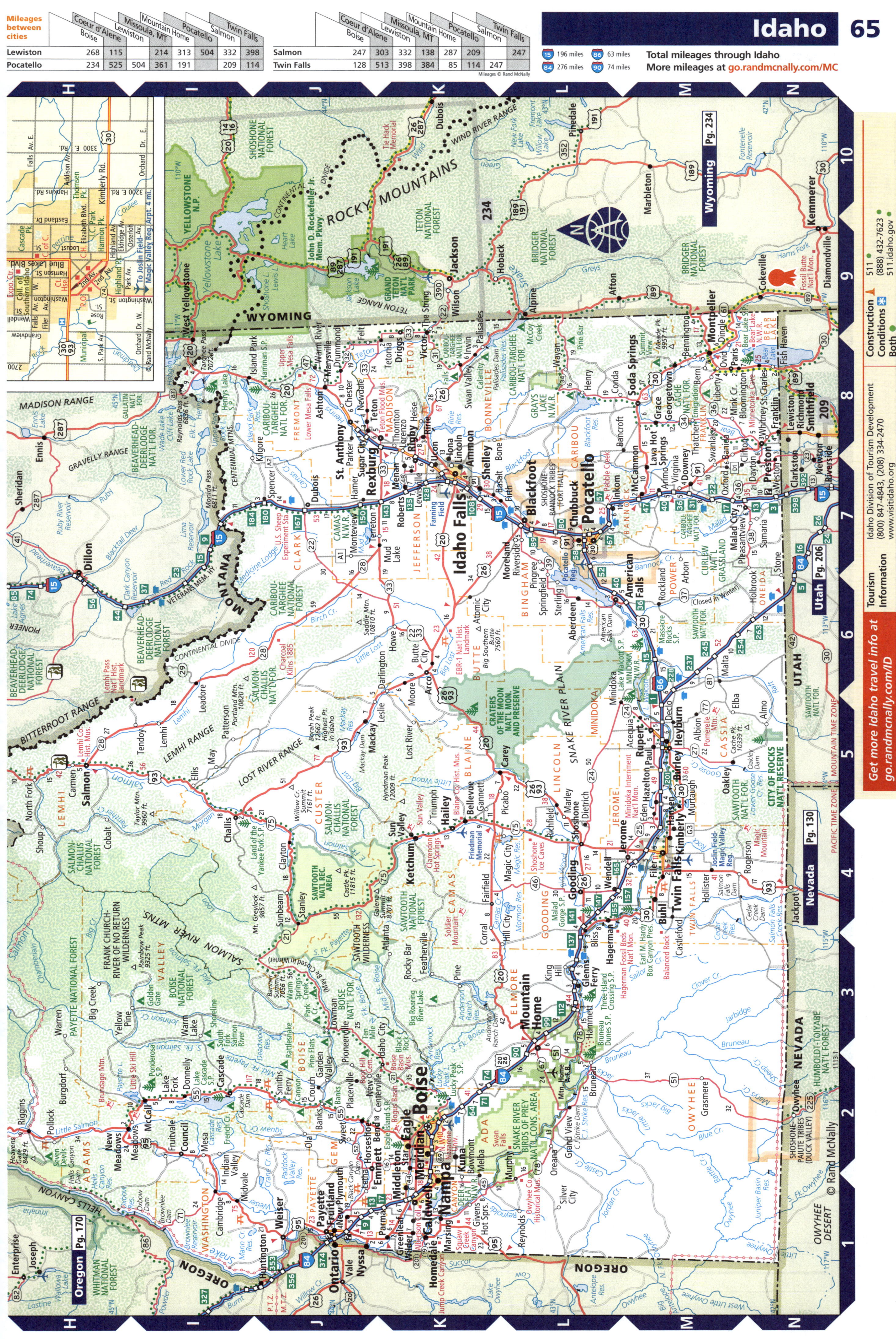

Index of cities Pg. 241 | Map legend Pg. 7

Mileages between cities

| | Bloomington | Carbondale | Champaign | Chicago | Decatur | Dubuque, IA | Kankakee | Lawrenceville | Moline | Mt. Vernon | Peoria | Quincy | Rockford | St. Louis, MO | Springfield | Waukegan |
|---|---|---|---|---|---|---|---|---|---|---|---|---|---|---|---|---|
| Carbondale | 245 | | 200 | 330 | 176 | 406 | 272 | 146 | 332 | 57 | 240 | 240 | 379 | 104 | 170 | 374 |
| Champaign | 51 | 200 | | 135 | 48 | 256 | 78 | 130 | 182 | 147 | 89 | 194 | 185 | 180 | 85 | 180 |

Mileages © Rand McNally

Illinois state facts

Nickname: Land of Lincoln
Capital: Springfield, J-8

Population: 12,901,563 (rank: 5th)
Largest city: Chicago, 2,853,114, C-13

Land area: 55,584 sq. mi. (rank: 24th)
Highest point: Charles Mound, 1,235 ft., A-6

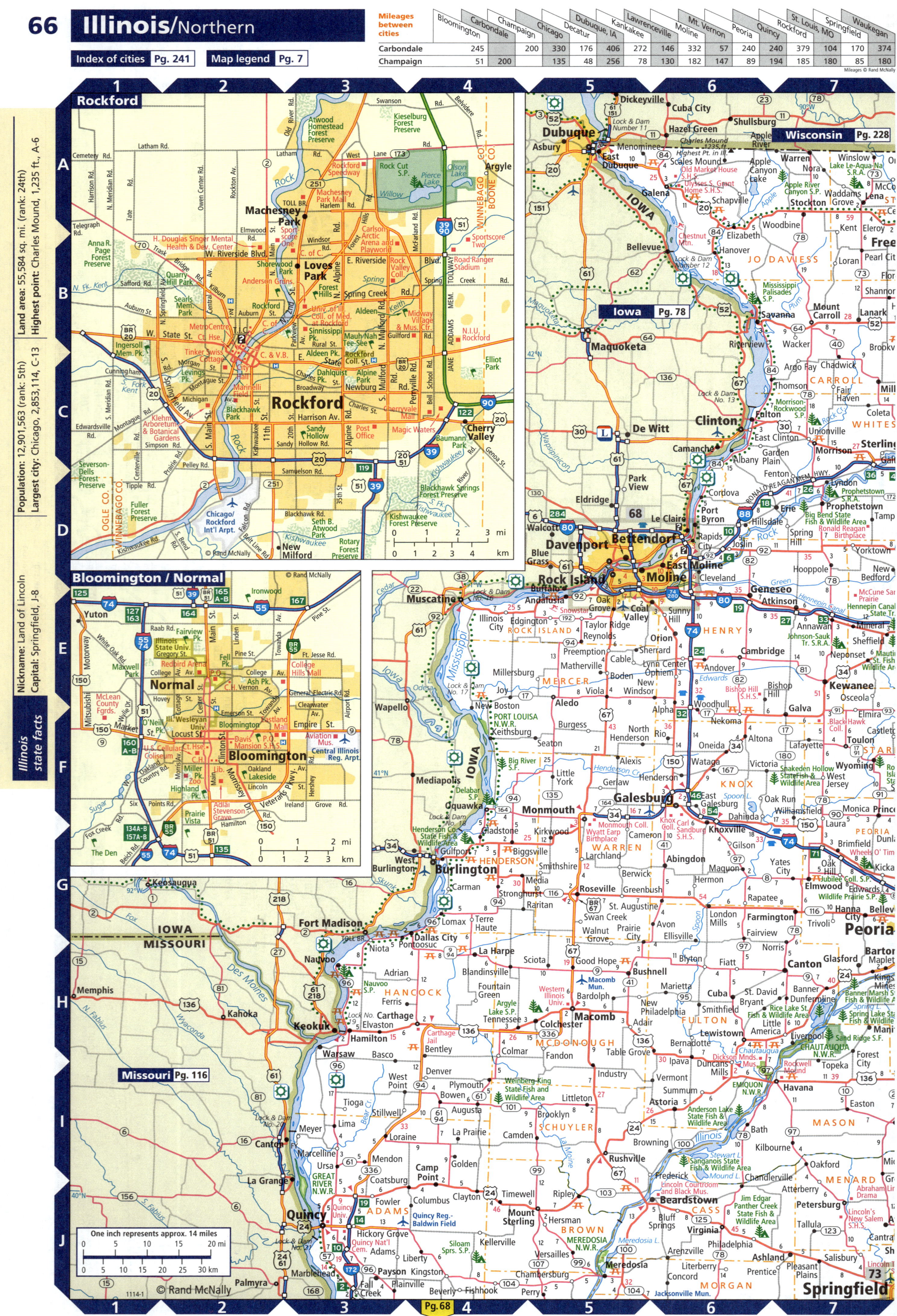


Pg. 68

| Mileages between cities | Bloomington | Carbondale | Champaign | Chicago | Decatur | Dubuque, IA | Kankakee | Lawrenceville | Moline | Mt. Vernon | Peoria | Quincy | Rockford | St. Louis, MO | Springfield | Waukegan |
|---|---|---|---|---|---|---|---|---|---|---|---|---|---|---|---|---|
| Chicago | 132 | 330 | 135 | | 179 | 177 | 58 | 247 | 166 | 277 | 154 | 309 | 84 | 296 | 198 | 38 |
| Moline | 131 | 332 | 182 | 166 | 171 | 75 | 158 | 307 | | 308 | 93 | 148 | 120 | 261 | 164 | 190 |

continued p. 68

Mileages © Rand McNally

55: 313 miles; 80: 164 miles; 70: 136 miles; 90: 124 miles

Total mileages through Illinois
More mileages at go.randmcnally.com/MC

Index of cities Pg. 241 | Map legend Pg. 7

Mileages between cities

| | Bloomington | Carbondale | Champaign | Chicago | Decatur | Dubuque, IA | Kankakee | Lawrenceville | Moline | Mt. Vernon | Peoria | Quincy | Rockford | St. Louis, MO | Springfield | Waukegan |
|---|---|---|---|---|---|---|---|---|---|---|---|---|---|---|---|---|
| Peoria | 38 | 240 | 89 | 154 | 78 | 167 | 108 | 214 | 93 | 215 | | 130 | 138 | 168 | 71 | 184 |
| Rockford | 132 | 379 | 185 | 84 | 180 | 93 | 139 | 309 | 120 | 326 | 138 | 268 | | 294 | 197 | 73 |



Illinois state facts

Nickname: Land of Lincoln
Capital: Springfield, J-8
Population: 12,901,563 (rank: 5th)
Largest city: Chicago, 2,853,114, C-13
Land area: 55,584 sq. mi. (rank: 24th)
Highest point: Charles Mound, 1,235 ft., A-6

Mileages between cities

| | Bloomington | Carbondale | Champaign | Chicago | Decatur | Dubuque, IA | Kankakee | Lawrenceville | Moline | Mt. Vernon | Peoria | Quincy | Rockford | St. Louis, MO | Springfield | Waukegan |
|---|---|---|---|---|---|---|---|---|---|---|---|---|---|---|---|---|
| St. Louis, MO | 162 | 104 | 180 | 296 | 135 | 335 | 252 | 144 | 261 | 79 | 168 | 139 | 294 | | 98 | 326 |
| Springfield | 66 | 170 | 85 | 198 | 38 | 238 | 157 | 153 | 164 | 138 | 71 | 112 | 197 | 98 | | 229 |



55 313 miles 80 164 miles 70 136 miles 90 124 miles

Total mileages through Illinois

More mileages at go.randmcnally.com/MC

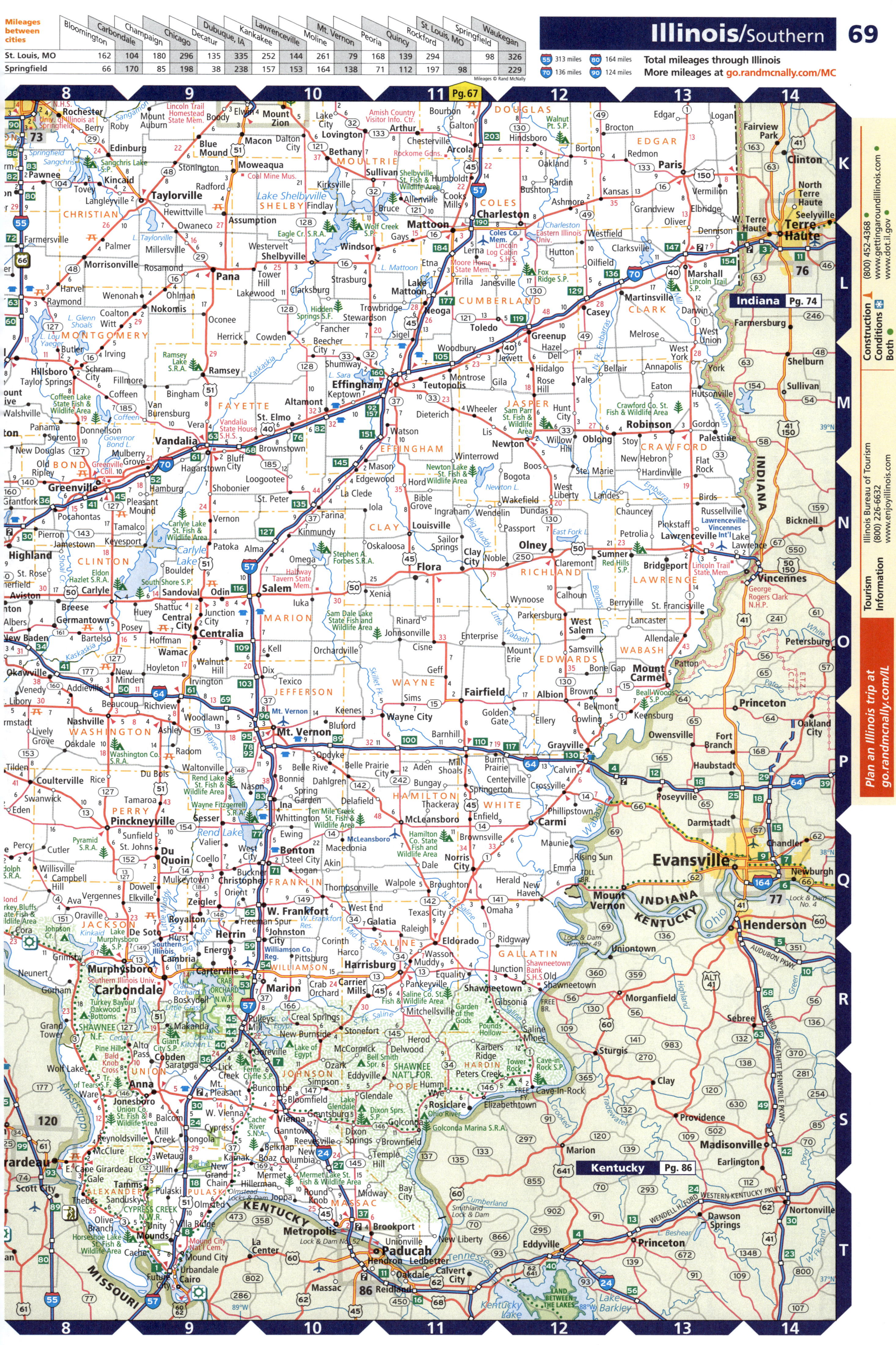

## City sights to see

- Adler Planetarium, Chicago, G-15
- Art Institute of Chicago, Chicago, E-13
- Baha'i Temple, Wilmette, E-8
- Chicago Botanic Garden, Glencoe, D-8
- Field Museum, Chicago, G-14
- Frank Lloyd Wright Home & Studio, Oak Park, H-8
- Illinois Holocaust Museum, Skokie, F-8

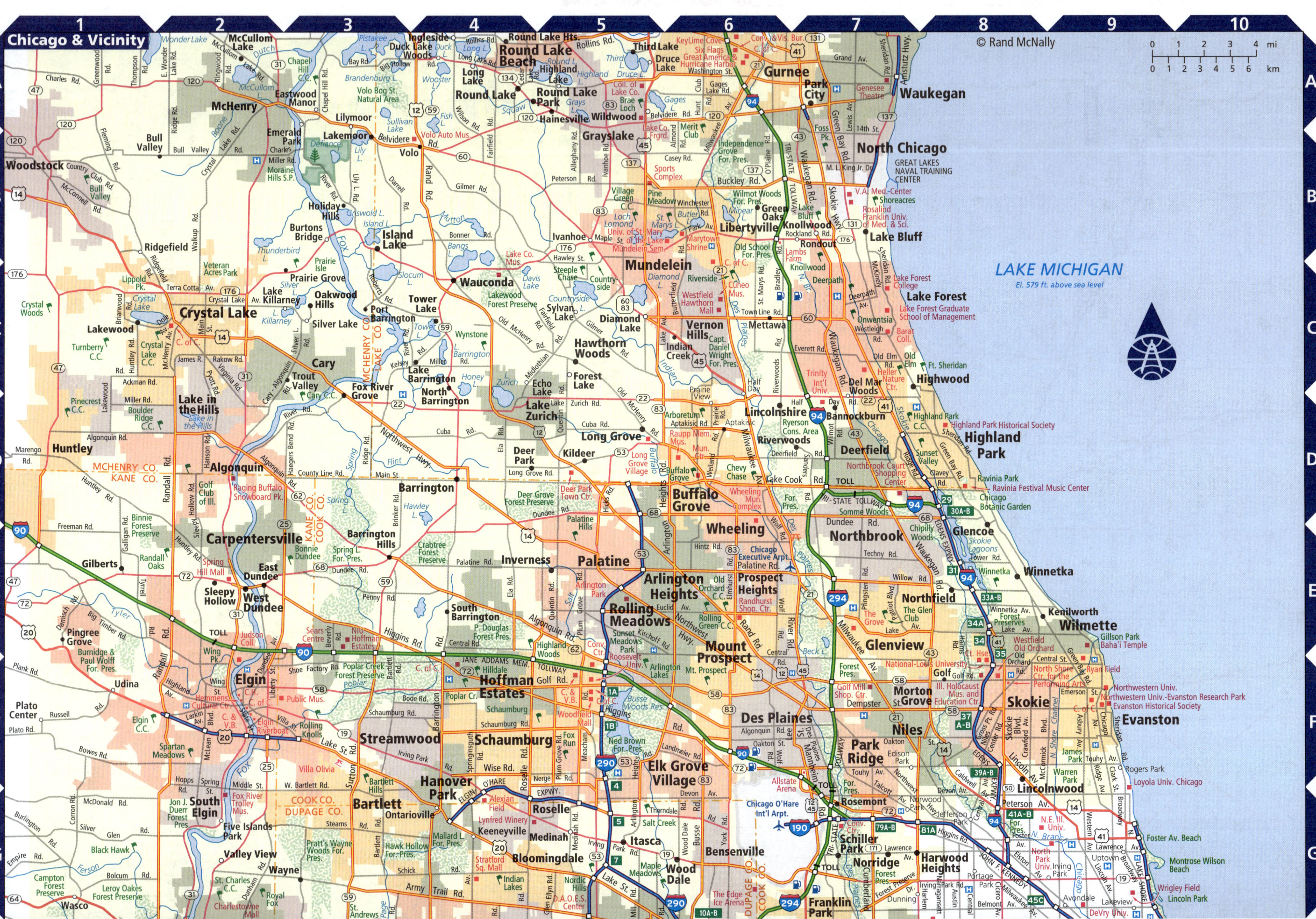

- John G. Shedd Aquarium, Chicago, G-14
- John Hancock Center, Chicago, C-13
- Lincoln Park Zoo, Chicago, H-9
- Millennium Park, Chicago, E-13
- Museum of Science & Industry, Chicago, J-10
- Navy Pier, Chicago, D-14
- Willis Tower, Chicago, E-12
- Wrigley Field, Chicago, G-9

**City sights to see**

- Abraham Lincoln Presidential Library & Museum, Springfield, M-16
- Children's Museum of Indianapolis, Indianapolis, D-18
- Eiteljorg Museum, Indianapolis, E-17
- Fort Wayne Childraen's Zoo, Fort Wayne, L-19
- Illinois State Capitol Complex, Springfield, M-16
- Indiana State Capitol, Indianapolis, E-18

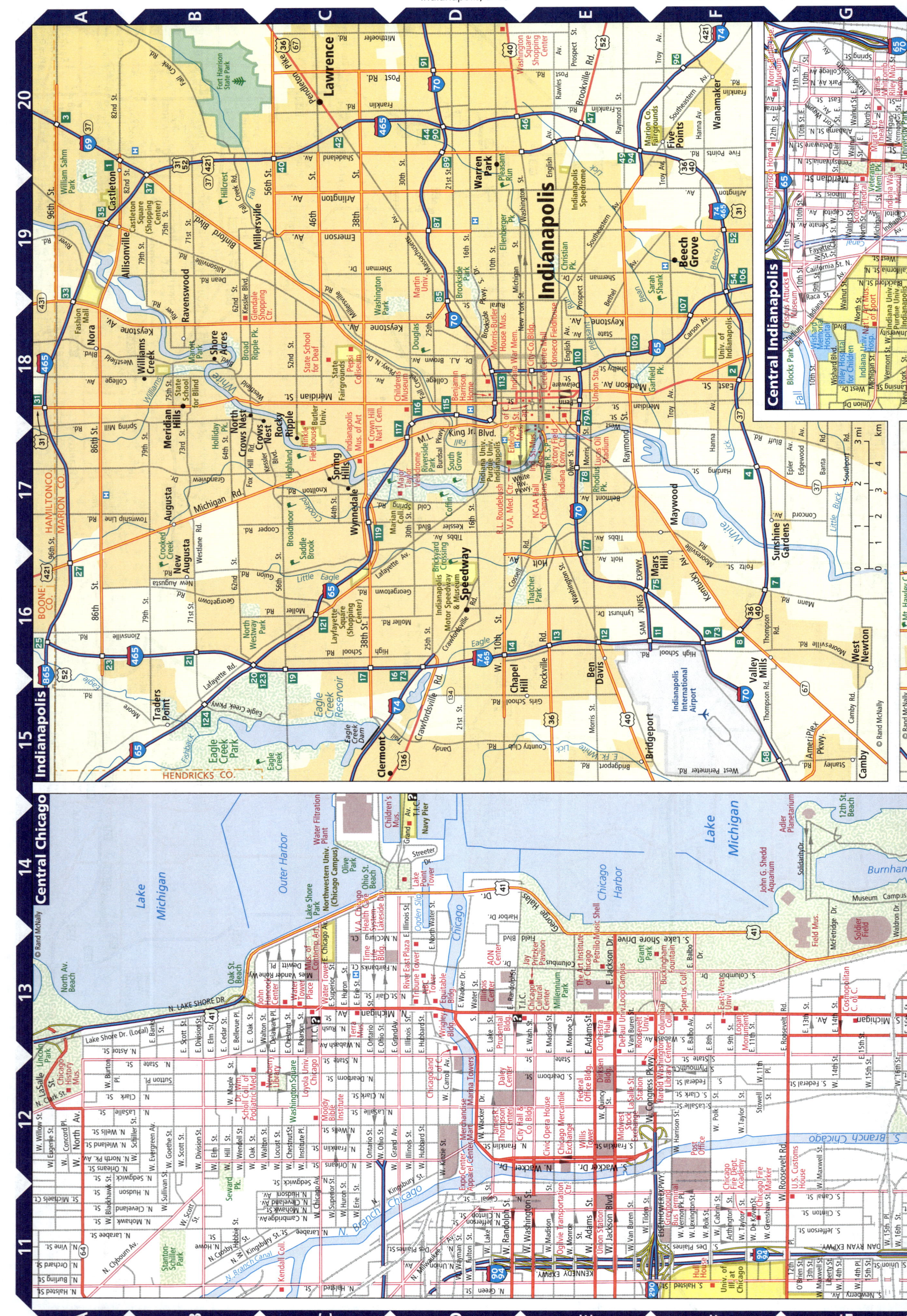

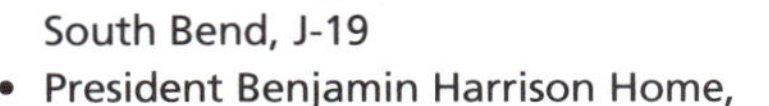

- Indiana State Museum, Indianapolis, E-17
- Indianapolis Motor Speedway and Hall of Fame Museum, Indianapolis, D-16
- NCAA Hall of Champions, Indianapolis, H-18
- National College Football Hall of Fame, South Bend, J-19
- President Benjamin Harrison Home, Indianapolis, D-18

South Bend

Fort Wayne

Peoria

Springfield

LAKE MICHIGAN

El. 579 ft. above sea level

Pg. 71

Index of cities Pg. 242 | Map legend Pg. 7

Mileages between cities

| | Angola | Bloomington | Chicago, IL | Crawfordsville | Evansville | Fort Wayne | Gary | Greensburg | Indianapolis | Kokomo | Lafayette | Muncie | New Albany | Richmond | South Bend | Terre Haute |
|---|---|---|---|---|---|---|---|---|---|---|---|---|---|---|---|---|
| Evansville | 347 | 120 | 289 | 178 | | 309 | 273 | 202 | 180 | 234 | 198 | 244 | 112 | 255 | 320 | 109 |
| Fort Wayne | 39 | 178 | 160 | 162 | 309 | | 132 | 147 | 129 | 86 | 117 | 72 | 238 | 92 | 89 | 205 |

Mileages © Rand McNally

Indiana state facts

Nickname: The Hoosier State
Capital: Indianapolis, J-9
Population: 6,376,792 (rank: 16th)
Largest city: Indianapolis, 798,382, J-9
Land area: 35,867 sq. mi. (rank: 38th)
Highest point: Hoosier Hill, 1,257 ft., I-14

**Mileages between cities**

| | Angola | Bloomington | Chicago, IL | Crawfordsville | Evansville | Fort Wayne | Gary | Greensburg | Indianapolis | Kokomo | Lafayette | Muncie | New Albany | Richmond | South Bend | Terre Haute |
|---|---|---|---|---|---|---|---|---|---|---|---|---|---|---|---|---|
| Gary | 135 | 200 | 30 | 118 | 273 | 132 | | 203 | 151 | 127 | 91 | 196 | 266 | 222 | 64 | 164 |
| Indianapolis | 166 | 52 | 181 | 49 | 180 | 129 | 151 | 50 | | 51 | 63 | 61 | 114 | 73 | 145 | 76 |

continued p. 76



Total mileages through Indiana
65: 261 miles; 70: 157 miles; 74: 172 miles; 90: 156 miles
More mileages at go.randmcnally.com/MC

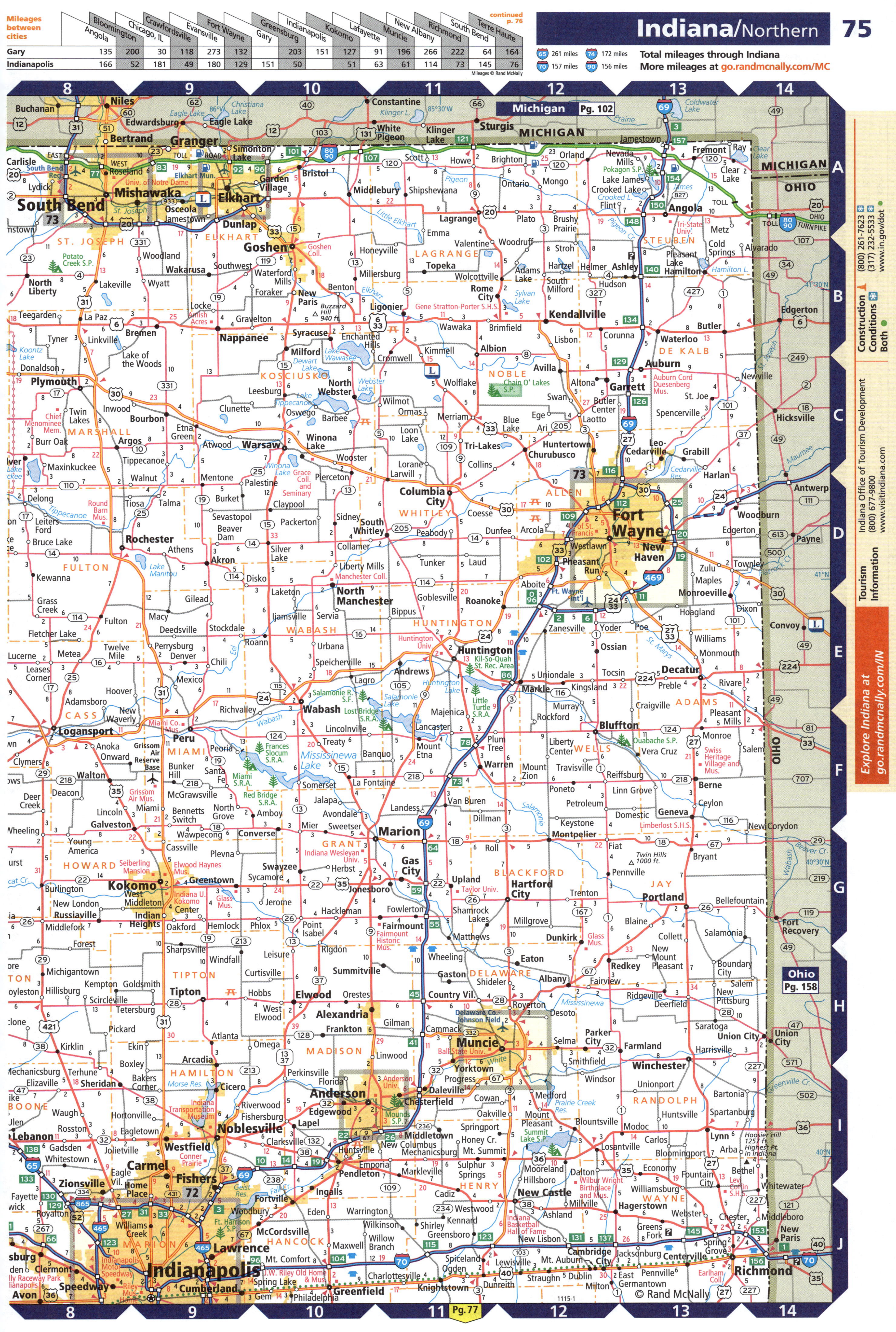

Index of cities Pg. 242
Map legend Pg. 7

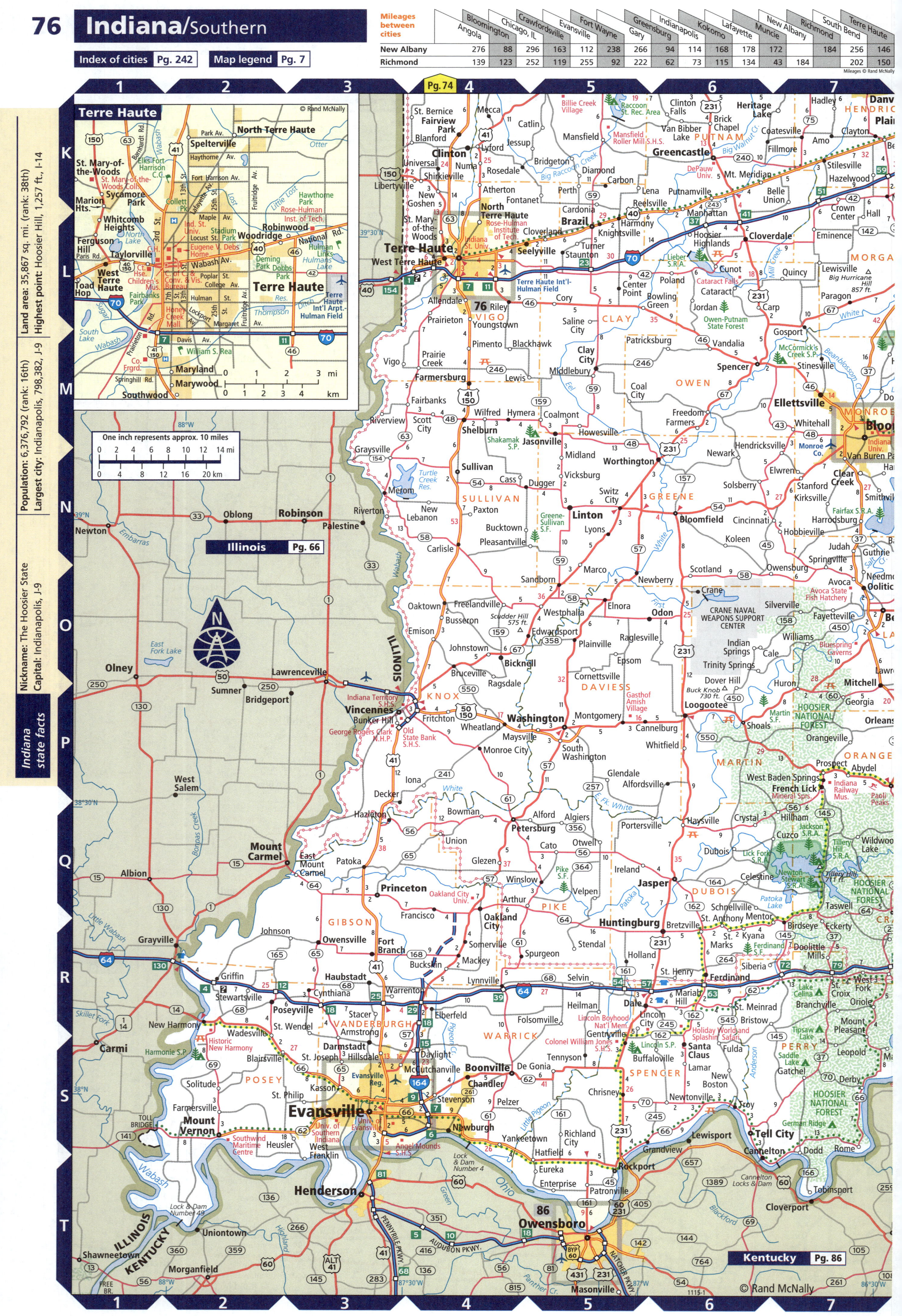


| Mileages between cities | Angola | Bloomington | Chicago, IL | Crawfordsville | Evansville | Fort Wayne | Gary | Greensburg | Indianapolis | Kokomo | Lafayette | Muncie | New Albany | Richmond | South Bend | Terre Haute |
|---|---|---|---|---|---|---|---|---|---|---|---|---|---|---|---|---|
| New Albany | 276 | 88 | 296 | 163 | 112 | 238 | 266 | 94 | 114 | 168 | 178 | 172 | | 184 | 256 | 146 |
| Richmond | 139 | 123 | 252 | 119 | 255 | 92 | 222 | 62 | 73 | 115 | 134 | 43 | 184 | | 202 | 150 |

Total mileages through Indiana
65: 261 miles | 70: 157 miles | 74: 172 miles | 90: 156 miles
More mileages at go.randmcnally.com/MC

Mileages between cities

| | Angola | Bloomington | Chicago, IL | Crawfordsville | Evansville | Fort Wayne | Gary | Greensburg | Indianapolis | Kokomo | Lafayette | Muncie | New Albany | Richmond | South Bend | Terre Haute |
|---|---|---|---|---|---|---|---|---|---|---|---|---|---|---|---|---|
| South Bend | 77 | 195 | 93 | 135 | 320 | 89 | 64 | 183 | 145 | 87 | 106 | 143 | 256 | 202 | | 216 |
| Terre Haute | 242 | 58 | 180 | 58 | 109 | 205 | 164 | 123 | 76 | 129 | 89 | 139 | 146 | 150 | 216 | |

Mileages © Rand McNally

Index of cities Pg. 243 | Map legend Pg. 7

**Mileages between cities**

| | Ames | Burlington | Cedar Rapids | Council Bluffs | Davenport | Decorah | Des Moines | Dubuque | Iowa City | Mason City | Ottumwa | Sioux City | Sioux Falls, SD | Spirit Lake | Storm Lake | Waterloo |
|---|---|---|---|---|---|---|---|---|---|---|---|---|---|---|---|---|
| Burlington | 209 | | 100 | 294 | 77 | 206 | 167 | 150 | 77 | 238 | 78 | 366 | 451 | 355 | 312 | 155 |
| Cedar Rapids | 108 | 100 | | 253 | 82 | 105 | 126 | 70 | 28 | 136 | 110 | 268 | 357 | 252 | 212 | 53 |



*Iowa state facts*

Nickname: The Hawkeye State
Capital: Des Moines, I-10
Population: 3,002,555 (rank: 30th)
Largest city: Des Moines, 197,052, I-10
Land area: 55,869 sq. mi. (rank: 23rd)
Highest point: Hawkeye Point, 1,670 ft., B-4

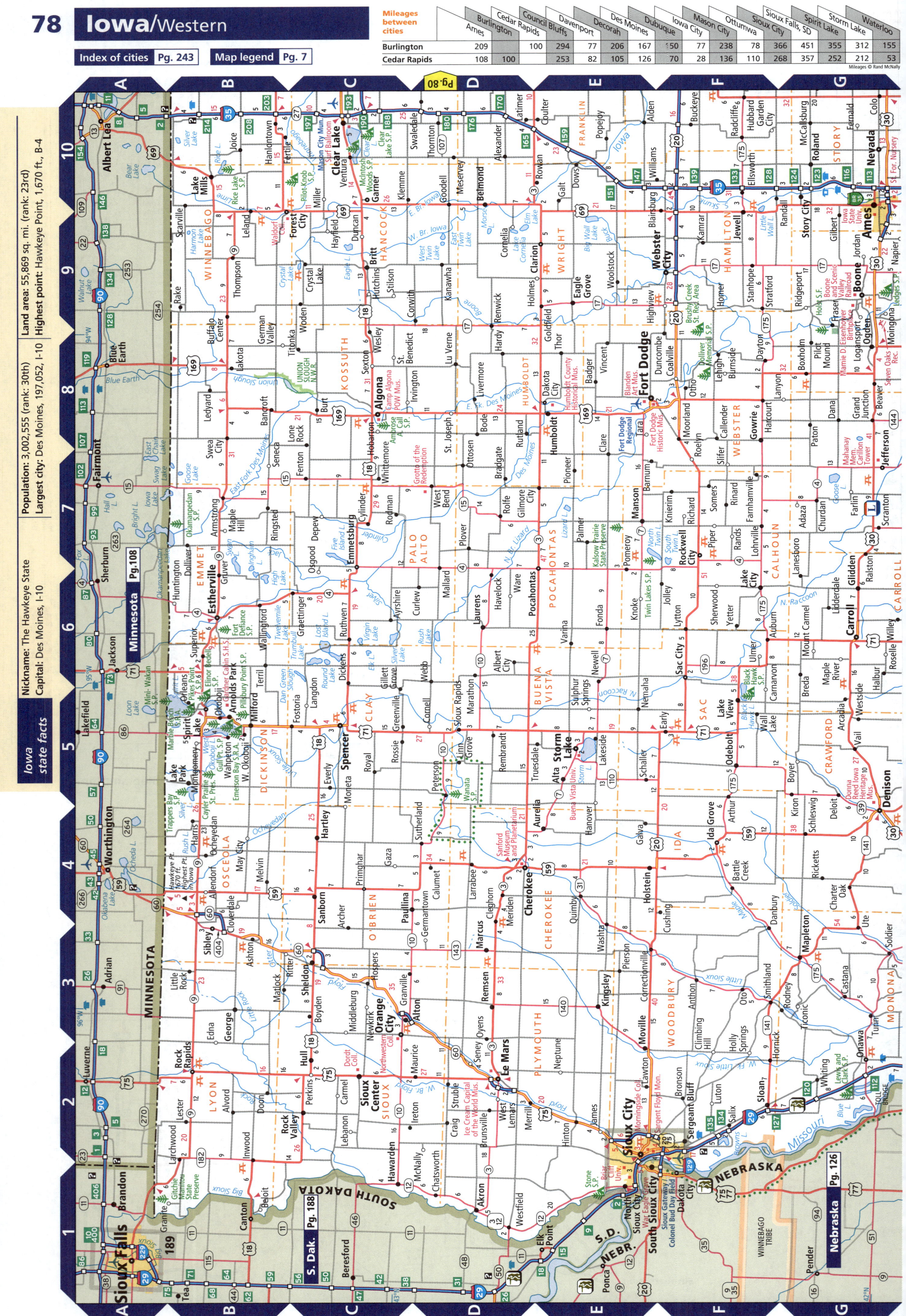

155 miles (I-29) · 303 miles (I-80) · 218 miles (I-35) · 257 miles (US 218)

**Total mileages through Iowa**
**More mileages at go.randmcnally.com/MC**

**Mileages between cities**

| | Ames | Burlington | Cedar Rapids | Council Bluffs | Davenport | Decorah | Des Moines | Dubuque | Iowa City | Mason City | Ottumwa | Sioux City | Sioux Falls, SD | Spirit Lake | Storm Lake | Waterloo |
|---|---|---|---|---|---|---|---|---|---|---|---|---|---|---|---|---|
| **Council Bluffs** | 160 | 294 | 253 | | 295 | 328 | 127 | 327 | 241 | 246 | 213 | 94 | 180 | 176 | 122 | 253 |
| **Davenport** | 191 | 77 | 82 | 295 | | 167 | 167 | 71 | 57 | 220 | 133 | 366 | 441 | 336 | 294 | 136 |

Mileages © Rand McNally

continued p. 80

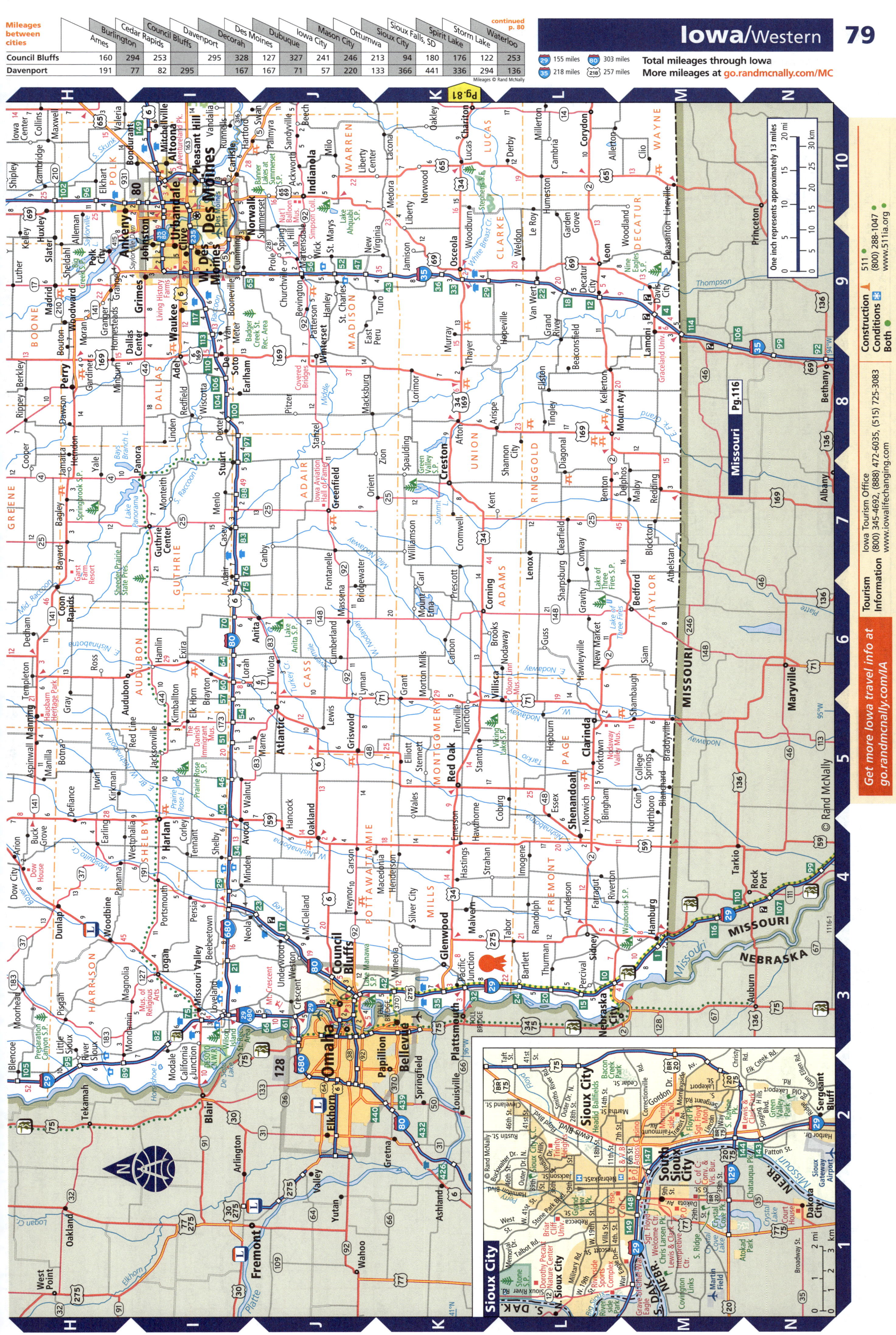

Index of cities Pg. 243 | Map legend Pg. 7

**Mileages between cities**

| | Ames | Burlington | Cedar Rapids | Council Bluffs | Davenport | Decorah | Des Moines | Dubuque | Iowa City | Mason City | Ottumwa | Sioux City | Sioux Falls, SD | Spirit Lake | Storm Lake | Waterloo |
|---|---|---|---|---|---|---|---|---|---|---|---|---|---|---|---|---|
| Des Moines | 33 | 167 | 126 | 127 | 167 | 201 | | 199 | 114 | 119 | 86 | 198 | 283 | 200 | 154 | 126 |
| Dubuque | 185 | 150 | 70 | 327 | 71 | 96 | 199 | | 84 | 174 | 184 | 305 | 395 | 290 | 249 | 91 |



**Iowa state facts**

Nickname: The Hawkeye State
Capital: Des Moines, I-10
Population: 3,002,555 (rank: 30th)
Largest city: Des Moines, 197,052, I-10
Land area: 55,869 sq. mi. (rank: 23rd)
Highest point: Hawkeye Point, 1,670 ft., B-4

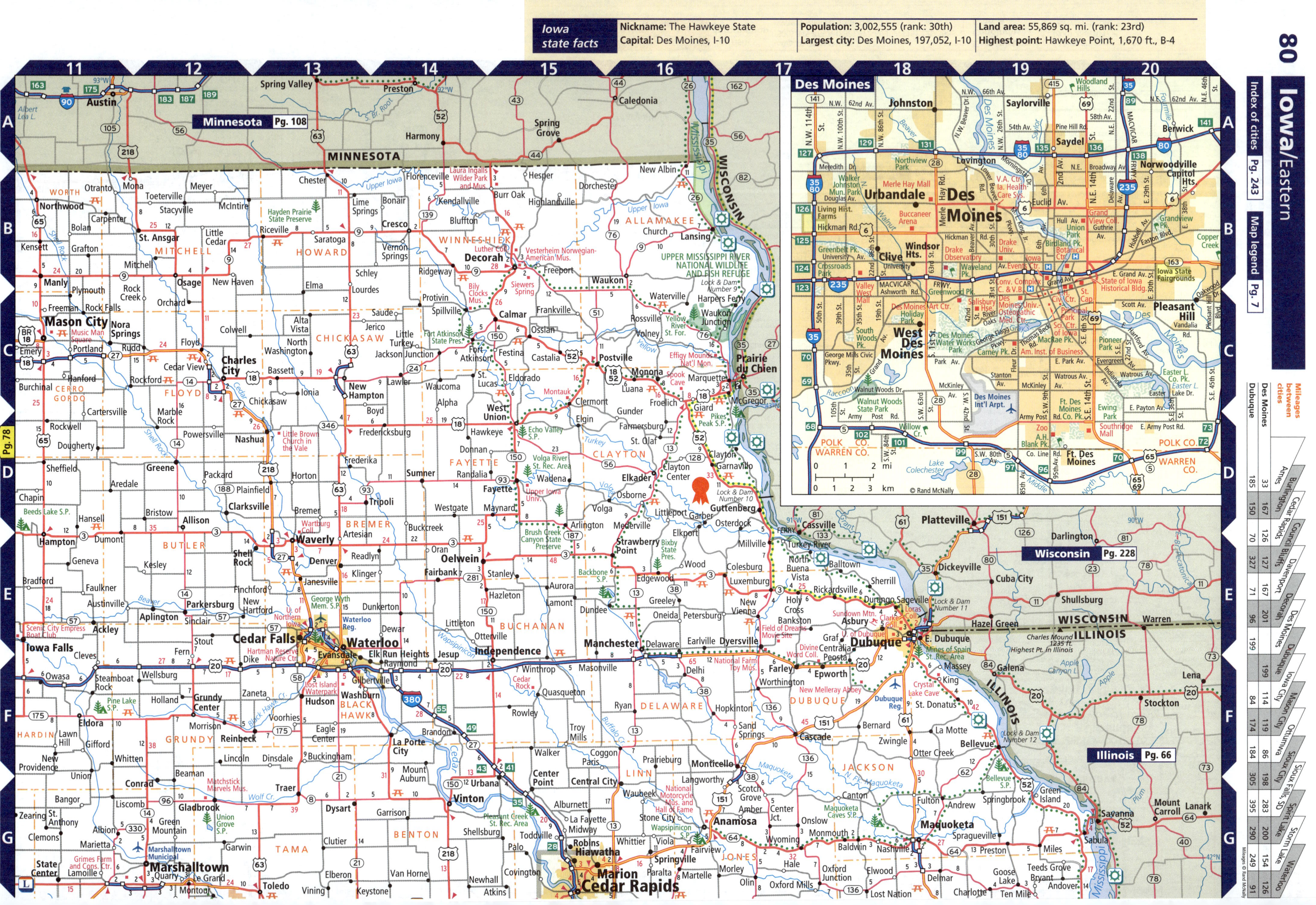

**Mileages between cities**

| | Ames | Burlington | Cedar Rapids | Council Bluffs | Davenport | Decorah | Des Moines | Dubuque | Iowa City | Mason City | Ottumwa | Sioux City | Sioux Falls, SD | Spirit Lake | Storm Lake | Waterloo |
|---|---|---|---|---|---|---|---|---|---|---|---|---|---|---|---|---|
| **Mason City** | 91 | 238 | 136 | 246 | 220 | 88 | 119 | 174 | 165 | | 203 | 200 | 222 | 118 | 135 | 83 |
| **Sioux City** | 175 | 366 | 268 | 94 | 366 | 304 | 198 | 305 | 312 | 200 | 285 | | 85 | 109 | 78 | 218 |



I-29 155 miles | I-80 303 miles
I-35 218 miles | US-218 257 miles

**Total mileages through Iowa**
**More mileages at go.randmcnally.com/MC**

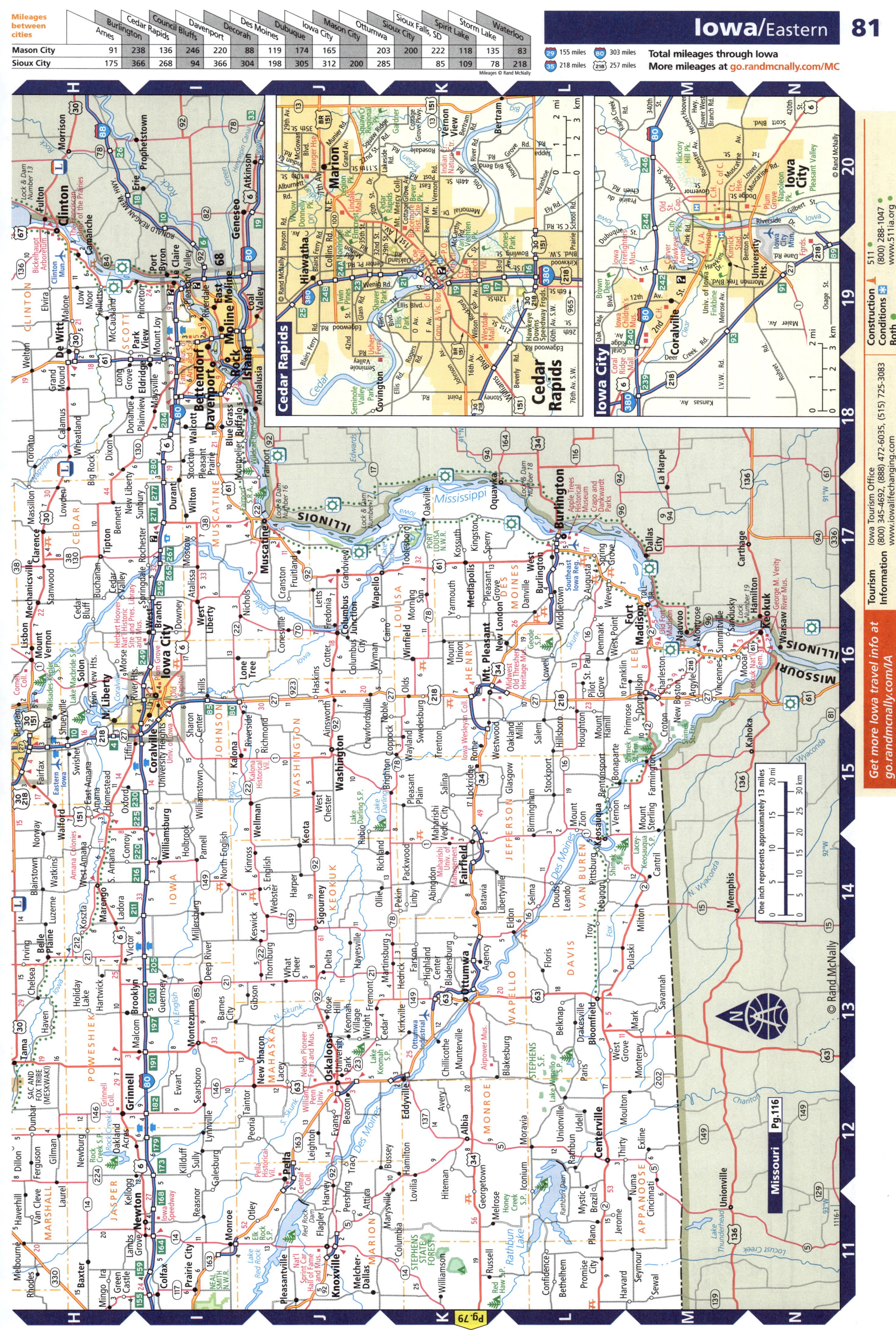

Index of cities Pg. 244 | Map legend Pg. 7

**Kansas state facts**

**Nickname:** The Sunflower State
**Capital:** Topeka, D-16
**Population:** 2,802,134 (rank: 33rd)
**Largest city:** Wichita, 366,046, H-13
**Land area:** 81,815 sq. mi. (rank: 13th)
**Highest point:** Mount Sunflower, 4,039 ft., D-1

**Mileages between cities**

| | Arkansas City | Atchison | Coffeyville | Dodge City | Emporia | Fort Scott | Goodland | Hays | Hutchinson | Joplin, MO | Kansas City | Liberal | Manhattan | Salina | Topeka | Wichita |
|---|---|---|---|---|---|---|---|---|---|---|---|---|---|---|---|---|
| Dodge City | 212 | 323 | 288 | | 240 | 304 | 192 | 104 | 122 | 337 | 333 | 82 | 227 | 164 | 273 | 154 |
| Goodland | 384 | 395 | 455 | 192 | 349 | 472 | | 144 | 268 | 505 | 406 | 209 | 299 | 235 | 344 | 323 |

Mileages © Rand McNally

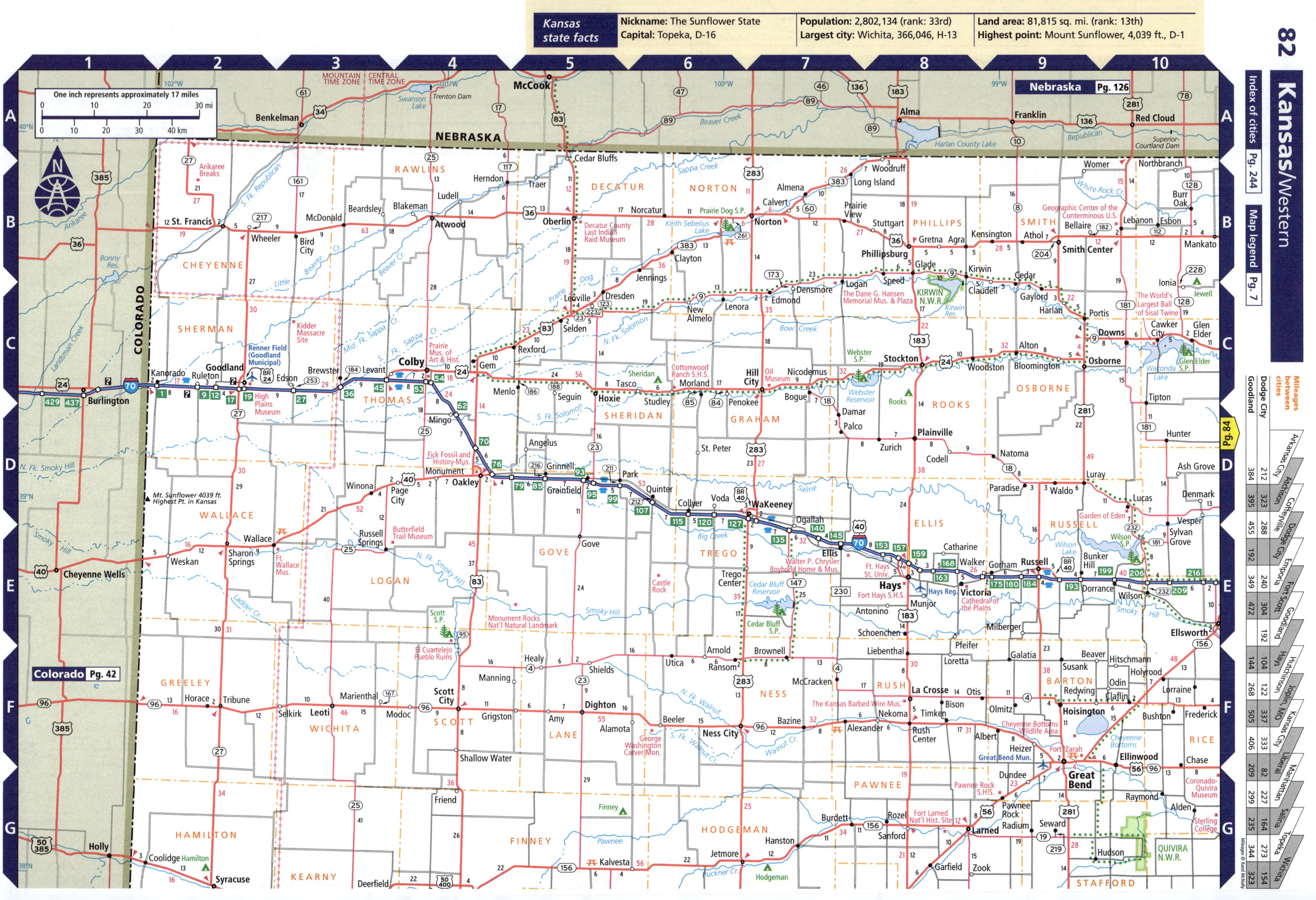

**Mileages between cities**

continued p. 84

| | Arkansas City | Atchison | Coffeyville | Dodge City | Emporia | Fort Scott | Goodland | Hays | Hutchinson | Joplin, MO | Kansas City | Liberal | Manhattan | Salina | Topeka | Wichita |
|---|---|---|---|---|---|---|---|---|---|---|---|---|---|---|---|---|
| Joplin, MO | 150 | 196 | 65 | 337 | 177 | 60 | 505 | 366 | 233 | | 154 | 395 | 252 | 274 | 196 | 183 |
| Kansas City | 228 | 58 | 172 | 333 | 109 | 94 | 406 | 266 | 220 | 154 | | 406 | 117 | 173 | 62 | 196 |



35 235 miles
70 424 miles
56 464 miles
81 220 miles

Total mileages through Kansas
More mileages at go.randmcnally.com/MC

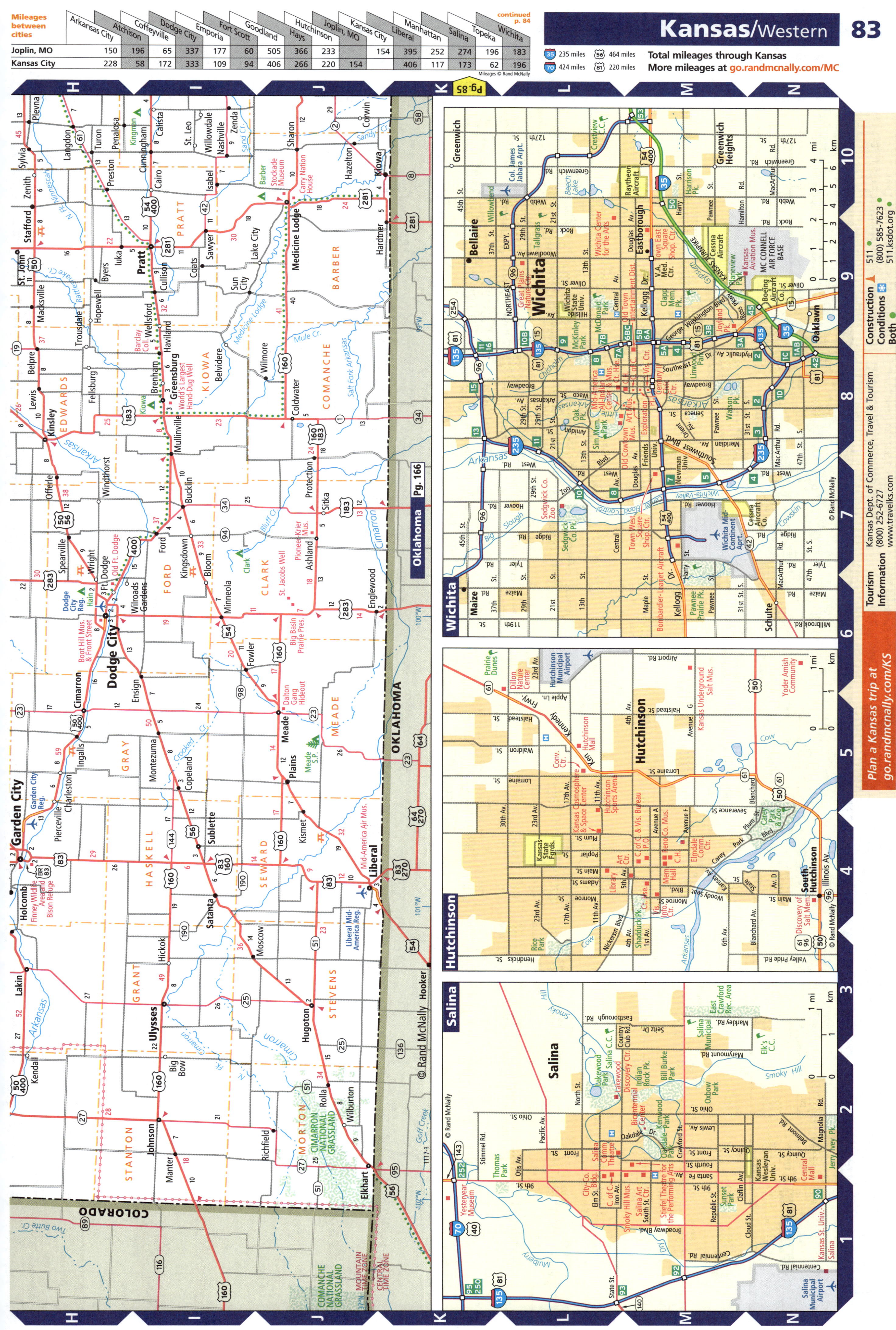

Index of cities Pg. 244 | Map legend Pg. 7

*Kansas state facts*

**Nickname:** The Sunflower State
**Capital:** Topeka, D-16
**Population:** 2,802,134 (rank: 33rd)
**Largest city:** Wichita, 366,046, H-13
**Land area:** 81,815 sq. mi. (rank: 13th)
**Highest point:** Mount Sunflower, 4,039 ft., D-1

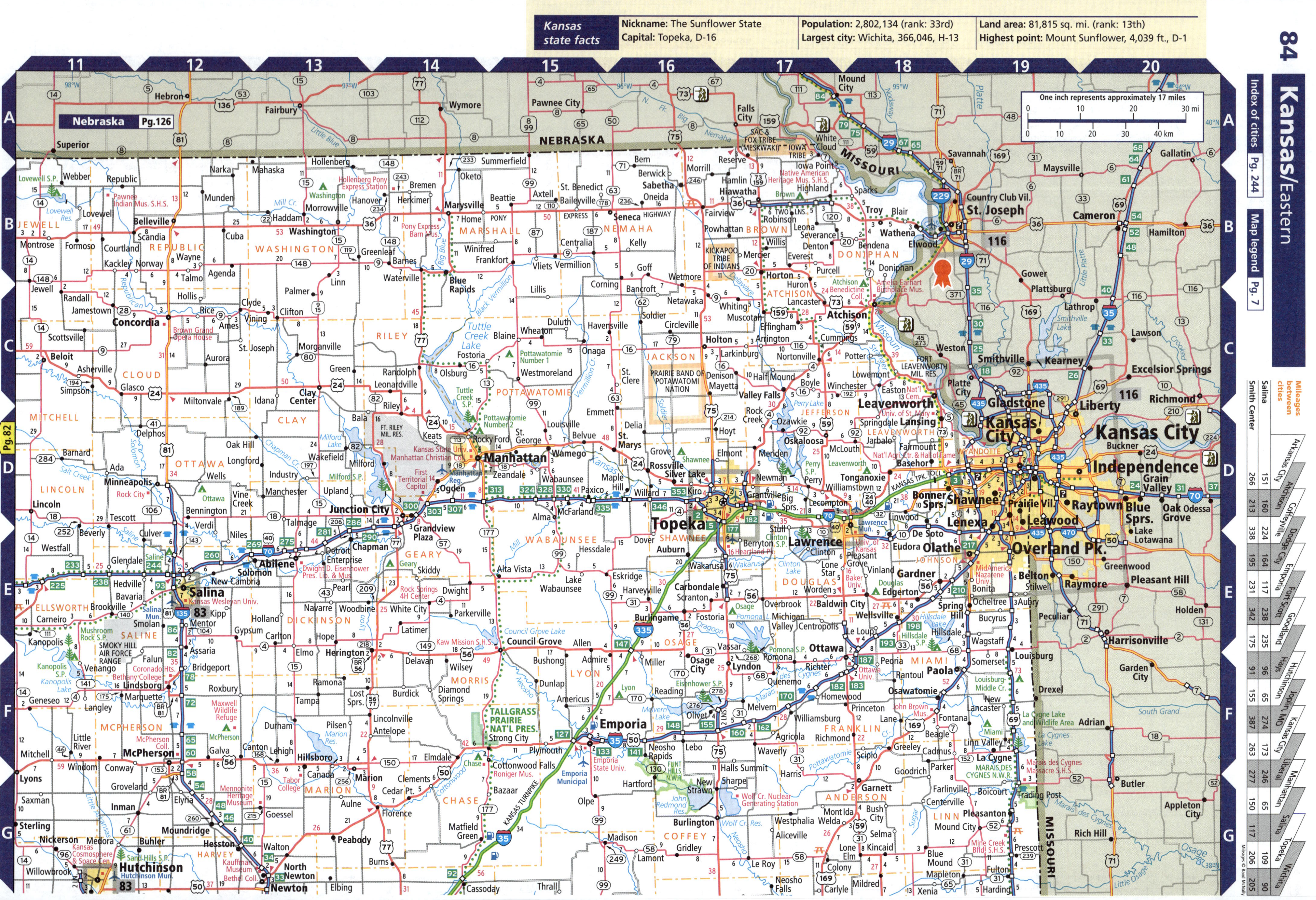


**Mileages between cities**

| | Arkansas City | Atchison | Coffeyville | Dodge City | Emporia | Fort Scott | Goodland | Hays | Hutchinson | Joplin, MO | Kansas City | Liberal | Manhattan | Salina | Topeka | Wichita |
|---|---|---|---|---|---|---|---|---|---|---|---|---|---|---|---|---|
| **Salina** | 151 | 160 | 224 | 164 | 117 | 238 | 235 | 96 | 65 | 274 | 173 | 246 | 65 | | 109 | 90 |
| **Smith Center** | 266 | 213 | 338 | 195 | 231 | 342 | 175 | 91 | 155 | 387 | 263 | 277 | 150 | 117 | 206 | 205 |

**Mileages between cities**

| | Arkansas City | Atchison | Coffeyville | Dodge City | Emporia | Fort Scott | Goodland | Hays | Hutchinson | Joplin, MO | Kansas City | Liberal | Manhattan | Salina | Topeka | Wichita |
|---|---|---|---|---|---|---|---|---|---|---|---|---|---|---|---|---|
| Topeka | 170 | 55 | 155 | 273 | 58 | 136 | 344 | 204 | 162 | 196 | 62 | 349 | 56 | 109 | | 137 |
| Wichita | 61 | 188 | 134 | 154 | 85 | 149 | 323 | 183 | 51 | 183 | 196 | 212 | 130 | 90 | 137 | |



35 235 miles
70 424 miles
56 464 miles
81 220 miles

Total mileages through Kansas
More mileages at go.randmcnally.com/MC

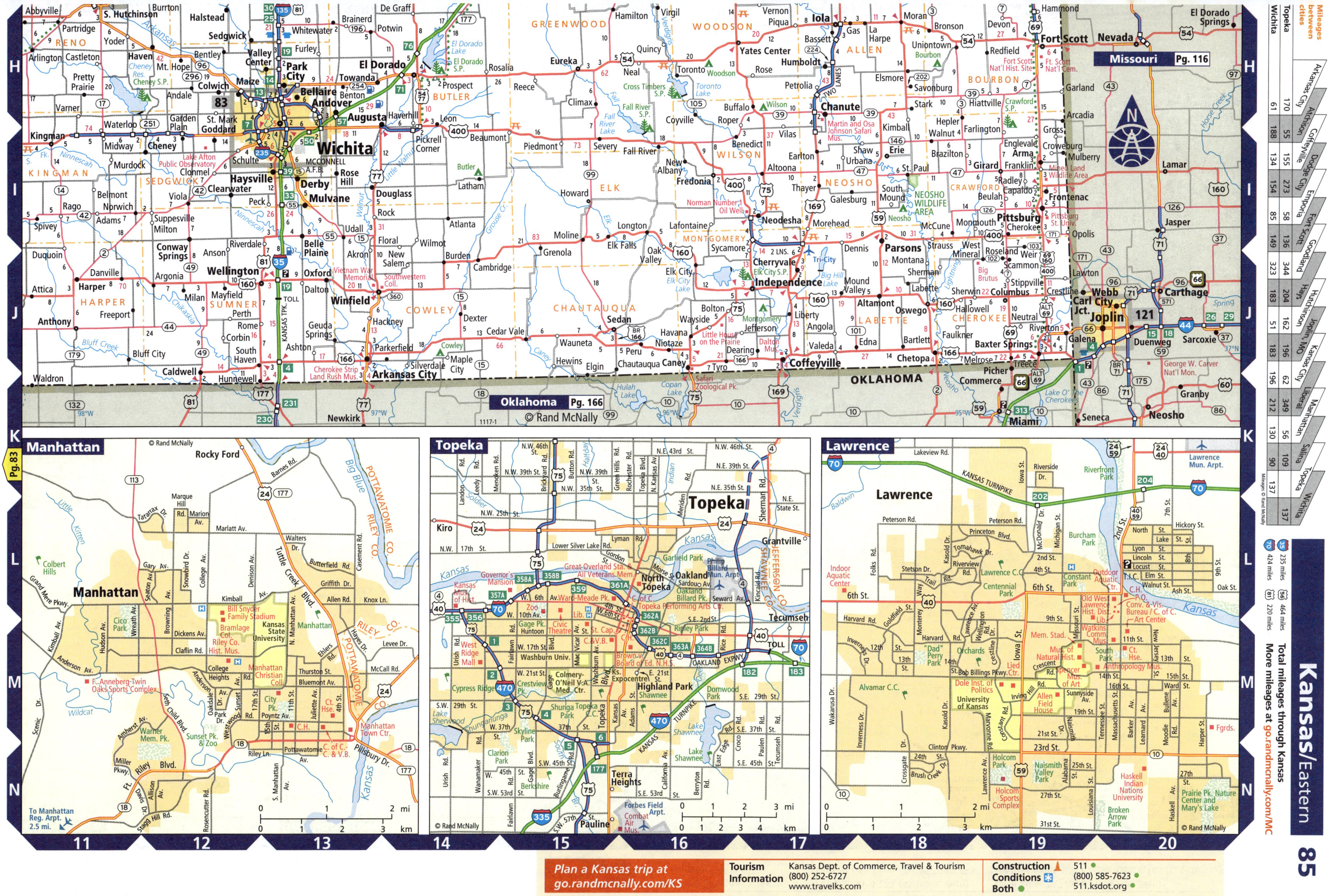


Plan a Kansas trip at go.randmcnally.com/KS

Tourism Information
Kansas Dept. of Commerce, Travel & Tourism
(800) 252-6727
www.travelks.com

Construction 511
Conditions (800) 585-7623
Both 511.ksdot.org

Index of cities Pg. 244 | Map legend Pg. 7

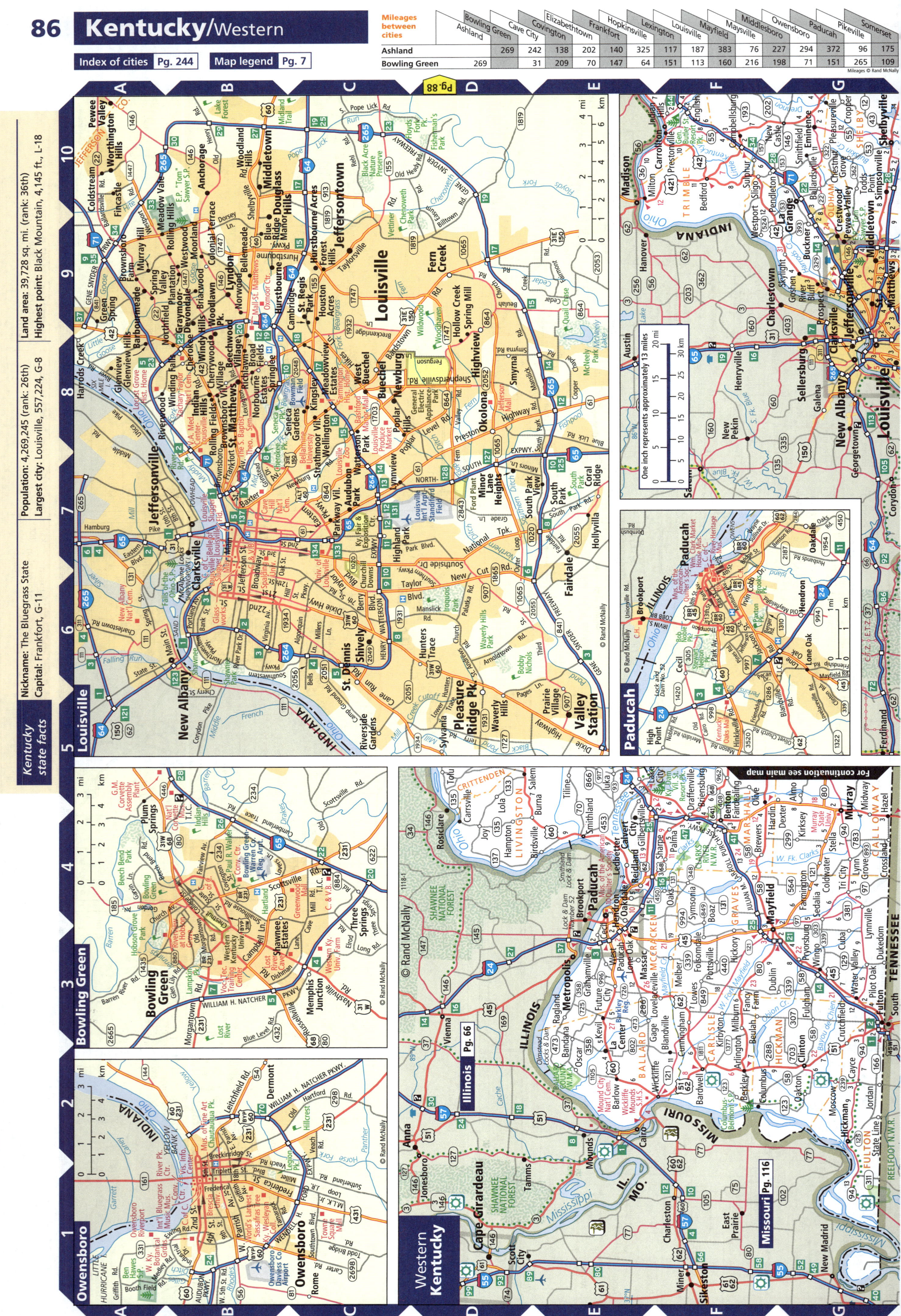

continued p. 88

| Mileages between cities | Ashland | Bowling Green | Cave City | Covington | Elizabethtown | Frankfort | Hopkinsville | Lexington | Louisville | Mayfield | Maysville | Middlesboro | Owensboro | Paducah | Pikeville | Somerset |
|---|---|---|---|---|---|---|---|---|---|---|---|---|---|---|---|---|
| Covington | 138 | 209 | 181 | | 140 | 78 | 265 | 81 | 97 | 322 | 59 | 208 | 203 | 312 | 216 | 157 |
| Lexington | 117 | 151 | 124 | 81 | 84 | 29 | 207 | | 76 | 266 | 63 | 130 | 177 | 256 | 140 | 78 |

Mileages © Rand McNally

64 185 miles · 65 137 miles · 71 97 miles · 75 192 miles

Total mileages through Kentucky
More mileages at go.randmcnally.com/MC

Index of cities Pg. 244 | Map legend Pg. 7

**Mileages between cities**

| | Ashland | Bowling Green | Cave City | Covington | Elizabethtown | Frankfort | Hopkinsville | Lexington | Louisville | Mayfield | Maysville | Middlesboro | Owensboro | Paducah | Pikeville | Somerset |
|---|---|---|---|---|---|---|---|---|---|---|---|---|---|---|---|---|
| Louisville | 187 | 113 | 85 | 97 | 44 | 50 | 170 | 76 | | 227 | 133 | 203 | 106 | 216 | 211 | 124 |
| Middlesboro | 227 | 198 | 176 | 208 | 182 | 157 | 265 | 130 | 203 | 363 | 191 | | 275 | 353 | 125 | 88 |



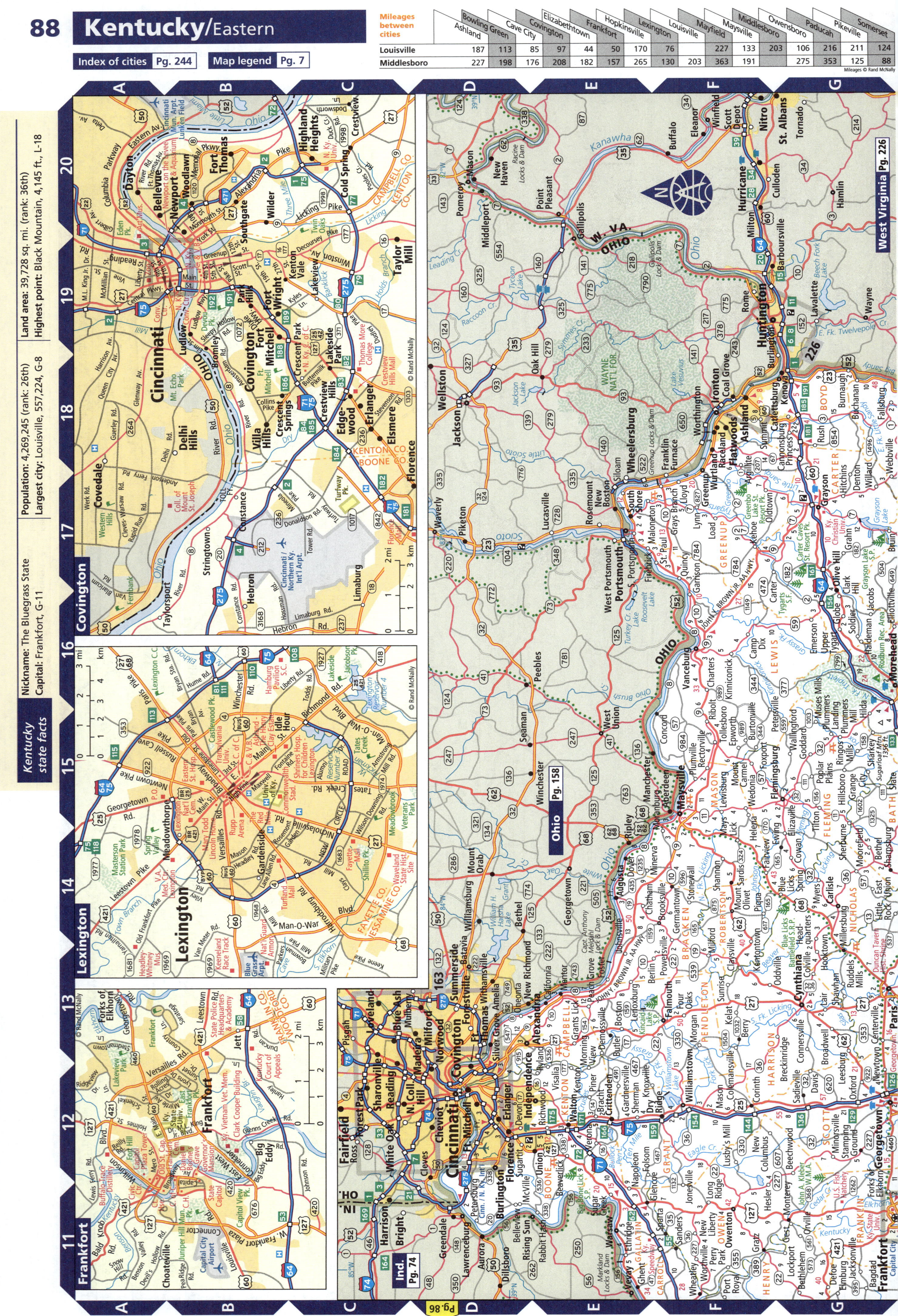

**Mileages between cities**

| | Ashland | Bowling Green | Cave City | Covington | Elizabethtown | Frankfort | Hopkinsville | Lexington | Louisville | Mayfield | Maysville | Middlesboro | Owensboro | Paducah | Pikeville | Somerset |
|---|---|---|---|---|---|---|---|---|---|---|---|---|---|---|---|---|
| Owensboro | 294 | 71 | 108 | 203 | 94 | 159 | 96 | 177 | 106 | 154 | 242 | 275 | | 143 | 318 | 187 |
| Paducah | 372 | 151 | 186 | 312 | 172 | 250 | 72 | 256 | 216 | 24 | 319 | 353 | 143 | | 396 | 265 |



# Kentucky/Eastern

64 185 miles · 71 97 miles
65 137 miles · 75 192 miles

**Total mileages through Kentucky**
**More mileages at go.randmcnally.com/MC**

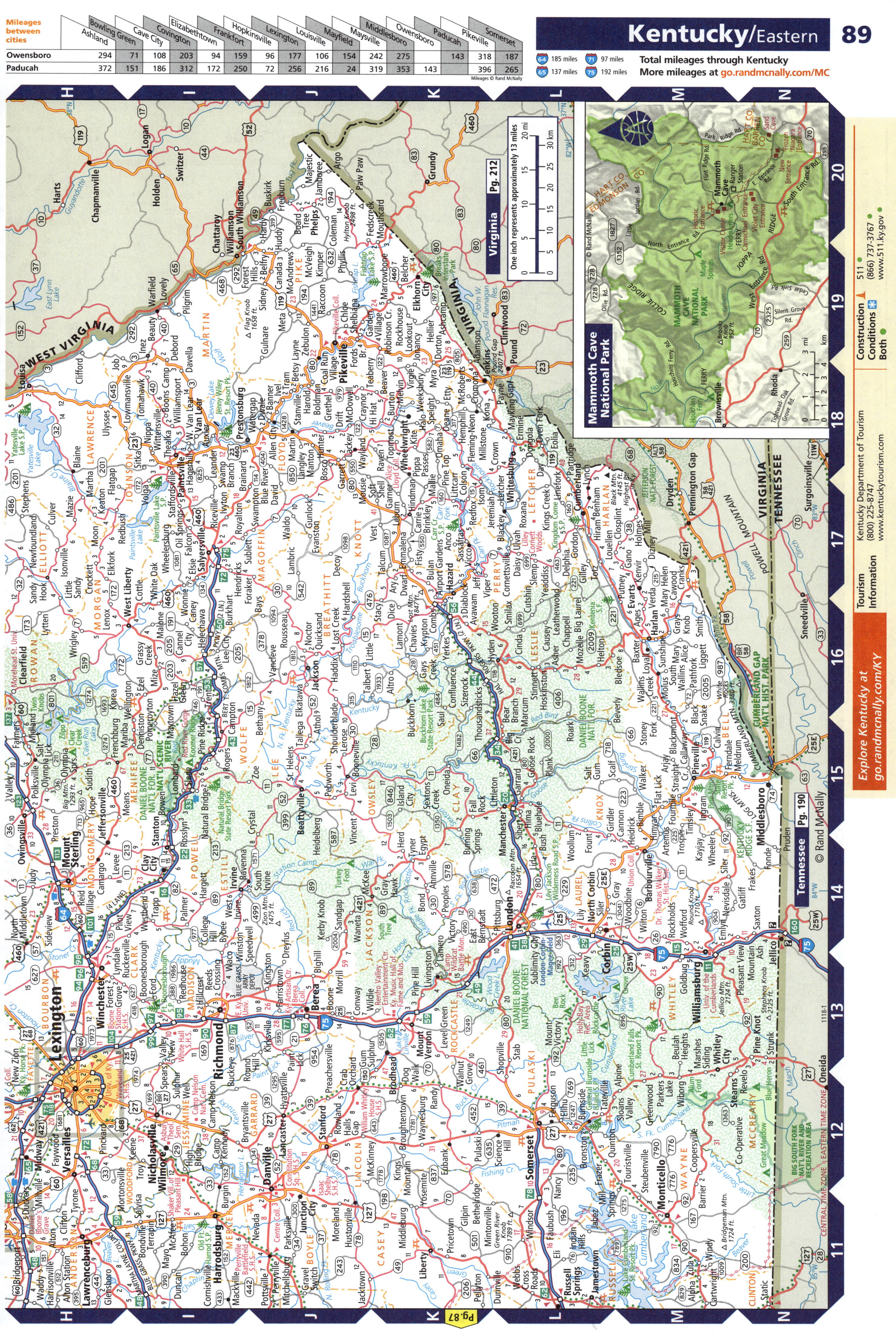

Index of cities Pg. 245 | Map legend Pg. 7

Mileages between cities

| | Baton Rouge | Beaumont, TX | Houma | Lake Charles | Monroe | New Orleans | Shreveport | Vicksburg, MS |
|---|---|---|---|---|---|---|---|---|
| Alexandria | 125 | 155 | 190 | 97 | 95 | 218 | 123 | 147 |
| Baton Rouge | | 183 | 85 | 124 | 186 | 79 | 250 | 157 |
| Gulfport, MS | 134 | 318 | 131 | 258 | 276 | 78 | 375 | 201 |
| Lafayette | 55 | 133 | 102 | 73 | 182 | 134 | 211 | 212 |



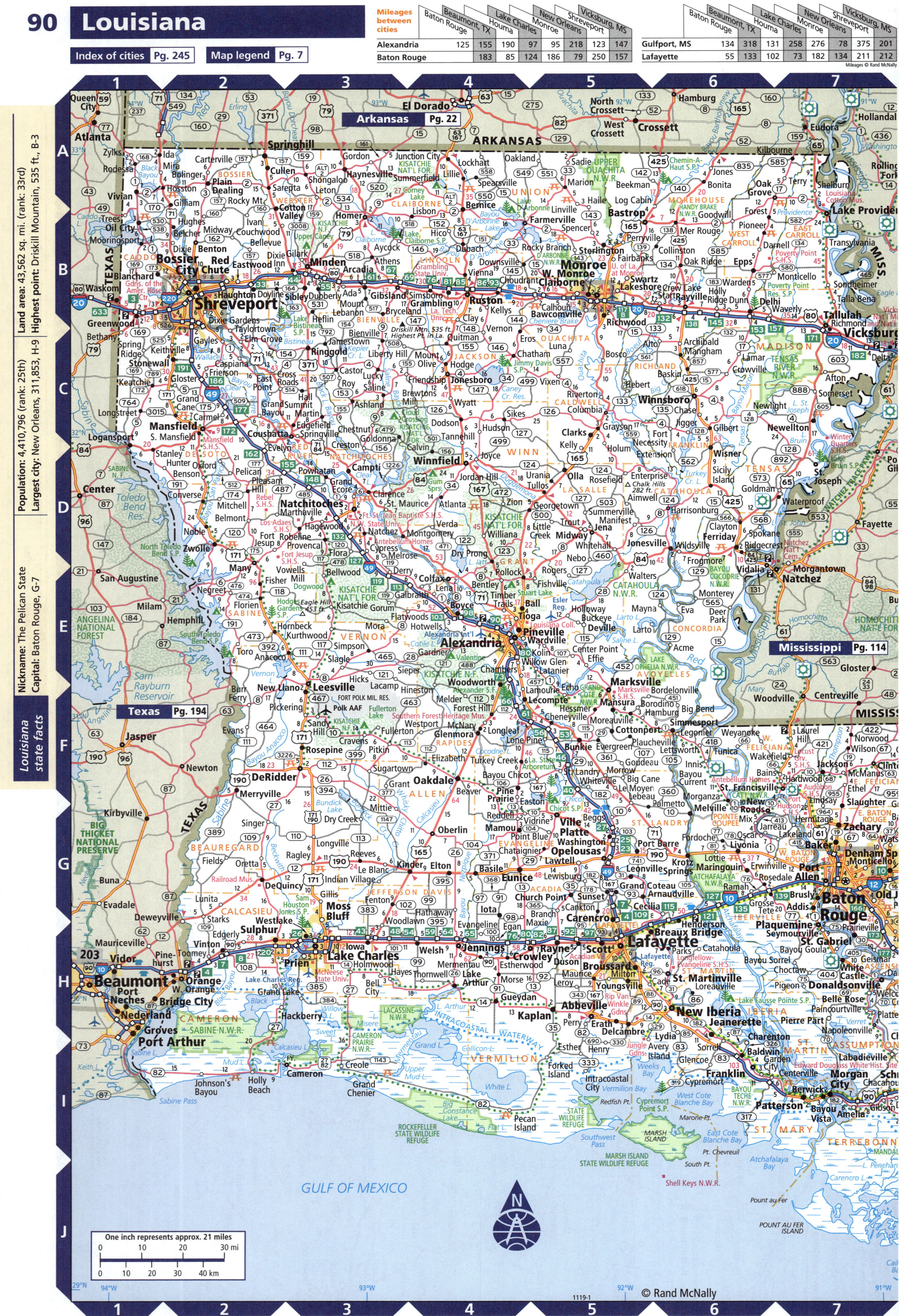

Mileages between cities

| | Baton Rouge | Beaumont, TX | Houma | Lake Charles | Monroe | New Orleans | Shreveport | Vicksburg, MS |
|---|---|---|---|---|---|---|---|---|
| Lake Charles | 124 | 60 | 177 | | 190 | 203 | 184 | 243 |
| New Orleans | 79 | 262 | 56 | 203 | 281 | | 340 | 207 |
| Shreveport | 250 | 206 | 314 | 184 | 98 | 340 | | 171 |
| Vicksburg, MS | 157 | 301 | 234 | 243 | 74 | 207 | 171 | |



| Interstate | Miles | Interstate | Miles |
|---|---|---|---|
| 10 | 274 miles | 49 | 208 miles |
| 20 | 190 miles | 55 | 66 miles |

Total mileages through Louisiana
More mileages at go.randmcnally.com/MC

Construction
Conditions
Both
511
(888) 762-3511
www.511la.org

Tourism Information
Louisiana Office of Tourism
(800) 334-8626
www.louisianatravel.com

Get more Louisiana travel info at go.randmcnally.com/LA

Index of cities Pg. 246 | Map legend Pg. 7

**Mileages between cities**

| | Auburn | Bangor | Bar Harbor | Eastport | Houlton | Millinocket | Portland | Rangeley |
|---|---|---|---|---|---|---|---|---|
| Bangor | 107 | | 47 | 120 | 118 | 72 | 128 | 120 |
| Eastport | 226 | 120 | 118 | | 115 | 125 | 247 | 242 |
| Houlton | 225 | 118 | 167 | 115 | | 69 | 246 | 238 |
| Madawaska | 326 | 219 | 267 | 218 | 102 | 170 | 347 | 339 |

Mileages © Rand McNally

**Maine state facts**

Nickname: The Pine Tree State
Capital: Augusta, F-4
Population: 1,316,456 (rank: 40th)
Largest city: Portland, 62,561, H-3
Land area: 30,862 sq. mi. (rank: 39th)
Highest point: Mount Katahdin, 5,268 ft., A-6

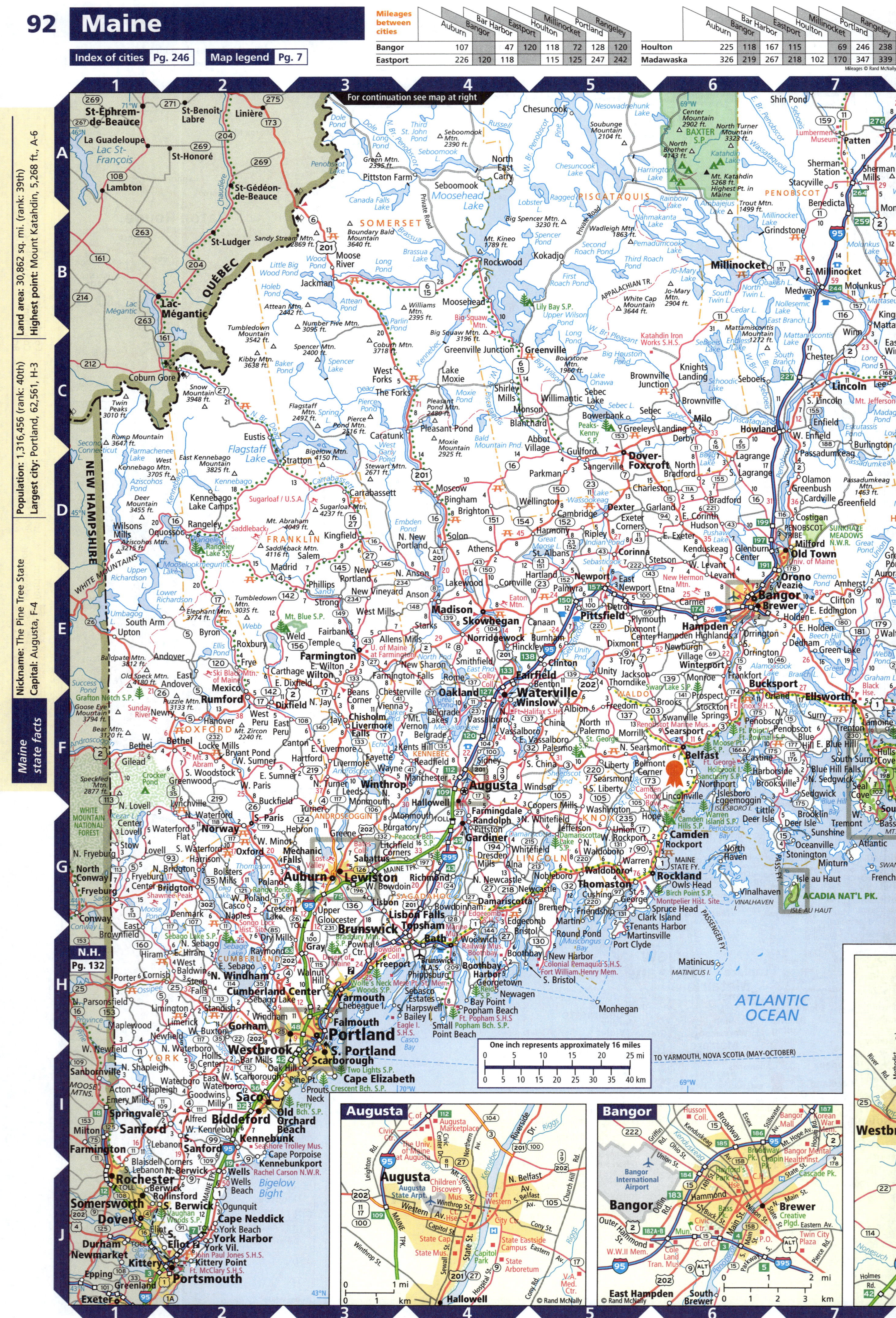

**Mileages between cities**

| | Auburn | Bangor | Bar Harbor | Eastport | Houlton | Millinocket | Portland | Rangeley |
|---|---|---|---|---|---|---|---|---|
| Portland | 35 | 128 | 174 | 247 | 246 | 181 | | 118 |
| Portsmouth, NH | 81 | 180 | 225 | 301 | 298 | 231 | 51 | 165 |
| Rangeley | 84 | 120 | 165 | 242 | 238 | 153 | 118 | |
| Waterville | 53 | 55 | 101 | 174 | 173 | 107 | 75 | 77 |

Mileages © Rand McNally

95 299 miles; 2 273 miles; 1 315 miles; 201 164 miles

Total mileages through Maine
More mileages at go.randmcnally.com/MC

**Road Conditions**
511
Construction (866) 282-7578, Conditions (207) 624-3595, Both www.511maine.gov

**Tourism Information**
Maine Office of Tourism
(888) 624-6345
www.visitmaine.com

Plan a Maine trip at go.randmcnally.com/ME

Index of cities Pg. 246 | Map legend Pg. 7

**Mileages between cities**

| | Aberdeen | Annapolis | Baltimore | Cambridge | Chestertown | Cumberland | Frederick | Hagerstown | Lexington Park | Ocean City | Pocomoke City | Rockville | St. Charles | Salisbury | Washington, DC | Wilmington, DE |
|---|---|---|---|---|---|---|---|---|---|---|---|---|---|---|---|---|
| **Aberdeen** | | 58 | 31 | 113 | 65 | 171 | 83 | 107 | 122 | 134 | 152 | 74 | 90 | 122 | 70 | 42 |
| **Annapolis** | 58 | | 28 | 57 | 47 | 157 | 68 | 93 | 73 | 108 | 120 | 42 | 41 | 89 | 30 | 96 |



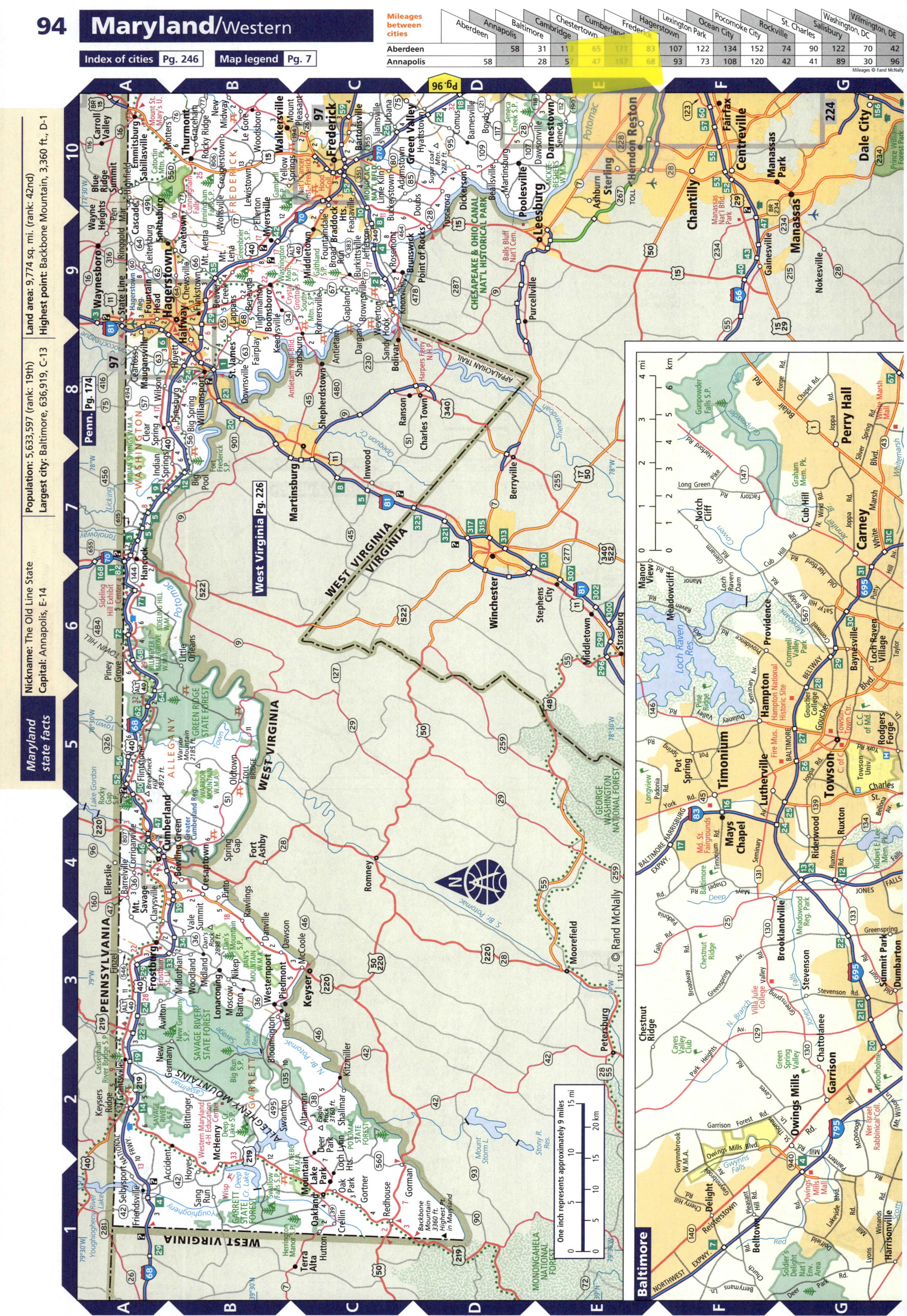

**Mileages between cities**

| | Aberdeen | Annapolis | Baltimore | Cambridge | Chestertown | Cumberland | Frederick | Hagerstown | Lexington Park | Ocean City | Pocomoke City | Rockville | St. Charles | Salisbury | Washington, DC | Wilmington, DE |
|---|---|---|---|---|---|---|---|---|---|---|---|---|---|---|---|---|
| Baltimore | 31 | 28 | | 84 | 73 | 136 | 47 | 72 | 93 | 136 | 146 | 42 | 59 | 116 | 39 | 70 |
| Cumberland | 171 | 157 | 136 | 212 | 203 | | 88 | 67 | 200 | 263 | 275 | 116 | 166 | 244 | 134 | 209 |

Mileages © Rand McNally

continued p. 96

68 81 miles · 70 94 miles · 81 12 miles · 95 110 miles

Total mileages through Maryland
More mileages at go.randmcnally.com/MC

Index of cities Pg. 246 | Map legend Pg. 7

| Mileages between cities | Aberdeen | Annapolis | Baltimore | Cambridge | Chestertown | Cumberland | Frederick | Hagerstown | Lexington Park | Ocean City | Pocomoke City | Rockville | St. Charles | Salisbury | Washington, DC | Wilmington, DE |
|---|---|---|---|---|---|---|---|---|---|---|---|---|---|---|---|---|
| Hagerstown | 107 | 93 | 72 | 149 | 139 | 67 | 25 | | 136 | 200 | 212 | 52 | 102 | 180 | 70 | 145 |
| Lexington Park | 122 | 73 | 93 | 127 | 118 | 200 | 113 | 136 | | 178 | 190 | 84 | 37 | 159 | 67 | 161 |



**Maryland state facts**

Nickname: The Old Line State
Capital: Annapolis, E-14
Population: 5,633,597 (rank: 19th)
Largest city: Baltimore, 636,919, C-13
Land area: 9,774 sq. mi. (rank: 42nd)
Highest point: Backbone Mountain, 3,360 ft., D-1

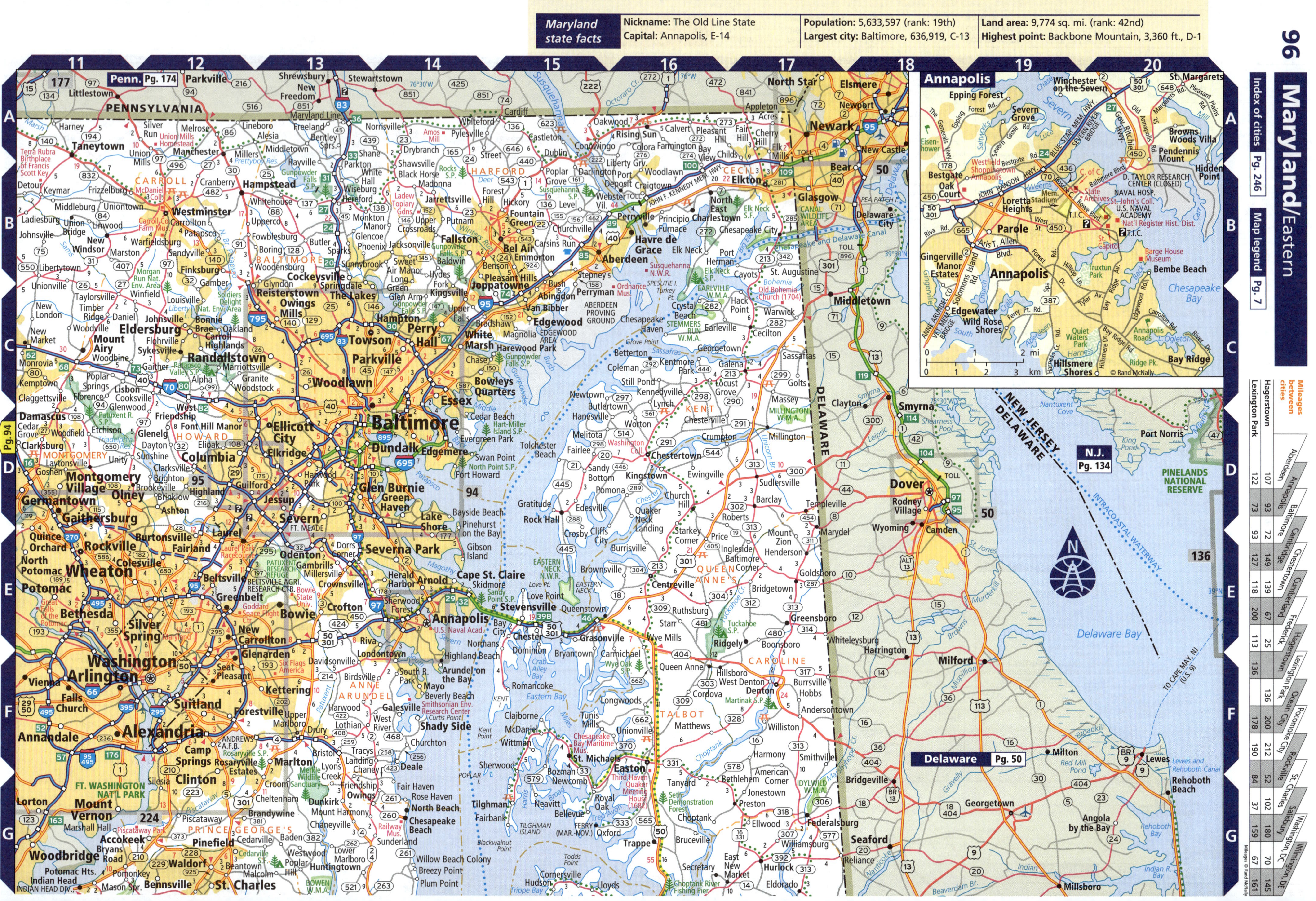

Mileages between cities

| | Aberdeen | Annapolis | Baltimore | Cambridge | Chestertown | Cumberland | Frederick | Hagerstown | Lexington Park | Ocean City | Pocomoke City | Rockville | St. Charles | Salisbury | Washington, DC | Wilmington, DE |
|---|---|---|---|---|---|---|---|---|---|---|---|---|---|---|---|---|
| Salisbury | 122 | 89 | 116 | 32 | 78 | 244 | 156 | 180 | 159 | 29 | 26 | 130 | 128 | | 118 | 107 |
| Washington, DC | 70 | 30 | 39 | 86 | 76 | 134 | 48 | 70 | 67 | 139 | 148 | 19 | 30 | 118 | | 109 |

Mileages © Rand McNally

68 81 miles · 81 12 miles · 70 94 miles · 95 110 miles

Total mileages through Maryland

More mileages at go.randmcnally.com/MC

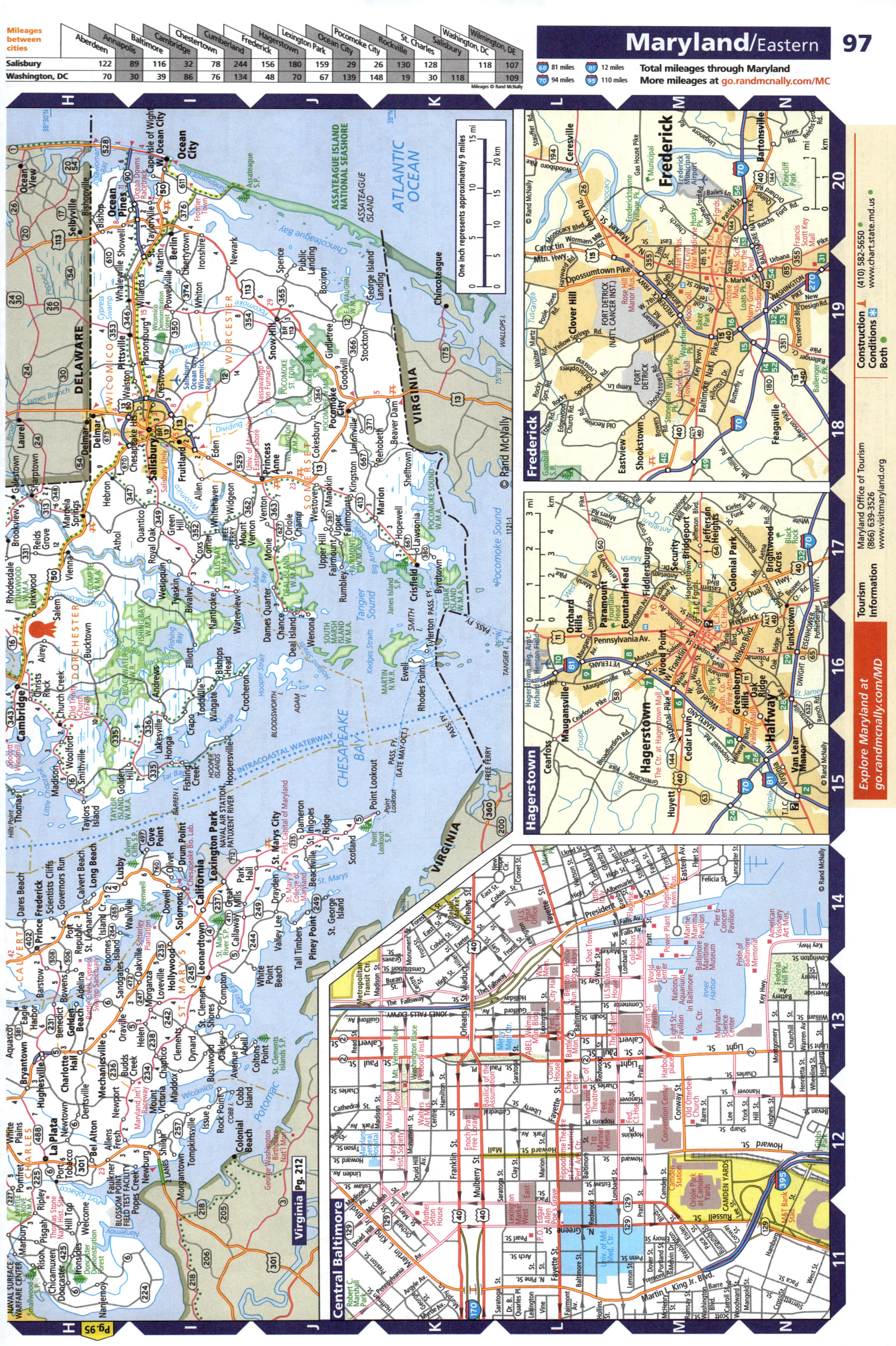

Index of cities Pg. 246 | Map legend Pg. 7

Mileages between cities

*via ferry

| | Boston | Brockton | Falmouth | Fitchburg | Gloucester | Greenfield | Lowell | Nantucket | New Bedford | North Adams | Pittsfield | Plymouth | Providence, RI | Provincetown | Springfield | Worcester |
|---|---|---|---|---|---|---|---|---|---|---|---|---|---|---|---|---|
| Boston | | 24 | 76 | 47 | 39 | 94 | 29 | 101* | 58 | 157 | 136 | 40 | 50 | 116 | 90 | 43 |
| Gloucester | 39 | 63 | 114 | 74 | | 120 | 47 | 140* | 97 | 157 | 169 | 78 | 90 | 154 | 122 | 75 |

Mileages © Rand McNally

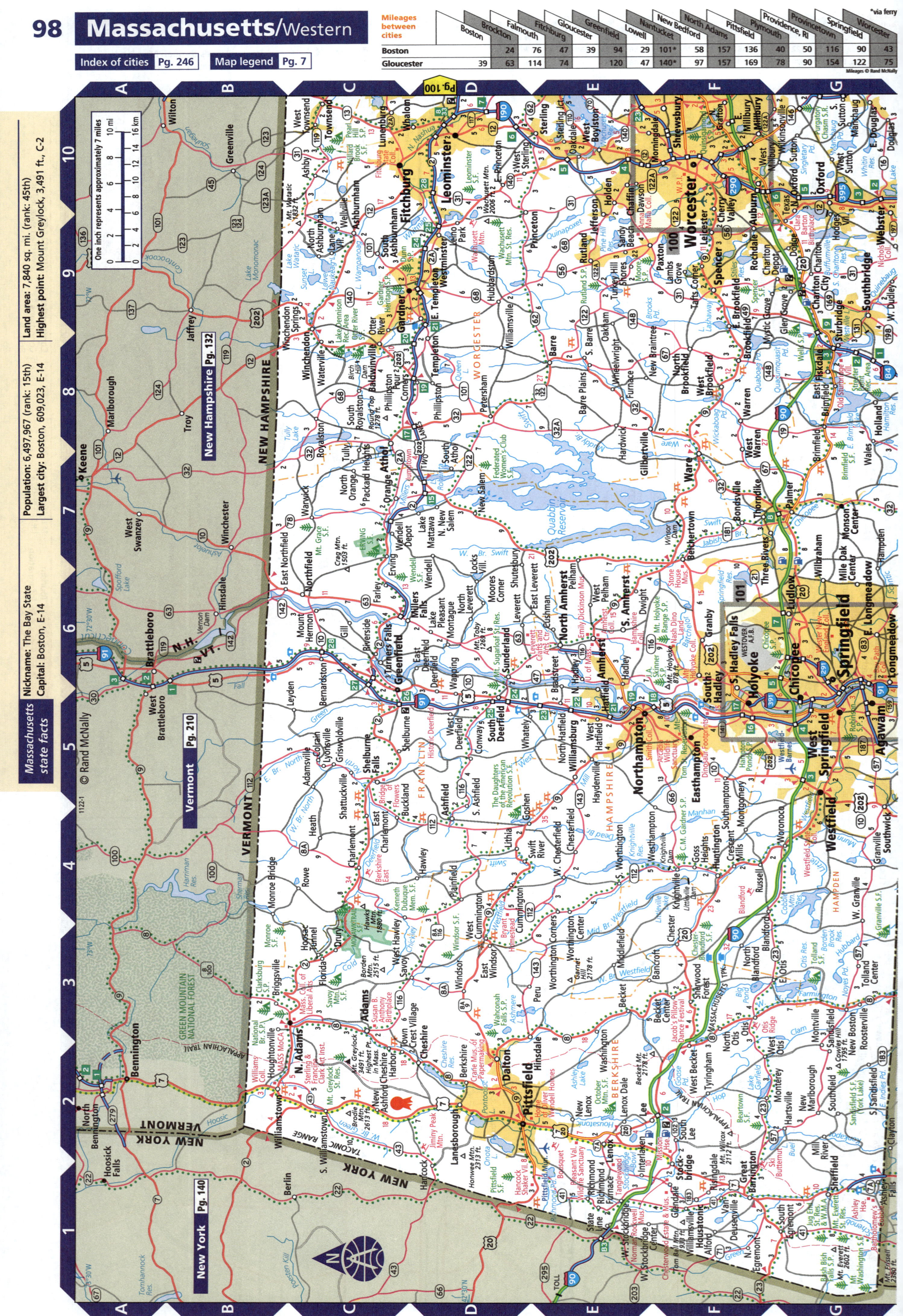

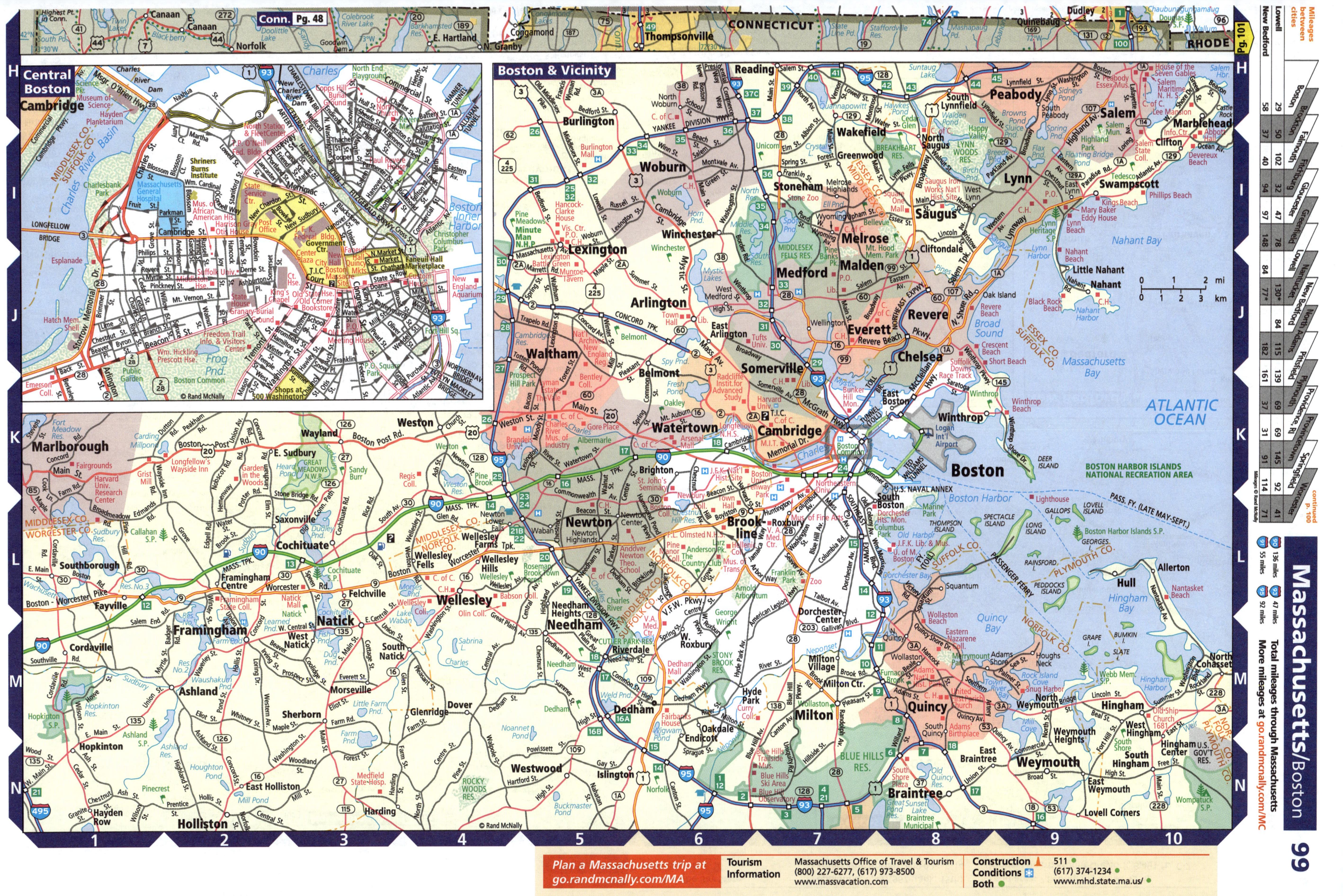


| Mileages between cities | Boston | Brockton | Falmouth | Fitchburg | Gloucester | Greenfield | Lowell | Nantucket | New Bedford | North Adams | Pittsfield | Plymouth | Providence, RI | Provincetown | Springfield | Worcester |
|---|---|---|---|---|---|---|---|---|---|---|---|---|---|---|---|---|
| Lowell | 29 | 50 | 102 | 32 | 47 | 78 | | 130* | 84 | 115 | 139 | 69 | 69 | 145 | 92 | 41 |
| New Bedford | 58 | 37 | 40 | 94 | 97 | 148 | 84 | 77* | | 182 | 161 | 37 | 31 | 91 | 114 | 71 |

Index of cities Pg. 246 | Map legend Pg. 7

Mileages between cities

*via ferry

| | Boston | Brockton | Falmouth | Fitchburg | Gloucester | Greenfield | Lowell | Nantucket | New Bedford | North Adams | Pittsfield | Plymouth | Providence, RI | Provincetown | Springfield | Worcester |
|---|---|---|---|---|---|---|---|---|---|---|---|---|---|---|---|---|
| Pittsfield | 136 | 150 | 189 | 124 | 169 | 79 | 139 | 226* | 161 | 22 | | 167 | 130 | 240 | 51 | 98 |
| Provincetown | 116 | 106 | 69 | 162 | 154 | 208 | 145 | 78* | 91 | 262 | 240 | 77 | 119 | | 194 | 146 |

Mileages © Rand McNally

Massachusetts state facts

Nickname: The Bay State
Capital: Boston, E-14
Population: 6,497,967 (rank: 15th)
Largest city: Boston, 609,023, E-14
Land area: 7,840 sq. mi. (rank: 45th)
Highest point: Mount Greylock, 3,491 ft., C-2

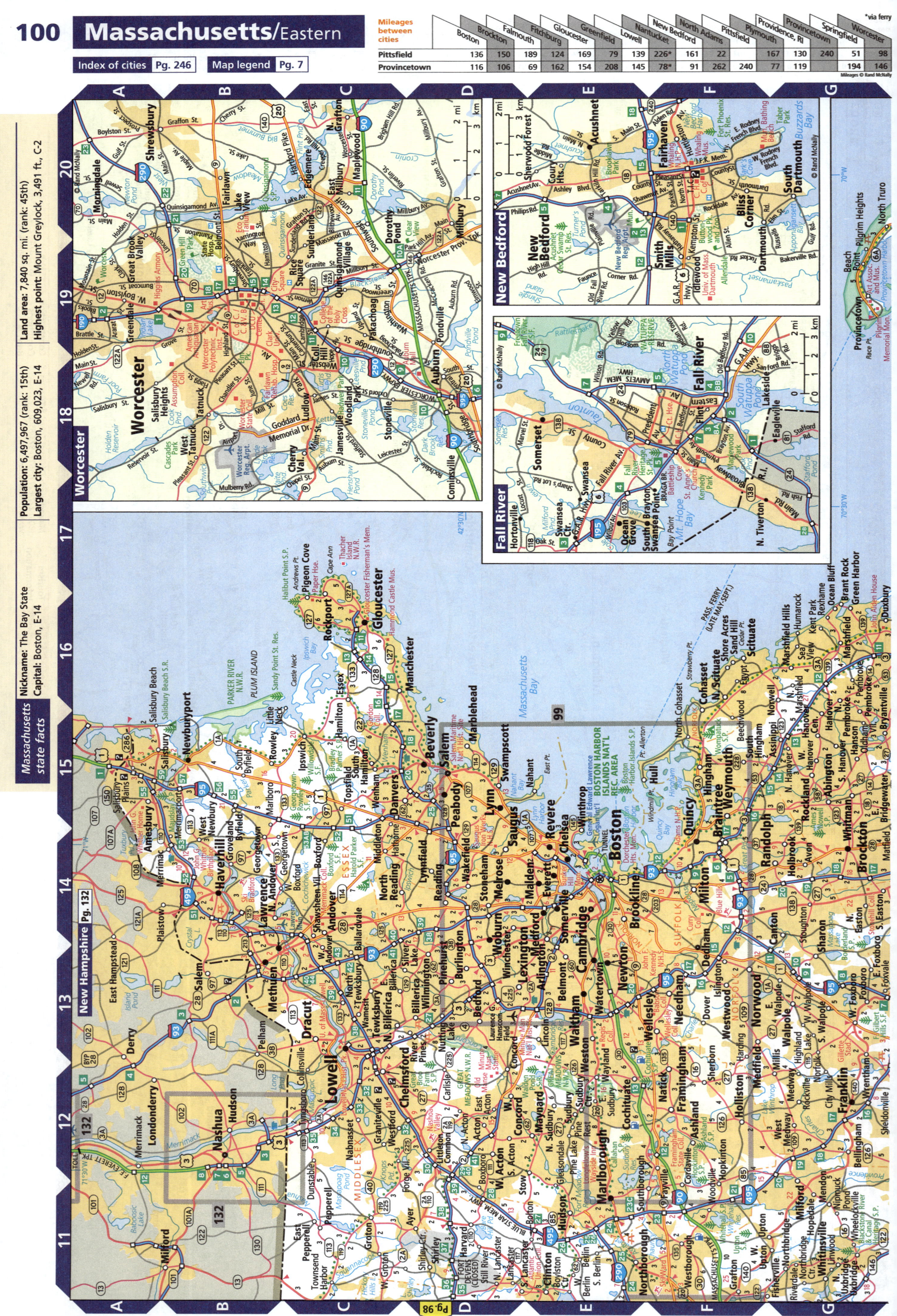

Mileages between cities

| | Boston | Brockton | Falmouth | Fitchburg | Gloucester | Greenfield | Lowell | Nantucket | New Bedford | North Adams | Pittsfield | Plymouth | Providence, RI | Provincetown | Springfield | Worcester |
|---|---|---|---|---|---|---|---|---|---|---|---|---|---|---|---|---|
| Springfield | 90 | 103 | 143 | 77 | 122 | 38 | 92 | 180* | 114 | 73 | 51 | 121 | 83 | 194 | | 51 |
| Worcester | 43 | 56 | 96 | 26 | 75 | 72 | 41 | 133* | 71 | 120 | 98 | 74 | 40 | 146 | 51 | |

*via ferry

Mileages © Rand McNally

90 136 miles
91 55 miles
93 47 miles
95 92 miles

Total mileages through Massachusetts
More mileages at go.randmcnally.com/MC

Plan a Massachusetts trip at go.randmcnally.com/MA

Tourism Information
Massachusetts Office of Travel & Tourism
(800) 227-6277, (617) 973-8500
www.massvacation.com

Construction Conditions Both
511
(617) 374-1234
www.mhd.state.ma.us/

Index of cities Pg. 247 Map legend Pg. 7

**Mileages between cities**

| | Alpena | Chicago, IL | Detroit | Grand Rapids | Houghton | Ironwood | Kalamazoo | Ludington | Mackinaw City | Menominee | Muskegon | Port Huron | Saginaw | Sault Ste. Marie | Toledo, OH | Traverse City |
|---|---|---|---|---|---|---|---|---|---|---|---|---|---|---|---|---|
| Ann Arbor | 227 | 240 | 43 | 132 | 538 | 584 | 98 | 228 | 272 | 473 | 172 | 102 | 86 | 329 | 51 | 238 |
| Detroit | 244 | 280 | | 157 | 553 | 599 | 140 | 252 | 290 | 488 | 197 | 62 | 102 | 345 | 59 | 255 |



**Michigan state facts**

Nickname: The Great Lake State
Capital: Lansing, Q-9
Population: 10,003,422 (rank: 8th)
Largest city: Detroit, 912,062, R-12
Land area: 56,804 sq. mi. (rank: 22nd)
Highest point: Mount Arvon, 1,979 ft., B-13

Mileages between cities

| | Alpena | Chicago, IL | Detroit | Grand Rapids | Houghton | Ironwood | Kalamazoo | Ludington | Mackinaw City | Menominee | Muskegon | Port Huron | Saginaw | Sault Ste. Marie | Toledo, OH | Traverse City |
|---|---|---|---|---|---|---|---|---|---|---|---|---|---|---|---|---|
| Flint | 178 | 271 | 68 | 113 | 489 | 534 | 130 | 186 | 224 | 423 | 152 | 66 | 37 | 280 | 107 | 188 |
| Grand Rapids | 249 | 177 | 157 | | 502 | 552 | 50 | 97 | 236 | 438 | 41 | 180 | 115 | 292 | 185 | 140 |

Mileages © Rand McNally

continued p. 104

69 199 miles · 94 275 miles · 75 396 miles · 96 192 miles

Total mileages through Michigan
More mileages at go.randmcnally.com/MC

Index of cities Pg. 247 | Map legend Pg. 7

**Mileages between cities**

| | Alpena | Chicago, IL | Detroit | Grand Rapids | Houghton | Ironwood | Kalamazoo | Ludington | Mackinaw City | Menominee | Muskegon | Port Huron | Saginaw | Sault Ste. Marie | Toledo, OH | Traverse City |
|---|---|---|---|---|---|---|---|---|---|---|---|---|---|---|---|---|
| Ironwood | 405 | 403 | 599 | 552 | 109 | | 544 | 319 | 311 | 195 | 586 | 600 | 499 | 307 | 636 | 413 |
| Kalamazoo | 298 | 145 | 140 | 50 | 556 | 544 | | 146 | 287 | 408 | 91 | 197 | 161 | 344 | 150 | 190 |



**Michigan state facts**

Nickname: The Great Lake State
Capital: Lansing, Q-9
Population: 10,003,422 (rank: 8th)
Largest city: Detroit, 912,062, R-12
Land area: 56,804 sq. mi. (rank: 22nd)
Highest point: Mount Arvon, 1,979 ft., B-13

Mileages between cities

| | Alpena | Chicago, IL | Detroit | Grand Rapids | Houghton | Ironwood | Kalamazoo | Ludington | Mackinaw City | Menominee | Muskegon | Port Huron | Saginaw | Sault Ste. Marie | Toledo, OH | Traverse City |
|---|---|---|---|---|---|---|---|---|---|---|---|---|---|---|---|---|
| Lansing | 228 | 216 | 90 | 68 | 494 | 539 | 75 | 162 | 228 | 429 | 107 | 122 | 88 | 284 | 118 | 180 |
| Mackinaw City | 94 | 412 | 290 | 236 | 266 | 311 | 287 | 218 | | 200 | 251 | 290 | 188 | 56 | 327 | 102 |



69 199 miles 94 275 miles
75 396 miles 96 192 miles

Total mileages through Michigan
More mileages at go.randmcnally.com/MC

One inch represents approx. 15 miles

Construction ▲ Conditions ❄ Both ● (800) 381-8477 www.michigan.gov/drive

Tourism Information: Travel Michigan (800) 644-2489 www.michigan.org

Get more Michigan info at go.randmcnally.com/MI

Pg. 103
Ohio Pg. 158

**City sights to see**

- Arab American National Museum, Dearborn, K-6
- Cranbrook Art Museum, Bloomfield Hills, G-5
- Detroit Zoo, Royal Oak, H-6
- Edsel & Eleanor Ford House, Grosse Pointe Shores, I-9
- Frederik Meijer Gardens, Grand Rapids, A-3
- Gerald R. Ford Museum, Grand Rapids, B-2

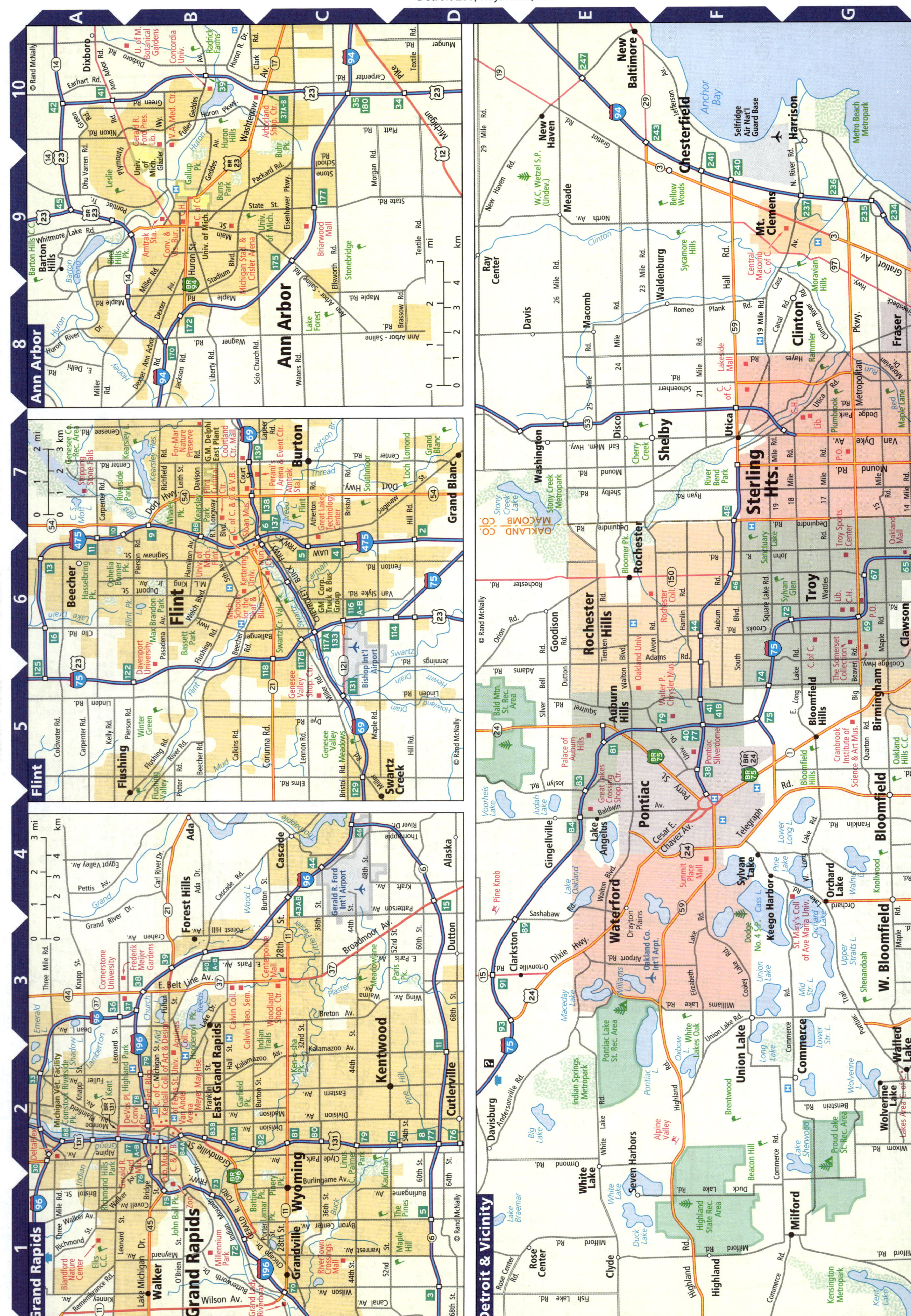

- Gerald R. Ford Presidential Library, Ann Arbor, B-10
- Henry Ford Museum, Dearborn, K-5
- Motown Historical Museum, Detroit, J-7
- New Detroit Science Center, Detroit, J-7
- Renaissance Center, Detroit, N-10
- Sloan Musuem, Flint, B-7
- University of Michigan, Ann Arbor, B-9

Index of cities Pg. 248 | Map legend Pg. 7

| Mileages between cities | Albert Lea | Bemidji | Brainerd | Duluth | Grand Forks, ND | Grand Marais | Hibbing | Int'l Falls | Mankato | Marshall | Minneapolis | Moorhead | Rochester | St. Cloud | Sioux Falls, SD | Willmar |
|---|---|---|---|---|---|---|---|---|---|---|---|---|---|---|---|---|
| Bemidji | 316 | | 97 | 151 | 114 | 259 | 105 | 112 | 290 | 258 | 222 | 135 | 306 | 151 | 380 | 188 |
| Duluth | 247 | 151 | 113 | | 266 | 110 | 76 | 162 | 233 | 273 | 152 | 250 | 226 | 141 | 390 | 204 |



**Minnesota state facts**

Nickname: The North Star State
Capital: St. Paul, O-10
Population: 5,220,393 (rank: 21st)
Largest city: Minneapolis, 382,605, O-9
Land area: 79,610 sq. mi. (rank: 14th)
Highest point: Eagle Mountain, 2,301 ft., B-11

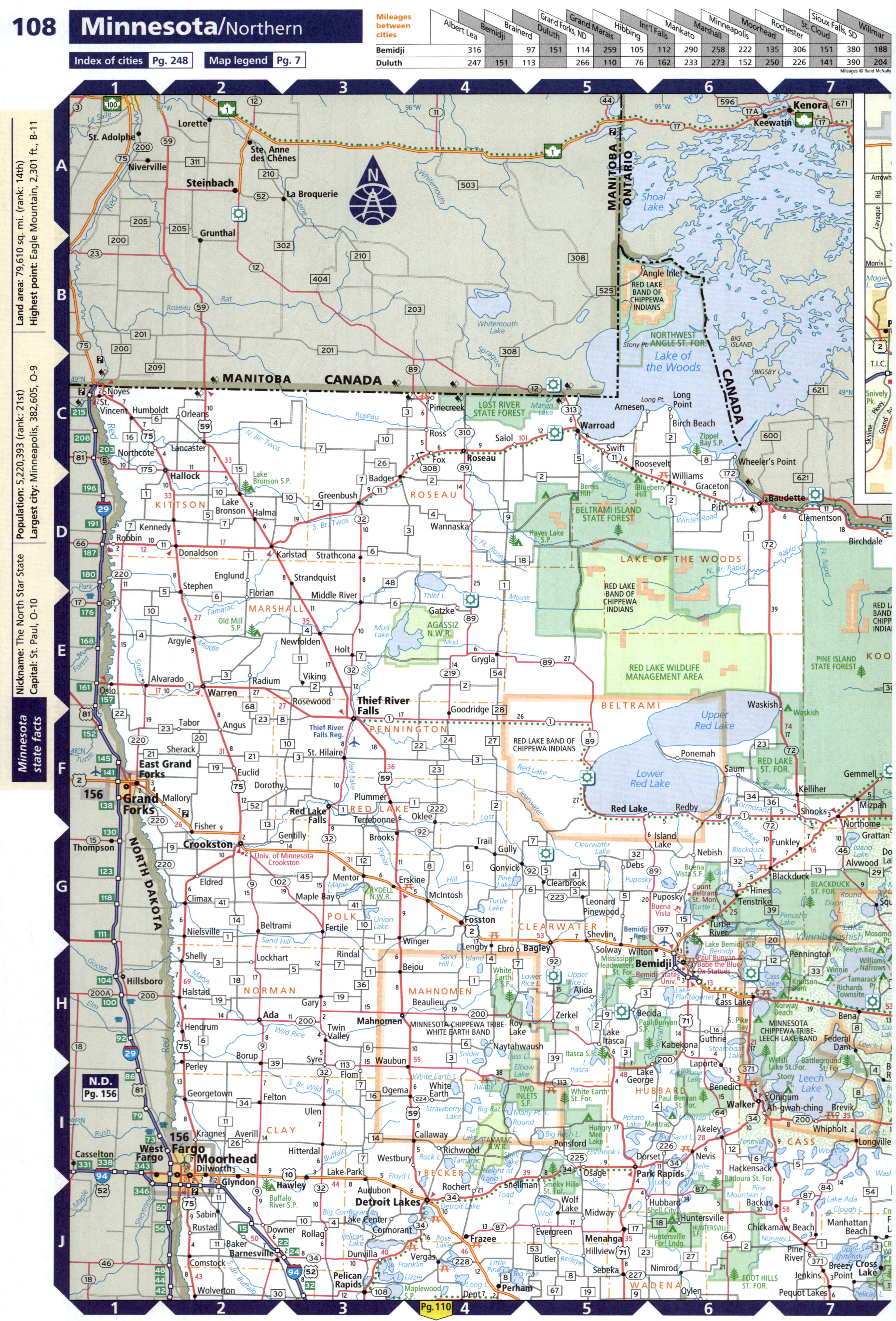

Mileages between cities

continued p. 110

| | Albert Lea | Bemidji | Brainerd | Duluth | Grand Forks, ND | Grand Marais | Hibbing | Int'l Falls | Mankato | Marshall | Minneapolis | Moorhead | Rochester | St. Cloud | Sioux Falls, SD | Willmar |
|---|---|---|---|---|---|---|---|---|---|---|---|---|---|---|---|---|
| Minneapolis | 96 | 222 | 130 | 152 | 314 | 262 | 208 | 293 | 80 | 153 | | 233 | 86 | 65 | 236 | 93 |
| Moorhead | 328 | 135 | 136 | 250 | 82 | 361 | 212 | 249 | 303 | 206 | 233 | | 321 | 170 | 244 | 172 |



35 260 miles
94 260 miles
90 276 miles
2 255 miles

Total mileages through Minnesota
More mileages at go.randmcnally.com/MC

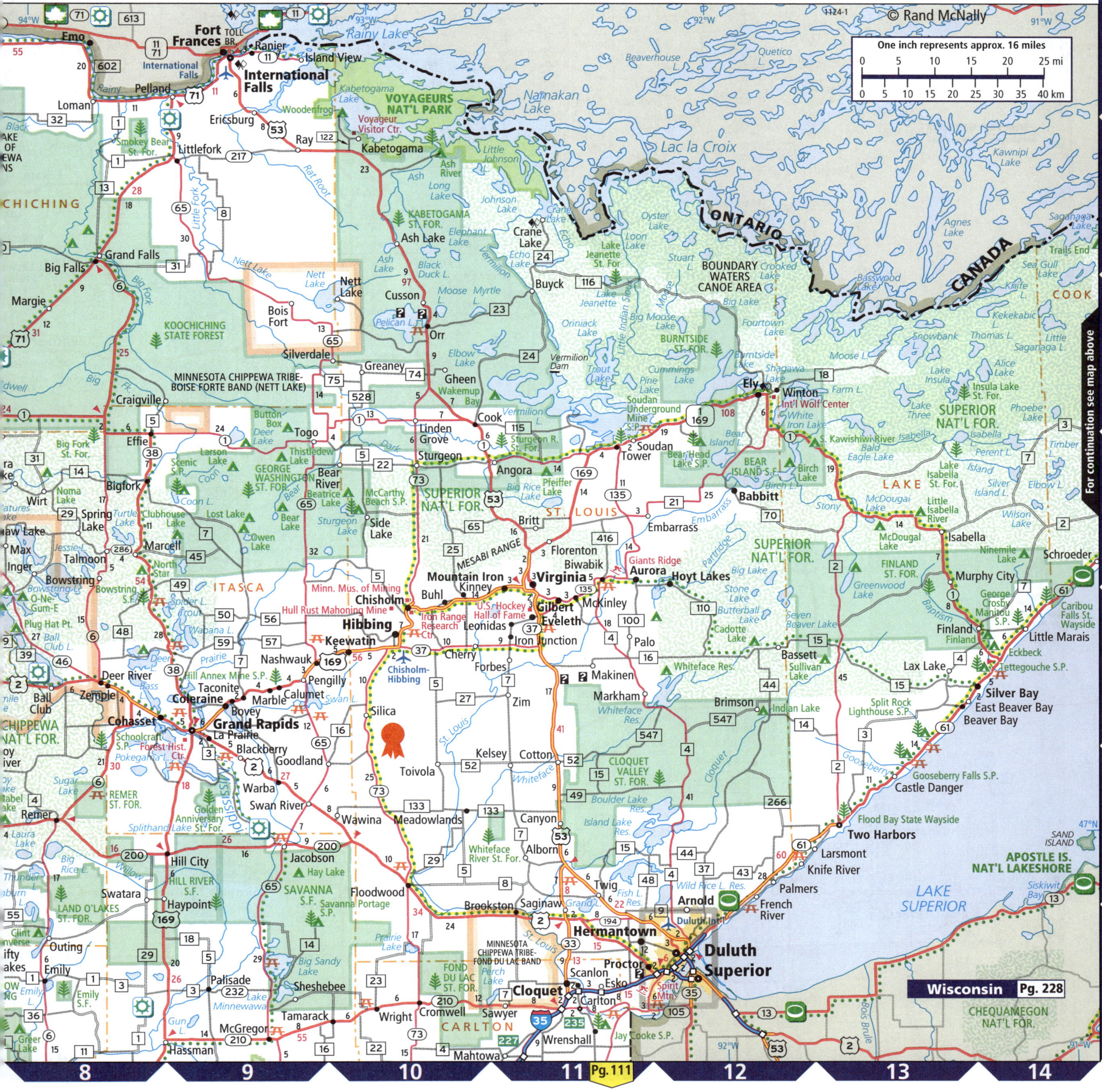

## Duluth / Superior

## Northeastern Minnesota

For continuation see main map

LAKE SUPERIOR



One inch represents approx. 16 miles

For continuation see map above

Wisconsin Pg. 228

Pg. 111

Explore Minnesota at go.randmcnally.com/MN

Tourism Information
Explore Minnesota Tourism
(888) 868-7476, (651) 296-5029
www.exploreminnesota.com

Construction ▲ Conditions ✱ Both ●
511 ● (651) 296-3000 ●
www.511mn.org ●
In MN: (800) 657-3774 ●

Index of cities Pg. 248 | Map legend Pg. 7

Mileages between cities

| | Albert Lea | Bemidji | Brainerd | Duluth | Grand Forks, ND | Grand Marais | Hibbing | Int'l Falls | Mankato | Marshall | Minneapolis | Moorhead | Rochester | St. Cloud | Sioux Falls, SD | Willmar |
|---|---|---|---|---|---|---|---|---|---|---|---|---|---|---|---|---|
| Rochester | 62 | 306 | 213 | 226 | 401 | 338 | 280 | 366 | 86 | 194 | 86 | 321 | | 153 | 236 | 178 |
| St. Cloud | 160 | 151 | 63 | 141 | 251 | 253 | 173 | 251 | 135 | 130 | 65 | 170 | 153 | | 220 | 62 |

Mileages © Rand McNally

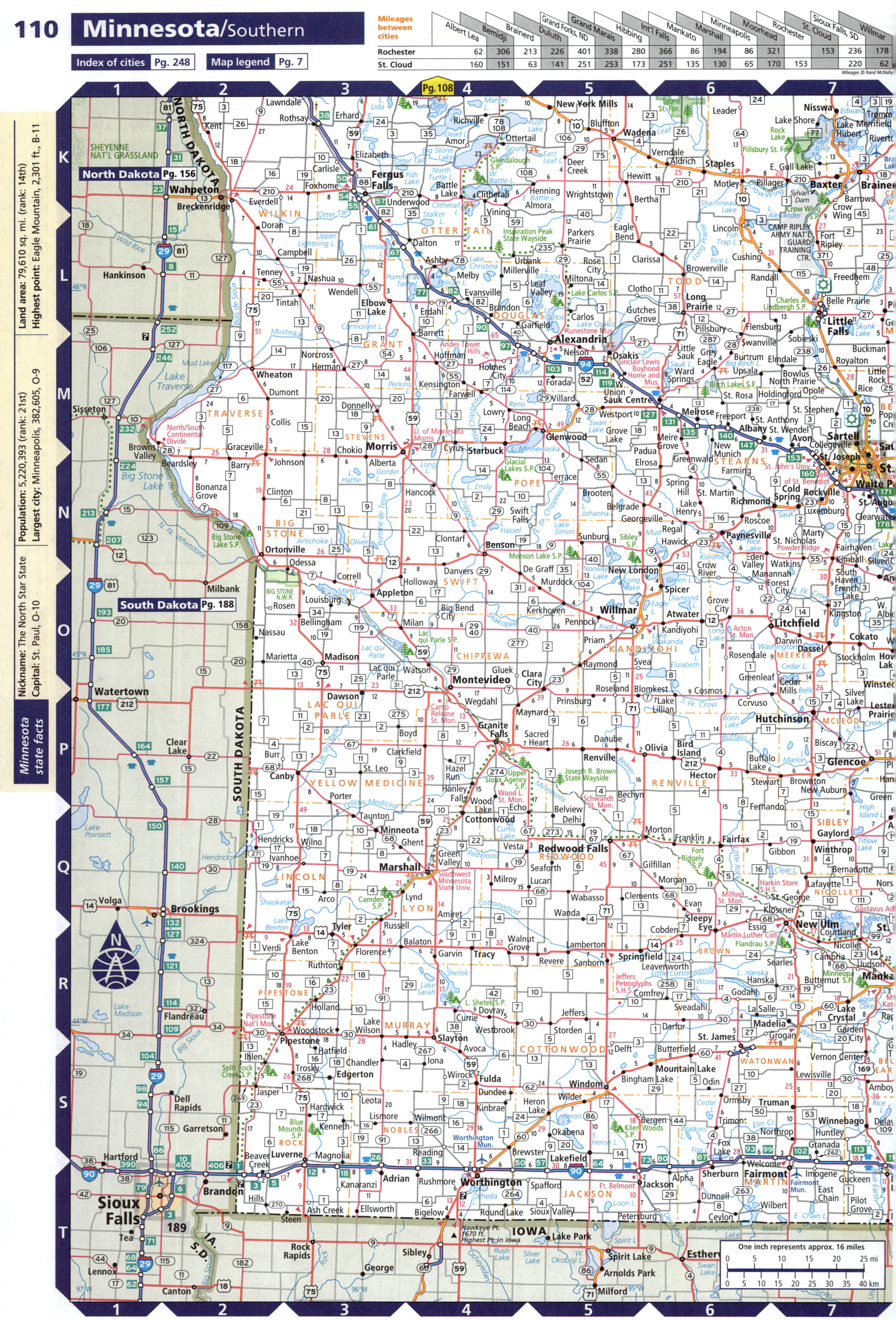

**Mileages between cities**

| | Albert Lea | Bemidji | Brainerd | Duluth | Grand Forks, ND | Grand Marais | Hibbing | Int'l Falls | Mankato | Marshall | Minneapolis | Moorhead | Rochester | St. Cloud | Sioux Falls, SD | Willmar |
|---|---|---|---|---|---|---|---|---|---|---|---|---|---|---|---|---|
| St. Paul | 98 | 230 | 137 | 149 | 325 | 260 | 204 | 290 | 87 | 159 | 9 | 243 | 78 | 75 | 241 | 102 |
| Sioux Falls, SD | 176 | 380 | 281 | 390 | 319 | 500 | 456 | 494 | 155 | 91 | 236 | 244 | 236 | 220 | | 158 |

Mileages © Rand McNally

Total mileages through Minnesota
35: 260 miles; 90: 276 miles; 94: 260 miles; 2: 255 miles
More mileages at go.randmcnally.com/MC

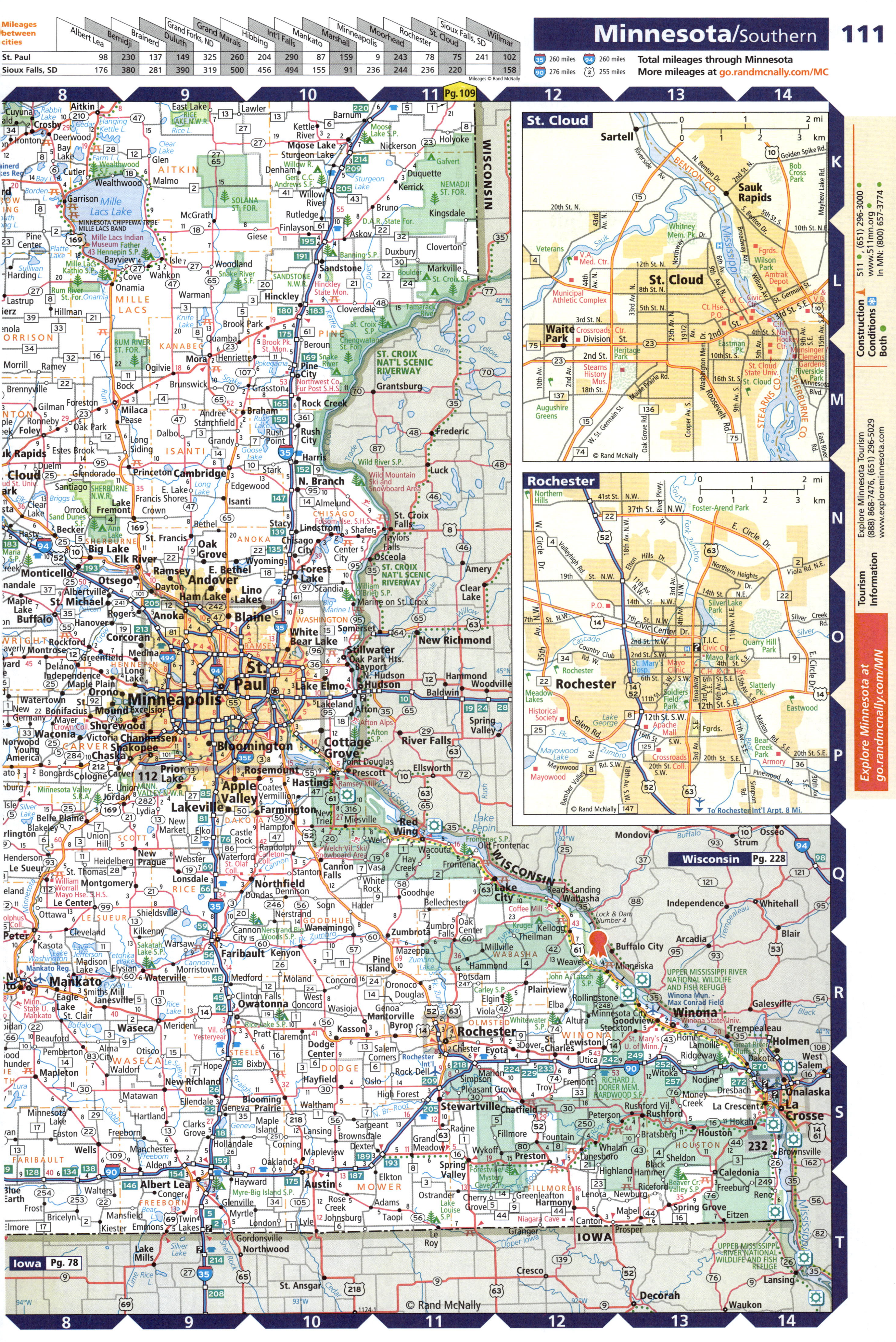

**City sights to see**

- Bell Mus. of Natural History, Minneapolis, L-4
- Cathedral of St. Paul, St. Paul, M-7
- Frederick R. Weisman Art Museum, Minneapolis, M-4
- Mall of America, Bloomington, I-5
- Mill City Museum, Minneapolis, L-3
- Minneapolis Institute of the Arts, Minneapolis, N-2

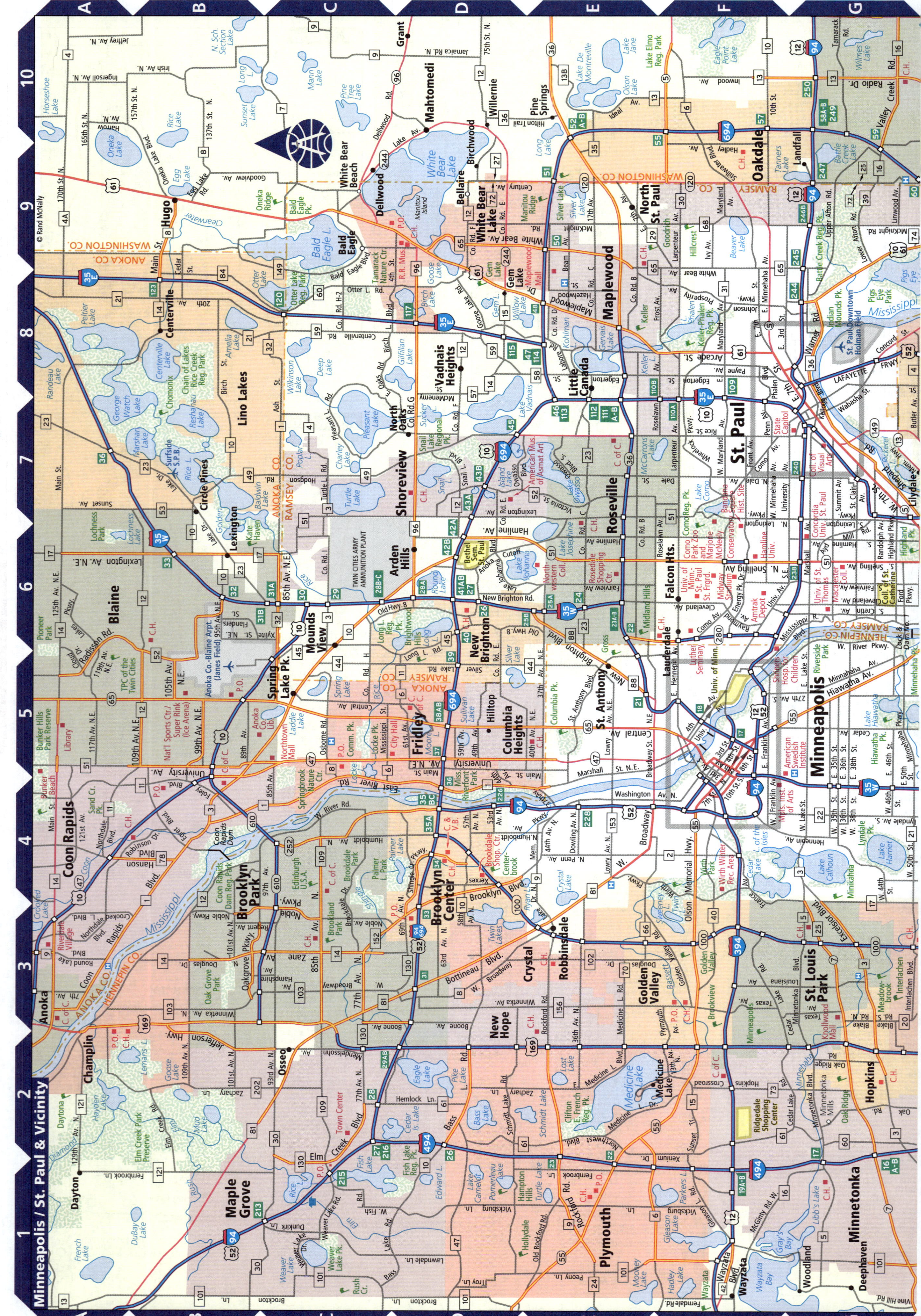

- Minneapolis Sculpture Garden, Minneapolis, M-1
- Minnesota History Center, St. Paul, M-7
- Minnesota State Capitol, St. Paul, L-7
- Ordway Center for the Performing Arts, St. Paul, M-7
- Science Museum of Minnesota, St. Paul, M-7
- Walker Art Center, Minneapolis, M-1

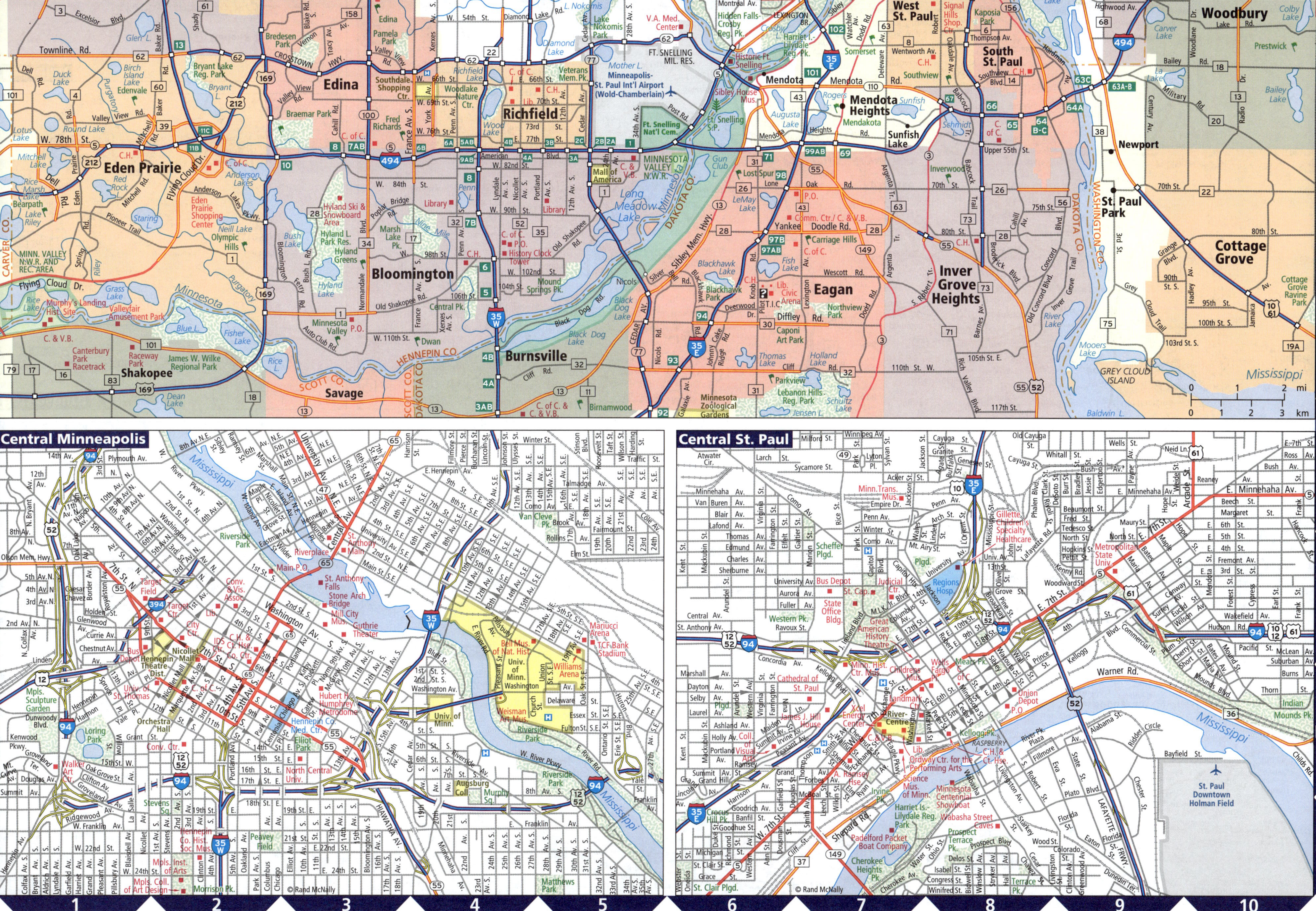

Index of cities Pg. 248 | Map legend Pg. 7

**Mileages between cities**

| | Batesville | Biloxi | Hattiesburg | Jackson | Memphis, TN | Natchez | Tupelo | Vicksburg |
|---|---|---|---|---|---|---|---|---|
| Biloxi | 320 | | 80 | 172 | 379 | 228 | 315 | 214 |
| Greenville | 112 | 293 | 210 | 121 | 152 | 152 | 177 | 91 |
| Jackson | 149 | 172 | 89 | | 209 | 103 | 190 | 44 |
| Memphis, TN | 61 | 379 | 297 | 209 | | 304 | 105 | 245 |



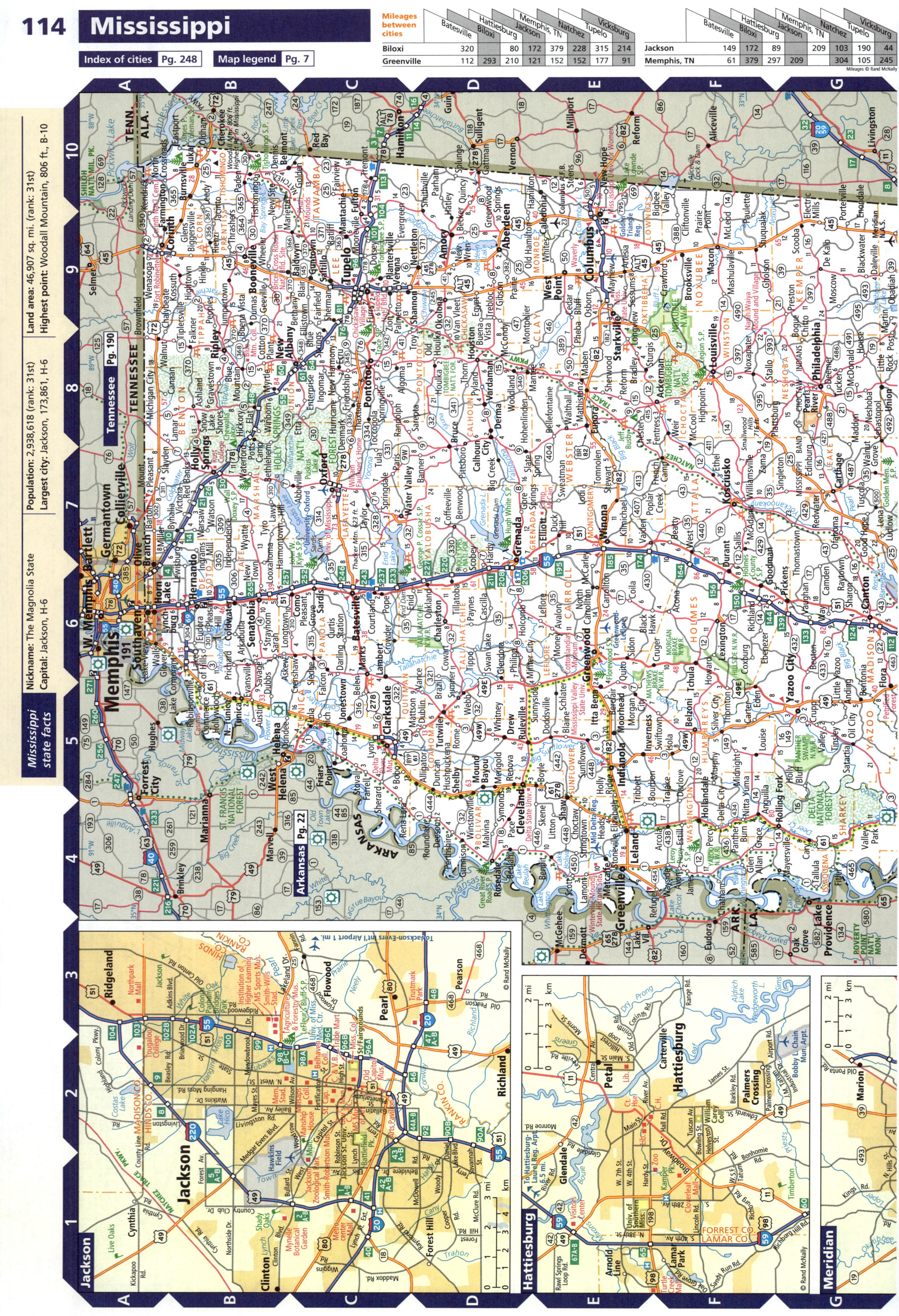

**Mileages between cities**

| | Batesville | Biloxi | Hattiesburg | Jackson | Memphis, TN | Natchez | Tupelo | Vicksburg |
|---|---|---|---|---|---|---|---|---|
| Meridian | 176 | 172 | 89 | 91 | 234 | 194 | 142 | 134 |
| New Orleans, LA | 335 | 90 | 109 | 183 | 394 | 171 | 340 | 207 |

| | Batesville | Biloxi | Hattiesburg | Jackson | Memphis, TN | Natchez | Tupelo | Vicksburg |
|---|---|---|---|---|---|---|---|---|
| Tupelo | 74 | 315 | 232 | 190 | 105 | 283 | | 225 |
| Vicksburg | 188 | 214 | 131 | 44 | 245 | 70 | 225 | |

Mileages © Rand McNally

**Total mileages through Mississippi**

| | |
|---|---|
| I-10 | 77 miles |
| I-20 | 169 miles |
| I-55 | 290 miles |
| I-59 | 172 miles |

More mileages at go.randmcnally.com/MC

Index of cities Pg. 249 | Map legend Pg. 7

**Missouri state facts**

Nickname: The Show Me State
Capital: Jefferson City, G-14
Population: 5,911,605 (rank: 18th)
Largest city: Kansas City, 451,572, F-9
Land area: 68,886 sq. mi. (rank: 18th)
Highest point: Taum Sauk Mtn., 1,772 ft., J-17

**Mileages between cities**

| | Branson | Cape Girardeau | Columbia | Hannibal | Hayti | Jefferson City | Joplin | Kansas City | Kirksville | Maryville | Osage Beach | Poplar Bluff | Rolla | St. Louis | Springfield | West Plains |
|---|---|---|---|---|---|---|---|---|---|---|---|---|---|---|---|---|
| Cape Girardeau | 295 | | 225 | 218 | 80 | 216 | 336 | 348 | 313 | 445 | 218 | 82 | 158 | 114 | 270 | 182 |
| Columbia | 205 | 225 | | 97 | 301 | 32 | 236 | 124 | 91 | 222 | 76 | 261 | 93 | 126 | 168 | 191 |

Mileages © Rand McNally

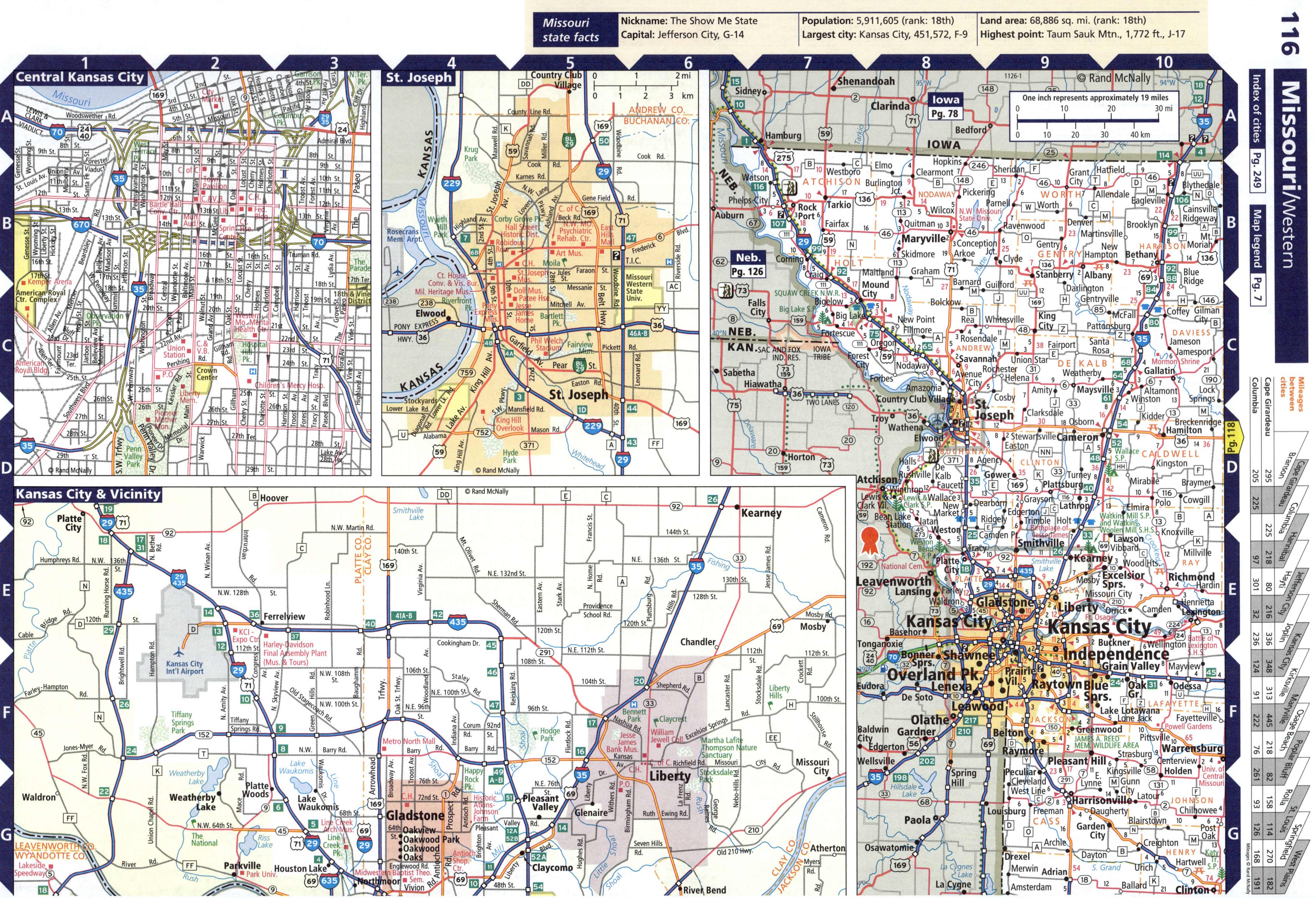

**Mileages between cities**

| | Branson | Cape Girardeau | Columbia | Hannibal | Hayti | Jefferson City | Joplin | Kansas City | Kirksville | Maryville | Osage Beach | Poplar Bluff | Rolla | St. Louis | Springfield | West Plains |
|---|---|---|---|---|---|---|---|---|---|---|---|---|---|---|---|---|
| Joplin | 109 | 336 | 236 | 312 | 319 | 206 | | 157 | 312 | 243 | 161 | 256 | 178 | 282 | 70 | 176 |
| Kansas City | 209 | 348 | 124 | 209 | 424 | 156 | 157 | | 157 | 93 | 164 | 356 | 219 | 250 | 166 | 275 |

Mileages © Rand McNally

continued p. 118

I-35 115 miles · I-55 210 miles · I-44 290 miles · I-70 252 miles

**Total mileages through Missouri**
**More mileages at go.randmcnally.com/MC**

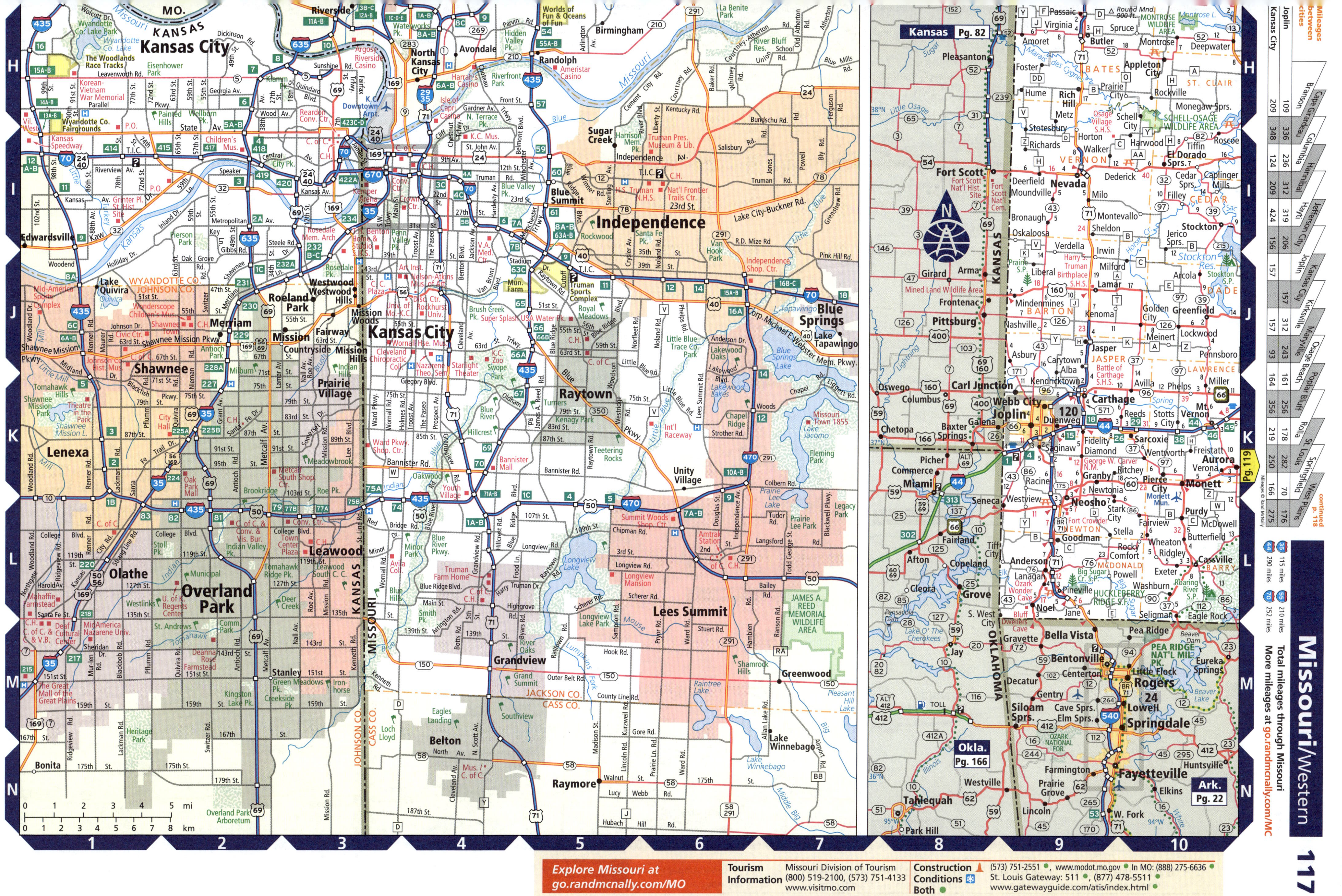


**Explore Missouri at go.randmcnally.com/MO**

**Tourism Information** Missouri Division of Tourism (800) 519-2100, (573) 751-4133 www.visitmo.com

**Construction** · **Conditions** · **Both**
(573) 751-2551, www.modot.mo.gov · In MO: (888) 275-6636
St. Louis Gateway: 511, (877) 478-5511
www.gatewayguide.com/atis/index.html

Index of cities Pg. 249 | Map legend Pg. 7

Mileages between cities

| | Branson | Cape Girardeau | Columbia | Hannibal | Hayti | Jefferson City | Joplin | Kansas City | Kirksville | Maryville | Osage Beach | Poplar Bluff | Rolla | St. Louis | Springfield | West Plains |
|---|---|---|---|---|---|---|---|---|---|---|---|---|---|---|---|---|
| Poplar Bluff | 215 | 82 | 261 | 255 | 62 | 223 | 256 | 356 | 350 | 457 | 224 | | 147 | 151 | 191 | 98 |
| St. Joseph | 270 | 405 | 182 | 191 | 481 | 214 | 203 | 53 | 141 | 43 | 222 | 416 | 276 | 308 | 225 | 336 |



**Missouri state facts**

Nickname: The Show Me State
Capital: Jefferson City, G-14
Population: 5,911,605 (rank: 18th)
Largest city: Kansas City, 451,572, F-9
Land area: 68,886 sq. mi. (rank: 18th)
Highest point: Taum Sauk Mtn., 1,772 ft., J-17

| Mileages between cities | Branson | Cape Girardeau | Columbia | Hannibal | Hayti | Jefferson City | Joplin | Kansas City | Kirksville | Maryville | Osage Beach | Poplar Bluff | Rolla | St. Louis | Springfield | West Plains |
|---|---|---|---|---|---|---|---|---|---|---|---|---|---|---|---|---|
| St. Louis | 249 | 114 | 126 | 120 | 192 | 124 | 282 | 250 | 217 | 347 | 164 | 151 | 104 | | 213 | 202 |
| Springfield | 42 | 270 | 168 | 242 | 253 | 136 | 70 | 166 | 259 | 266 | 91 | 191 | 108 | 213 | | 108 |



35 115 miles
44 290 miles
55 210 miles
70 252 miles

Total mileages through Missouri
More mileages at go.randmcnally.com/MC

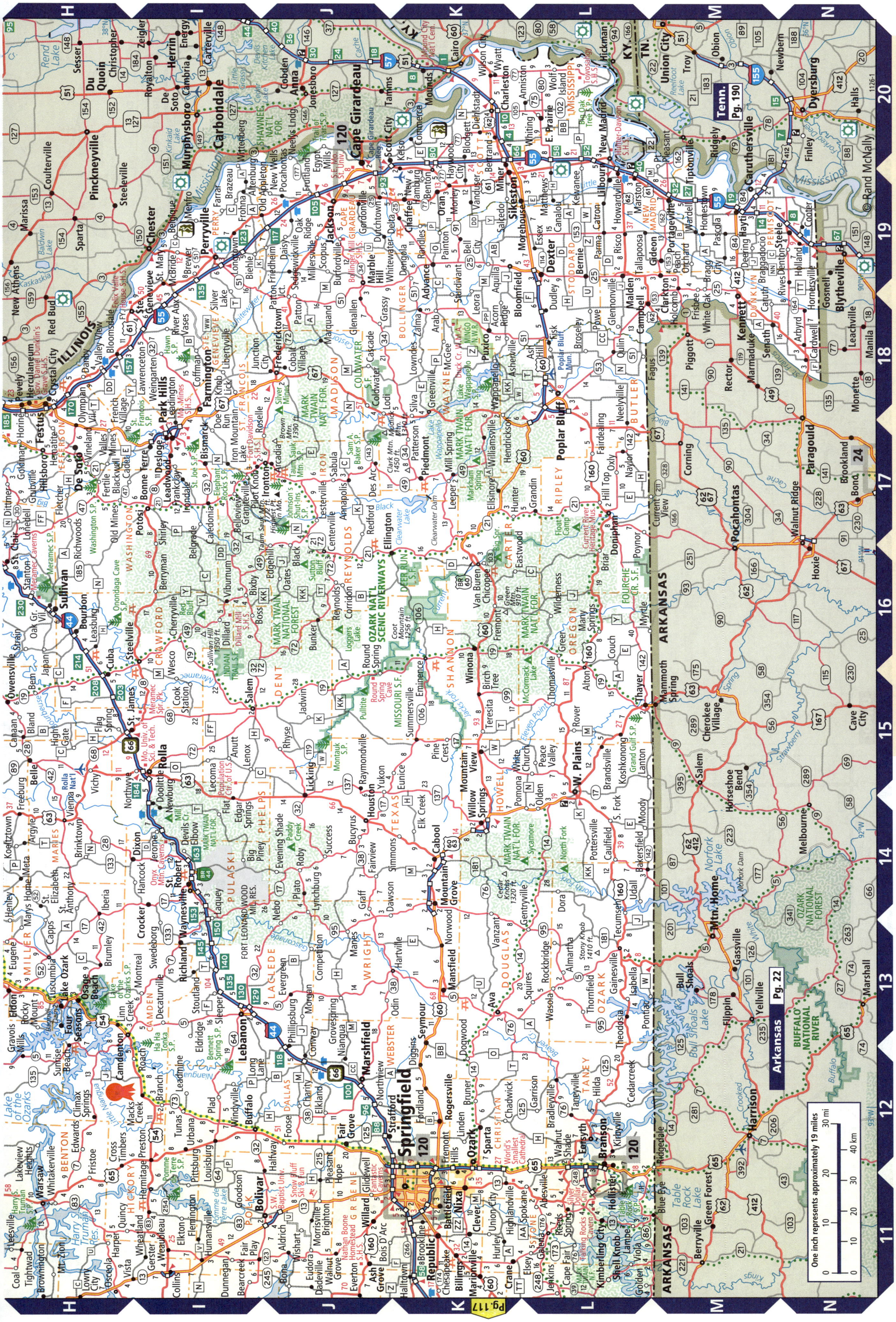

One inch represents approximately 19 miles

**Explore Missouri at go.randmcnally.com/MO**

**Tourism Information:** Missouri Division of Tourism (800) 519-2100, (573) 751-4133 www.visitmo.com

**Construction / Conditions / Both:** (573) 751-2551 • www.modot.mo.gov • In MO: (888) 275-6636
St. Louis Gateway: 511 • (877) 478-5511
www.gatewayguide.com/atis/index.html

**City sights to see**

- Andy Williams Moon River Theatre, Branson, M-8
- Anheuser-Busch Brewery, St. Louis, I-7
- Bass Pro Shops® Outdoor World®, Springfield, C-3
- Dolly Parton's Dixie Stampede, Branson, M-9
- Gateway Arch, St. Louis, L-4
- Laumeier Sculpture Park, St. Louis, J-4
- Magic House, Kirkwood, I-4

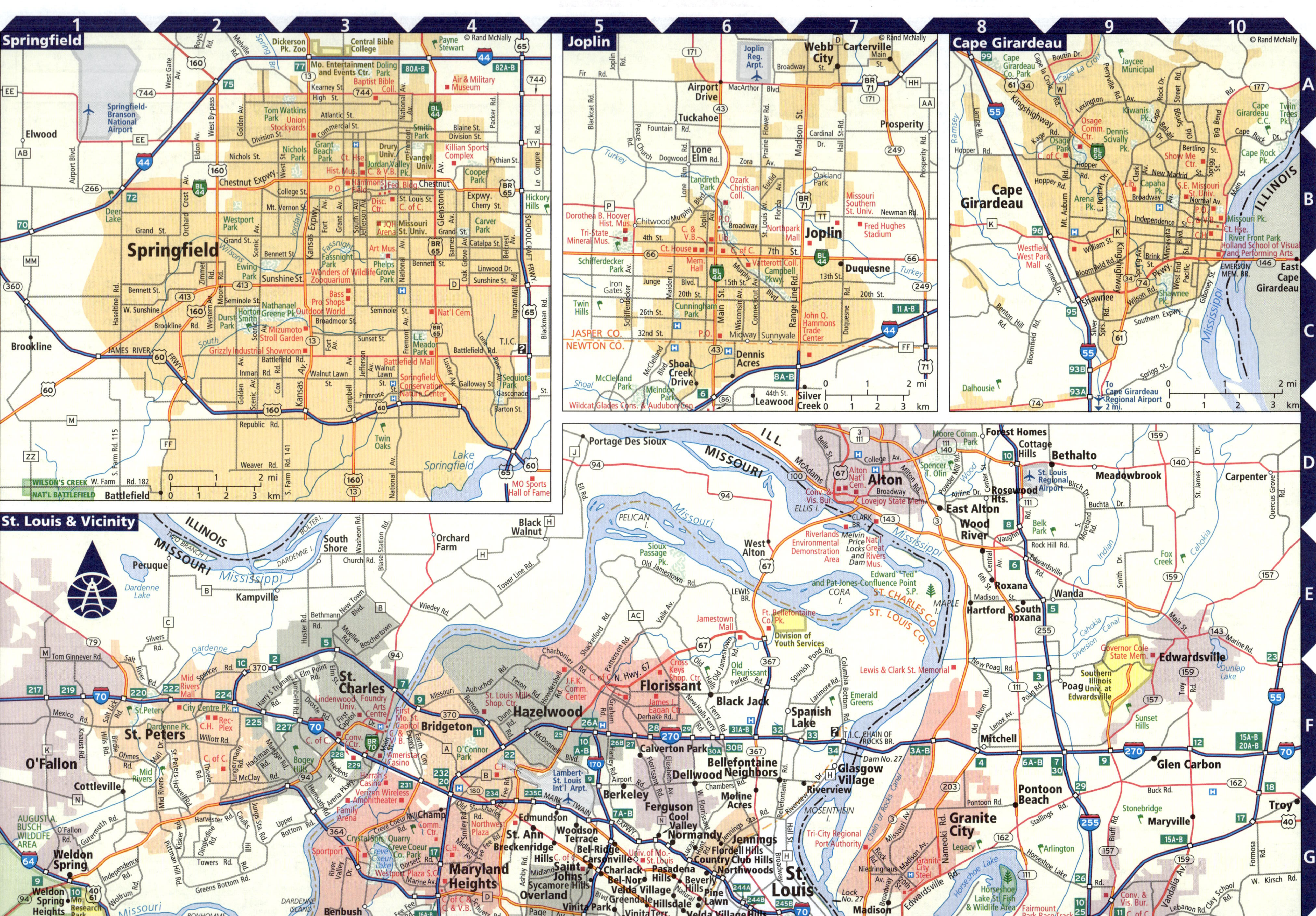

- Missouri Botanical Garden, St. Louis, I-6
- Shoji Tabuchi Theatre, Branson, L-7
- St. Louis Art Museum, St. Louis, H-6
- St. Louis Science Center, St. Louis, H-6
- St. Louis Zoo, St. Louis, H-6
- Shepherd of the Hills Homestead & Outdoor Theatre, Branson, K-6
- White Water, Branson, M-7

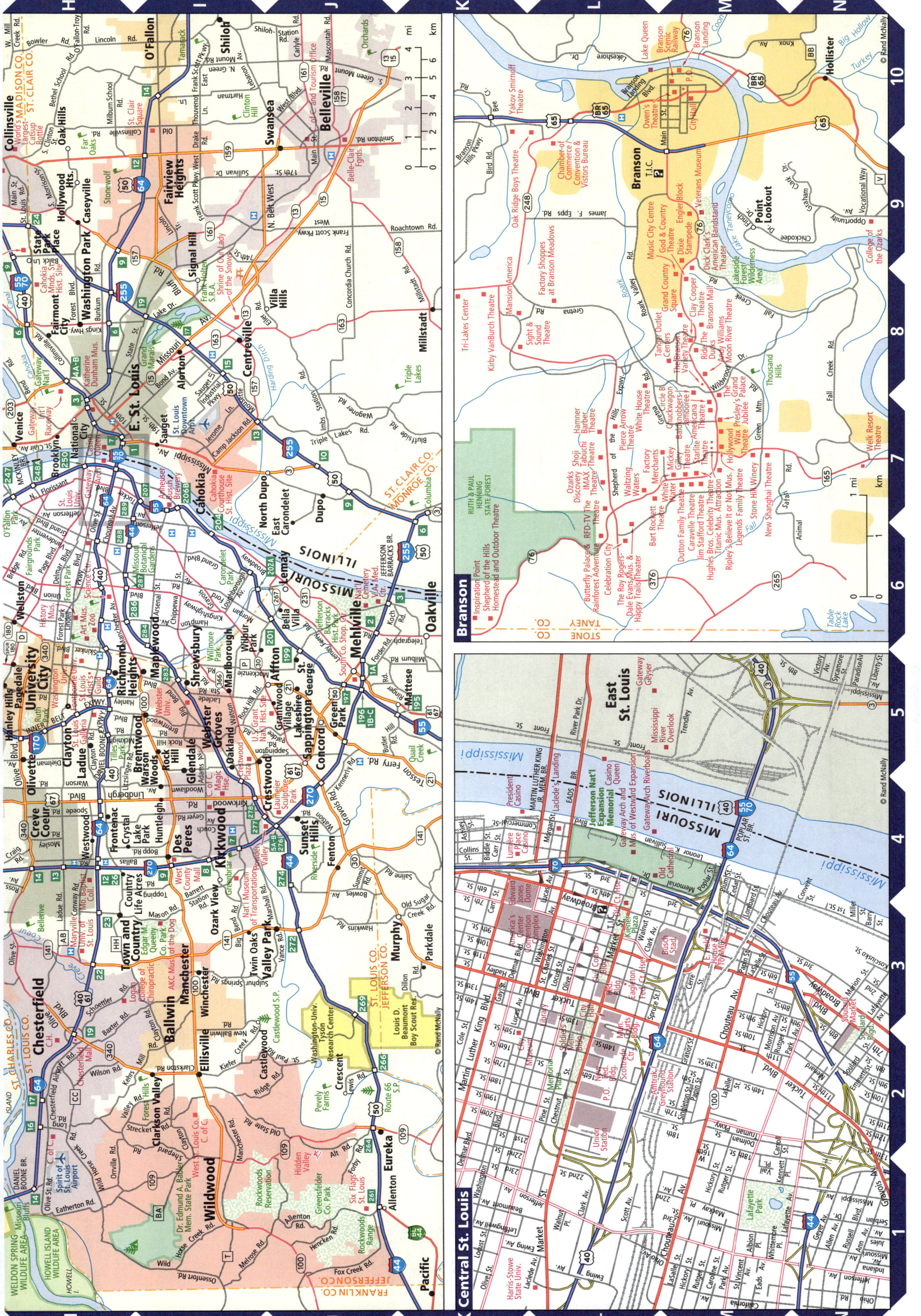

Index of cities Pg. 250
Map legend Pg. 7

**Mileages between cities**

| | Belle Fourche, SD | Billings | Bozeman | Butte | Dillon | Glasgow | Great Falls | Havre | Kalispell | Lewistown | Libby | Miles City | Missoula | St. Mary | Sidney | W. Yellowstone |
|---|---|---|---|---|---|---|---|---|---|---|---|---|---|---|---|---|
| Billings | 261 | | 143 | 223 | 256 | 276 | 218 | 247 | 451 | 125 | 536 | 144 | 343 | 375 | 269 | 232 |
| Butte | 486 | 223 | 82 | | 54 | 425 | 154 | 267 | 224 | 244 | 309 | 367 | 120 | 269 | 494 | 149 |

Mileages © Rand McNally

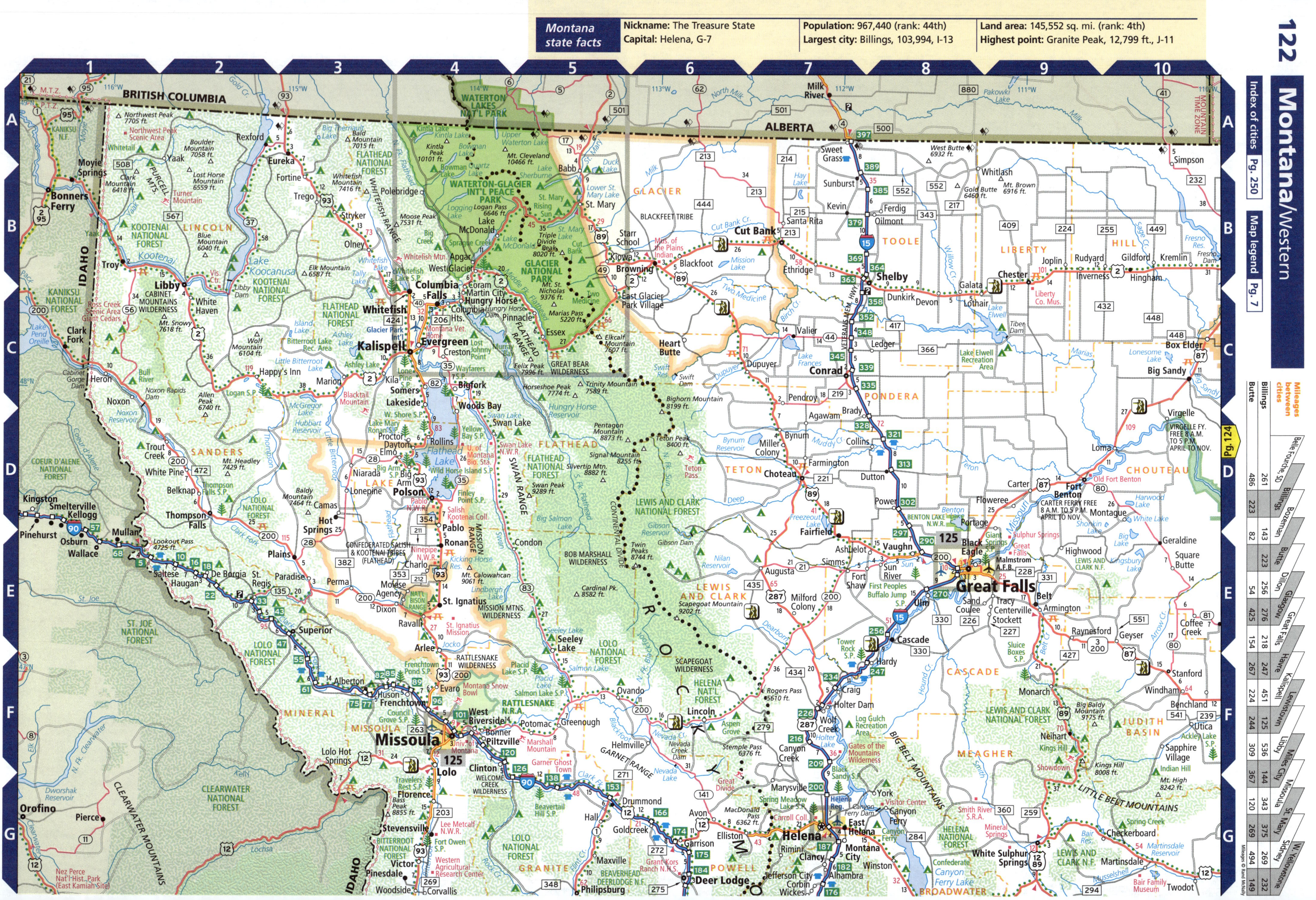

Mileages between cities

| | Belle Fourche, SD | Billings | Bozeman | Butte | Dillon | Glasgow | Great Falls | Havre | Kalispell | Lewistown | Libby | Miles City | Missoula | St. Mary | Sidney | W. Yellowstone |
|---|---|---|---|---|---|---|---|---|---|---|---|---|---|---|---|---|
| Great Falls | 481 | 218 | 186 | 154 | 219 | 271 | | 113 | 224 | 106 | 312 | 317 | 166 | 158 | 375 | 264 |
| Helena | 500 | 238 | 98 | 66 | 132 | 360 | 90 | 202 | 193 | 193 | 281 | 383 | 113 | 205 | 463 | 177 |

Mileages © Rand McNally

continued p. 124

15: 396 miles  
90: 552 miles  
94: 249 miles

Total mileages through Montana  
More mileages at go.randmcnally.com/MC

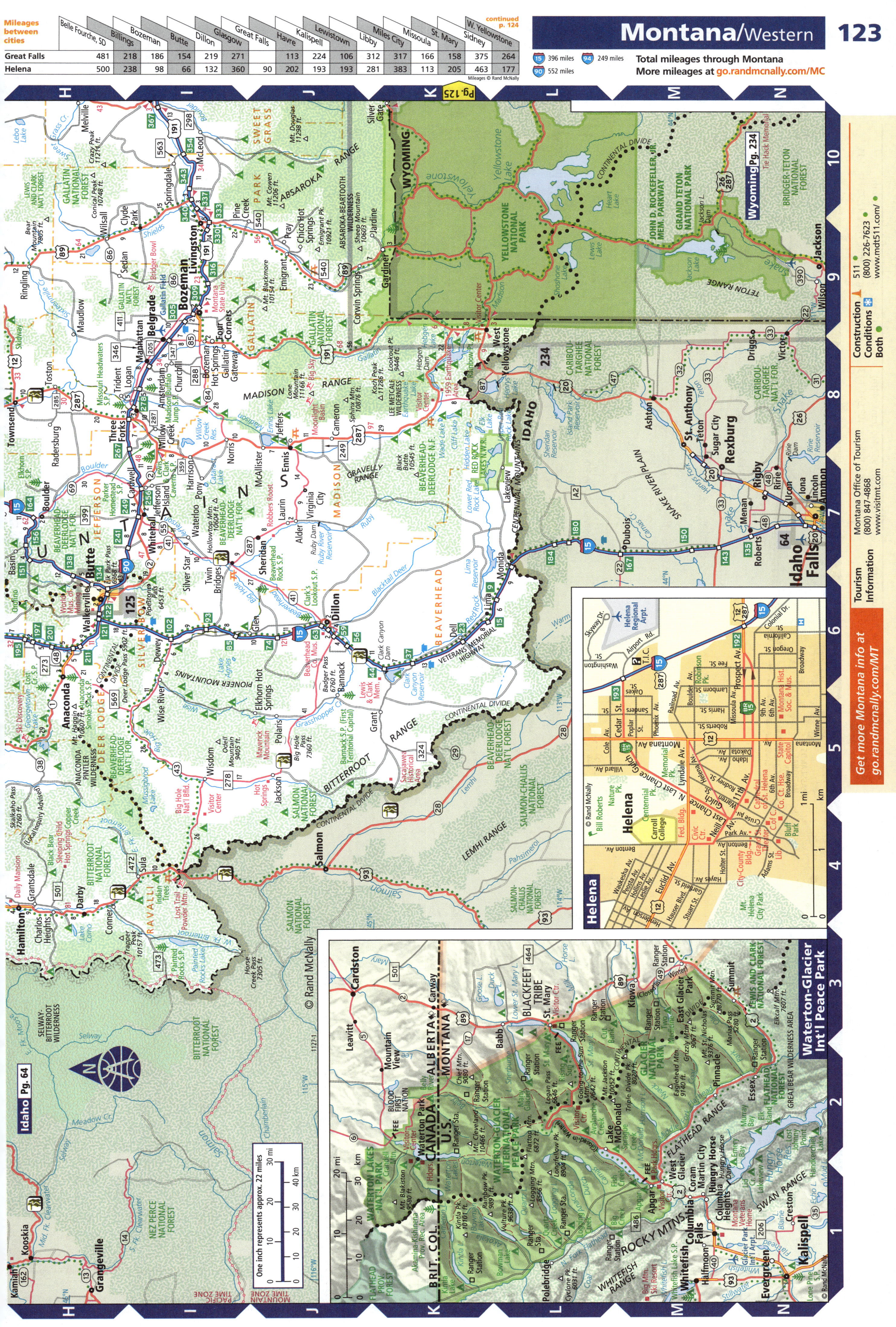

Index of cities Pg. 250 | Map legend Pg. 7

**Mileages between cities**

| | Belle Fourche, SD | Billings | Bozeman | Butte | Dillon | Glasgow | Great Falls | Havre | Kalispell | Lewistown | Libby | Miles City | Missoula | St. Mary | Sidney | W. Yellowstone |
|---|---|---|---|---|---|---|---|---|---|---|---|---|---|---|---|---|
| Kalispell | 711 | 451 | 308 | 224 | 278 | 419 | 224 | 261 | | 330 | 88 | 593 | 121 | 82 | 558 | 371 |
| Miles City | 174 | 144 | 285 | 367 | 399 | 195 | 317 | 333 | 593 | 211 | 678 | | 487 | 473 | 126 | 375 |

Mileages © Rand McNally

**Montana state facts**

Nickname: The Treasure State
Capital: Helena, G-7
Population: 967,440 (rank: 44th)
Largest city: Billings, 103,994, I-13
Land area: 145,552 sq. mi. (rank: 4th)
Highest point: Granite Peak, 12,799 ft., J-11

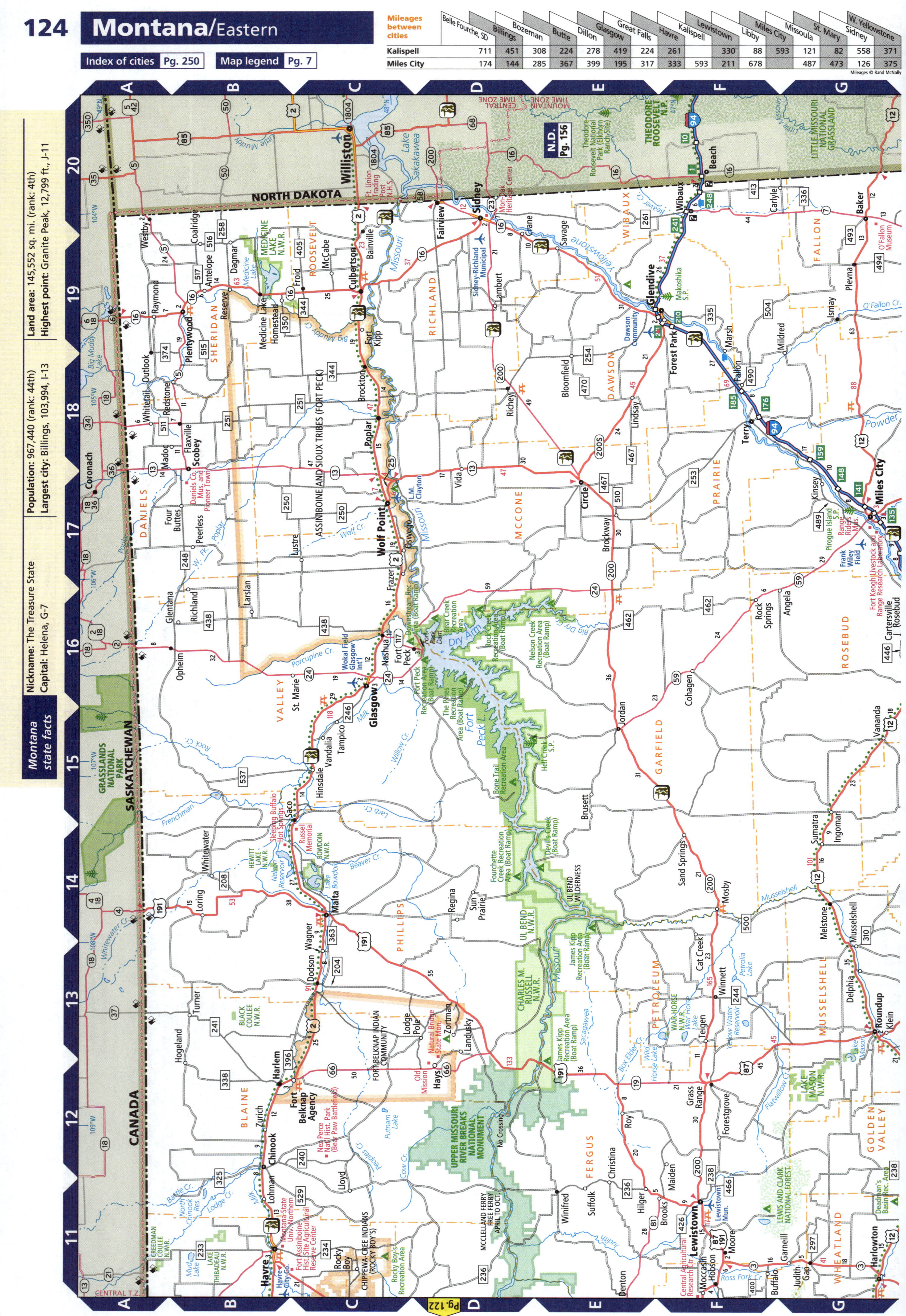

Mileages between cities

| | Belle Fourche, SD | Billings | Bozeman | Butte | Dillon | Glasgow | Great Falls | Havre | Kalispell | Lewistown | Libby | Miles City | Missoula | St. Mary | Sidney | W. Yellowstone |
|---|---|---|---|---|---|---|---|---|---|---|---|---|---|---|---|---|
| Missoula | 606 | 343 | 202 | 120 | 172 | 437 | 166 | 280 | 121 | 272 | 191 | 487 | | 203 | 614 | 267 |
| Sidney | 298 | 269 | 411 | 494 | 524 | 140 | 375 | 298 | 558 | 270 | 646 | 126 | 614 | 490 | | 501 |

Mileages © Rand McNally

15 396 miles 94 249 miles
90 552 miles

Total mileages through Montana
More mileages at go.randmcnally.com/MC

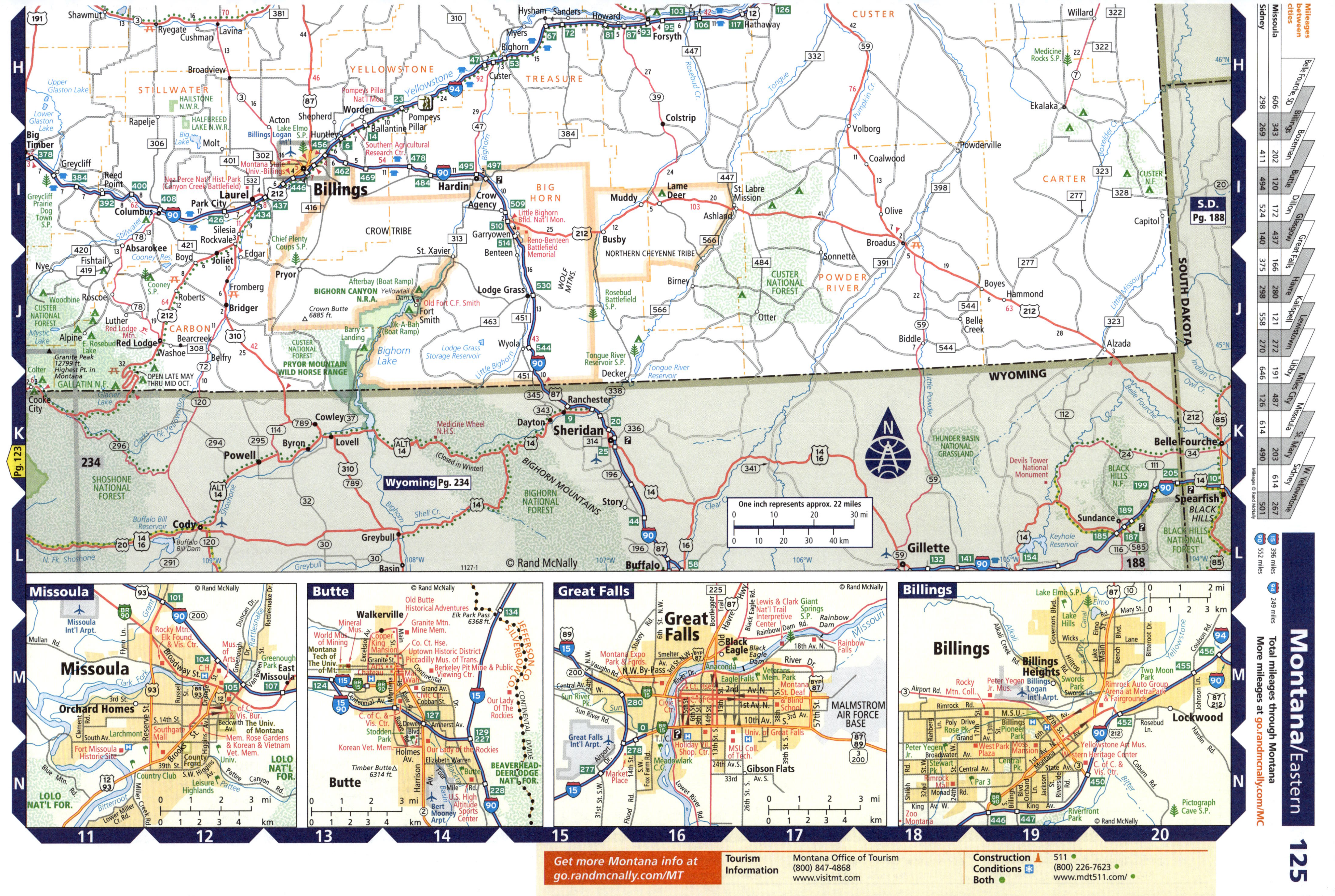

Index of cities Pg. 250 Map legend Pg. 7

Mileages between cities

| | Beatrice | Chadron | Columbus | Falls City | Grand Island | Kearney | Lincoln | McCook | Norfolk | North Platte | Ogallala | Omaha | O'Neill | Scottsbluff | Sioux City, IA | Valentine |
|---|---|---|---|---|---|---|---|---|---|---|---|---|---|---|---|---|
| Grand Island | 131 | 326 | 64 | 196 | | 50 | 93 | 152 | 105 | 145 | 194 | 147 | 112 | 323 | 187 | 210 |
| Lincoln | 41 | 450 | 79 | 102 | 93 | 129 | | 232 | 124 | 224 | 274 | 55 | 208 | 402 | 151 | 304 |

Mileages © Rand McNally

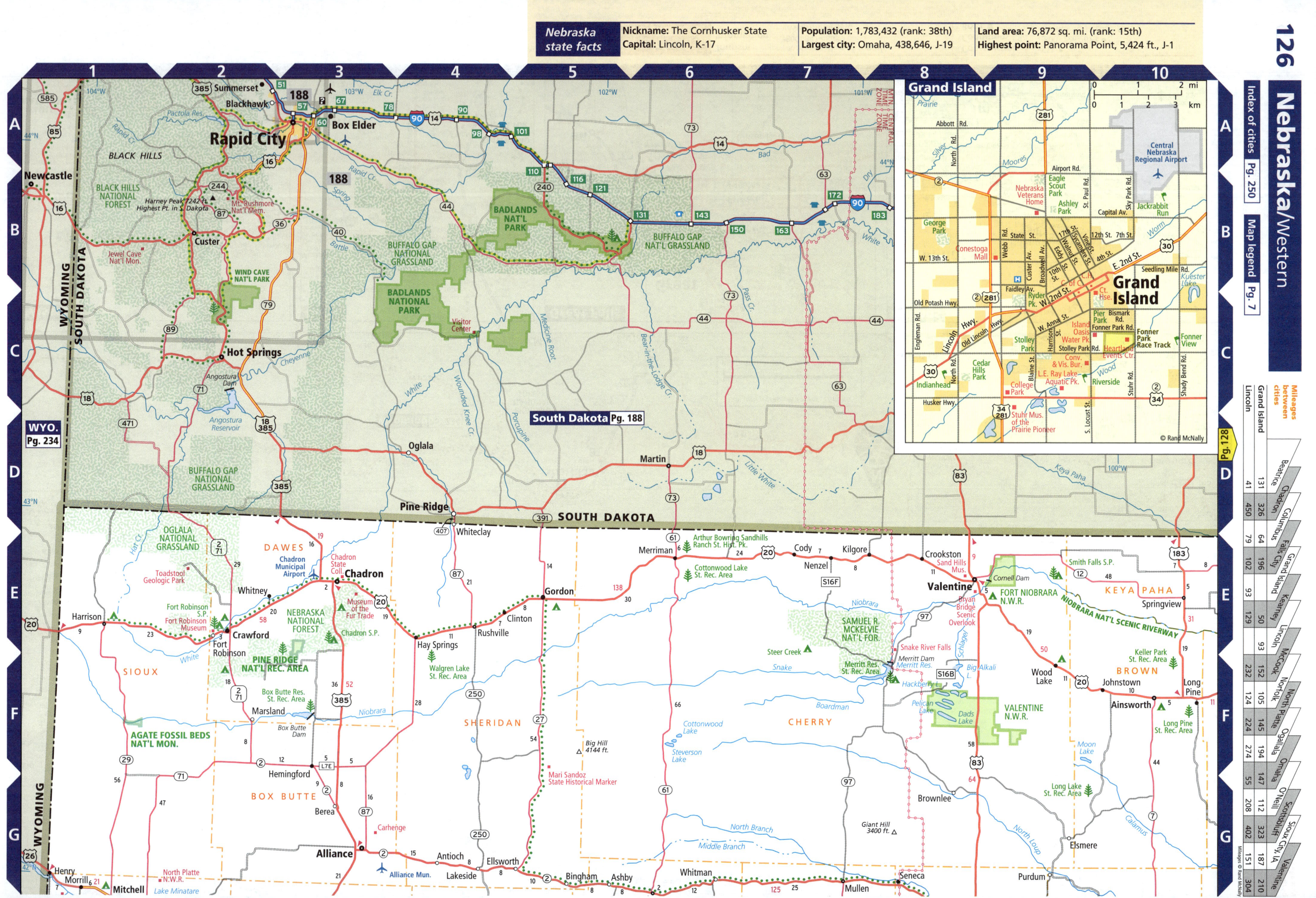

Mileages between cities

| | Beatrice | Chadron | Columbus | Falls City | Grand Island | Kearney | Lincoln | McCook | Norfolk | North Platte | Ogallala | Omaha | O'Neill | Scottsbluff | Sioux City, IA | Valentine |
|---|---|---|---|---|---|---|---|---|---|---|---|---|---|---|---|---|
| Norfolk | 162 | 322 | 45 | 218 | 105 | 155 | 124 | 259 | | 250 | 300 | 109 | 75 | 417 | 82 | 186 |
| North Platte | 262 | 229 | 210 | 327 | 145 | 99 | 224 | 67 | 250 | | 53 | 276 | 189 | 182 | 373 | 129 |

continued p. 128



80 455 miles | 83 226 miles
81 219 miles | 20 436 miles

Total mileages through Nebraska
More mileages at go.randmcnally.com/MC

Index of cities Pg. 250 | Map legend Pg. 7

**Nebraska state facts**

Nickname: The Cornhusker State
Capital: Lincoln, K-17
Population: 1,783,432 (rank: 38th)
Largest city: Omaha, 438,646, J-19
Land area: 76,872 sq. mi. (rank: 15th)
Highest point: Panorama Point, 5,424 ft., J-1

| Mileages between cities | Beatrice | Chadron | Columbus | Falls City | Grand Island | Kearney | Lincoln | McCook | Norfolk | North Platte | Ogallala | Omaha | O'Neill | Scottsbluff | Sioux City, IA | Valentine |
|---|---|---|---|---|---|---|---|---|---|---|---|---|---|---|---|---|
| Omaha | 95 | 431 | 83 | 104 | 147 | 181 | 55 | 283 | 109 | 276 | 325 | | 184 | 458 | 97 | 294 |
| Scottsbluff | 440 | 99 | 388 | 505 | 323 | 277 | 402 | 245 | 417 | 182 | 129 | 458 | 322 | | 467 | 216 |

Mileages © Rand McNally

| Mileages between cities | Beatrice | Chadron | Columbus | Falls City | Grand Island | Kearney | Lincoln | McCook | Norfolk | North Platte | Ogallala | Omaha | O'Neill | Scottsbluff | Sioux City, IA | Valentine |
|---|---|---|---|---|---|---|---|---|---|---|---|---|---|---|---|---|
| Sidney | 381 | 131 | 329 | 445 | 263 | 218 | 343 | 186 | 369 | 122 | 71 | 394 | 311 | 77 | 492 | 251 |
| Valentine | 342 | 137 | 230 | 406 | 210 | 195 | 304 | 197 | 186 | 129 | 182 | 294 | 111 | 216 | 236 | |

Mileages © Rand McNally

80 455 miles 83 226 miles
81 219 miles 20 436 miles

**Total mileages through Nebraska**
**More mileages at go.randmcnally.com/MC**

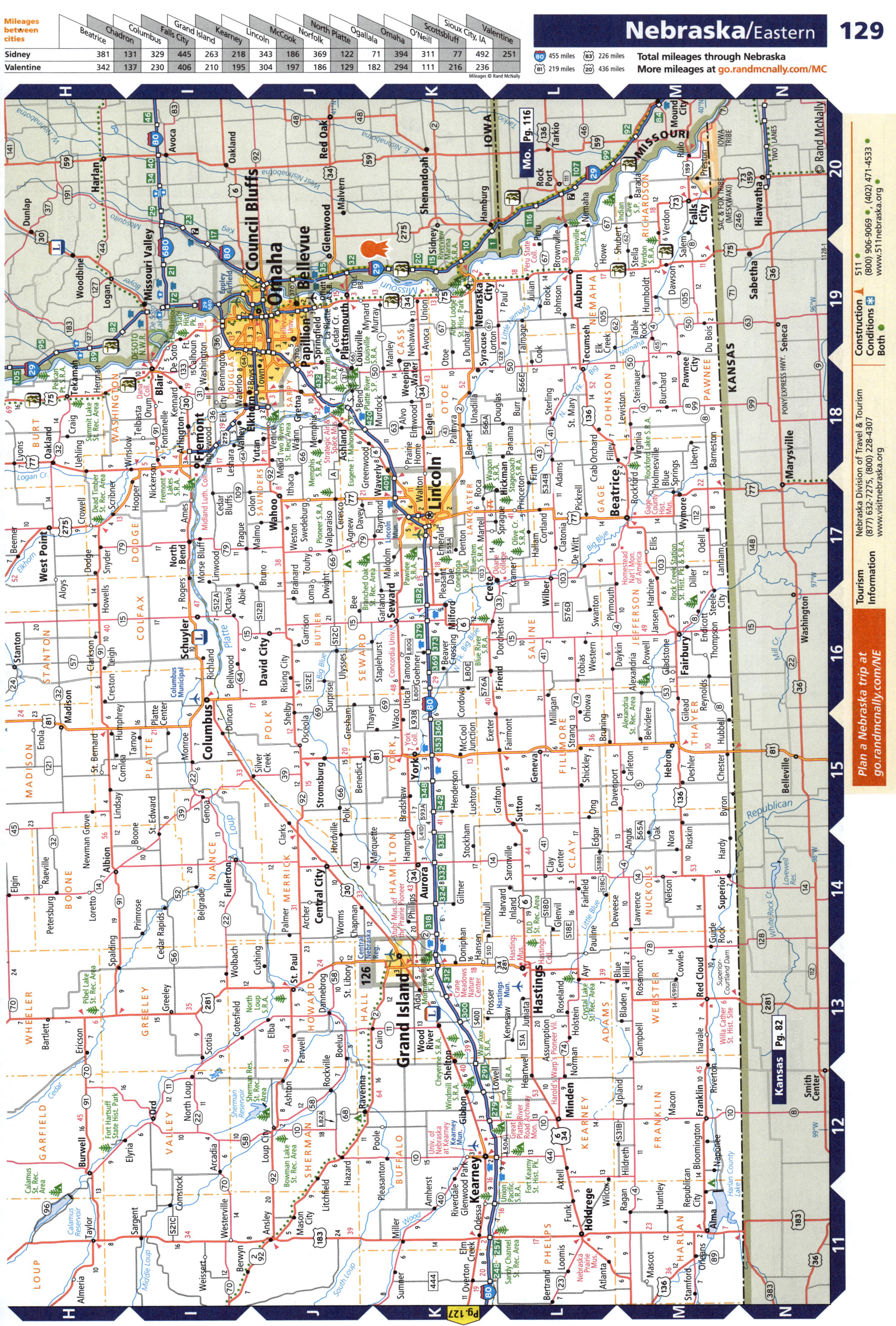

Index of cities Pg. 251 | Map legend Pg. 7

| Mileages between cities | Carson City | Elko | Ely | Jackpot | Las Vegas | Reno | Tonopah | Winnemucca |
|---|---|---|---|---|---|---|---|---|
| Elko | 304 | | 188 | 117 | 429 | 288 | 252 | 125 |
| Ely | 319 | 188 | | 205 | 241 | 319 | 167 | 271 |

| | Carson City | Elko | Ely | Jackpot | Las Vegas | Reno | Tonopah | Winnemucca |
|---|---|---|---|---|---|---|---|---|
| Las Vegas | 435 | 429 | 241 | 446 | | 447 | 210 | 472 |
| Reno | 32 | 288 | 319 | 405 | 447 | | 237 | 163 |

Mileages © Rand McNally

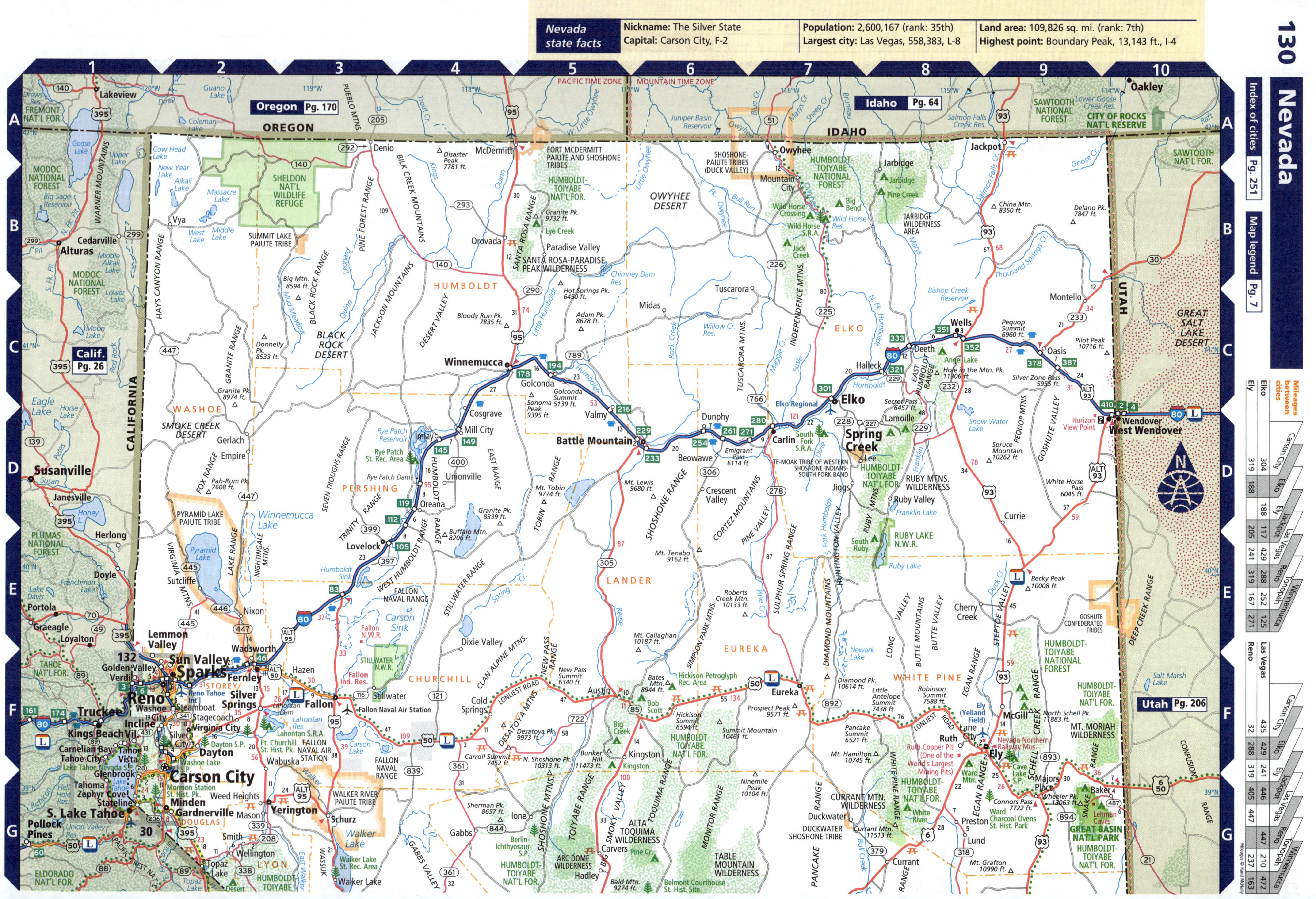

**Mileages between cities**

| | Carson City | Elko | Ely | Jackpot | Las Vegas | Reno | Tonopah | Winnemucca |
|---|---|---|---|---|---|---|---|---|
| S. Lake Tahoe, CA | 27 | 332 | 347 | 450 | 451 | 60 | 237 | 208 |
| Tonopah | 225 | 252 | 167 | 373 | 210 | 237 | | 261 |
| West Wendover | 414 | 109 | 120 | 125 | 361 | 397 | 288 | 232 |
| Winnemucca | 179 | 125 | 271 | 240 | 472 | 163 | 261 | |

Mileages © Rand McNally

Total mileages through Nevada: 15 — 124 miles; 80 — 411 miles; 6 — 307 miles; 95 — 652 miles

More mileages at go.randmcnally.com/MC

Index of cities Pg. 251 | Map legend Pg. 7

**Mileages between cities**

| | Colebrook | Concord | Conway | Keene | Laconia | Littleton | Nashua | Portsmouth |
|---|---|---|---|---|---|---|---|---|
| Berlin | 49 | 115 | 40 | 168 | 97 | 42 | 151 | 117 |
| Concord | 137 | | 77 | 51 | 27 | 87 | 36 | 44 |
| Keene | 181 | 51 | 130 | | 80 | 136 | 50 | 99 |
| Lebanon | 128 | 57 | 88 | 64 | 58 | 82 | 89 | 111 |



**New Hampshire state facts**

Nickname: The Granite State
Capital: Concord, K-7
Population: 1,315,809 (rank: 41st)
Largest city: Manchester, 108,586, L-7
Land area: 8,968 sq. mi. (rank: 44th)
Highest point: Mount Washington, 6,288 ft., F-8

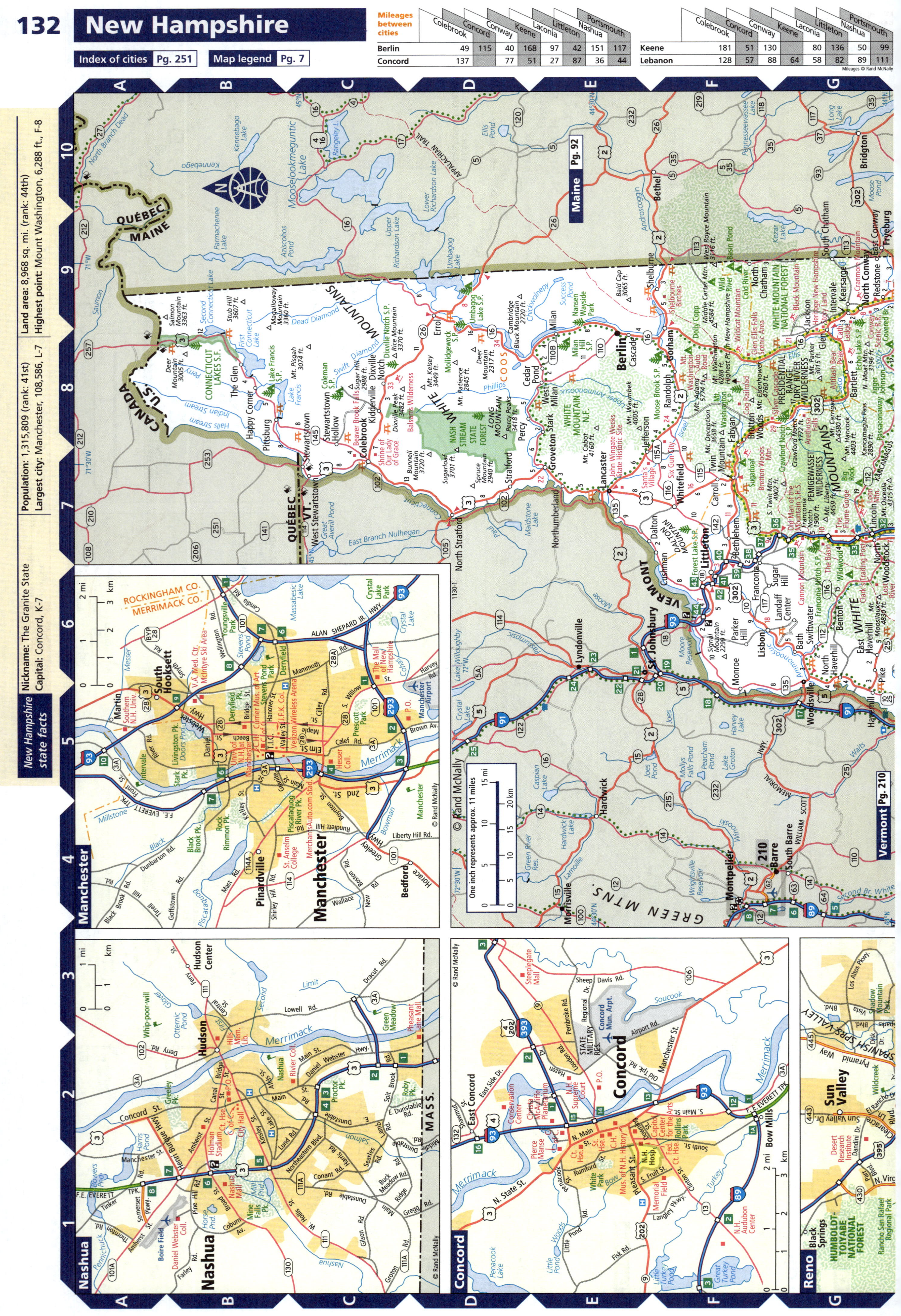

| Mileages between cities | Colebrook | Concord | Conway | Keene | Laconia | Littleton | Nashua | Portsmouth |
|---|---|---|---|---|---|---|---|---|
| Littleton | 56 | 87 | 54 | 136 | 66 | | 121 | 129 |
| Manchester | 155 | 18 | 95 | 55 | 45 | 105 | 18 | 43 |

| | Colebrook | Concord | Conway | Keene | Laconia | Littleton | Nashua | Portsmouth |
|---|---|---|---|---|---|---|---|---|
| Nashua | 172 | 36 | 113 | 50 | 63 | 121 | | 54 |
| Portsmouth | 180 | 44 | 77 | 99 | 57 | 129 | 54 | |

Mileages © Rand McNally

89 61 miles
95 16 miles
93 132 miles
2 36 miles

Total mileages through New Hampshire
More mileages at go.randmcnally.com/MC

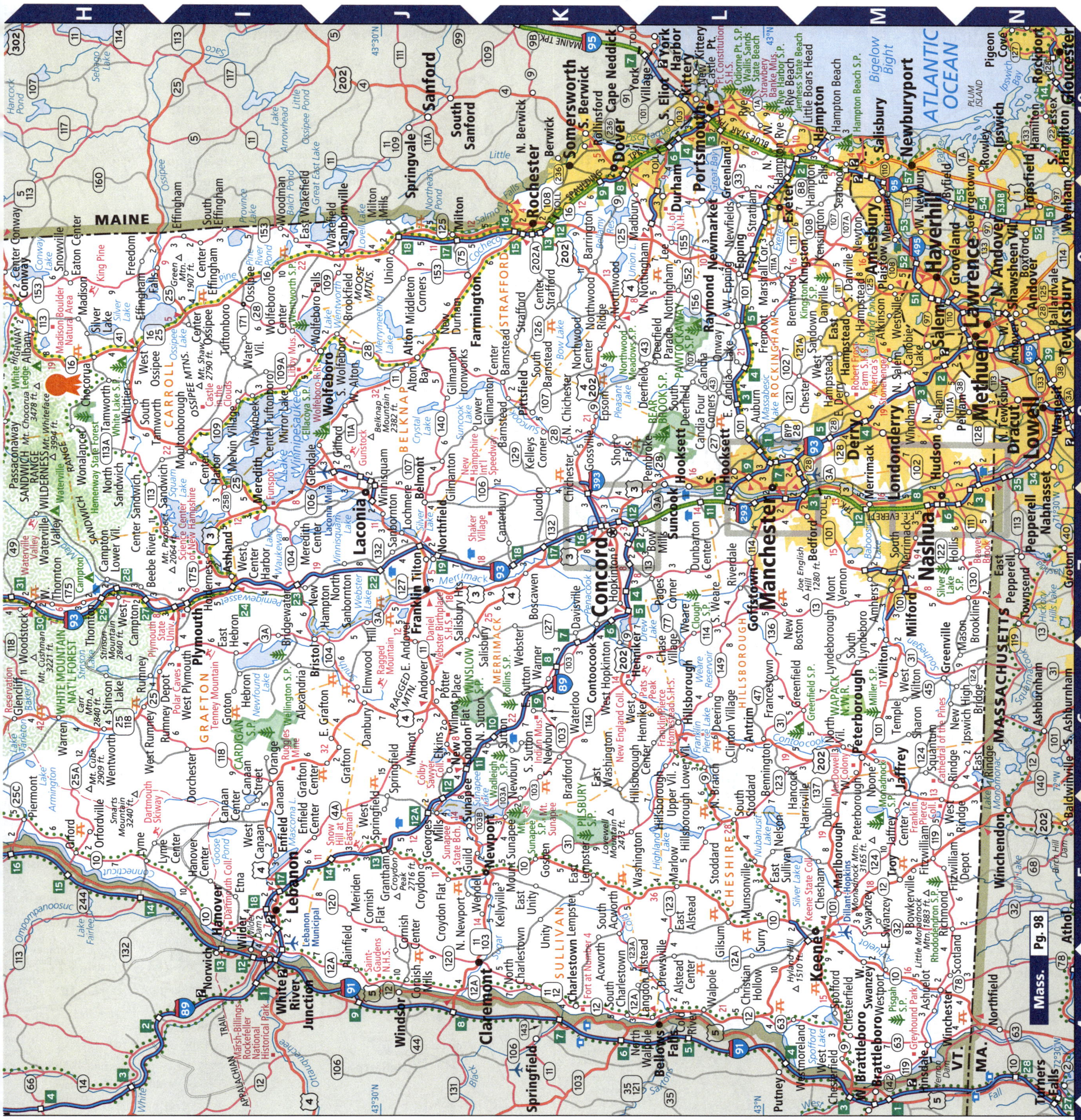

Plan a New Hampshire trip at go.randmcnally.com/NH

Tourism Information: New Hampshire Division of Travel & Tourism Development (800) 386-4664 www.visitnh.gov

Construction / Conditions / Both: 511 (866) 282-7579 www.nh.gov/dot/511

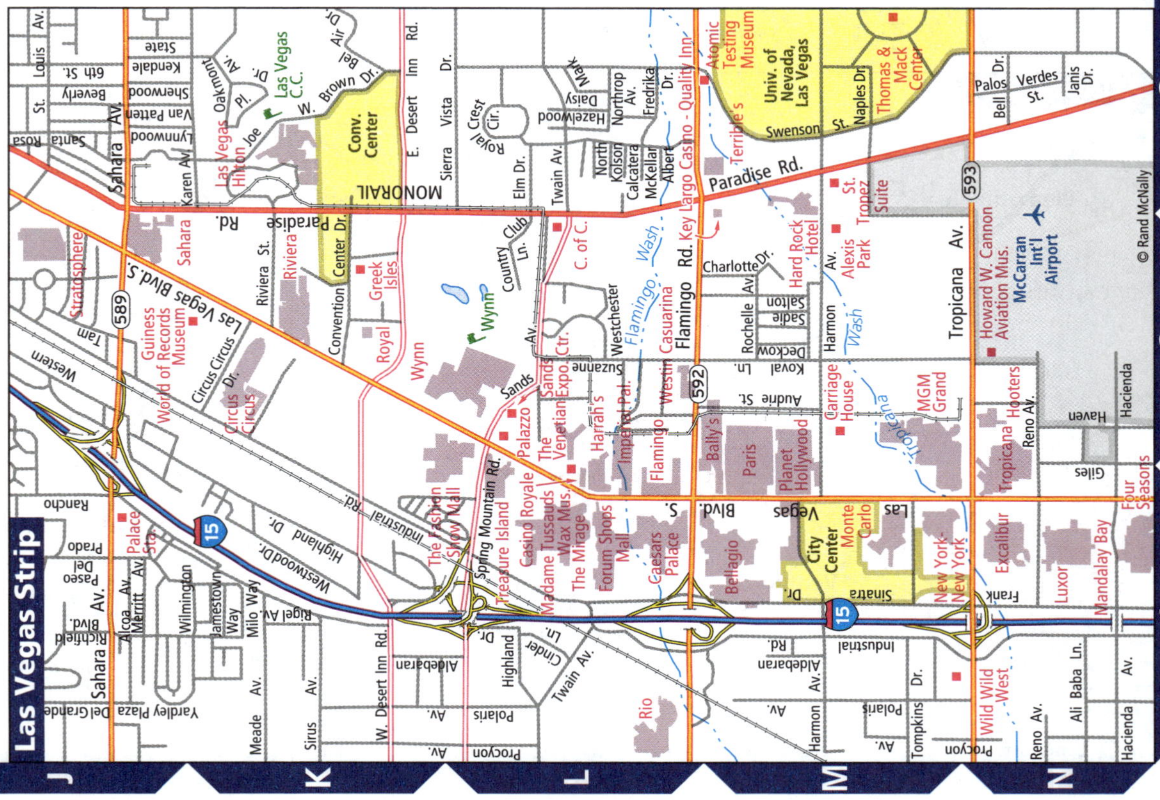

Reno

Sparks

Las Vegas Strip

© Rand McNally

Index of cities Pg. 251 | Map legend Pg. 7

Mileages between cities

| | Atlantic City | Camden | Cape May | Jersey City | Long Branch | Newark | New Brunswick | New York, NY | Paterson | Phillipsburg | Port Jervis, NY | Princeton | Toms River | Trenton | Vineland | Wilmington, DE |
|---|---|---|---|---|---|---|---|---|---|---|---|---|---|---|---|---|
| Atlantic City | | 58 | 47 | 120 | 82 | 115 | 94 | 126 | 129 | 138 | 182 | 99 | 52 | 90 | 36 | 82 |
| Camden | 58 | | 88 | 86 | 76 | 80 | 61 | 96 | 94 | 80 | 143 | 45 | 55 | 34 | 36 | 31 |



New Jersey state facts

Nickname: The Garden State
Capital: Trenton, J-8
Population: 8,682,661 (rank: 11th)
Largest city: Newark, 278,980, F-12
Land area: 7,417 sq. mi. (rank: 46th)
Highest point: High Point, 1,803 ft., A-9

**Mileages between cities**

| | Atlantic City | Camden | Cape May | Jersey City | Long Branch | Newark | New Brunswick | New York, NY | Paterson | Phillipsburg | Port Jervis, NY | Princeton | Toms River | Trenton | Vineland | Wilmington, DE |
|---|---|---|---|---|---|---|---|---|---|---|---|---|---|---|---|---|
| Cape May | 47 | 88 | | 151 | 114 | 147 | 126 | 157 | 161 | 170 | 214 | 131 | 84 | 121 | 48 | 98 |
| Newark | 115 | 80 | 147 | 6 | 43 | | 25 | 10 | 15 | 58 | 74 | 41 | 63 | 55 | 114 | 112 |

continued p. 136



78: 68 miles
80: 68 miles
95: 98 miles

Total mileages through New Jersey
More mileages at go.randmcnally.com/MC

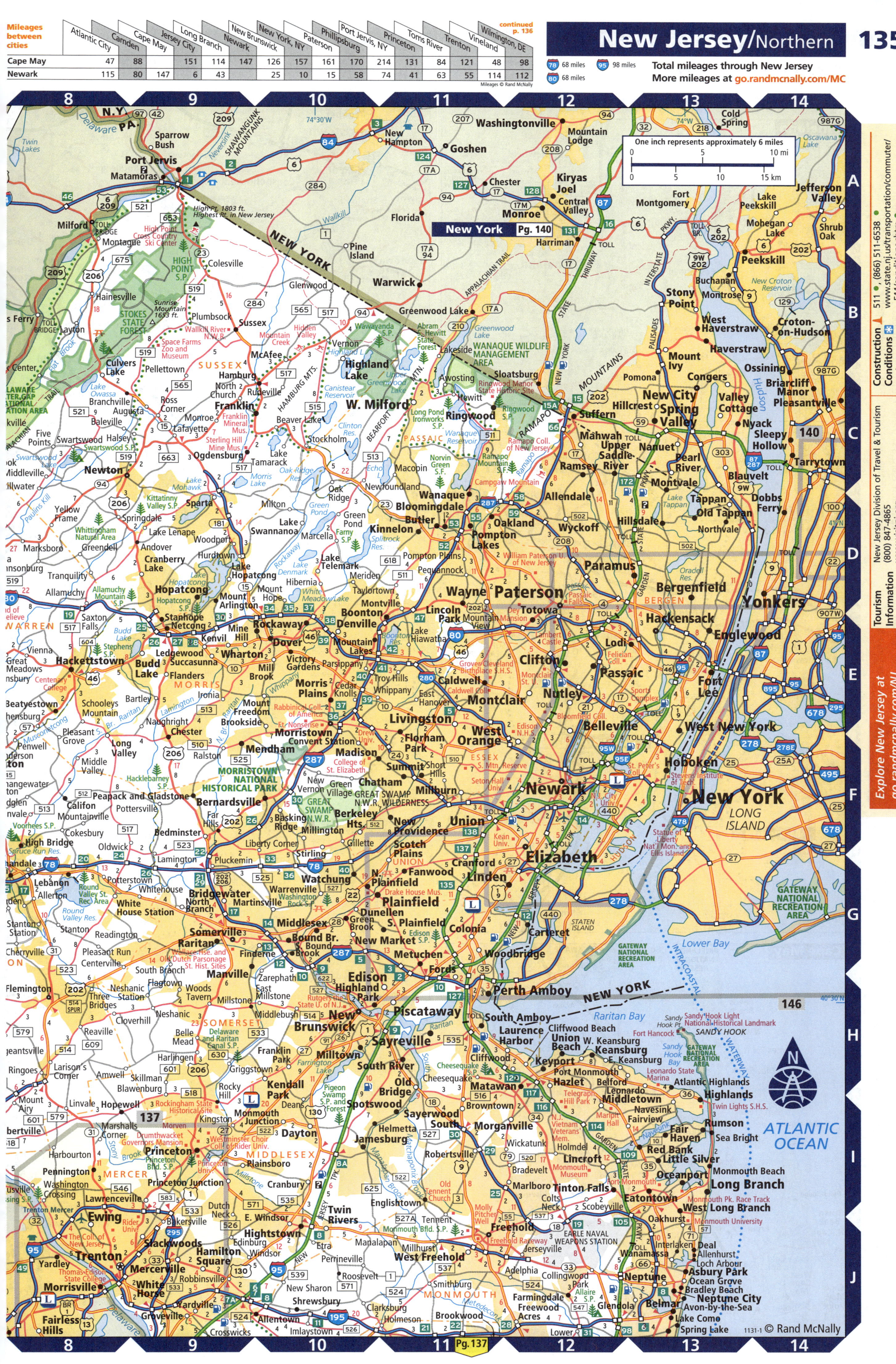


Tourism Information
New Jersey Division of Travel & Tourism
(800) 847-4865
www.visitnj.org

Construction / Conditions / Both
511, (866) 511-6538
www.state.nj.us/transportation/commuter/511/conditions.shtm

Explore New Jersey at go.randmcnally.com/NJ

Index of cities Pg. 251 | Map legend Pg. 7

Mileages between cities

| | Atlantic City | Camden | Cape May | Jersey City | Long Branch | Newark | New Brunswick | New York, NY | Paterson | Phillipsburg | Port Jervis, NY | Princeton | Toms River | Trenton | Vineland | Wilmington, DE |
|---|---|---|---|---|---|---|---|---|---|---|---|---|---|---|---|---|
| New Brunswick | 94 | 61 | 126 | 30 | 34 | 25 | | 36 | 39 | 48 | 92 | 16 | 43 | 26 | 95 | 93 |
| Phillipsburg | 138 | 80 | 170 | 64 | 81 | 58 | 48 | 68 | 67 | | 74 | 54 | 101 | 54 | 118 | 95 |

Mileages © Rand McNally

**New Jersey state facts**

Nickname: The Garden State
Capital: Trenton, J-8
Population: 8,682,661 (rank: 11th)
Largest city: Newark, 278,980, F-12
Land area: 7,417 sq. mi. (rank: 46th)
Highest point: High Point, 1,803 ft., A-9

**Mileages between cities**

| | Atlantic City | Camden | Cape May | Jersey City | Long Branch | Newark | New Brunswick | New York, NY | Paterson | Phillipsburg | Port Jervis, NY | Princeton | Toms River | Trenton | Vineland | Wilmington, DE |
|---|---|---|---|---|---|---|---|---|---|---|---|---|---|---|---|---|
| Port Jervis | 182 | 143 | 214 | 89 | 110 | 74 | 92 | 95 | 73 | 74 | | 94 | 130 | 122 | 180 | 158 |
| Trenton | 90 | 34 | 121 | 61 | 52 | 55 | 26 | 66 | 69 | 54 | 122 | 11 | 47 | | 69 | 61 |

Mileages © Rand McNally

Total mileages through New Jersey
78 68 miles · 95 98 miles · 80 68 miles
More mileages at go.randmcnally.com/MC

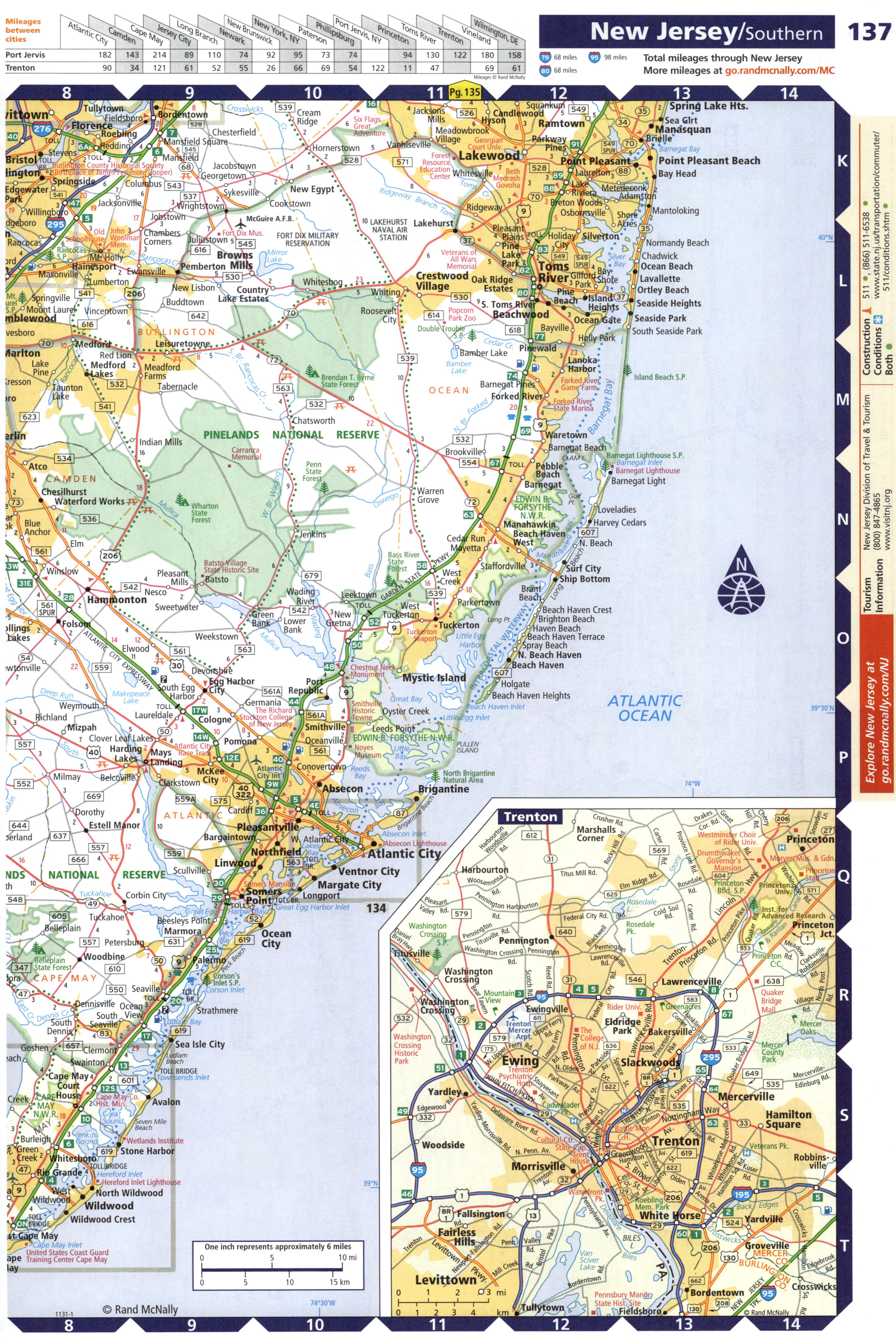

Index of cities Pg. 251 | Map legend Pg. 7

**New Mexico state facts**

Nickname: Land of Enchantment
Capital: Santa Fe, D-6
Population: 1,984,356 (rank: 36th)
Largest city: Albuquerque, 521,999, E-4
Land area: 121,356 sq. mi. (rank: 5th)
Highest point: Wheeler Peak, 13,161 ft., B-6

**Mileages between cities**

| | Albuquerque | Carlsbad | Clayton | Gallup | Las Cruces | Socorro | Taos | Tucumcari |
|---|---|---|---|---|---|---|---|---|
| Albuquerque | | 277 | 270 | 137 | 222 | 78 | 128 | 173 |
| Carlsbad | 277 | | 374 | 412 | 206 | 241 | 336 | 263 |
| Clayton | 270 | 374 | | 407 | 415 | 347 | 163 | 111 |
| Clovis | 219 | 180 | 168 | 356 | 292 | 248 | 246 | 83 |

Mileages © Rand McNally

Mileages between cities

| | Albuquerque | Carlsbad | Clayton | Gallup | Las Cruces | Socorro | Taos | Tucumcari |
|---|---|---|---|---|---|---|---|---|
| Farmington | 180 | 455 | 418 | 121 | 404 | 258 | 202 | 354 |
| Las Cruces | 222 | 206 | 415 | 338 | | 146 | 351 | 303 |
| Roswell | 199 | 76 | 293 | 336 | 184 | 165 | 260 | 182 |
| Santa Fe | 58 | 268 | 215 | 197 | 282 | 136 | 68 | 166 |

Mileages © Rand McNally

10 164 miles
25 462 miles
40 374 miles

Total mileages through New Mexico
More mileages at go.randmcnally.com/MC

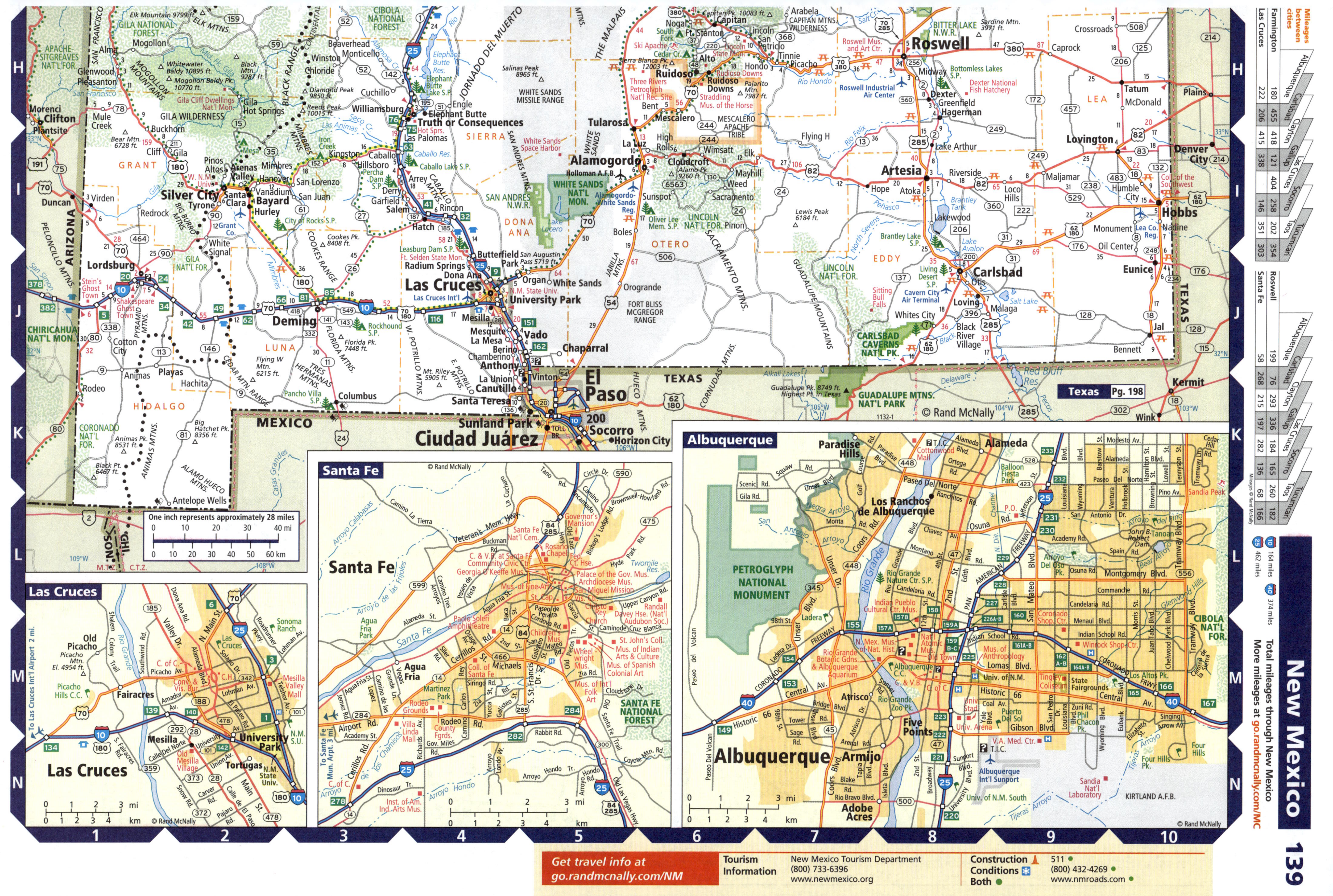

Index of cities Pg. 252 | Map legend Pg. 7

Mileages between cities

| | Albany | Buffalo | Hempstead | Newburgh | New York | Poughkeepsie | Riverhead | White Plains |
|---|---|---|---|---|---|---|---|---|
| Albany | | 289 | 167 | 87 | 156 | 75 | 219 | 138 |
| Buffalo | 289 | | 423 | 361 | 395 | 362 | 471 | 394 |
| Hempstead | 167 | 423 | | 78 | 12 | 92 | 59 | 34 |
| Kingston | 55 | 339 | 116 | 37 | 106 | 19 | 168 | 87 |

Mileages © Rand McNally

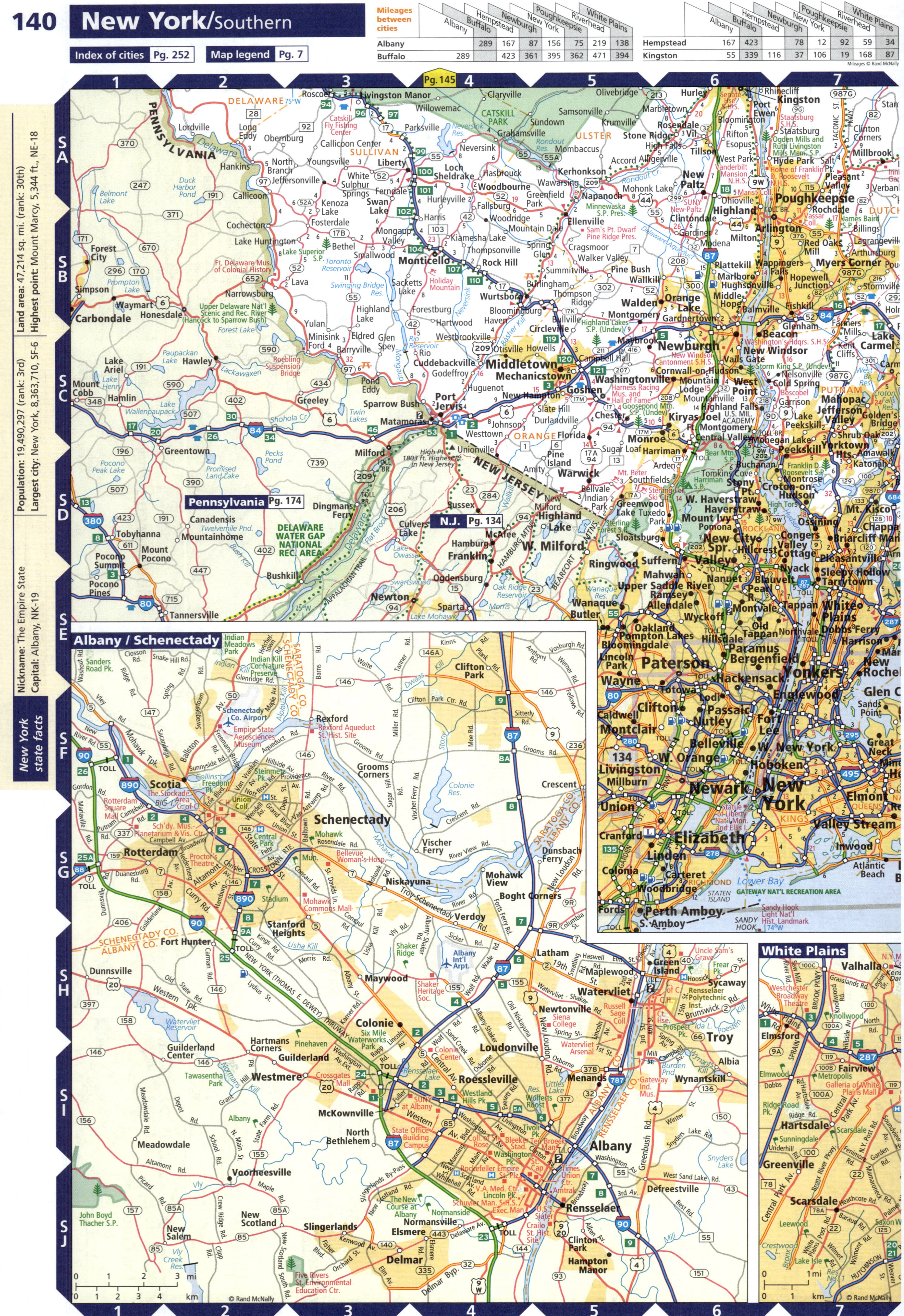

Mileages between cities

| | Albany | Buffalo | Hempstead | Newburgh | New York | Poughkeepsie | Riverhead | White Plains |
|---|---|---|---|---|---|---|---|---|
| Montauk | 260 | 513 | 97 | 172 | 107 | 184 | 42 | 126 |
| Newburgh | 87 | 361 | 78 | | 72 | 19 | 130 | 49 |
| New York | 156 | 395 | 12 | 72 | | 84 | 66 | 26 |
| Poughkeepsie | 75 | 362 | 92 | 19 | 84 | | 143 | 60 |

Mileages © Rand McNally

Total mileages through New York

| Route | Miles |
|---|---|
| 84 | 72 miles |
| 87 | 334 miles |
| 95 | 24 miles |
| 495 | 66 miles |

More mileages at go.randmcnally.com/MC

Index of cities Pg. 252 | Map legend Pg. 7

**Mileages between cities**

| | Albany | Binghamton | Buffalo | Elmira | Glens Falls | Jamestown | Kingston | Lake Placid | Massena | New York | Niagara Falls | Plattsburgh | Rochester | Syracuse | Utica | Watertown |
|---|---|---|---|---|---|---|---|---|---|---|---|---|---|---|---|---|
| **Albany** | | 140 | 289 | 195 | 53 | 356 | 55 | 140 | 217 | 156 | 302 | 160 | 226 | 145 | 94 | 175 |
| **Binghamton** | 140 | | 222 | 56 | 179 | 218 | 130 | 266 | 231 | 176 | 235 | 287 | 159 | 73 | 89 | 143 |



*New York state facts*

Nickname: The Empire State
Capital: Albany, NK-19
Population: 19,490,297 (rank: 3rd)
Largest city: New York, 8,363,710, SF-6
Land area: 47,214 sq. mi. (rank: 30th)
Highest point: Mount Marcy, 5,344 ft., NE-18

# New York/Northwestern

**Mileages between cities**

| | Albany | Binghamton | Buffalo | Elmira | Glens Falls | Jamestown | Kingston | Lake Placid | Massena | New York | Niagara Falls | Plattsburgh | Rochester | Syracuse | Utica | Watertown |
|---|---|---|---|---|---|---|---|---|---|---|---|---|---|---|---|---|
| Buffalo | 289 | 222 | | 148 | 313 | 71 | 339 | 337 | 305 | 395 | 21 | 373 | 73 | 150 | 198 | 212 |
| Jamestown | 356 | 218 | 71 | 163 | 395 | | 349 | 404 | 370 | 392 | 92 | 436 | 139 | 214 | 263 | 278 |

Mileages © Rand McNally

continued p. 144

I-81 184 miles · I-87 334 miles · I-86 176 miles · I-90 385 miles — Total mileages through New York

More mileages at go.randmcnally.com/MC

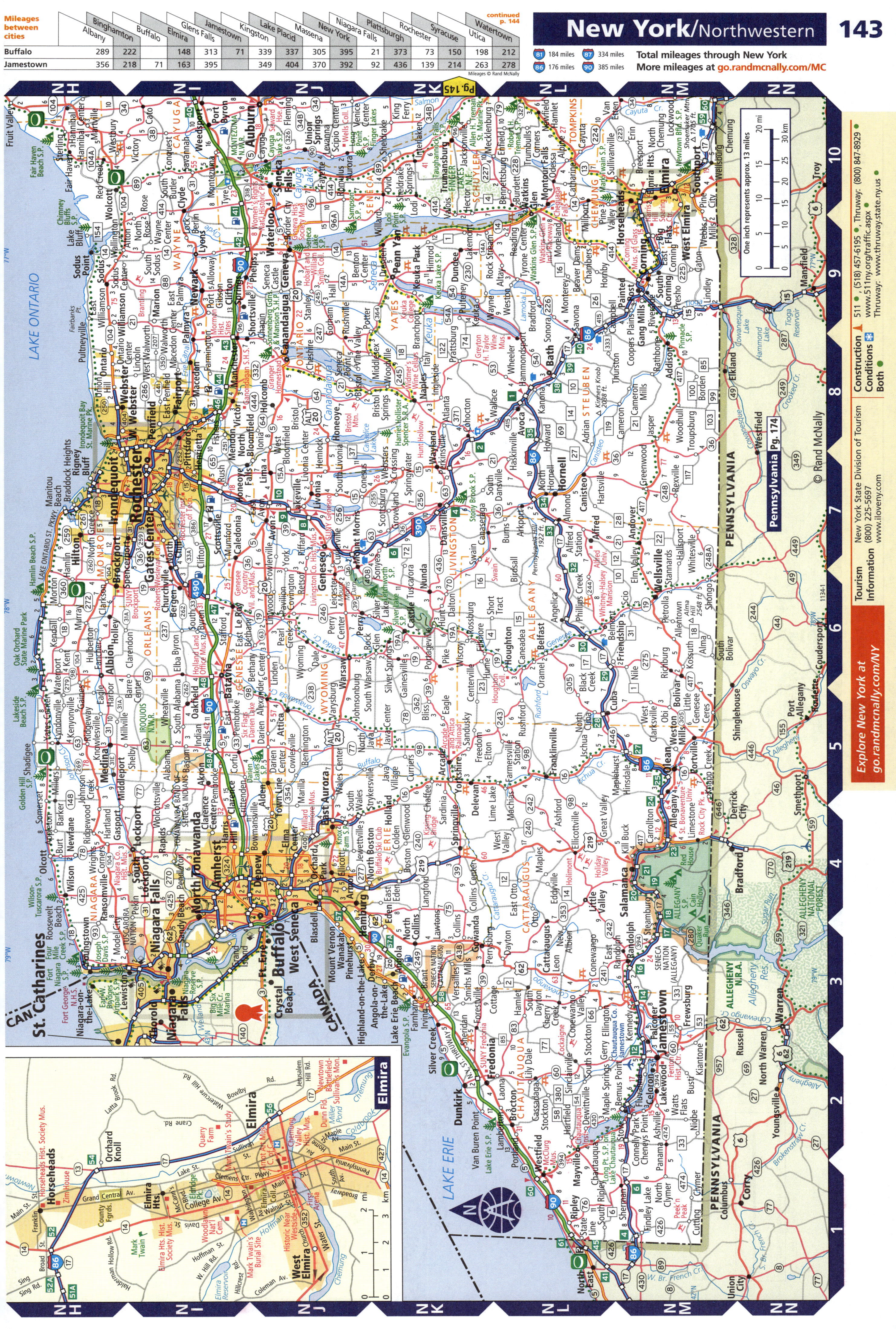

Index of cities Pg. 252 | Map legend Pg. 7

| Mileages between cities | Albany | Binghamton | Buffalo | Elmira | Glens Falls | Jamestown | Kingston | Lake Placid | Massena | New York | Niagara Falls | Plattsburgh | Rochester | Syracuse | Utica | Watertown |
|---|---|---|---|---|---|---|---|---|---|---|---|---|---|---|---|---|
| Plattsburgh | 160 | 287 | 373 | 342 | 110 | 436 | 214 | 50 | 82 | 317 | 384 | | 308 | 227 | 183 | 165 |
| Rochester | 226 | 159 | 73 | 120 | 248 | 139 | 277 | 275 | 242 | 332 | 87 | 308 | | 86 | 135 | 149 |



**New York state facts**

Nickname: The Empire State
Capital: Albany, NK-19
Population: 19,490,297 (rank: 3rd)
Largest city: New York, 8,363,710, SF-6
Land area: 47,214 sq. mi. (rank: 30th)
Highest point: Mount Marcy, 5,344 ft., NE-18

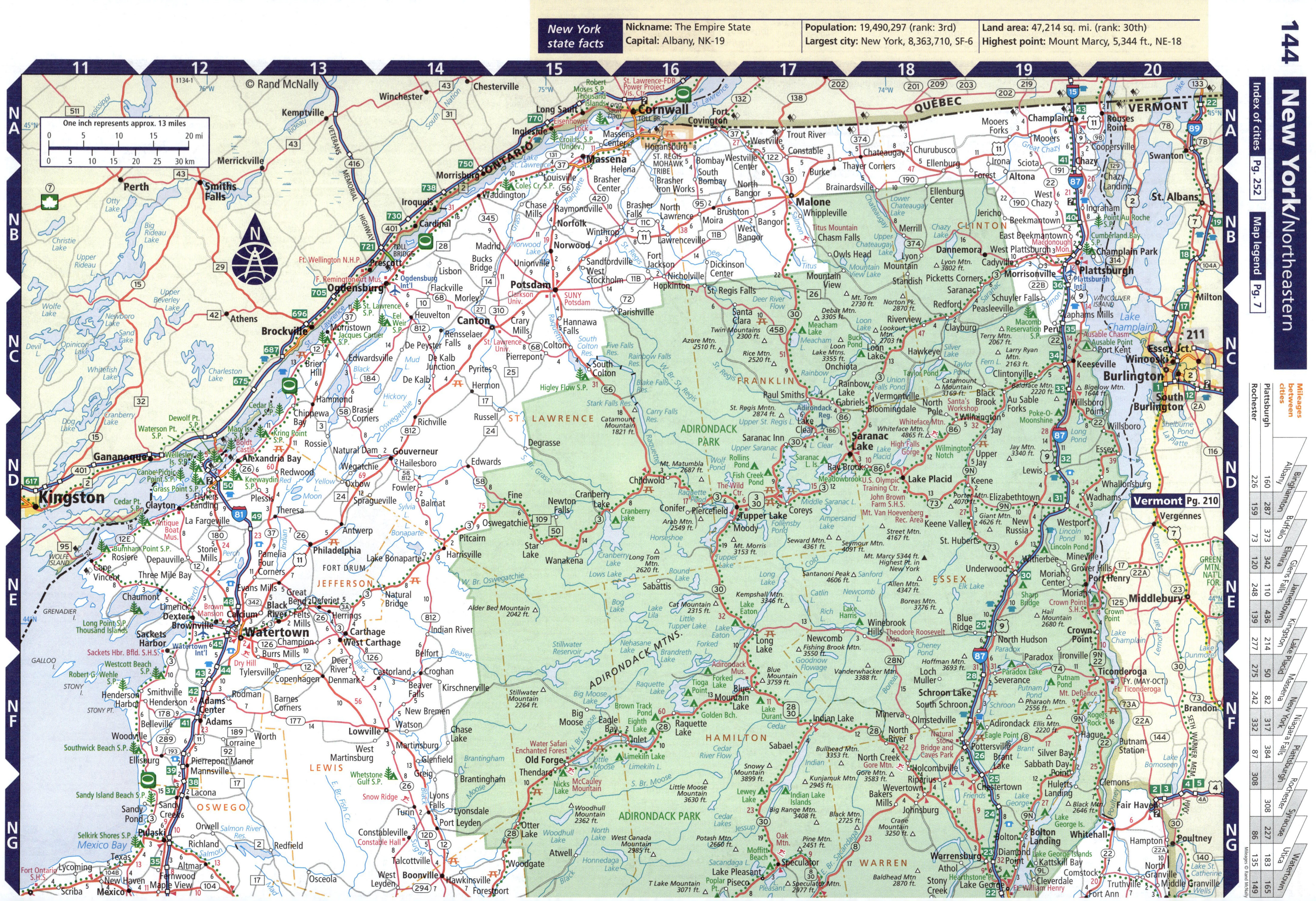

Mileages between cities

| | Albany | Binghamton | Buffalo | Elmira | Glens Falls | Jamestown | Kingston | Lake Placid | Massena | New York | Niagara Falls | Plattsburgh | Rochester | Syracuse | Utica | Watertown |
|---|---|---|---|---|---|---|---|---|---|---|---|---|---|---|---|---|
| Syracuse | 145 | 73 | 150 | 90 | 160 | 214 | 195 | 195 | 159 | 246 | 162 | 227 | 86 | | 53 | 70 |
| Watertown | 175 | 143 | 212 | 160 | 179 | 278 | 226 | 125 | 89 | 316 | 225 | 165 | 149 | 70 | 80 | |

Mileages © Rand McNally

81 184 miles 87 334 miles
86 176 miles 90 385 miles

Total mileages through New York
More mileages at go.randmcnally.com/MC

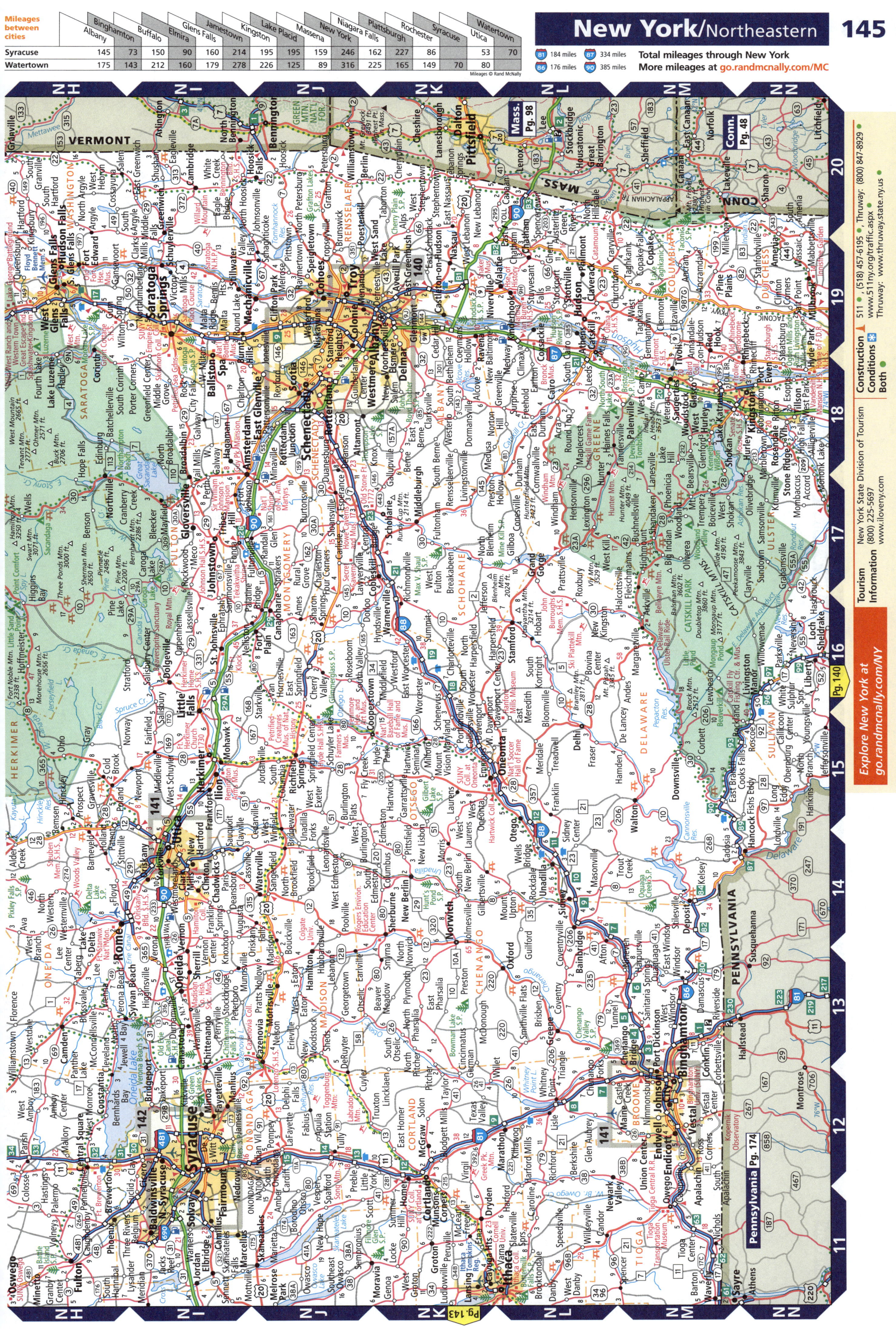


Explore New York at go.randmcnally.com/NY

Tourism Information: New York State Division of Tourism (800) 225-5697 www.iloveny.com

Construction ▲ Conditions ❄ Both ●: 511 ●, (518) 457-6195 ●, Thruway: (800) 847-8929 ● www.511ny.org/traffic.aspx ● Thruway: www.thruway.state.ny.us ●

## City sights to see

- American Museum of Natural History, Manhattan, A-4
- Battery Park, Manhattan, I-1
- Belmont Park Race Track, Elmont, H-16
- Bronx Zoo, Bronx, E-12
- Brooklyn Bridge, New York, H-2
- Carnegie Hall, Manhattan, C-4
- Central Park, Manhattan, B-4

- Chrysler Building, Manhattan, D-4
- Coney Island, Brooklyn, L-10
- Edison Nat'l Historic Site, W. Orange, N.J., F-5
- Ellis Island, Jersey City, N.J./Manhattan, I-9
- Empire State Building, Manhattan, D-3
- Greenwich Village, Manhattan, H-10
- Grand Central Terminal, Manhattan, D-4

continued p.148

**City sights to see–continued**

- Guggenheim Museum, Manhattan, A-5
- Intrepid Sea-Air Space Mus., Manhattan, C-2
- Lincoln Center, Manhattan, B-3
- Madison Square Garden, Manhattan, D-2
- Meadowlands Sports Complex, East Rutherford, N.J., F-8
- Metropolitan Museum of Art, Manhattan, B-5

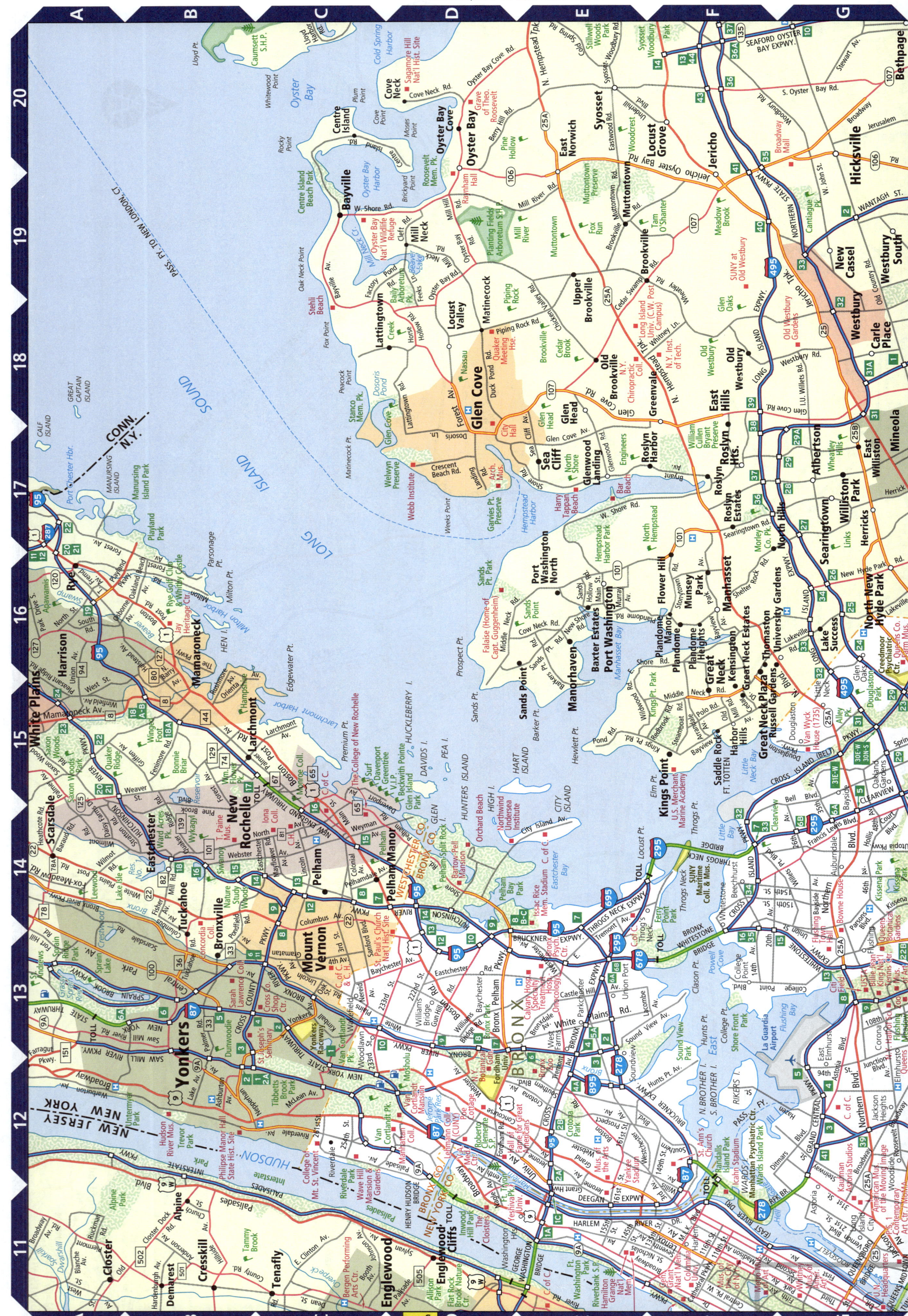

- New York Stock Exchange and Wall Street, Manhattan, H-1
- Rockefeller Center, Manhattan, C-4
- Staten Island Ferry, New York, J-9
- Statue of Liberty, Manhattan, I-9
- Times Square, Manhattan, D-3
- World Trade Center Site, Manhattan, H-1
- Yankee Stadium, Bronx, E-11

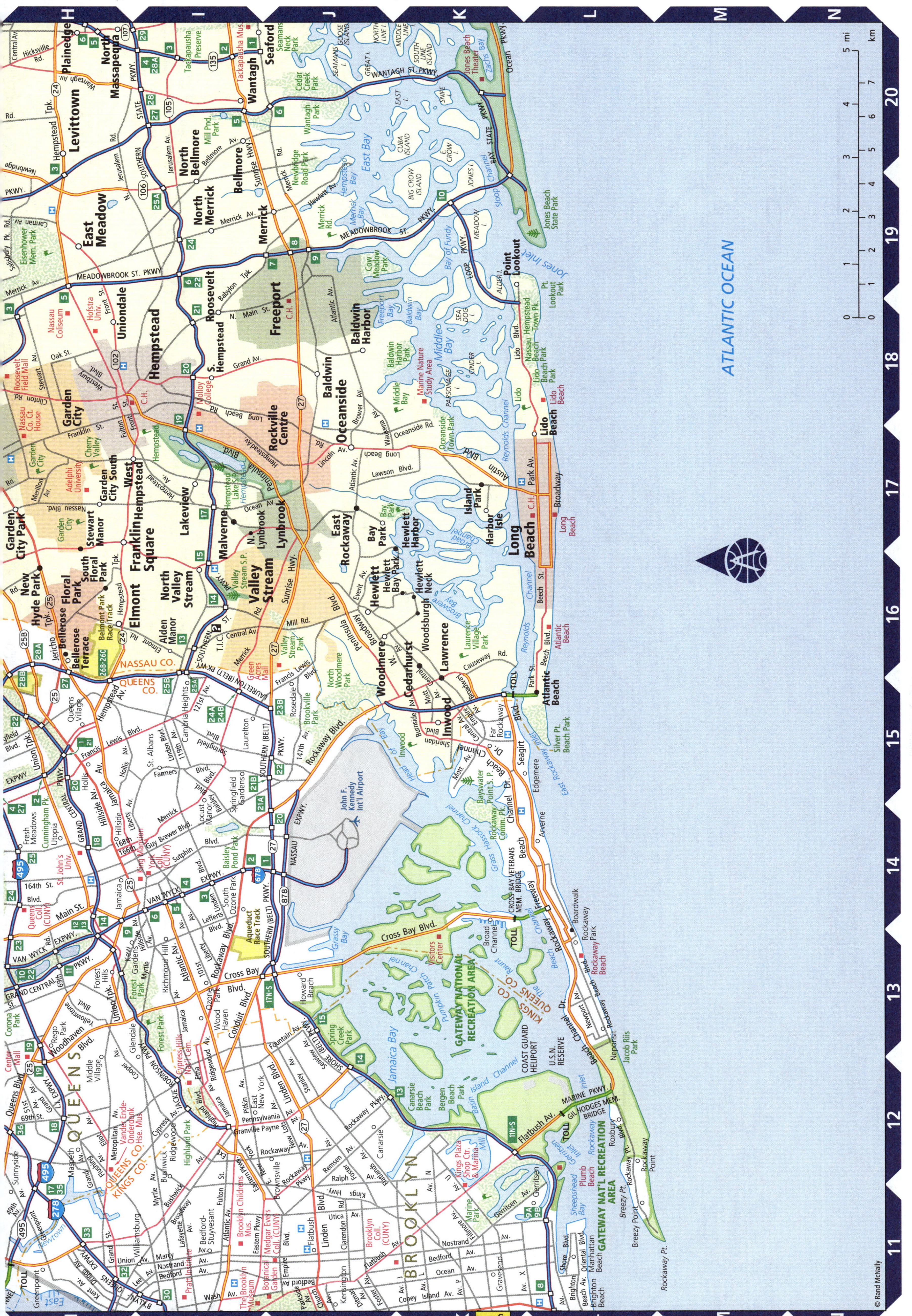

# North Carolina/Western

Index of cities Pg. 253 | Map legend Pg. 7

Mileages between cities

| | Asheville | Boone | Charlotte | Durham | Elizabeth City | Greensboro | Hickory | Morehead City | Murphy | Nags Head | New Bern | Raleigh | Roanoke Rapids | Rockingham | Wilmington | Winston-Salem |
|---|---|---|---|---|---|---|---|---|---|---|---|---|---|---|---|---|
| Asheville | | 94 | 128 | 224 | 412 | 172 | 77 | 393 | 110 | 444 | 358 | 251 | 308 | 200 | 327 | 145 |
| Charlotte | 128 | 100 | | 144 | 332 | 93 | 57 | 313 | 223 | 364 | 278 | 168 | 231 | 71 | 197 | 77 |



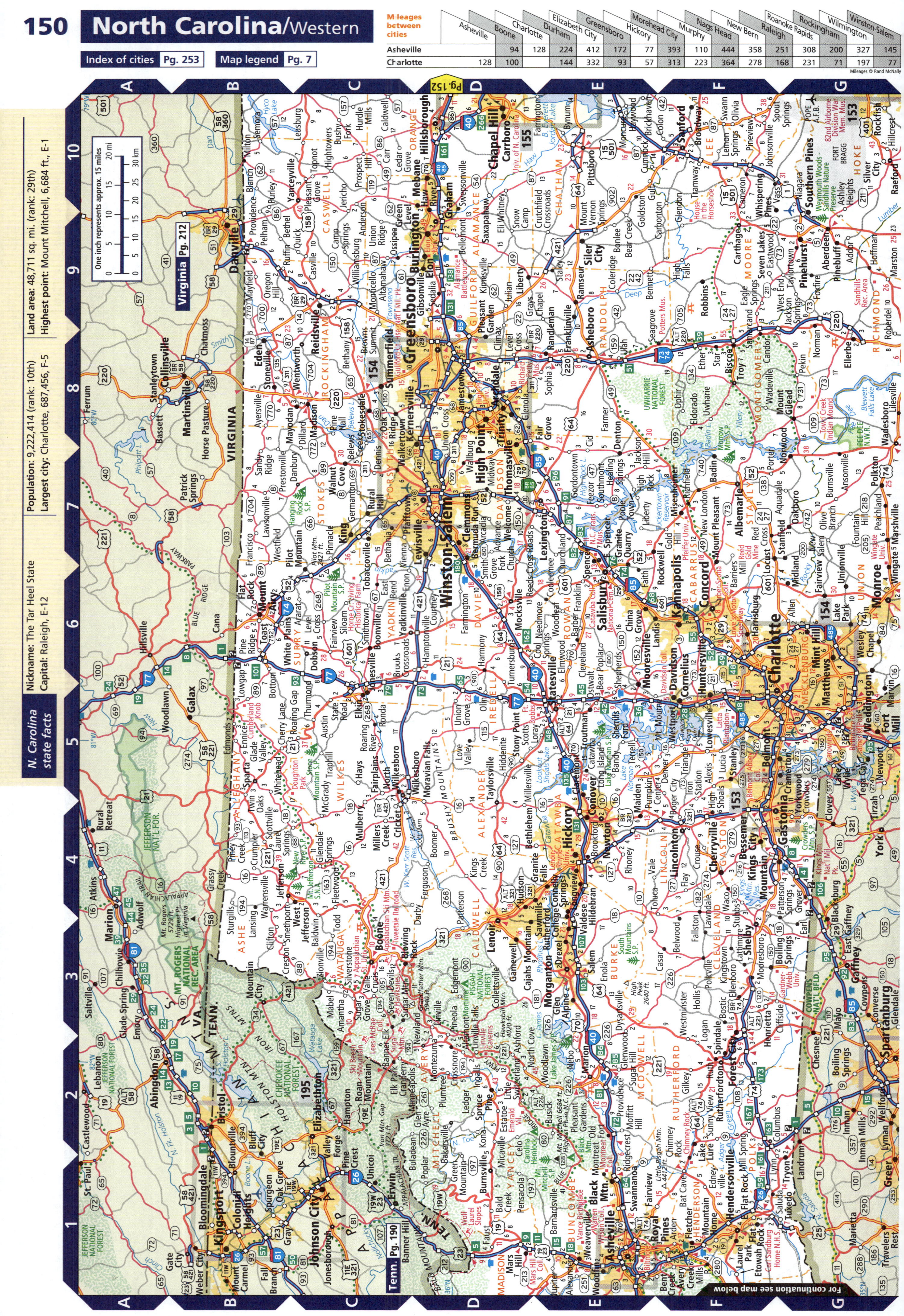

Mileages between cities

continued p. 152

| | Asheville | Boone | Charlotte | Durham | Elizabeth City | Greensboro | Hickory | Morehead City | Murphy | Nags Head | New Bern | Raleigh | Roanoke Rapids | Rockingham | Wilmington | Winston-Salem |
|---|---|---|---|---|---|---|---|---|---|---|---|---|---|---|---|---|
| Elizabeth City | 412 | 354 | 332 | 185 | | 241 | 338 | 152 | 520 | 56 | 119 | 164 | 97 | 259 | 208 | 269 |
| Fayetteville | 261 | 202 | 137 | 89 | 203 | 94 | 189 | 138 | 369 | 234 | 130 | 63 | 127 | 64 | 89 | 119 |



40 419 miles · 85 233 miles · 77 102 miles · 95 182 miles

Total mileages through North Carolina
More mileages at go.randmcnally.com/MC

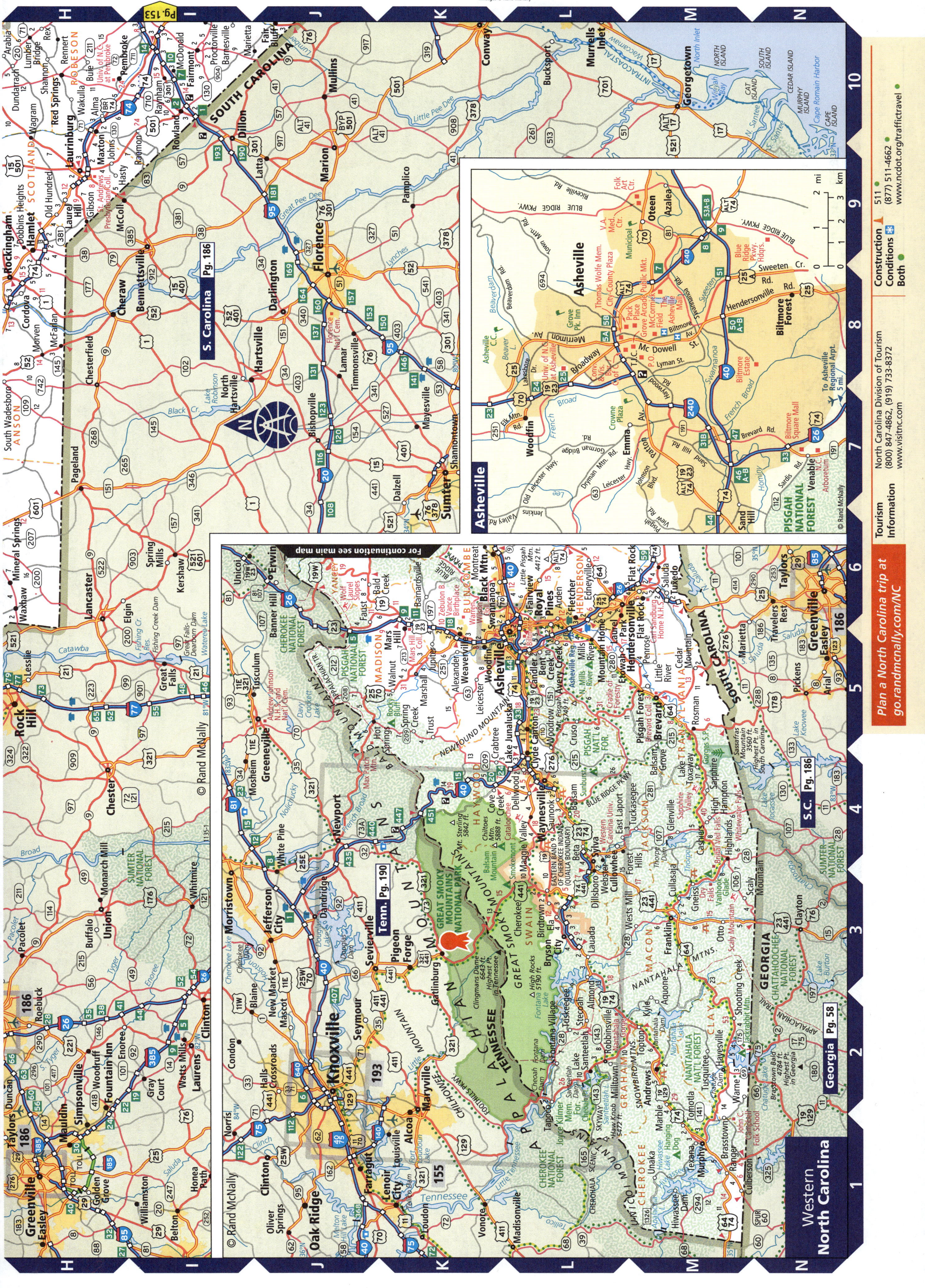

Western North Carolina
Asheville
S. Carolina Pg. 186
Tenn. Pg. 190
Georgia Pg. 58
S.C. Pg. 186
Pg. 153
For continuation see main map
Great Smoky Mountains National Park
Knoxville
Greenville
Rock Hill
Florence
Sumter
Tourism Information
North Carolina Division of Tourism
(800) 847-4862, (919) 733-8372
www.visitnc.com
Construction Conditions Both
511
(877) 511-4662
www.ncdot.org/traffictravel
Plan a North Carolina trip at go.randmcnally.com/NC
© Rand McNally

Index of cities Pg. 253 | Map legend Pg. 7

| Mileages between cities | Asheville | Boone | Charlotte | Durham | Elizabeth City | Greensboro | Hickory | Morehead City | Murphy | Nags Head | New Bern | Raleigh | Roanoke Rapids | Rockingham | Wilmington | Winston-Salem |
|---|---|---|---|---|---|---|---|---|---|---|---|---|---|---|---|---|
| Greensboro | 172 | 113 | 93 | 53 | 241 | | 98 | 223 | 279 | 271 | 188 | 80 | 138 | 83 | 207 | 29 |
| Greenville | 332 | 273 | 250 | 101 | 97 | 156 | 258 | 79 | 440 | 129 | 44 | 82 | 86 | 176 | 116 | 188 |



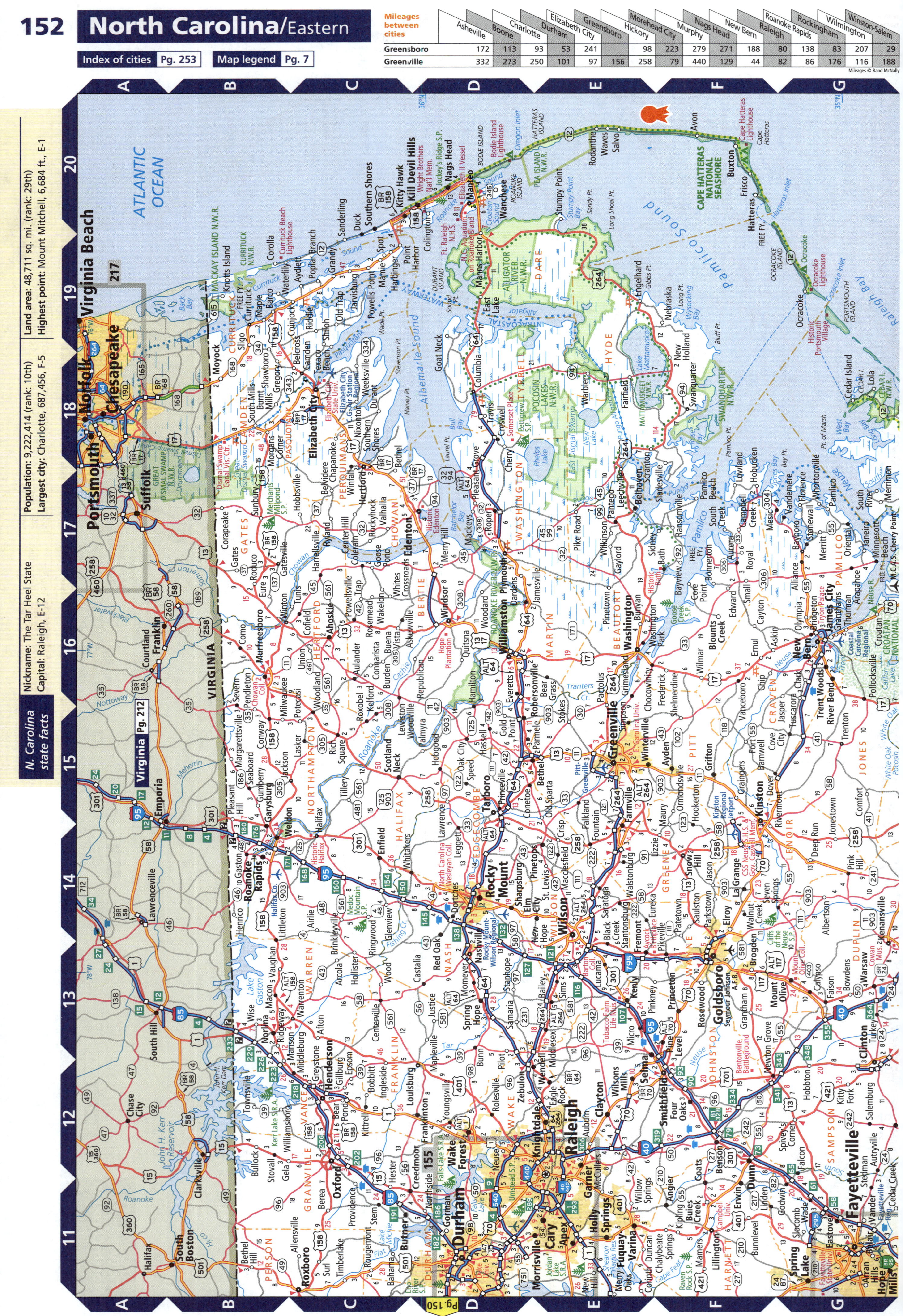

**Mileages between cities**

| | Asheville | Boone | Charlotte | Durham | Elizabeth City | Greensboro | Hickory | Morehead City | Murphy | Nags Head | New Bern | Raleigh | Roanoke Rapids | Rockingham | Wilmington | Winston-Salem |
|---|---|---|---|---|---|---|---|---|---|---|---|---|---|---|---|---|
| Raleigh | 251 | 192 | 168 | 22 | 164 | 80 | 177 | 146 | 358 | 195 | 111 | | 89 | 98 | 130 | 107 |
| Wilmington | 327 | 319 | 197 | 156 | 208 | 207 | 259 | 91 | 428 | 230 | 90 | 130 | 178 | 127 | | 236 |



40 419 miles · 85 233 miles · 77 102 miles · 95 182 miles

Total mileages through North Carolina
More mileages at go.randmcnally.com/MC

## City sights to see

- Dollywood, Pigeon Forge, Tenn., B-11
- Discovery Place, Charlotte, H-4
- Duke Homestead State Historic Site, Durham, F-9
- Guilford Courthouse National Military Park, Greensboro, B-6
- Historic Bethabara Park, Winston-Salem, A-1
- Mint Museum of Art, Charlotte, H-5

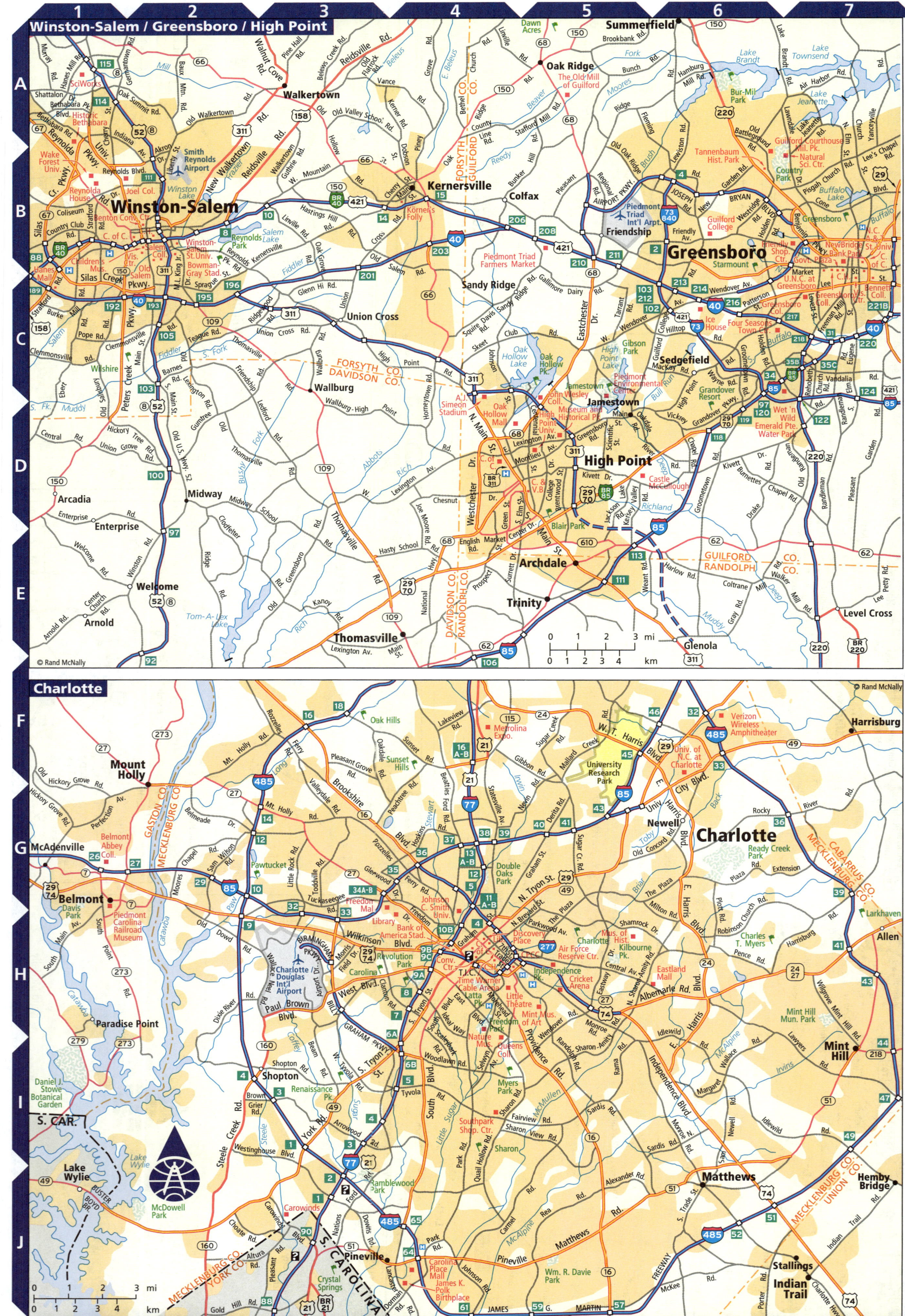

- Morehead Planetarium and Science Center, Chapel Hill, H-8
- North Carolina Museum of Life and Science, Durham, F-10
- North Carolina Museum of History, Raleigh, I-12
- North Carolina State Capitol, Raleigh, I-13
- Old Salem, Winston-Salem, B-2
- Reynolda House, Winston-Salem, B-1

Index of cities Pg. 254 | Map legend Pg. 7

**Mileages between cities**

| | Bismarck | Bowman | Fargo | Garrison | Grand Forks | Jamestown | Williston | Winnepeg, MB |
|---|---|---|---|---|---|---|---|---|
| Bismarck | | 174 | 195 | 75 | 272 | 102 | 228 | 413 |
| Devils Lake | 180 | 354 | 165 | 167 | 89 | 99 | 245 | 230 |
| Dickinson | 97 | 78 | 292 | 149 | 368 | 198 | 132 | 509 |
| Fargo | 195 | 368 | | 266 | 80 | 94 | 422 | 222 |



**N. Dakota state facts**

Nickname: The Peace Garden State
Capital: Bismarck, H-7

Population: 641,481 (rank: 48th)
Largest city: Fargo, 93,531, H-13

Land area: 68,976 sq. mi. (rank: 17th)
Highest point: White Butte, 3,506 ft., I-3

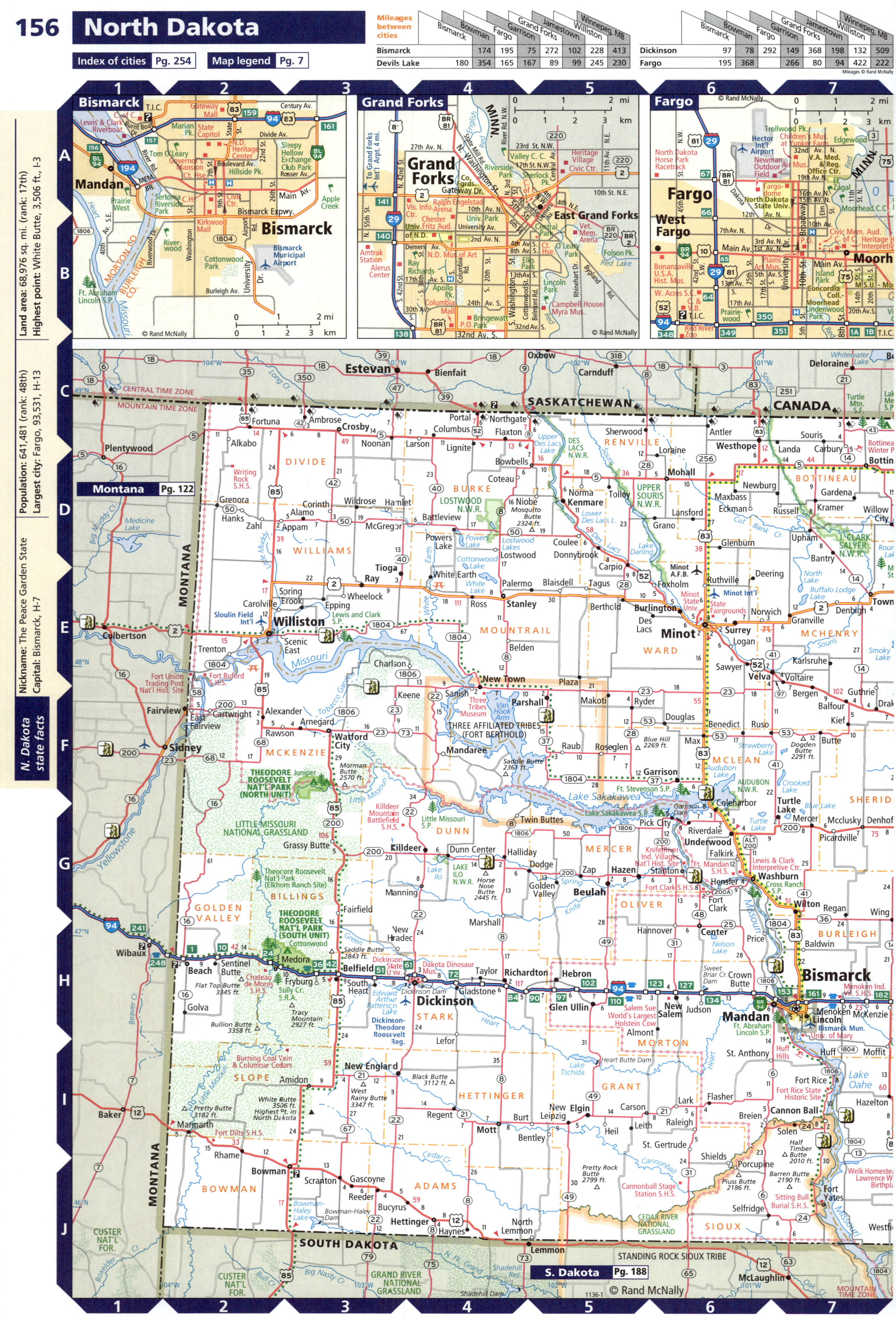

Mileages between cities

| | Bismarck | Bowman | Fargo | Garrison | Grand Forks | Jamestown | Williston | Winnepeg, MB |
|---|---|---|---|---|---|---|---|---|
| Grand Forks | 272 | 444 | 80 | 256 | | 171 | 334 | 146 |
| Minot | 110 | 260 | 268 | 47 | 210 | 170 | 124 | 299 |

| | Bismarck | Bowman | Fargo | Garrison | Grand Forks | Jamestown | Williston | Winnepeg, MB |
|---|---|---|---|---|---|---|---|---|
| Wahpeton | 243 | 416 | 54 | 315 | 131 | 142 | 470 | 273 |
| Williston | 228 | 170 | 422 | 144 | 334 | 293 | | 424 |



29 218 miles 2 359 miles
94 352 miles 83 265 miles

Total mileages through North Dakota
More mileages at go.randmcnally.com/MC

One inch represents approx. 22 miles
0 10 20 30 mi
0 10 20 30 40 km

Construction 511 (866) 696-3511 www.dot.nd.gov/travel-info/travel-info.htm
Conditions
Both

Tourism Information North Dakota Tourism Division (800) 435-5663 www.ndtourism.com

Plan a North Dakota trip at go.randmcnally.com/ND

Minnesota Pg. 108

MANITOBA
MINNESOTA
SOUTH DAKOTA
Winnipeg
Grand Forks
Fargo
Jamestown
Devils Lake
Valley City
Moorhead
Wahpeton

Index of cities Pg. 254 | Map legend Pg. 7

Mileages between cities

| | Akron | Ashtabula | Canton | Cincinnati | Cleveland | Columbus | Coshocton | Findlay | Lima | Mansfield | New Philadelphia | Pittsburgh, PA | Sandusky | Steubenville | Toledo | Youngstown |
|---|---|---|---|---|---|---|---|---|---|---|---|---|---|---|---|---|
| Akron | | 81 | 20 | 232 | 39 | 124 | 80 | 132 | 154 | 62 | 47 | 107 | 85 | 82 | 133 | 48 |
| Cleveland | 39 | 58 | 58 | 248 | | 142 | 102 | 121 | 156 | 80 | 85 | 131 | 62 | 124 | 111 | 72 |

Mileages © Rand McNally

**Mileages between cities**

| | Akron | Ashtabula | Canton | Cincinnati | Cleveland | Columbus | Coshocton | Findlay | Lima | Mansfield | New Philadelphia | Pittsburgh, PA | Sandusky | Steubenville | Toledo | Youngstown |
|---|---|---|---|---|---|---|---|---|---|---|---|---|---|---|---|---|
| Columbus | 124 | 194 | 126 | 106 | 142 | | 71 | 96 | 91 | 66 | 118 | 184 | 112 | 150 | 142 | 172 |
| Defiance | 180 | 214 | 185 | 169 | 157 | 135 | 177 | 51 | 44 | 123 | 190 | 274 | 98 | 246 | 57 | 214 |

continued p. 160

Mileages © Rand McNally

71 248 miles 80 237 miles
75 211 miles 90 245 miles

Total mileages through Ohio
More mileages at go.randmcnally.com/MC

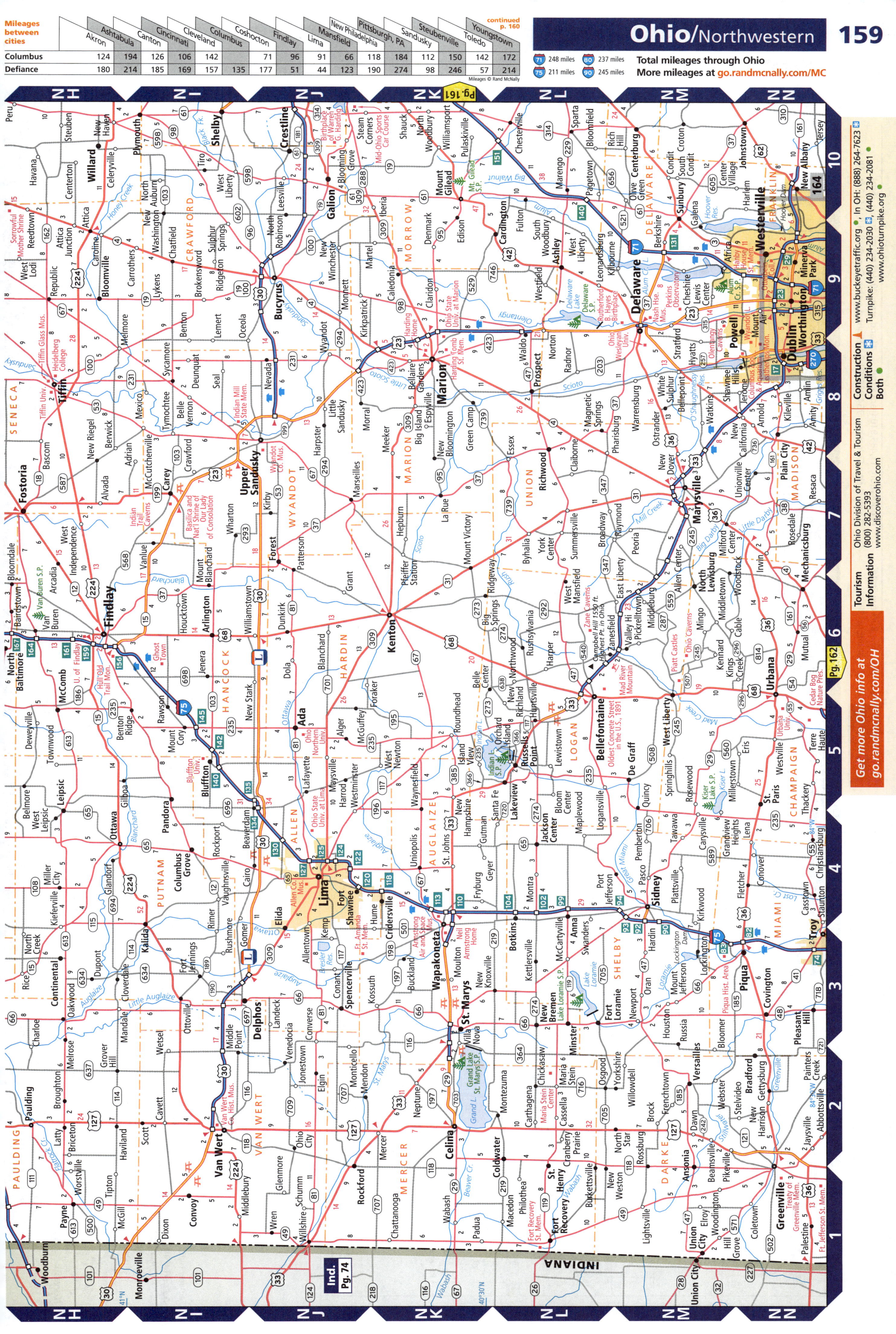

Index of cities Pg. 254 | Map legend Pg. 7

**Mileages between cities**

| | Akron | Ashtabula | Canton | Cincinnati | Cleveland | Columbus | Coshocton | Findlay | Lima | Mansfield | New Philadelphia | Pittsburgh, PA | Sandusky | Steubenville | Toledo | Youngstown |
|---|---|---|---|---|---|---|---|---|---|---|---|---|---|---|---|---|
| Lima | 154 | 216 | 156 | 124 | 156 | 91 | 134 | 34 | | 94 | 162 | 261 | 96 | 217 | 77 | 202 |
| Mansfield | 62 | 132 | 64 | 172 | 80 | 66 | 62 | 72 | 94 | | 67 | 170 | 53 | 124 | 99 | 110 |



**Ohio state facts**

Nickname: The Buckeye State
Capital: Columbus, SB-9
Population: 11,485,910 (rank: 7th)
Largest city: Columbus, 754,885, SB-9
Land area: 40,948 sq. mi. (rank: 35th)
Highest point: Campbell Hill, 1,550 ft., NL-6

Mileages between cities

| | Akron | Ashtabula | Canton | Cincinnati | Cleveland | Columbus | Coshocton | Findlay | Lima | Mansfield | New Philadelphia | Pittsburgh, PA | Sandusky | Steubenville | Toledo | Youngstown |
|---|---|---|---|---|---|---|---|---|---|---|---|---|---|---|---|---|
| Toledo | 133 | 171 | 152 | 200 | 111 | 142 | 152 | 44 | 77 | 99 | 179 | 228 | 58 | 221 | | 169 |
| Youngstown | 48 | 57 | 57 | 279 | 72 | 172 | 117 | 180 | 202 | 110 | 84 | 67 | 122 | 66 | 169 | |



71 248 miles · 75 211 miles · 80 237 miles · 90 245 miles

Total mileages through Ohio
More mileages at go.randmcnally.com/MC

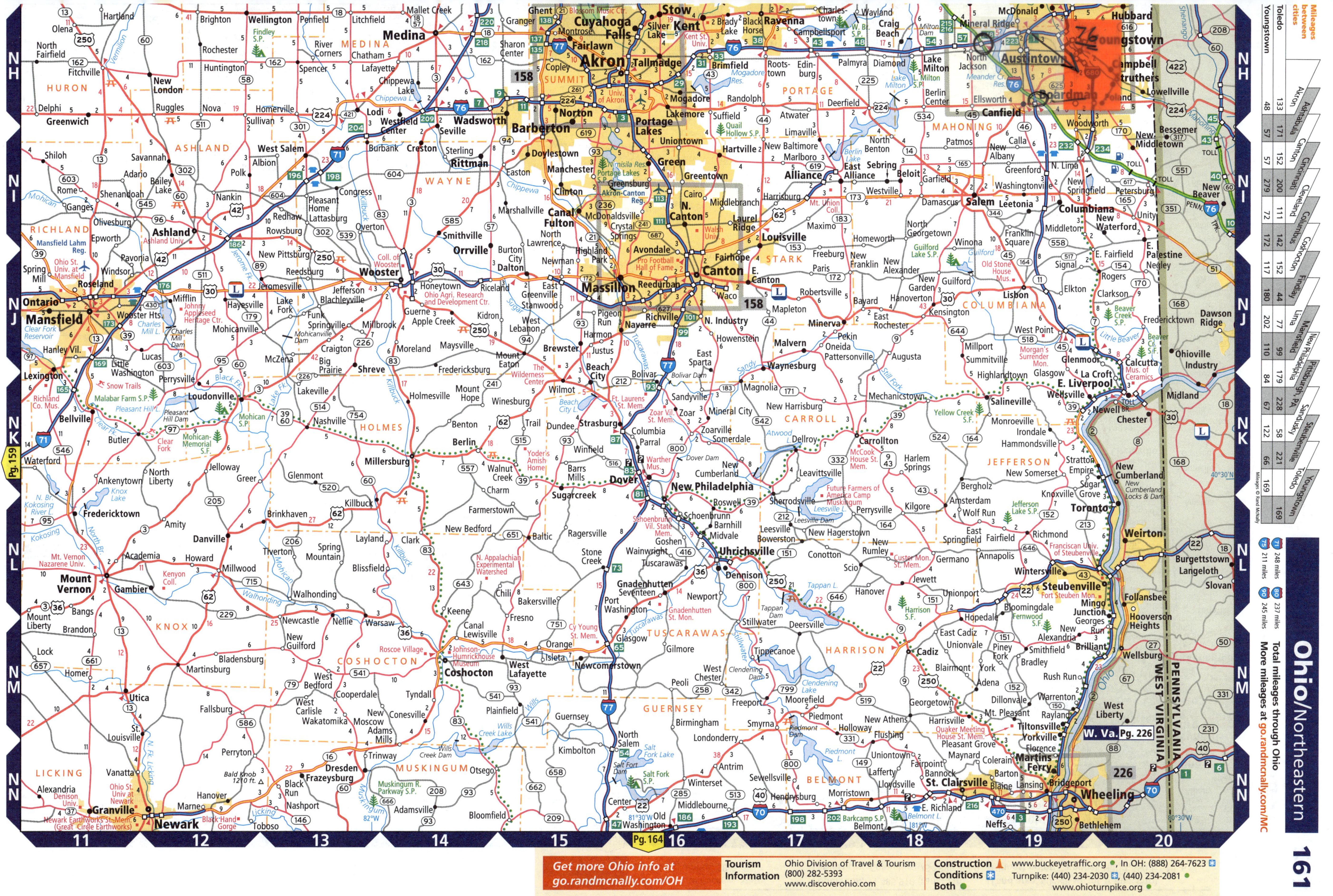


Get more Ohio info at go.randmcnally.com/OH

Tourism Information
Ohio Division of Travel & Tourism
(800) 282-5393
www.discoverohio.com

Construction
Conditions
Both

www.buckeyetraffic.org, In OH: (888) 264-7623
Turnpike: (440) 234-2030, (440) 234-2081
www.ohioturnpike.org

Index of cities Pg. 254
Map legend Pg. 7

Mileages between cities

| | Athens | Cambridge | Chillicothe | Cincinnati | Cleveland | Columbus | Dayton | Gallipolis | Huntington, WV | Lancaster | Marietta | Maysville, KY | Portsmouth | Wheeling, WV | Wilmington | Zanesville |
|---|---|---|---|---|---|---|---|---|---|---|---|---|---|---|---|---|
| Cincinnati | 160 | 183 | 106 | | 248 | 106 | 50 | 153 | 148 | 133 | 210 | 61 | 110 | 230 | 51 | 158 |
| Columbus | 74 | 79 | 47 | 106 | 142 | | 71 | 106 | 137 | 30 | 124 | 112 | 91 | 126 | 62 | 55 |



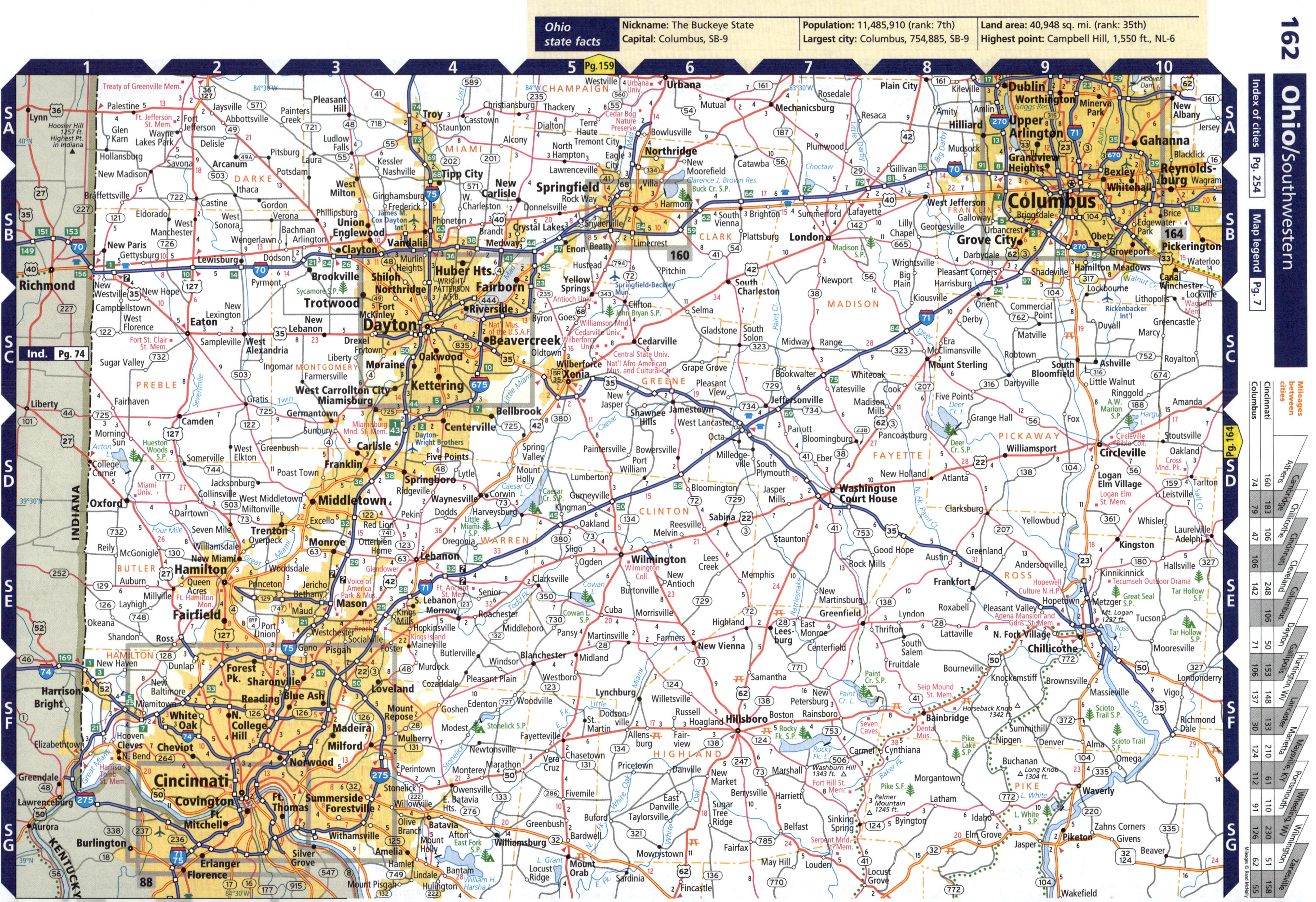

**Mileages between cities**

| | Athens | Cambridge | Chillicothe | Cincinnati | Cleveland | Columbus | Dayton | Gallipolis | Huntington, WV | Lancaster | Marietta | Maysville, KY | Portsmouth | Wheeling, WV | Wilmington | Zanesville |
|---|---|---|---|---|---|---|---|---|---|---|---|---|---|---|---|---|
| **Dayton** | 134 | 149 | 77 | 50 | 212 | 71 | | 137 | 168 | 101 | 195 | 108 | 122 | 197 | 34 | 126 |
| **Gallipolis** | 42 | 114 | 60 | 153 | 235 | 106 | 137 | | 39 | 86 | 66 | 111 | 55 | 162 | 112 | 94 |

continued p. 164



70 225 miles 75 211 miles
71 248 miles 77 160 miles

**Total mileages through Ohio**
**More mileages at go.randmcnally.com/MC**

Pg. 165

Kentucky Pg. 86

One inch represents approximately 9 miles

**Central Cincinnati**

**Cincinnati**

**Dayton**

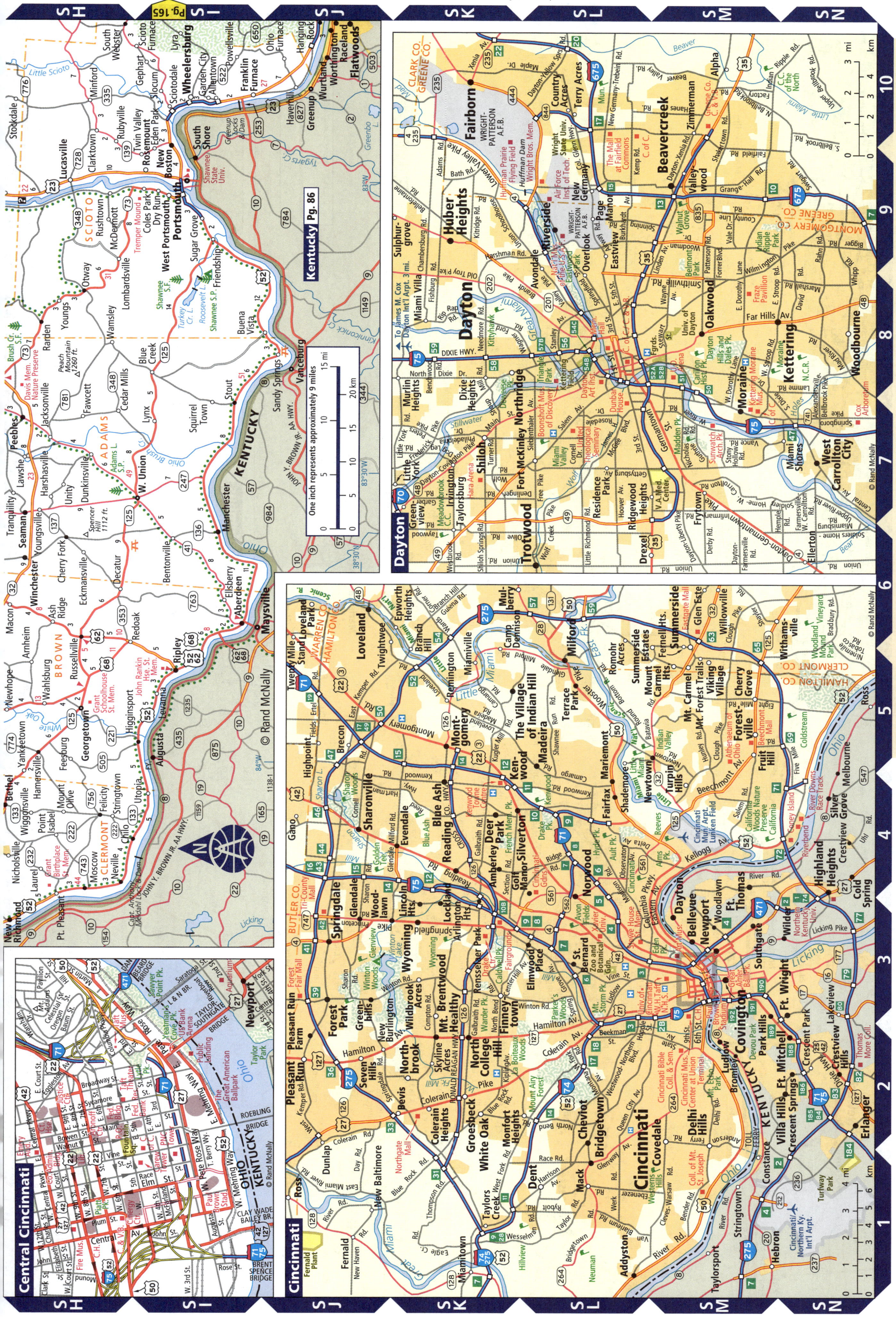

*Explore Ohio at go.randmcnally.com/OH*

| | |
|---|---|
| **Tourism Information** | Ohio Division of Travel & Tourism (800) 282-5393 www.discoverohio.com |
| **Construction** / **Conditions** / **Both** | www.buckeyetraffic.org, In OH: (888) 264-7623; Cincinnati/northern Kentucky area: 511, (513) 333-3333, www.artimis.org |

Index of cities Pg. 254 | Map legend Pg. 7

| Mileages between cities | Athens | Cambridge | Chillicothe | Cincinnati | Cleveland | Columbus | Dayton | Gallipolis | Huntington, WV | Lancaster | Marietta | Maysville, KY | Portsmouth | Wheeling, WV | Wilmington | Zanesville |
|---|---|---|---|---|---|---|---|---|---|---|---|---|---|---|---|---|
| Marietta | 44 | 48 | 104 | 210 | 164 | 124 | 195 | 66 | 106 | 82 | | 165 | 128 | 90 | 156 | 69 |
| Portsmouth | 81 | 162 | 44 | 110 | 233 | 91 | 122 | 55 | 46 | 80 | 128 | 52 | | 210 | 79 | 138 |

Mileages © Rand McNally

**Ohio state facts**

Nickname: The Buckeye State
Capital: Columbus, SB-9

Population: 11,485,910 (rank: 7th)
Largest city: Columbus, 754,885, SB-9

Land area: 40,948 sq. mi. (rank: 35th)
Highest point: Campbell Hill, 1,550 ft., NL-6

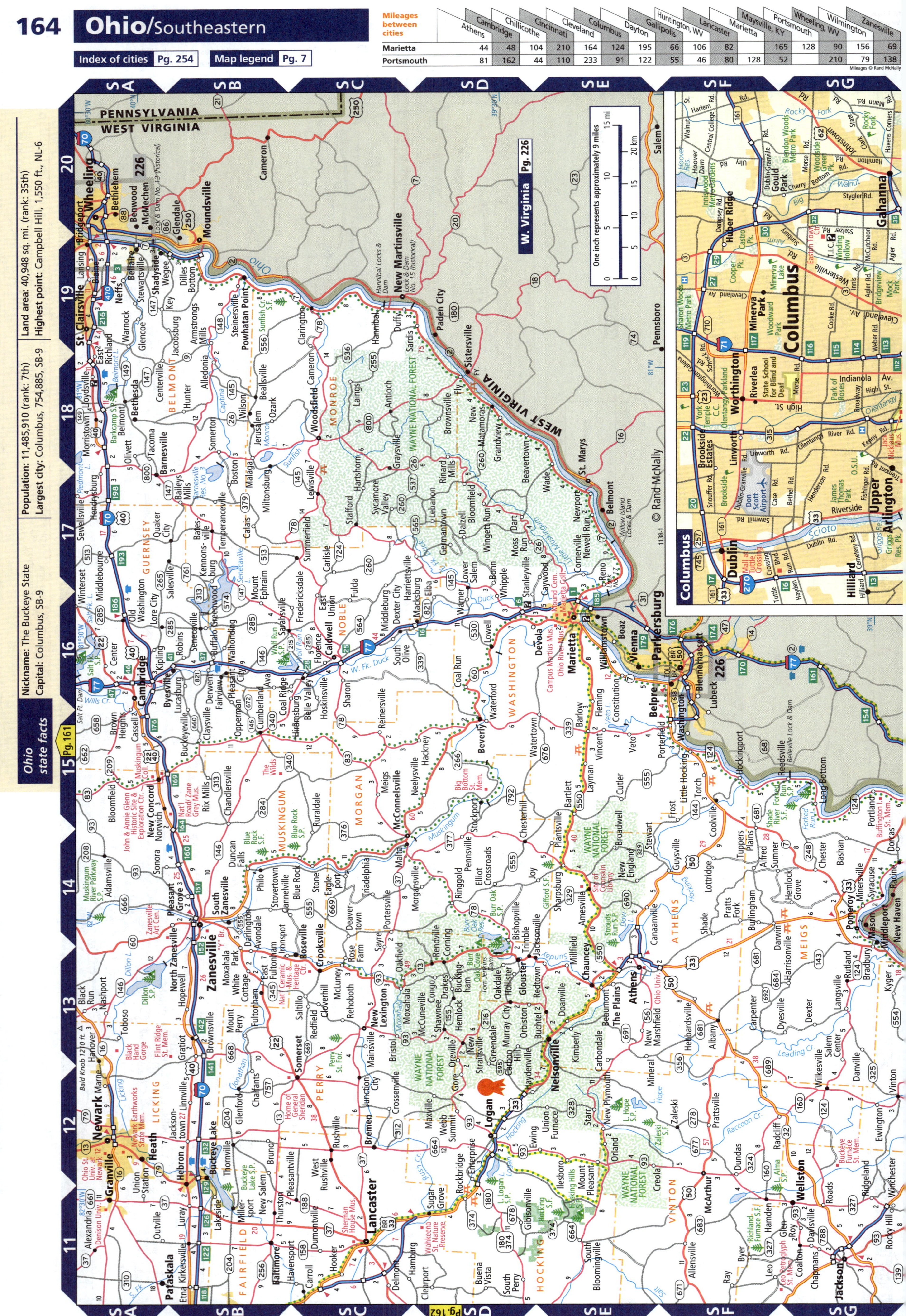

**Mileages between cities**

| | Athens | Cambridge | Chillicothe | Cincinnati | Cleveland | Columbus | Dayton | Gallipolis | Huntington, WV | Lancaster | Marietta | Maysville, KY | Portsmouth | Wheeling, WV | Wilmington | Zanesville |
|---|---|---|---|---|---|---|---|---|---|---|---|---|---|---|---|---|
| **Springfield** | 118 | 123 | 69 | 77 | 185 | 45 | 27 | 129 | 160 | 74 | 168 | 102 | 114 | 171 | 38 | 99 |
| **Zanesville** | 52 | 24 | 94 | 158 | 145 | 55 | 126 | 94 | 134 | 45 | 69 | 164 | 138 | 72 | 114 | |

Mileages © Rand McNally

70 225 miles | 75 211 miles | 71 248 miles | 77 160 miles

**Total mileages through Ohio**
**More mileages at go.randmcnally.com/MC**

Index of cities Pg. 256 | Map legend Pg. 7

**Mileages between cities**

| | Ardmore | Bartlesville | Dallas, TX | Elk City | Enid | Ft. Smith, AR | Guymon | Joplin, MO | Lawton | McAlester | Muskogee | Oklahoma City | Ponca City | Tulsa | Wichita Falls, TX | Woodward |
|---|---|---|---|---|---|---|---|---|---|---|---|---|---|---|---|---|
| Ardmore | | 246 | 109 | 208 | 195 | 223 | 360 | 312 | 99 | 116 | 180 | 97 | 200 | 201 | 86 | 236 |
| Elk City | 208 | 260 | 303 | | 148 | 292 | 184 | 327 | 108 | 240 | 249 | 112 | 216 | 215 | 143 | 77 |



**Oklahoma state facts**

Nickname: The Sooner State
Capital: Oklahoma City, F-13
Population: 3,642,361 (rank: 28th)
Largest city: Oklahoma City, 551,789, F-13
Land area: 68,667 sq. mi. (rank: 19th)
Highest point: Black Mesa, 4,973 ft., B-1

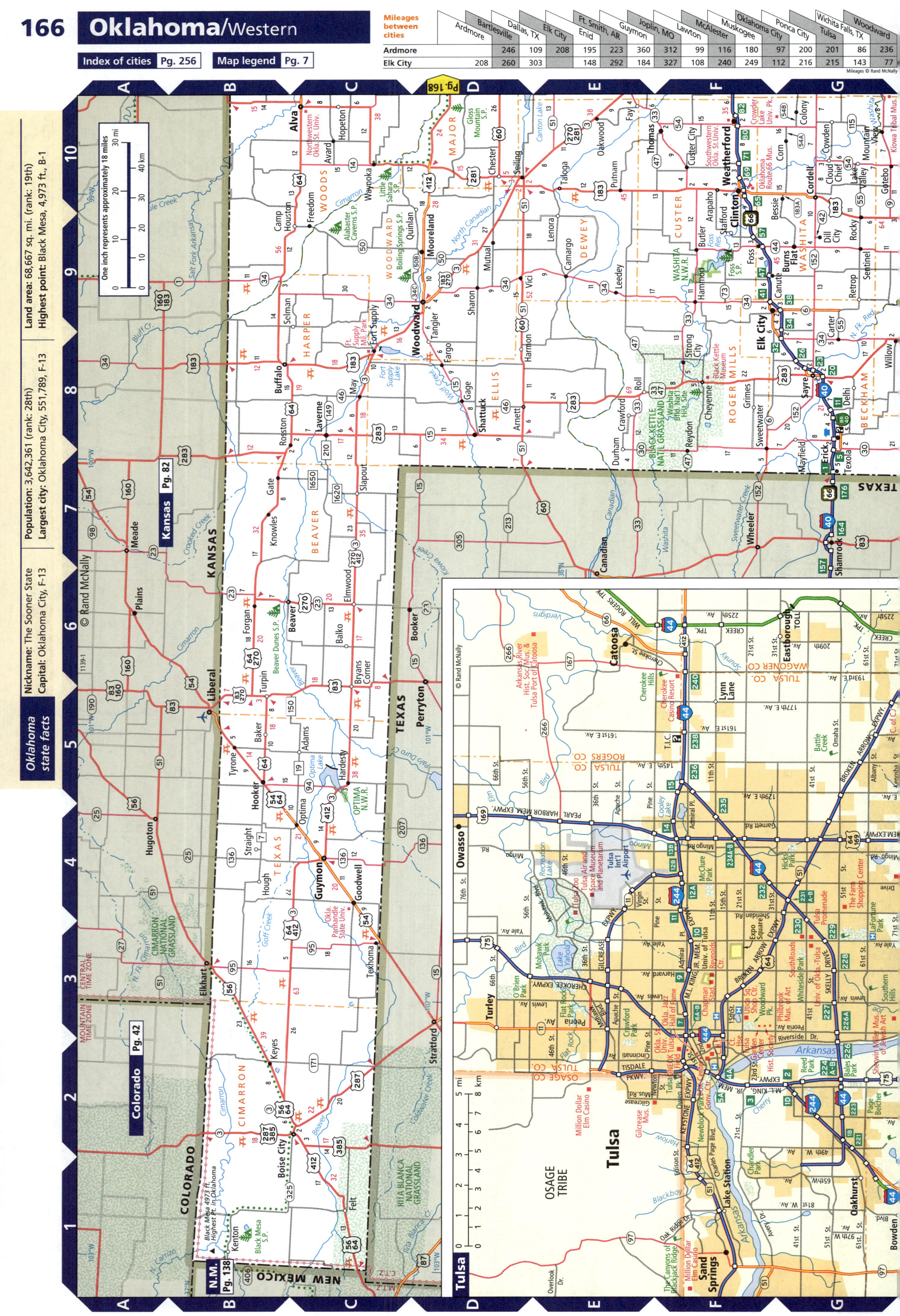

**Mileages between cities**

| | Ardmore | Bartlesville | Dallas, TX | Elk City | Enid | Ft. Smith, AR | Guymon | Joplin, MO | Lawton | McAlester | Muskogee | Oklahoma City | Ponca City | Tulsa | Wichita Falls, TX | Woodward |
|---|---|---|---|---|---|---|---|---|---|---|---|---|---|---|---|---|
| **Enid** | 195 | 134 | 302 | 148 | | 232 | 211 | 227 | 142 | 204 | 164 | 99 | 67 | 114 | 196 | 87 |
| **Guymon** | 360 | 344 | 459 | 184 | 211 | 443 | | 438 | 294 | 391 | 375 | 263 | 278 | 326 | 317 | 124 |




continued p. 168


35: 236 miles
44: 329 miles
40: 331 miles
75: 227 miles

**Total mileages through Oklahoma**
**More mileages at go.randmcnally.com/MC**

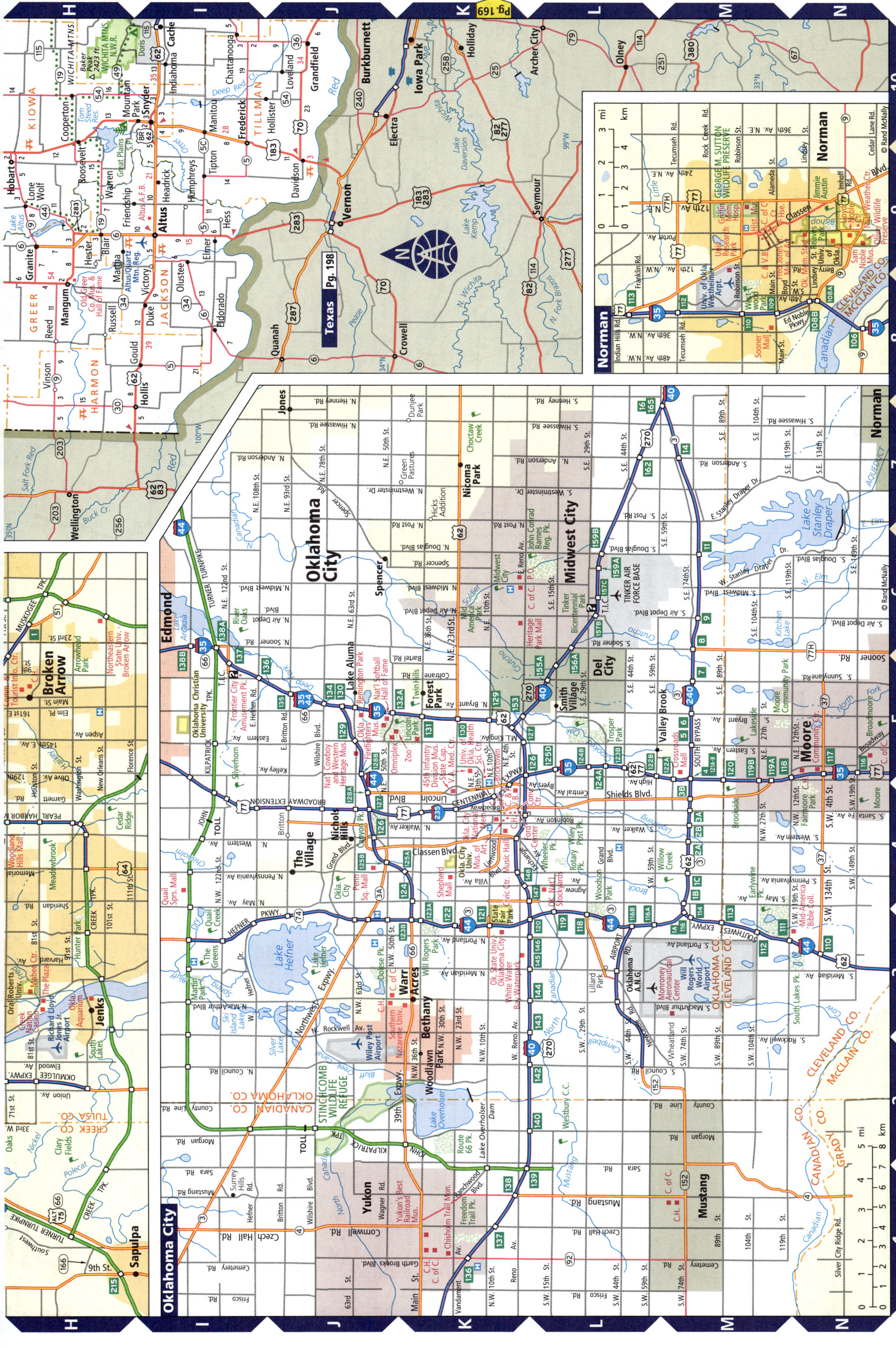

Plan an Oklahoma trip at go.randmcnally.com/OK

**Tourism Information**: Oklahoma Tourism & Recreation Department (800) 652-6552 www.travelok.com

**Construction / Conditions / Both**: (405) 425-2385 www.okladot.state.ok.us In OK: (888) 425-2385

Index of cities Pg. 256 | Map legend Pg. 7

| Mileages between cities | Ardmore | Bartlesville | Dallas, TX | Elk City | Enid | Ft. Smith, AR | Guymon | Joplin, MO | Lawton | McAlester | Muskogee | Oklahoma City | Ponca City | Tulsa | Wichita Falls, TX | Woodward |
|---|---|---|---|---|---|---|---|---|---|---|---|---|---|---|---|---|
| Idabel | 149 | 248 | 171 | 352 | 316 | 136 | 504 | 295 | 245 | 116 | 180 | 240 | 293 | 203 | 238 | 380 |
| Muskogee | 180 | 91 | 236 | 249 | 164 | 70 | 375 | 117 | 218 | 65 | | 137 | 142 | 50 | 272 | 251 |



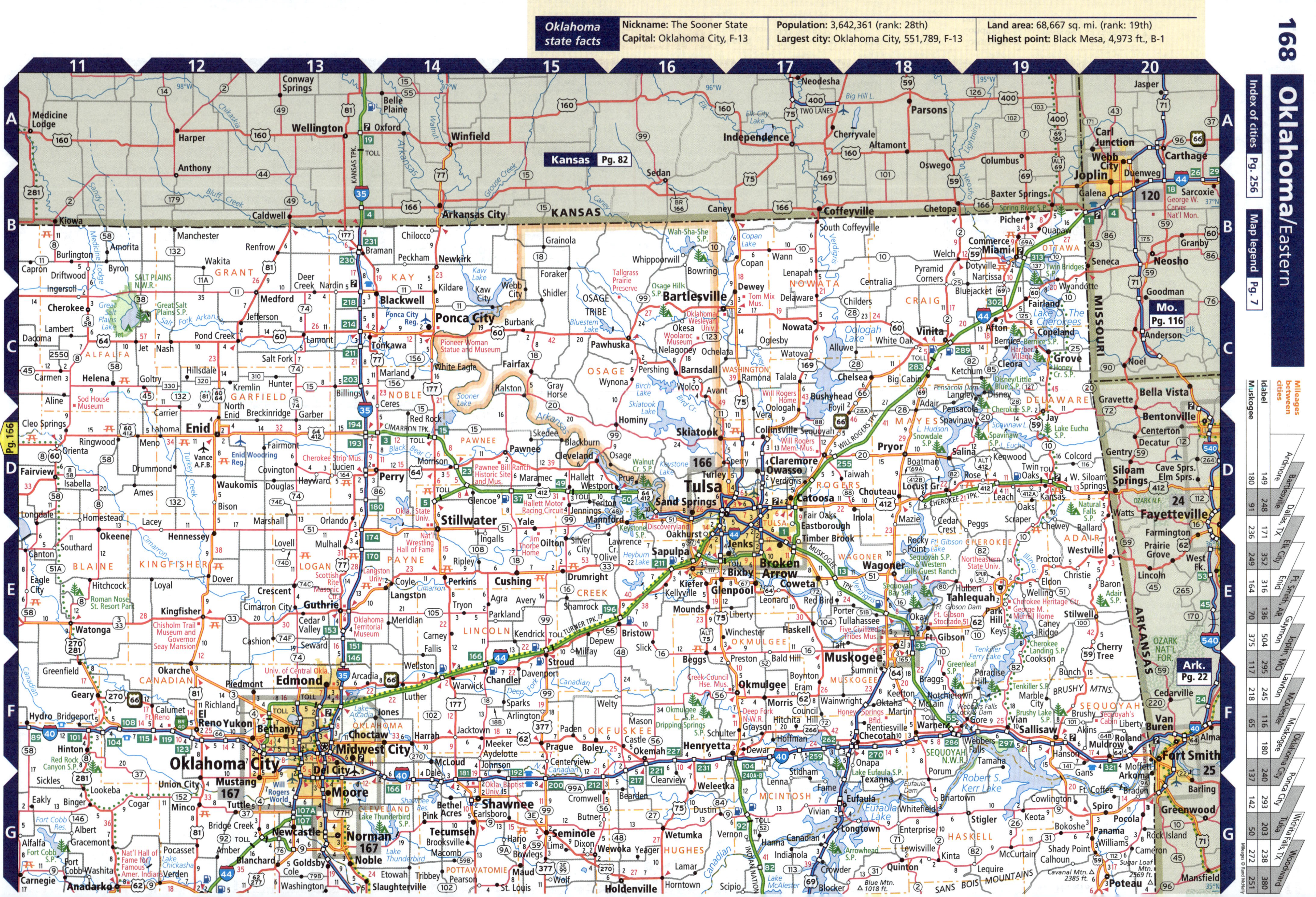

Mileages between cities

| | Ardmore | Bartlesville | Dallas, TX | Elk City | Enid | Ft. Smith, AR | Guymon | Joplin, MO | Lawton | McAlester | Muskogee | Oklahoma City | Ponca City | Tulsa | Wichita Falls, TX | Woodward |
|---|---|---|---|---|---|---|---|---|---|---|---|---|---|---|---|---|
| Oklahoma City | 97 | 149 | 204 | 112 | 99 | 180 | 263 | 216 | 86 | 128 | 137 | | 105 | 104 | 140 | 139 |
| Tulsa | 201 | 45 | 258 | 215 | 114 | 118 | 326 | 113 | 191 | 91 | 50 | 104 | 91 | | 244 | 202 |



35 236 miles 44 329 miles
40 331 miles 75 227 miles

Total mileages through Oklahoma
More mileages at go.randmcnally.com/MC

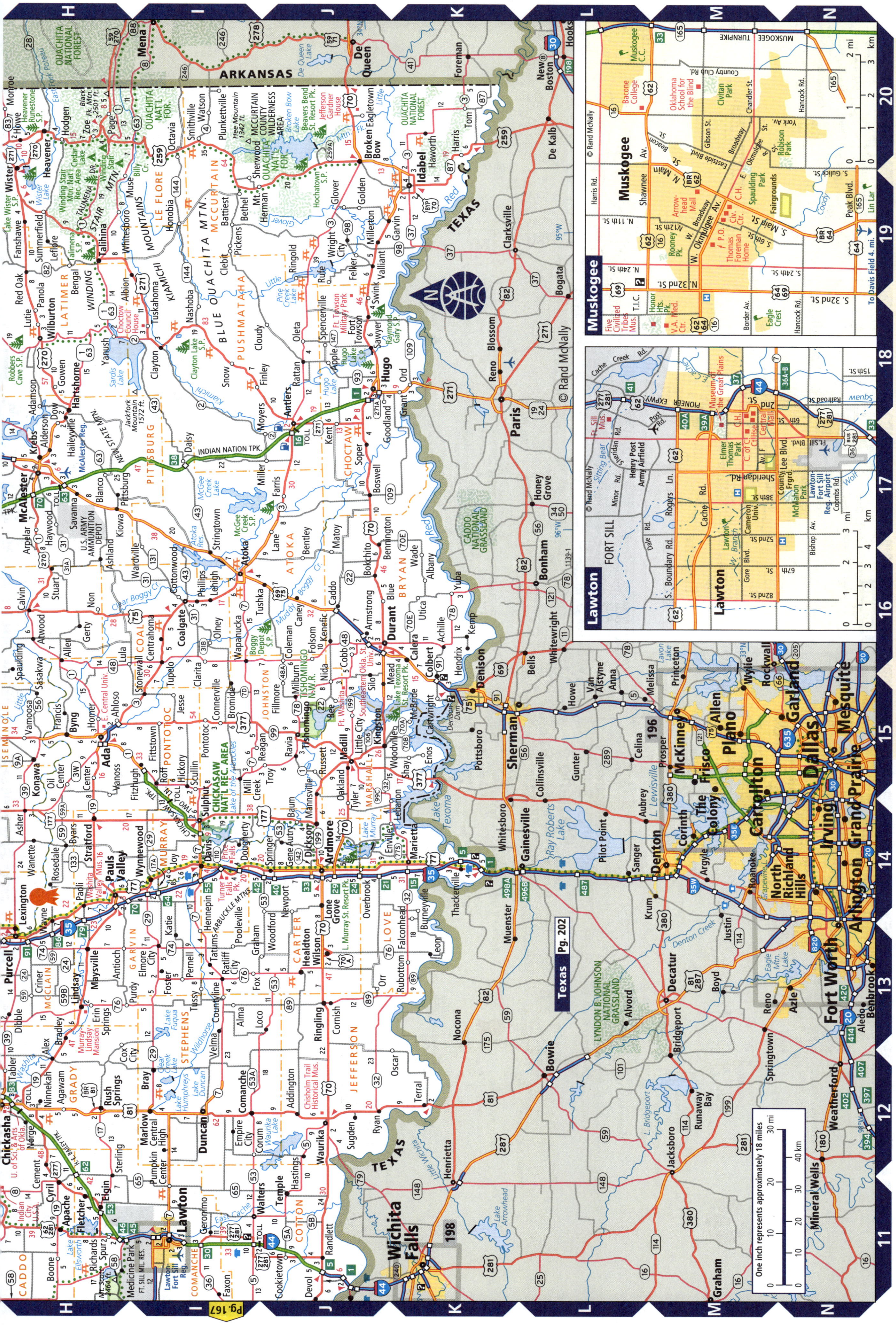

Plan an Oklahoma trip at go.randmcnally.com/OK

Tourism Information: Oklahoma Tourism & Recreation Department (800) 652-6552 www.travelok.com

Construction Conditions: (405) 425-2385 www.okladot.state.ok.us

Both: In OK: (888) 425-2385

One inch represents approximately 18 miles

Index of cities Pg. 256 | Map legend Pg. 7

| Mileages between cities | Astoria | Bend | Brookings | Burns | Coos Bay | Crater Lake N.P. | Eugene | Gov't Camp | John Day | Lakeview | Medford | Ontario | Pendleton | Portland | Salem | The Dalles |
|---|---|---|---|---|---|---|---|---|---|---|---|---|---|---|---|---|
| Bend | 250 | | 287 | 130 | 228 | 107 | 115 | 106 | 151 | 175 | 172 | 260 | 242 | 161 | 131 | 129 |
| Corvallis | 166 | 127 | 280 | 257 | 132 | 187 | 47 | 126 | 260 | 284 | 210 | 387 | 290 | 82 | 37 | 165 |



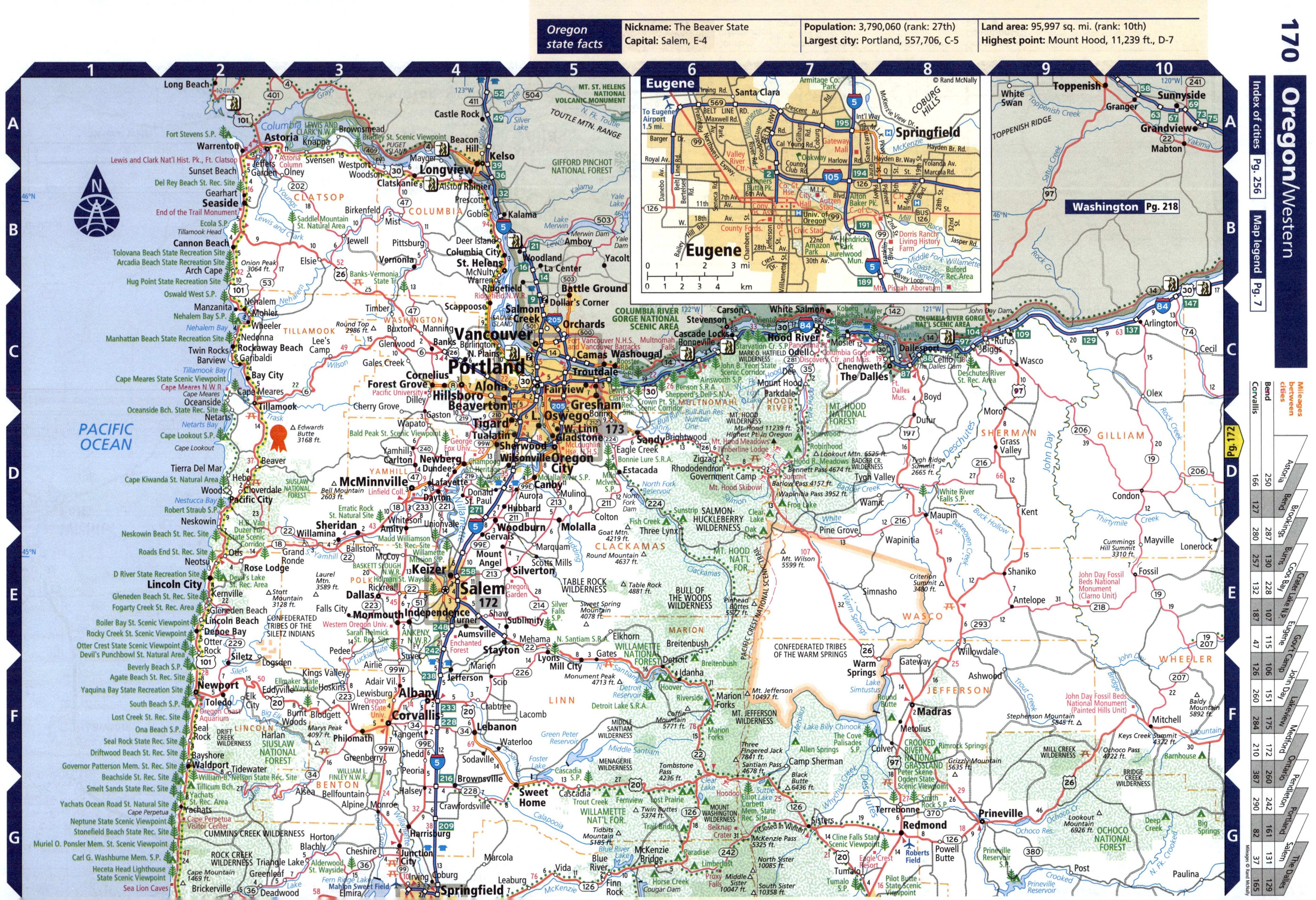

Mileages between cities

| | Astoria | Bend | Brookings | Burns | Coos Bay | Crater Lake N.P. | Eugene | Gov't Camp | John Day | Lakeview | Medford | Ontario | Pendleton | Portland | Salem | The Dalles |
|---|---|---|---|---|---|---|---|---|---|---|---|---|---|---|---|---|
| Eugene | 193 | 115 | 234 | 245 | 109 | 142 | | 154 | 249 | 241 | 166 | 375 | 318 | 110 | 66 | 193 |
| McDermitt, NV | 525 | 277 | 525 | 147 | 505 | 356 | 392 | 380 | 218 | 222 | 400 | 187 | 354 | 436 | 408 | 405 |

continued p. 172

Mileages © Rand McNally

5 308 miles · 84 375 miles · 82 11 miles · 101 348 miles

Total mileages through Oregon
More mileages at go.randmcnally.com/MC

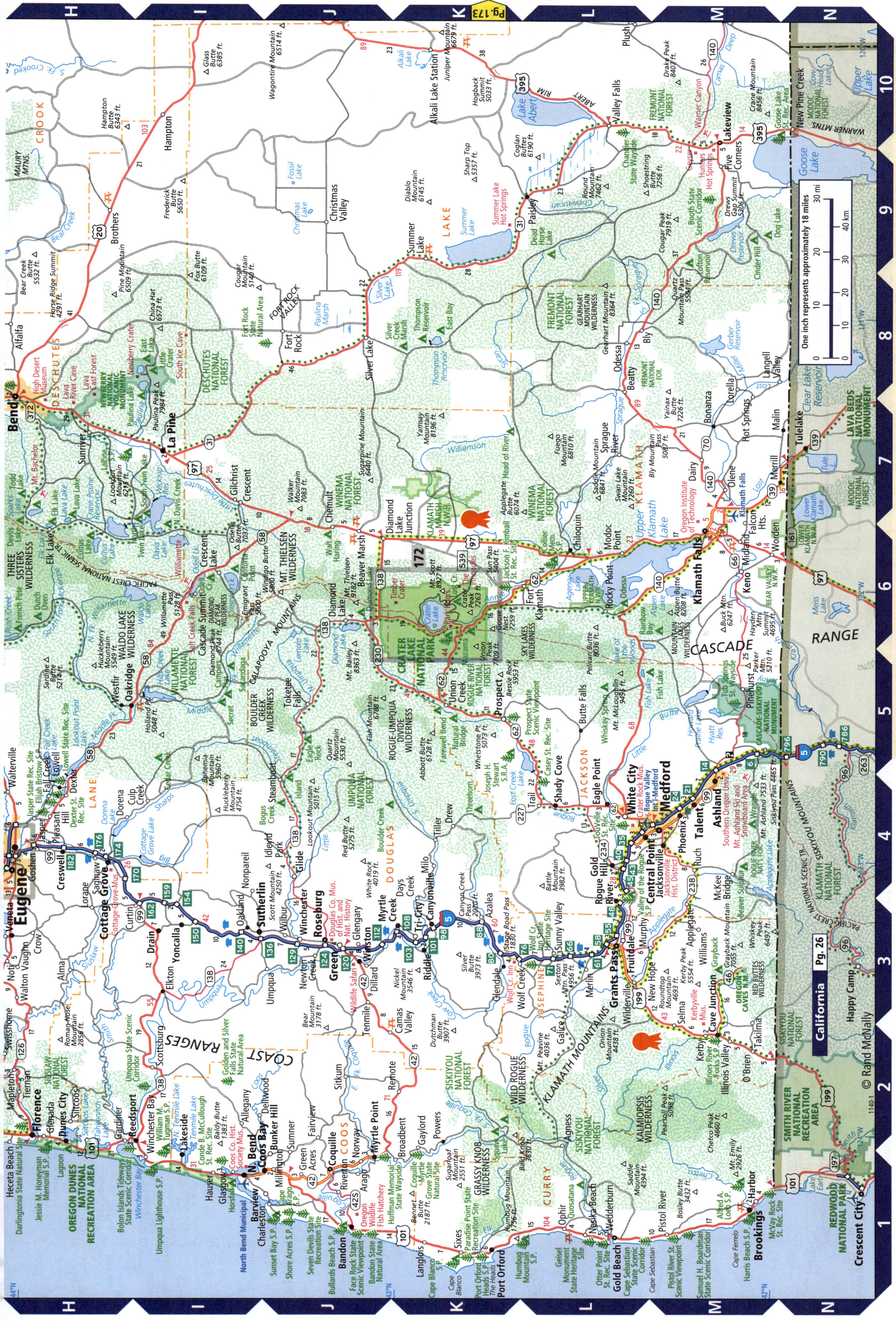

One inch represents approximately 18 miles

Pg. 173

Pg. 26

Get more Oregon info at go.randmcnally.com/OR

Tourism Information — Travel Oregon (800) 547-7842 www.traveloregon.com

Construction / Conditions / Both — 511 (800) 977-6368 (503) 588-2941 www.tripcheck.com

Index of cities Pg. 256 | Map legend Pg. 7

**Mileages between cities**

| | Astoria | Bend | Brookings | Burns | Coos Bay | Crater Lake N.P. | Eugene | Gov't Camp | John Day | Lakeview | Medford | Ontario | Pendleton | Portland | Salem | The Dalles |
|---|---|---|---|---|---|---|---|---|---|---|---|---|---|---|---|---|
| Medford | 356 | 172 | 125 | 305 | 169 | 74 | 166 | 317 | 328 | 171 | | 432 | 481 | 273 | 228 | 356 |
| Ontario | 464 | 260 | 547 | 130 | 488 | 367 | 375 | 354 | 131 | 269 | 432 | | 167 | 374 | 420 | 291 |



**Oregon state facts**

Nickname: The Beaver State
Capital: Salem, E-4
Population: 3,790,060 (rank: 27th)
Largest city: Portland, 557,706, C-5
Land area: 95,997 sq. mi. (rank: 10th)
Highest point: Mount Hood, 11,239 ft., D-7

Mileages between cities

| | Astoria | Bend | Brookings | Burns | Coos Bay | Crater Lake N.P. | Eugene | Gov't Camp | John Day | Lakeview | Medford | Ontario | Pendleton | Portland | Salem | The Dalles |
|---|---|---|---|---|---|---|---|---|---|---|---|---|---|---|---|---|
| Pendleton | 298 | 242 | 550 | 196 | 428 | 349 | 318 | 188 | 126 | 335 | 481 | 167 | | 208 | 254 | 125 |
| Portland | 96 | 161 | 342 | 281 | 220 | 250 | 110 | 55 | 265 | 336 | 273 | 374 | 208 | | 47 | 83 |

Mileages © Rand McNally

Total mileages through Oregon

| | | | |
|---|---|---|---|
| 5 | 308 miles | 84 | 375 miles |
| 82 | 11 miles | 101 | 348 miles |

More mileages at go.randmcnally.com/MC

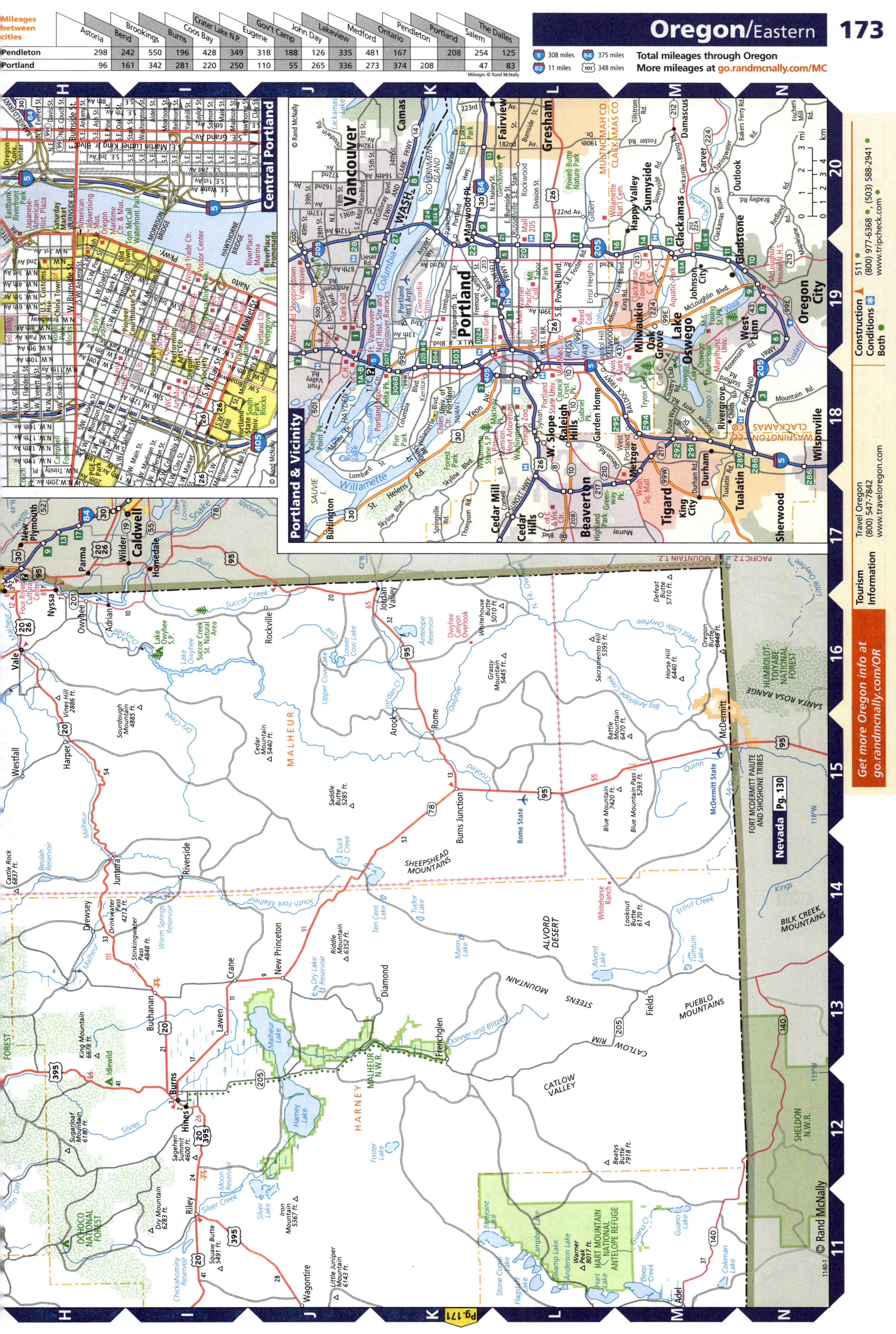

Index of cities Pg. 257 | Map legend Pg. 7

Mileages between cities

| | Altoona | Chambersburg | Cumberland, MD | Du Bois | Erie | Galeton | Harrisburg | Johnstown | Kittanning | Meadville | New Castle | Philadelphia | Pittsburgh | State College | Uniontown | Warren |
|---|---|---|---|---|---|---|---|---|---|---|---|---|---|---|---|---|
| Altoona | | 90 | 66 | 71 | 202 | 135 | 134 | 46 | 79 | 165 | 127 | 234 | 96 | 41 | 112 | 130 |
| Chambersburg | 90 | | 87 | 153 | 282 | 215 | 54 | 94 | 160 | 246 | 206 | 157 | 160 | 101 | 149 | 218 |

Mileages © Rand McNally

**Pennsylvania state facts**

Nickname: The Keystone State
Capital: Harrisburg, EN-5

Population: 12,448,279 (rank: 6th)
Largest city: Philadelphia, 1,447,395, EP-12

Land area: 44,817 sq. mi. (rank: 32nd)
Highest point: Mount Davis, 3,213 ft., WQ-7

**Mileages between cities**

continued p. 176

| | Altoona | Chambersburg | Cumberland, MD | Du Bois | Erie | Galeton | Harrisburg | Johnstown | Kittanning | Meadville | New Castle | Philadelphia | Pittsburgh | State College | Uniontown | Warren |
|---|---|---|---|---|---|---|---|---|---|---|---|---|---|---|---|---|
| Erie | 202 | 282 | 232 | 148 | | 159 | 297 | 177 | 123 | 41 | 88 | 419 | 127 | 208 | 184 | 66 |
| Johnstown | 46 | 94 | 70 | 77 | 177 | 179 | 137 | | 53 | 141 | 102 | 238 | 67 | 85 | 80 | 135 |



Total mileages through Pennsylvania
70: 168 miles; 79: 183 miles; 80: 311 miles; 90: 46 miles
More mileages at go.randmcnally.com/MC

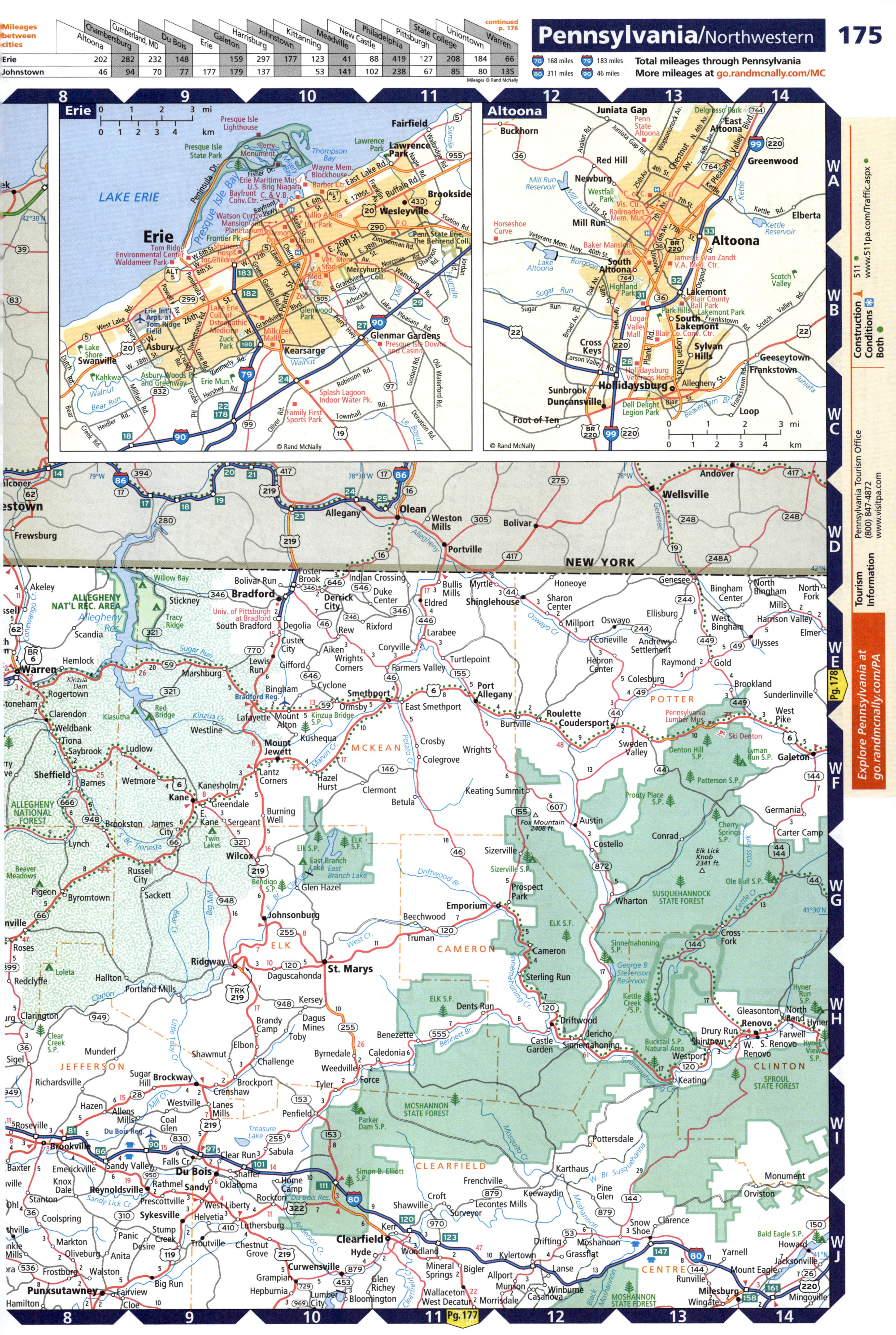


Tourism Information
Pennsylvania Tourism Office
(800) 847-4872
www.visitpa.com

Construction Conditions
Both
511
www.511pa.com/Traffic.aspx

Explore Pennsylvania at go.randmcnally.com/PA

Pg. 178
Pg. 177

Index of cities Pg. 257 | Map legend Pg. 7

Mileages between cities

| | Altoona | Chambersburg | Cumberland, MD | Du Bois | Erie | Galeton | Harrisburg | Johnstown | Kittanning | Meadville | New Castle | Philadelphia | Pittsburgh | State College | Uniontown | Warren |
|---|---|---|---|---|---|---|---|---|---|---|---|---|---|---|---|---|
| New Castle | 127 | 206 | 156 | 110 | 88 | 197 | 250 | 102 | 48 | 52 | | 350 | 52 | 171 | 108 | 120 |
| Pittsburgh | 96 | 160 | 111 | 101 | 127 | 200 | 203 | 67 | 42 | 91 | 52 | 304 | | 135 | 51 | 148 |



**Pennsylvania state facts**

Nickname: The Keystone State
Capital: Harrisburg, EN-5
Population: 12,448,279 (rank: 6th)
Largest city: Philadelphia, 1,447,395, EP-12
Land area: 44,817 sq. mi. (rank: 32nd)
Highest point: Mount Davis, 3,213 ft, WQ-7

Mileages between cities

| | Altoona | Chambersburg | Cumberland, MD | Du Bois | Erie | Galeton | Harrisburg | Johnstown | Kittanning | Meadville | New Castle | Philadelphia | Pittsburgh | State College | Uniontown | Warren |
|---|---|---|---|---|---|---|---|---|---|---|---|---|---|---|---|---|
| State College | 41 | 101 | 106 | 61 | 208 | 100 | 87 | 85 | 120 | 173 | 171 | 193 | 135 | | 152 | 119 |
| Williamsport | 100 | 132 | 166 | 110 | 257 | 72 | 83 | 146 | 168 | 220 | 219 | 176 | 196 | 63 | 212 | 171 |

Mileages © Rand McNally

Total mileages through Pennsylvania: 70 168 miles; 80 311 miles; 79 183 miles; 90 46 miles

More mileages at go.randmcnally.com/MC

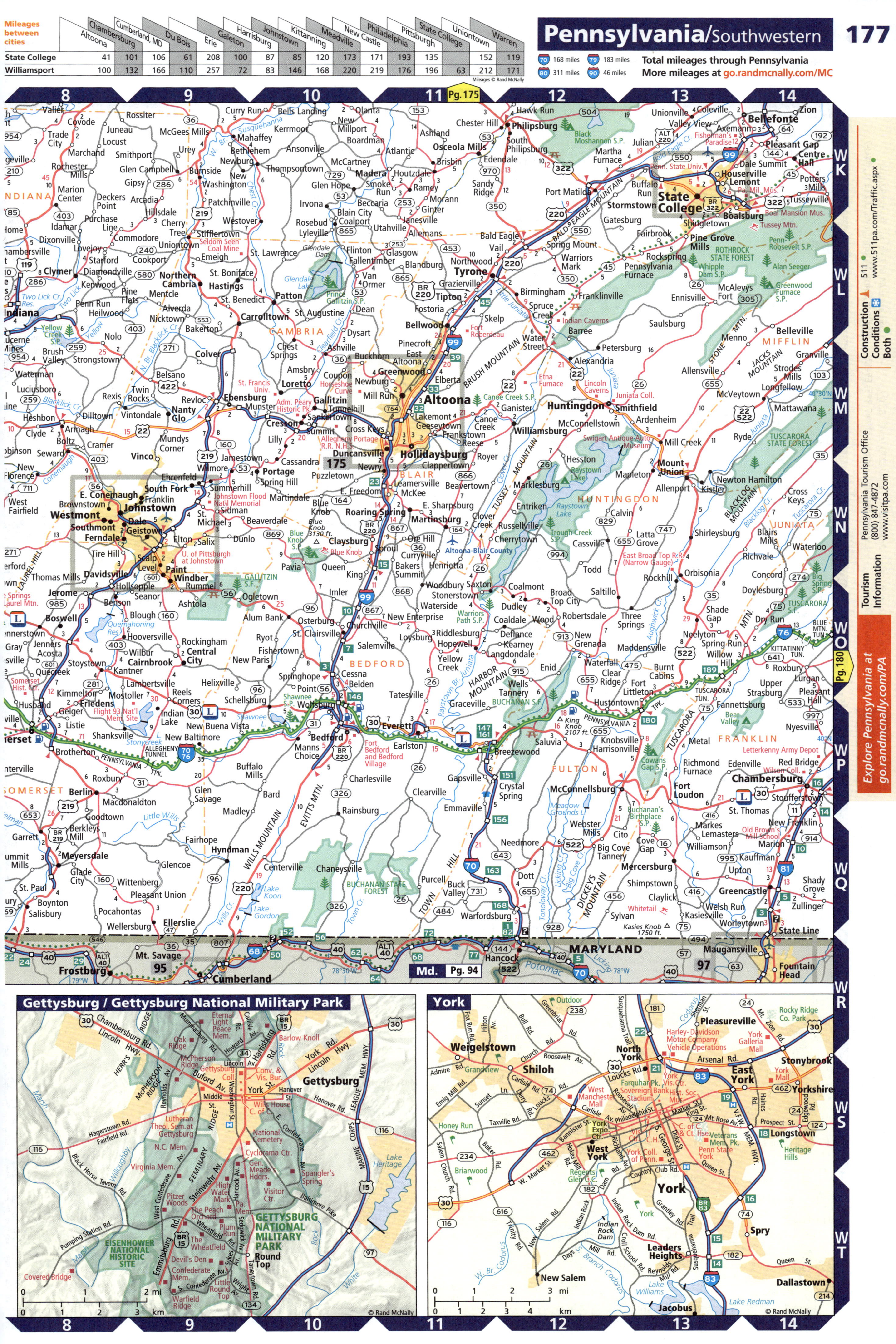

Index of cities Pg. 257 | Map legend Pg. 7

Mileages between cities

| | Allentown | Gettysburg | Harrisburg | Lancaster | Mansfield | Philadelphia | Pittsburgh | Port Jervis, NY | Scranton | State College | Stroudsburg | Towanda | Trenton, NJ | Wilkes Barre | Williamsport | York |
|---|---|---|---|---|---|---|---|---|---|---|---|---|---|---|---|---|
| Allentown | | 121 | 81 | 67 | 177 | 62 | 282 | 81 | 74 | 175 | 40 | 126 | 75 | 60 | 127 | 92 |
| Chambersburg | 132 | 25 | 54 | 91 | 182 | 157 | 160 | 227 | 171 | 101 | 170 | 188 | 177 | 154 | 132 | 74 |



Pennsylvania state facts

Nickname: The Keystone State
Capital: Harrisburg, EN-5

Population: 12,448,279 (rank: 6th)
Largest city: Philadelphia, 1,447,395, EP-12

Land area: 44,817 sq. mi. (rank: 32nd)
Highest point: Mount Davis, 3,213 ft., WQ-7

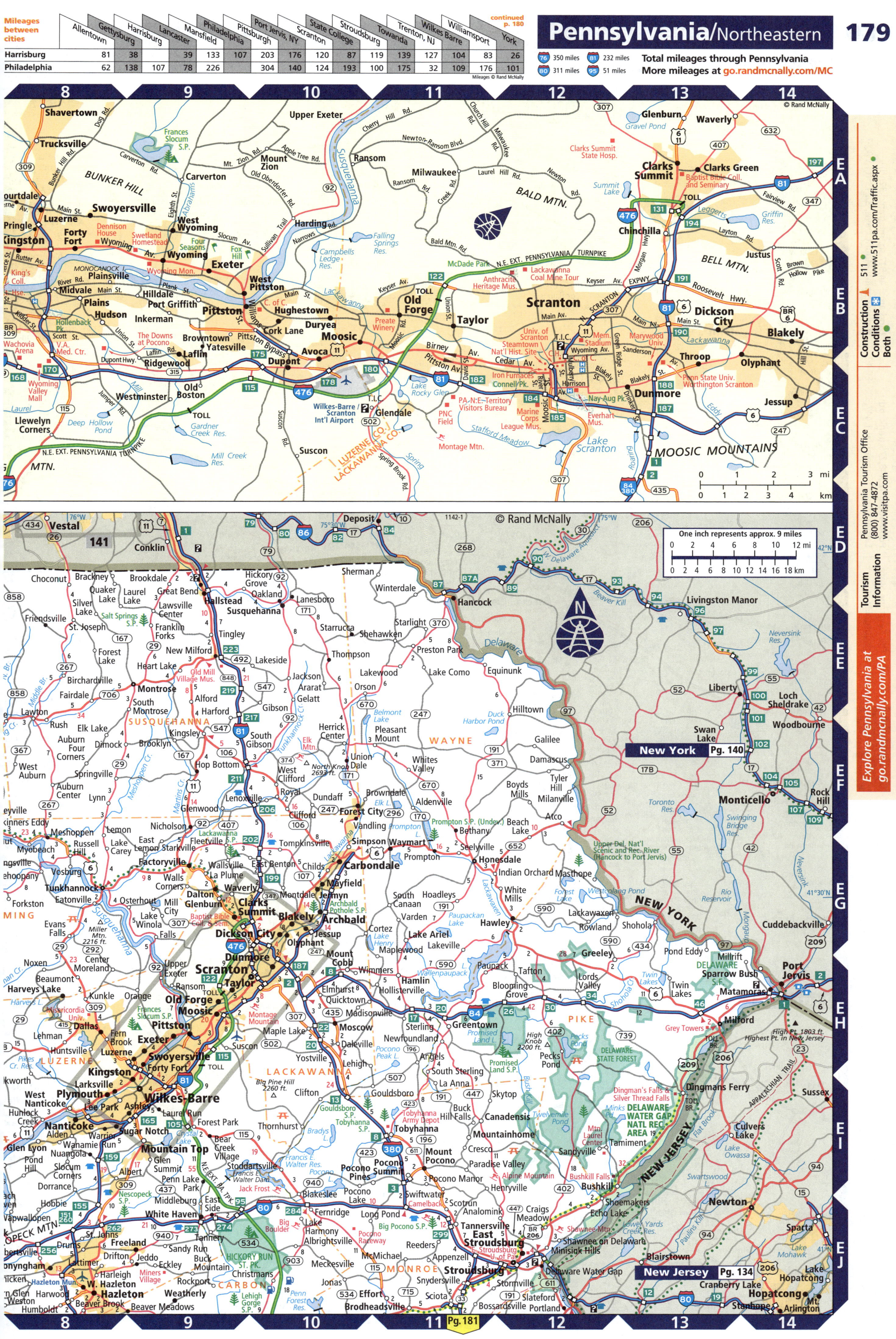

Mileages between cities

| | Allentown | Gettysburg | Harrisburg | Lancaster | Mansfield | Philadelphia | Pittsburgh | Port Jervis, NY | Scranton | State College | Stroudsburg | Towanda | Trenton, NJ | Wilkes Barre | Williamsport | York |
|---|---|---|---|---|---|---|---|---|---|---|---|---|---|---|---|---|
| Harrisburg | 81 | 38 | | 39 | 133 | 107 | 203 | 176 | 120 | 87 | 119 | 139 | 127 | 104 | 83 | 26 |
| Philadelphia | 62 | 138 | 107 | 78 | 226 | | 304 | 140 | 124 | 193 | 100 | 175 | 32 | 109 | 176 | 101 |



continued p. 180

76 350 miles 81 232 miles
80 311 miles 95 51 miles

Total mileages through Pennsylvania
More mileages at go.randmcnally.com/MC

Index of cities Pg. 257 | Map legend Pg. 7

| Mileages between cities | Allentown | Gettysburg | Harrisburg | Lancaster | Mansfield | Philadelphia | Pittsburgh | Port Jervis, NY | Scranton | State College | Stroudsburg | Towanda | Trenton, NJ | Wilkes Barre | Williamsport | York |
|---|---|---|---|---|---|---|---|---|---|---|---|---|---|---|---|---|
| Reading | 37 | 96 | 64 | 34 | 175 | 62 | 261 | 118 | 100 | 150 | 76 | 152 | 82 | 86 | 126 | 56 |
| Scranton | 74 | 160 | 120 | 132 | 102 | 124 | 279 | 59 | | 150 | 46 | 64 | 137 | 16 | 101 | 146 |

Mileages between cities

| | Allentown | Gettysburg | Harrisburg | Lancaster | Mansfield | Philadelphia | Pittsburgh | Port Jervis, NY | Scranton | State College | Stroudsburg | Towanda | Trenton, NJ | Wilkes Barre | Williamsport | York |
|---|---|---|---|---|---|---|---|---|---|---|---|---|---|---|---|---|
| State College | 175 | 129 | 87 | 126 | 107 | 193 | 135 | 205 | 150 | | 162 | 134 | 213 | 132 | 63 | 118 |
| Williamsport | 127 | 126 | 83 | 123 | 50 | 176 | 196 | 157 | 101 | 63 | 113 | 67 | 189 | 84 | | 115 |



76 350 miles · 81 232 miles · 80 311 miles · 95 51 miles

Total mileages through Pennsylvania

More mileages at go.randmcnally.com/MC

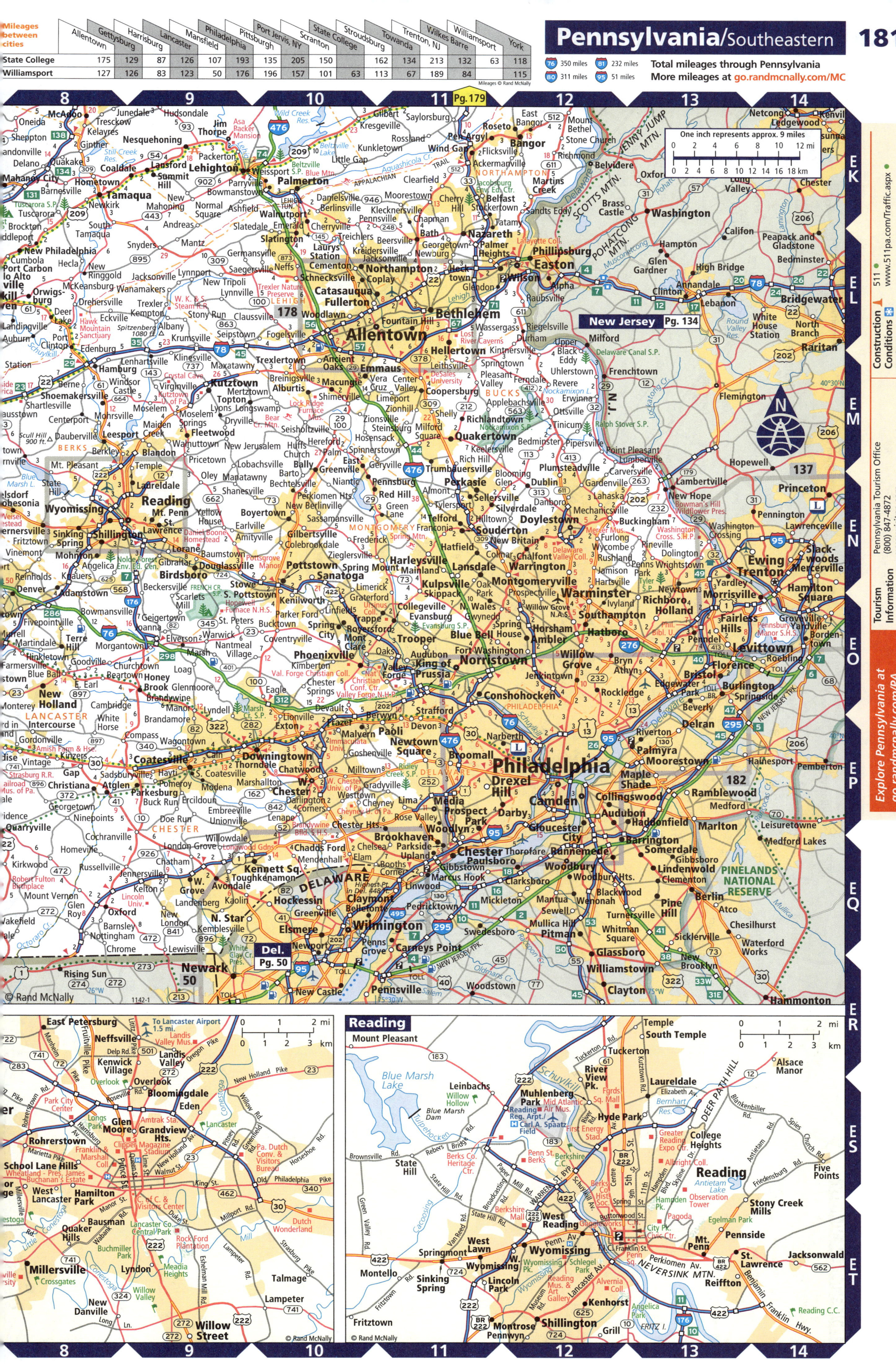


Construction ● Conditions ● Both ●

511 ● www.511pa.com/Traffic.aspx ●

Tourism Information

Pennsylvania Tourism Office
(800) 847-4872
www.visitpa.com

Explore Pennsylvania at go.randmcnally.com/PA

## City sights to see

- Adventure Aquarium, Camden, N.J., E-5
- The Andy Warhol Museum, Pittsburgh, L-2
- Betsy Ross House, Philadelphia, F-10
- Carnegie Science Center, Pittsburgh, L-1
- Duquesne & Monongahela Inclines, Pittsburgh, M-1 & N-2
- Franklin Institute Science Museum, Philadelphia, F-6

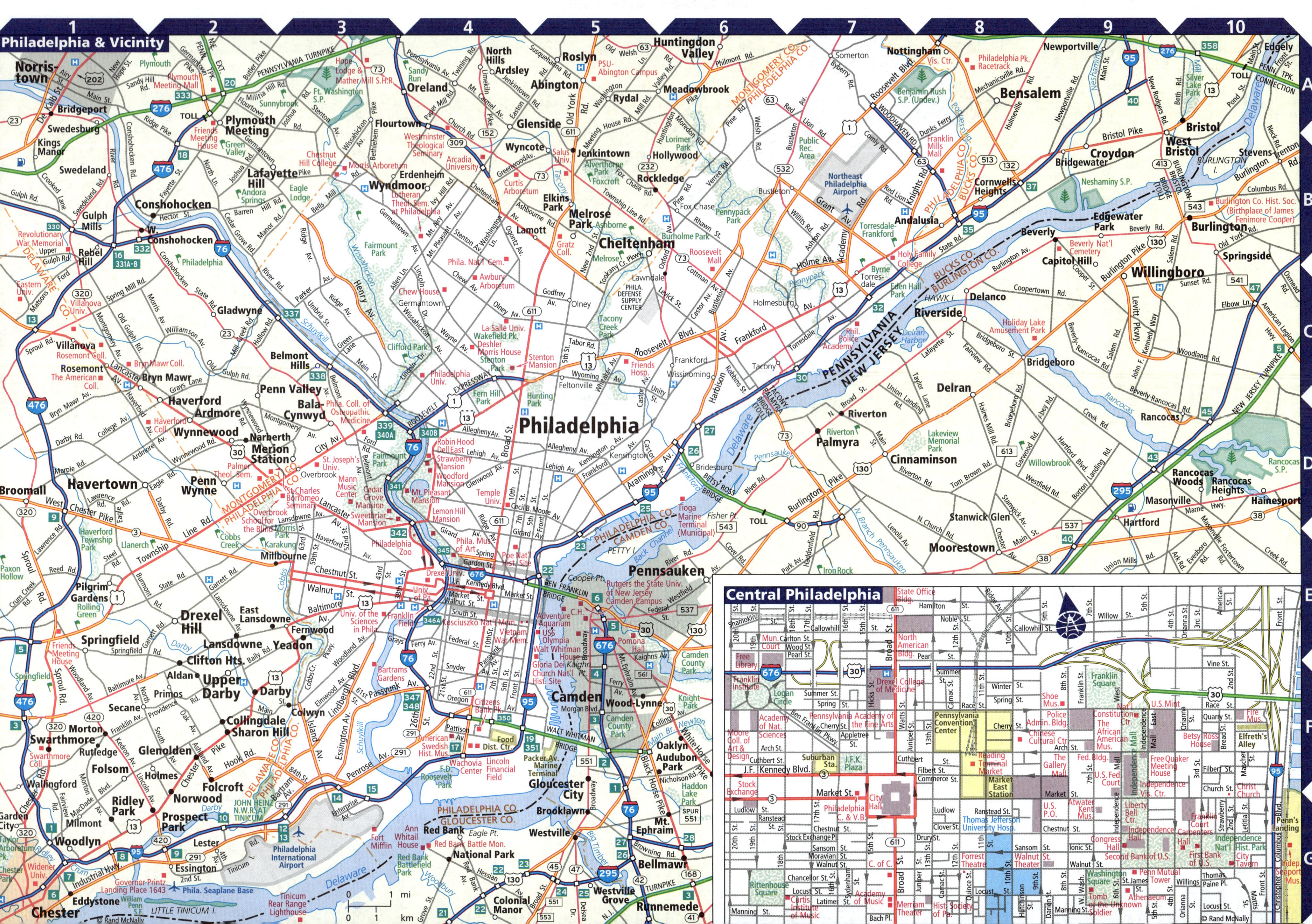

- Independence Hall, Philadelphia, G-9
- Liberty Bell, Philadelphia, G-9
- National Constitution Center, Philadelphia, F-9
- Philadelphia Museum of Art, Philadelphia, E-4
- Point State Park, Pittsburgh, M-1
- The Strip District, Pittsburgh, L-3
- The U.S. Mint, Philadelphia, F-9
- USS *Olympia*, Philadelphia, E-5

# Rhode Island

Index of cities Pg. 259 | Map legend Pg. 7

Mileages between cities

|  | Fall River, MA | Kingston | Newport | Providence | Warwick | Westerly | Woonsocket | Worcester, MA |
|---|---|---|---|---|---|---|---|---|
| Chepachet | 35 | 41 | 45 | 19 | 23 | 54 | 13 | 37 |
| Fall River, MA |  | 35 | 20 | 16 | 25 | 58 | 31 | 56 |
| Newport | 20 | 16 |  | 33 | 26 | 39 | 47 | 72 |
| Providence | 16 | 29 | 33 |  | 10 | 42 | 14 | 40 |

Mileages © Rand McNally

**Rhode Island state facts**

Nickname: The Ocean State
Capital: Providence, D-6
Population: 1,050,788 (rank: 43rd)
Largest city: Providence, 171,557, D-6
Land area: 1,045 sq. mi. (rank: 50th)
Highest point: Jerimoth Hill, 812 ft., C-3

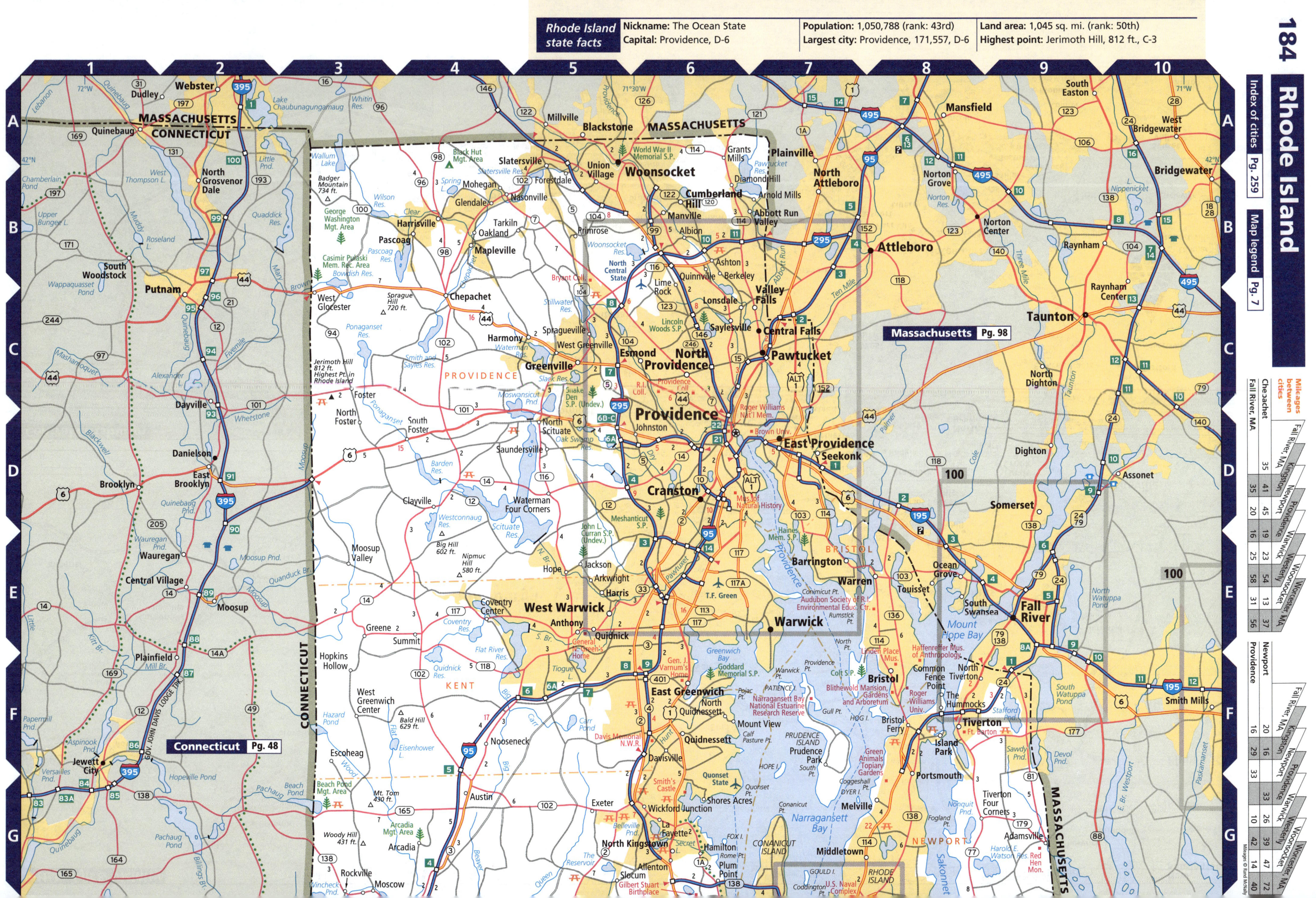

**Mileages between cities**

| | Fall River, MA | Kingston | Newport | Providence | Warwick | Westerly | Woonsocket | Worcester, MA |
|---|---|---|---|---|---|---|---|---|
| Warwick | 25 | 23 | 26 | 10 | | 37 | 24 | 50 |
| Westerly | 58 | 23 | 39 | 42 | 37 | | 56 | 82 |
| Woonsocket | 31 | 43 | 47 | 14 | 24 | 56 | | 27 |
| Worcester, MA | 56 | 68 | 72 | 40 | 50 | 82 | 27 | |

Mileages © Rand McNally

95 42 miles · 1 60 miles · 6 31 miles

**Total mileages through Rhode Island**
More mileages at go.randmcnally.com/MC

Plan a Rhode Island trip at go.randmcnally.com/RI

Tourism Information: Rhode Island Tourism Division, (800) 250-7384, (888) 886-9463, (800) 556-2484, www.visitrhodeisland.com

Construction · Conditions · Both: 511, (888) 401-4511, www2.tmc.state.ri.us

Index of cities Pg. 259 | Map legend Pg. 7

**Mileages between cities**

| | Anderson | Augusta, GA | Charleston | Charlotte, NC | Columbia | Hilton Head I. | Myrtle Beach | Spartanburg |
|---|---|---|---|---|---|---|---|---|
| Augusta, GA | 92 | | 175 | 160 | 72 | 151 | 216 | 120 |
| Charleston | 238 | 175 | | 207 | 112 | 104 | 95 | 201 |
| Charlotte, NC | 128 | 160 | 207 | | 93 | 253 | 176 | 72 |
| Columbia | 117 | 72 | 112 | 93 | | 158 | 148 | 93 |



**S. Carolina state facts**

Nickname: The Palmetto State
Capital: Columbia, D-7
Population: 4,479,800 (rank: 24th)
Largest city: Columbia, 127,029, D-7
Land area: 30,110 sq. mi. (rank: 40th)
Highest point: Sassafras Mountain, 3,560 ft., A-3

| Mileages between cities | Anderson | Augusta, GA | Charleston | Charlotte, NC | Columbia | Hilton Head I. | Myrtle Beach | Spartanburg |
|---|---|---|---|---|---|---|---|---|
| Florence | 206 | 148 | 130 | 104 | 81 | 177 | 67 | 169 |
| Myrtle Beach | 273 | 216 | 95 | 176 | 148 | 200 | | 237 |
| Savannah, GA | 282 | 134 | 106 | 251 | 156 | 34 | 202 | 246 |
| Spartanburg | 60 | 120 | 201 | 72 | 93 | 247 | 237 | |

Mileages © Rand McNally

| Interstate | Miles | Interstate | Miles |
|---|---|---|---|
| 20 | 142 miles | 85 | 106 miles |
| 26 | 221 miles | 95 | 199 miles |

Total mileages through South Carolina
More mileages at go.randmcnally.com/MC

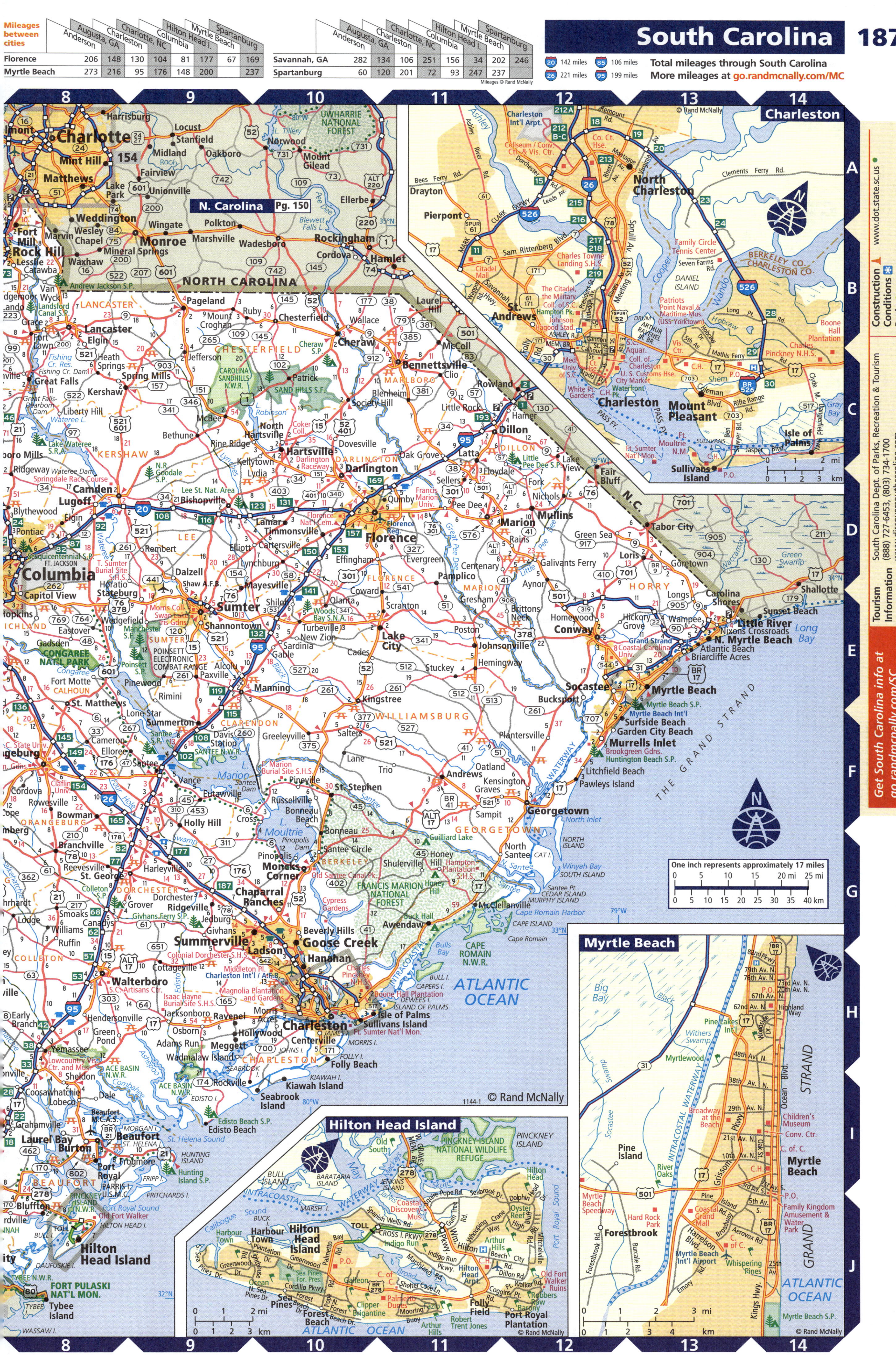

Index of cities Pg. 259
Map legend Pg. 7

**Mileages between cities**

| | Aberdeen | Mobridge | Pierre | Pine Ridge | Rapid City | Sioux Falls | Watertown | Yankton |
|---|---|---|---|---|---|---|---|---|
| Aberdeen | | 100 | 160 | 360 | 333 | 203 | 96 | 236 |
| Belle Fourche | 312 | 212 | 206 | 172 | 60 | 403 | 362 | 421 |
| Mobridge | 100 | | 108 | 308 | 243 | 303 | 196 | 332 |
| Pierre | 160 | 108 | | 200 | 173 | 224 | 188 | 242 |



**S. Dakota state facts**

Nickname: The Mount Rushmore State
Capital: Pierre, D-7
Population: 804,194 (rank: 46th)
Largest city: Sioux Falls, 154,997, F-13
Land area: 75,885 sq. mi. (rank: 16th)
Highest point: Harney Peak, 7,242 ft., E-2

**Mileages between cities**

| | Aberdeen | Mobridge | Pierre | Pine Ridge | Rapid City | Sioux Falls | Watertown | Yankton |
|---|---|---|---|---|---|---|---|---|
| Rapid City | 333 | 243 | 173 | 111 | | 347 | 403 | 365 |
| Sioux City, IA | 285 | 384 | 305 | 358 | 428 | 85 | 184 | 63 |
| Sioux Falls | 203 | 303 | 224 | 356 | 347 | | 103 | 81 |
| Watertown | 96 | 196 | 188 | 415 | 403 | 103 | | 155 |



29 253 miles
90 413 miles
12 317 miles
83 242 miles

**Total mileages through South Dakota**
**More mileages at go.randmcnally.com/MC**

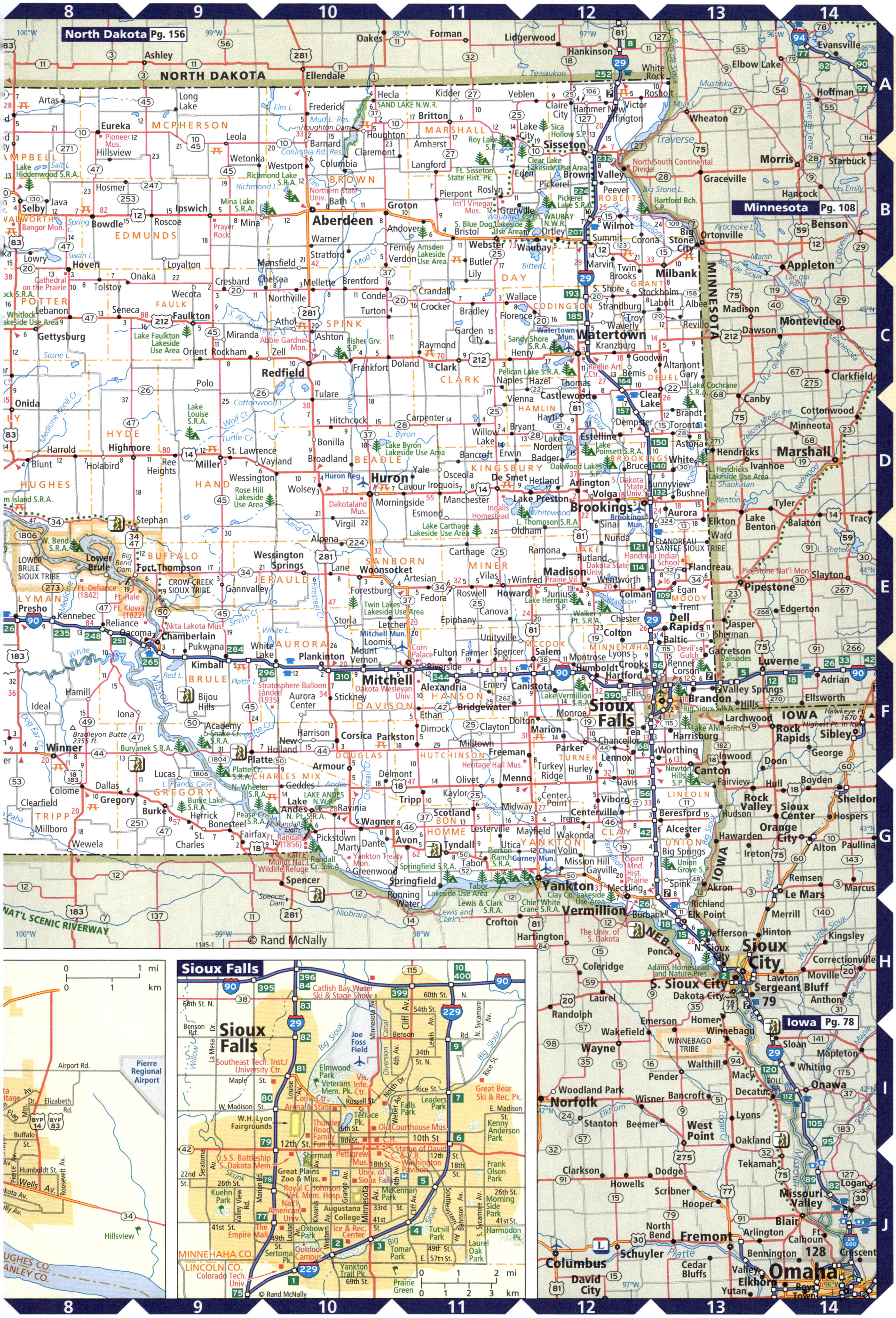

North Dakota Pg. 156

Minnesota Pg. 108

Iowa Pg. 78

**Sioux Falls**

**Tourism Information** South Dakota Office of Tourism (800) 732-5682 www.travelsd.com

**Construction** / **Conditions** / **Both** 511 (866) 697-3511 www.safetravelusa.com/sd

*Plan a South Dakota trip at go.randmcnally.com/SD*

Index of cities Pg. 259 | Map legend Pg. 7

**Mileages between cities**

| | Atlanta, GA | Bristol | Chattanooga | Clarksville | Cookeville | Dyersburg | Fayetteville | Gatlinburg | Jackson | Johnson City | Knoxville | Memphis | Morristown | Nashville | Oak Ridge | Union City |
|---|---|---|---|---|---|---|---|---|---|---|---|---|---|---|---|---|
| Chattanooga | 117 | 223 | | 177 | 98 | 303 | 94 | 151 | 260 | 215 | 110 | 314 | 158 | 131 | 108 | 311 |
| Clarksville | 293 | 337 | 177 | | 125 | 173 | 136 | 265 | 123 | 329 | 224 | 201 | 271 | 47 | 207 | 138 |

Mileages © Rand McNally

*Tennessee state facts*

Nickname: The Volunteer State
Capital: Nashville, C-11

Population: 6,214,888 (rank: 17th)
Largest city: Memphis, 669,651, G-2

Land area: 41,217 sq. mi. (rank: 34th)
Highest point: Clingmans Dome, 6,643 ft., M-15

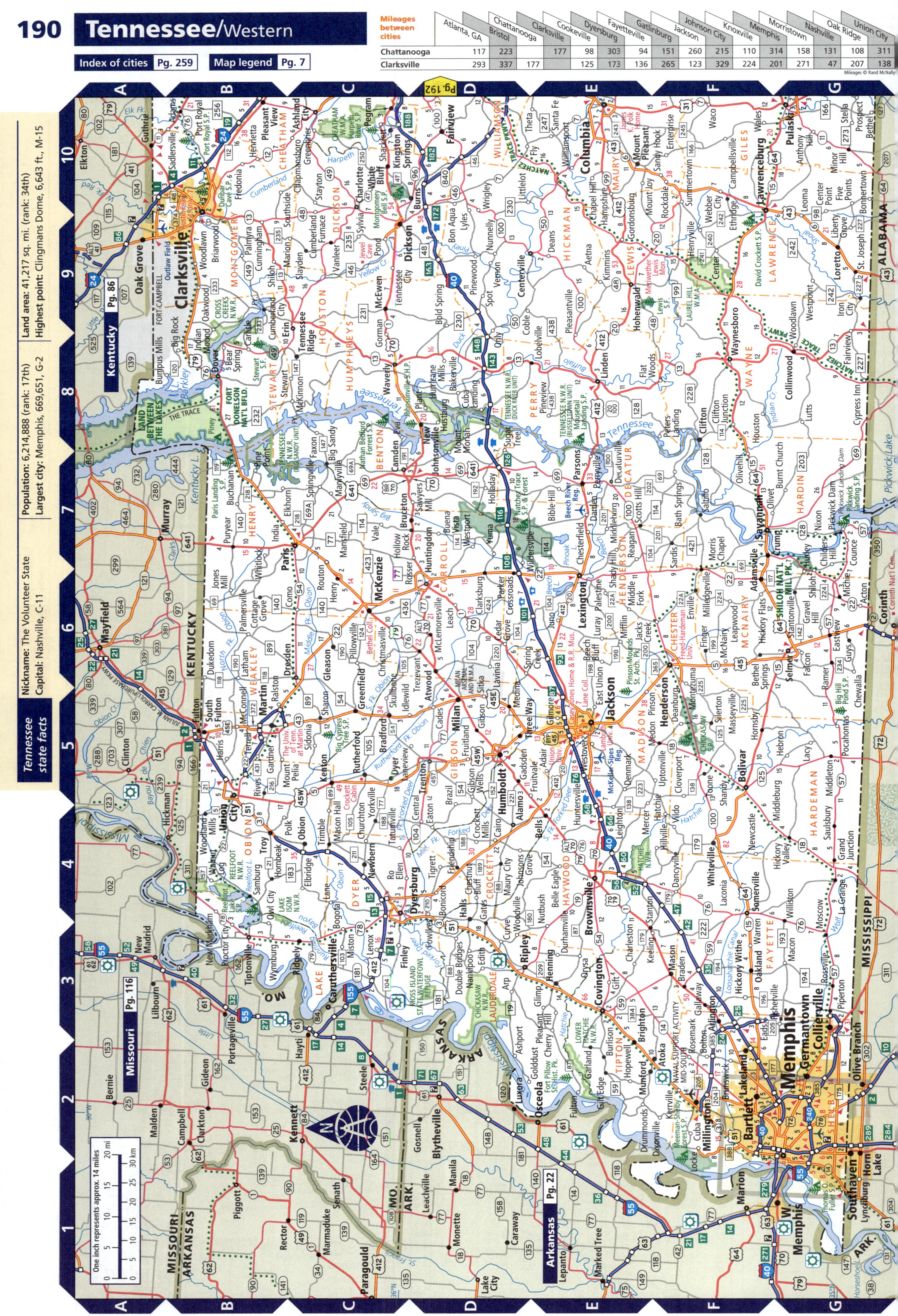

Mileages between cities

| | Atlanta, GA | Bristol | Chattanooga | Clarksville | Cookeville | Dyersburg | Fayetteville | Gatlinburg | Jackson | Johnson City | Knoxville | Memphis | Morristown | Nashville | Oak Ridge | Union City |
|---|---|---|---|---|---|---|---|---|---|---|---|---|---|---|---|---|
| Dyersburg | 418 | 463 | 303 | 173 | 252 | | 229 | 392 | 47 | 455 | 351 | 76 | 398 | 172 | 334 | 34 |
| Fayetteville | 211 | 317 | 94 | 136 | 109 | 229 | | 246 | 167 | 308 | 204 | 243 | 252 | 90 | 189 | 224 |

continued p. 192



40 455 miles · 65 121 miles · 75 161 miles · 81 76 miles

Total mileages through Tennessee
More mileages at go.randmcnally.com/MC

Nashville

Memphis & Vicinity

Alabama Pg. 10

Mississippi Pg. 114

Explore Tennessee at go.randmcnally.com/TN

Tourism Information: Tennessee Department of Tourist Development (800) 462-8366, (615) 741-2159 www.tnvacation.com

Construction / Conditions / Both: 511 (877) 244-0065 www.tn511.com

Index of cities Pg. 259 | Map legend Pg. 7

**Mileages between cities**

| | Atlanta, GA | Bristol | Chattanooga | Clarksville | Cookeville | Dyersburg | Fayetteville | Gatlinburg | Jackson | Johnson City | Knoxville | Memphis | Morristown | Nashville | Oak Ridge | Union City |
|---|---|---|---|---|---|---|---|---|---|---|---|---|---|---|---|---|
| Johnson City | 256 | 24 | 215 | 329 | 206 | 455 | 308 | 106 | 412 | | 104 | 495 | 65 | 283 | 128 | 463 |
| Knoxville | 202 | 113 | 110 | 224 | 102 | 351 | 204 | 41 | 308 | 104 | | 390 | 48 | 179 | 24 | 358 |



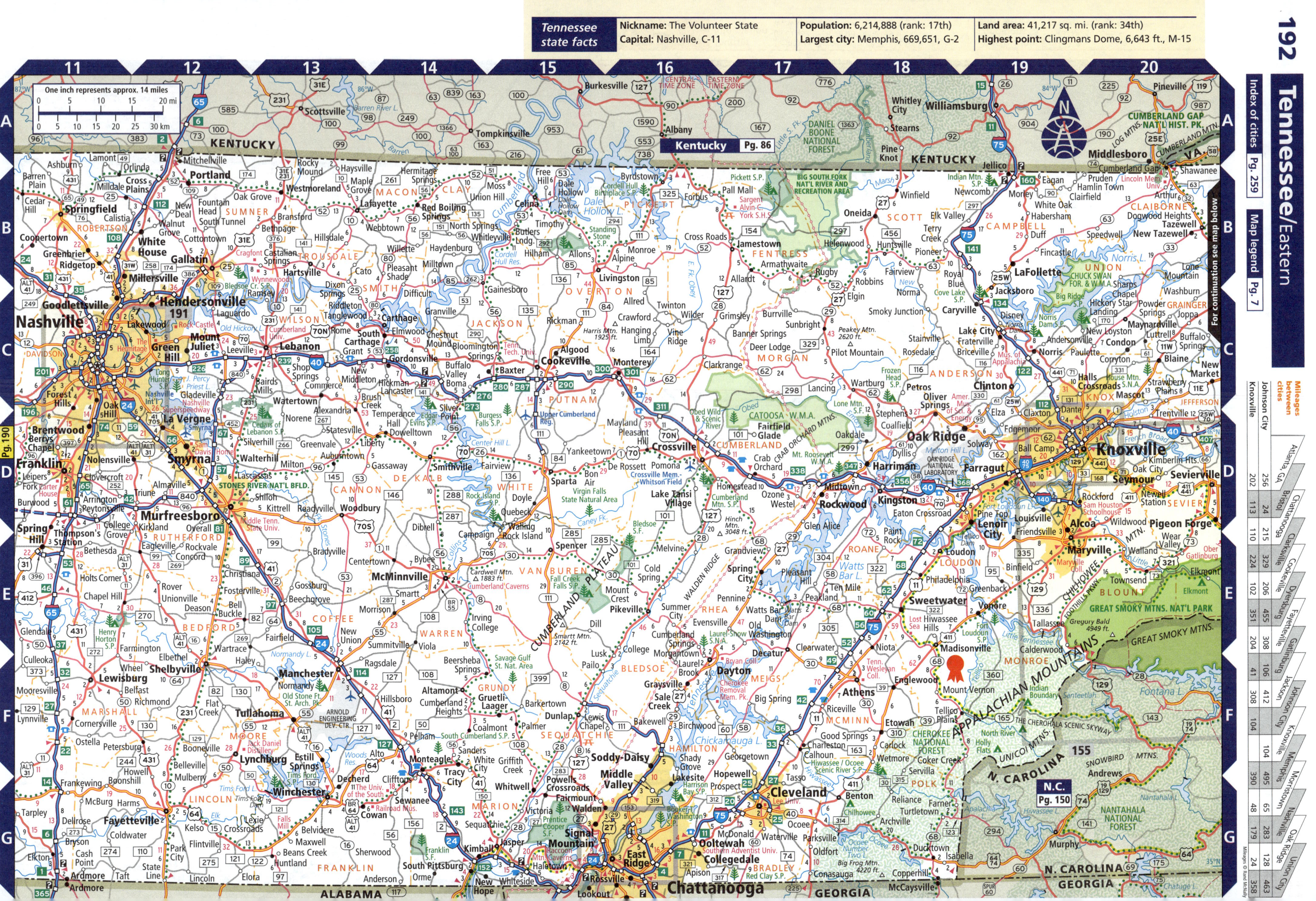

**Mileages between cities**

| | Atlanta, GA | Bristol | Chattanooga | Clarksville | Cookeville | Dyersburg | Fayetteville | Gatlinburg | Jackson | Johnson City | Knoxville | Memphis | Morristown | Nashville | Oak Ridge | Union City |
|---|---|---|---|---|---|---|---|---|---|---|---|---|---|---|---|---|
| Memphis | 380 | 502 | 314 | 201 | 291 | 76 | 243 | 431 | 87 | 495 | 390 | | 437 | 212 | 373 | 113 |
| Nashville | 249 | 292 | 131 | 47 | 80 | 172 | 90 | 220 | 129 | 283 | 179 | 212 | 226 | | 162 | 168 |



40 455 miles 75 161 miles
65 121 miles 81 76 miles

Total mileages through Tennessee
More mileages at go.randmcnally.com/MC

# Eastern Tennessee

Explore Tennessee at go.randmcnally.com/TN

Tourism Information
Tennessee Department of Tourist Development
(800) 462-8366, (615) 741-2159
www.tnvacation.com

Construction / Conditions / Both
511
(877) 244-0065
www.tn511.com

Knoxville

Chattanooga

Georgia Pg. 58
Alabama Pg. 10
Virginia Pg. 212
Ky. Pg. 86
North Carolina Pg. 150
For continuation see main map

## City sights to see

- Appalachian Caverns, Blountville, Tenn., K-3
- Battleship USS *Texas*, La Porte, D-9
- Bayou Place, Houston, K-8
- Bishop's Palace, Galveston, B-10
- Bristol Caverns, Bristol, Tenn., J-6
- Bristol Motor Speedway, Bristol, Tenn., K-4
- Contemporary Arts Museum, Houston, E-5
- Houston Fire Museum, Houston, E-5

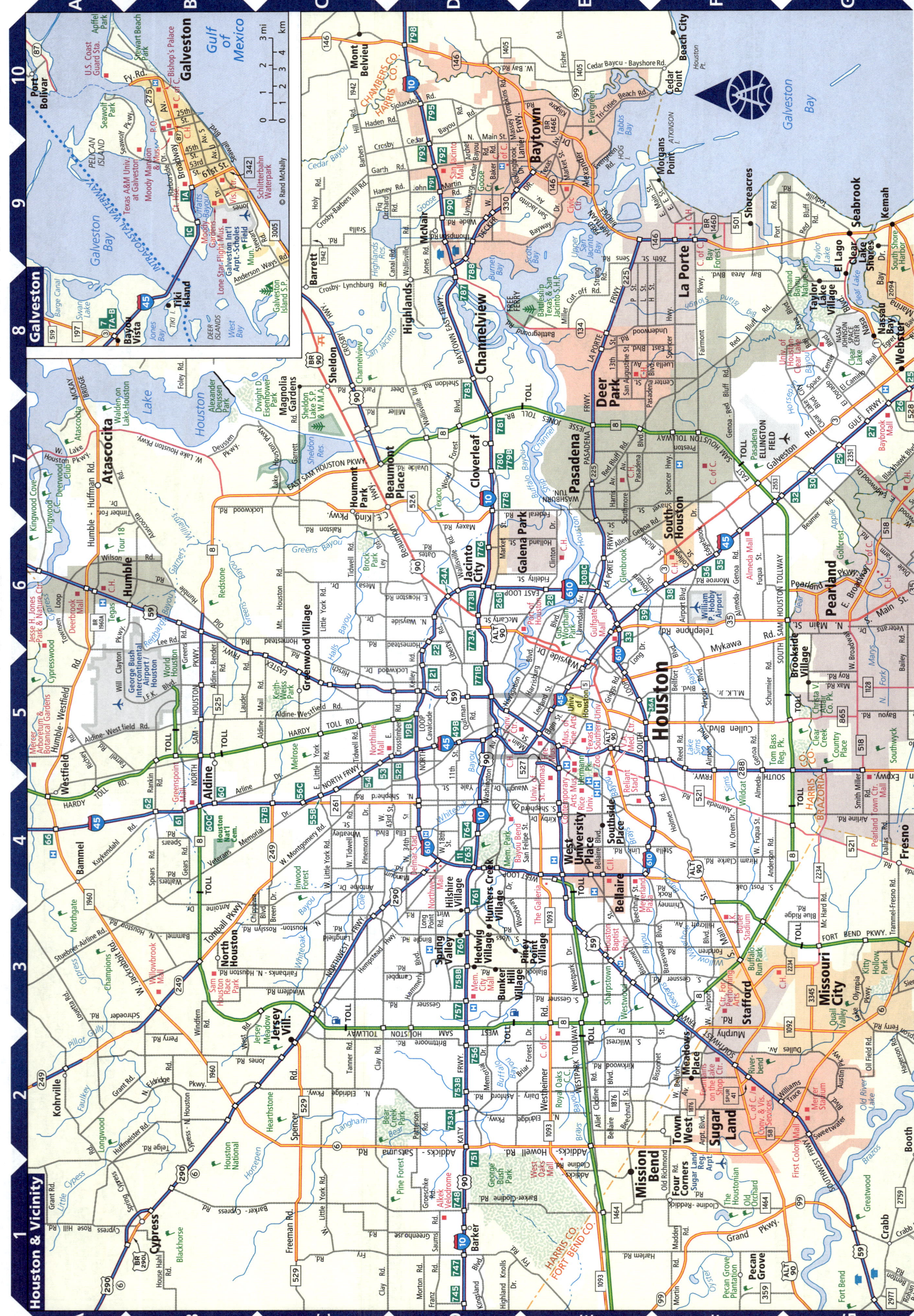

- Houston Zoo, Houston, E-5
- Minute Maid Park, L-9
- Moody Gardens, Galveston, B-9
- Museum of Fine Arts, Houston, E-5
- Reliant Stadium, Houston, E-5
- Rocky Mount Museum, Piney Flats, Tenn., L-3
- Space Center Houston, Houston, G-8
- Wortham Theatre Center, Houston, K-8

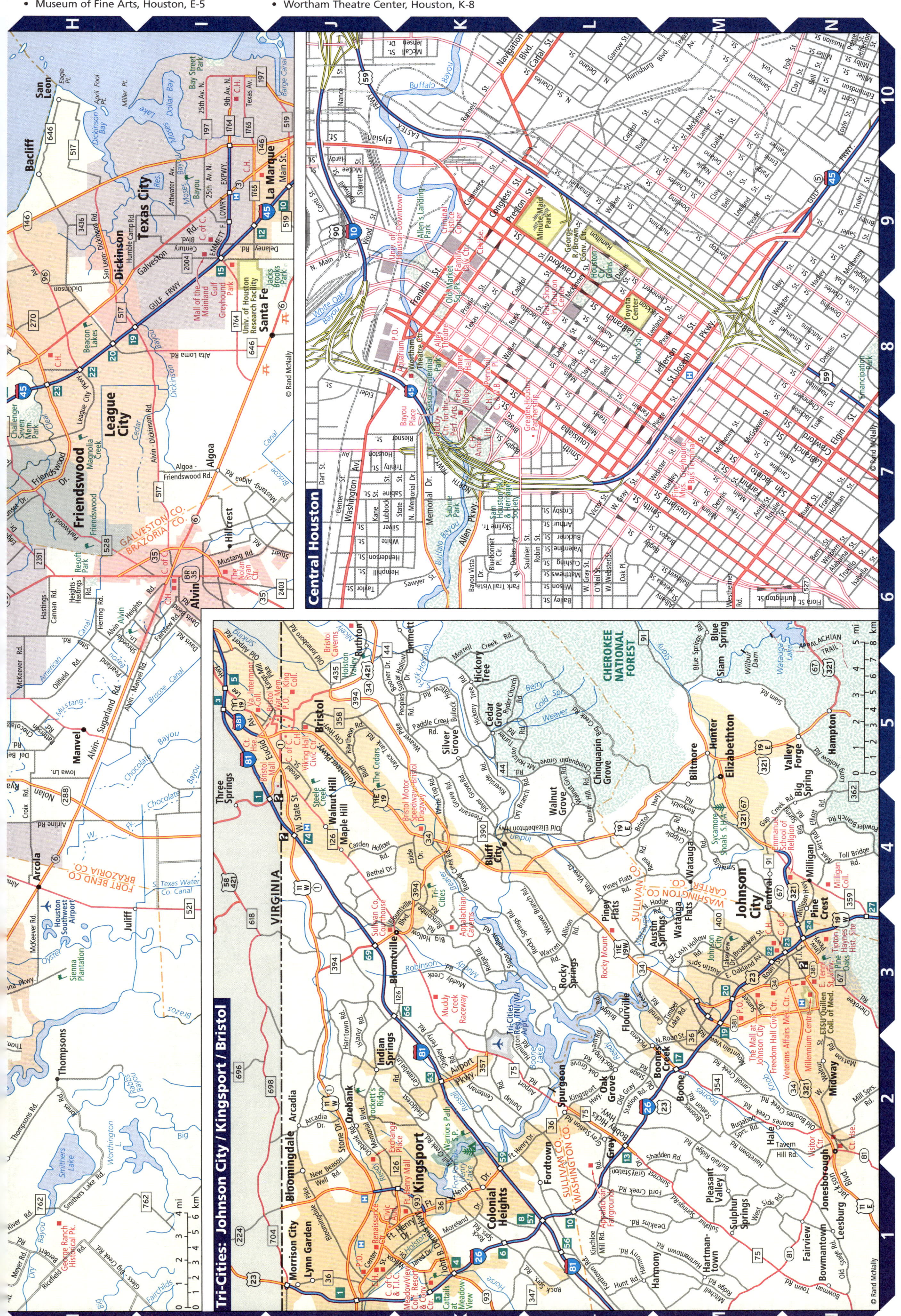

**City sights to see**

- Cowboys Stadium, Arlington, H-7
- Dallas Arts District, Dallas, A-5
- Dallas Museum of Art, Dallas, A-5
- Dallas Zoo, Dallas, H-10
- Fair Park (Cotton Bowl), Dallas, G-11
- Heard Natural Science Museum & Sanctuary, McKinney, B-10
- Old City Park, Dallas, C-6

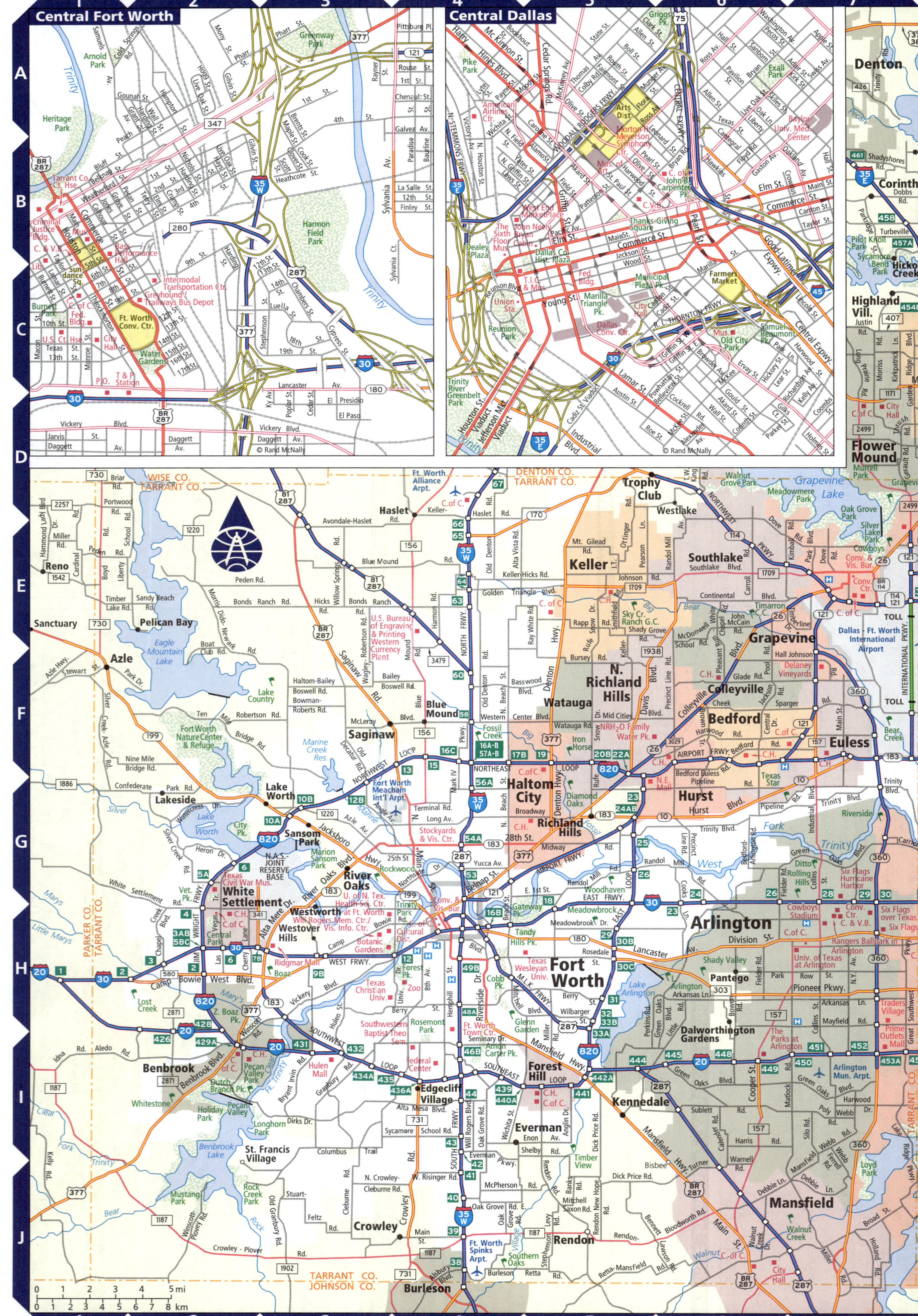

- Ripley's Believe It or Not! & Louis Tussaud's Palace of Wax, Grand Prairie, G-8
- The Sixth Floor Museum at Dealey Plaza, Dallas, B-4
- Stockyards Historic District, Fort Worth, G-4
- Sundance Square, Fort Worth, B-1
- Texas Civil War Museum, Fort Worth, G-2
- Will Rogers Memorial Center, Fort Worth, H-3



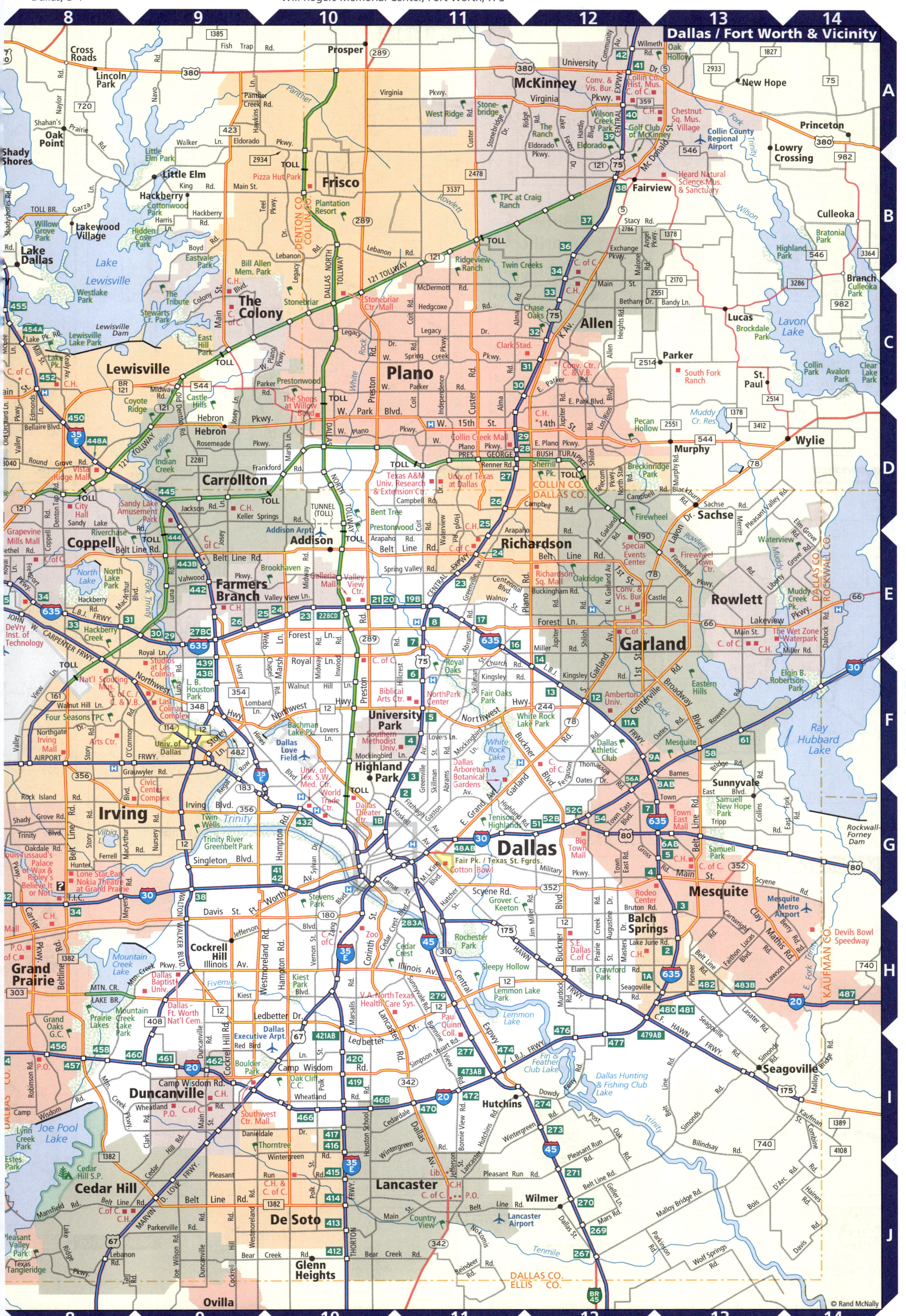

Index of cities Pg. 260 | Map legend Pg. 7

Mileages between cities

| | Abilene | Amarillo | Big Bend N.P. | Big Spring | Childress | Clovis, NM | Dallas | Eagle Pass | El Paso | Fort Stockton | Lubbock | Odessa | Perryton | San Angelo | San Antonio | Van Horn |
|---|---|---|---|---|---|---|---|---|---|---|---|---|---|---|---|---|
| Abilene | | 268 | 380 | 108 | 155 | 267 | 179 | 304 | 454 | 255 | 163 | 168 | 306 | 88 | 250 | 332 |
| Amarillo | 268 | | 470 | 226 | 112 | 104 | 363 | 510 | 407 | 344 | 120 | 258 | 115 | 318 | 510 | 423 |

Mileages © Rand McNally

Texas state facts

Nickname: The Lone Star State
Capital: Austin, EJ-6
Population: 24,326,974 (rank: 2nd)
Largest city: Houston, 2,242,193, EK-10
Land area: 261,797 sq. mi. (rank: 2nd)
Highest point: Guadalupe Peak, 8,749 ft, WK-4

| Mileages between cities | Abilene | Amarillo | Big Bend N.P. | Big Spring | Childress | Clovis, NM | Dallas | Eagle Pass | El Paso | Fort Stockton | Lubbock | Odessa | Perryton | San Angelo | San Antonio | Van Horn |
|---|---|---|---|---|---|---|---|---|---|---|---|---|---|---|---|---|
| Del Rio | 241 | 454 | 242 | 240 | 383 | 425 | 426 | 56 | 428 | 184 | 333 | 258 | 534 | 154 | 151 | 303 |
| El Paso | 454 | 407 | 325 | 346 | 482 | 301 | 635 | 484 | | 240 | 343 | 284 | 516 | 404 | 554 | 121 |

continued p. 200

Mileages © Rand McNally

10 881 miles 40 177 miles 20 636 miles

Total mileages through Texas
More mileages at go.randmcnally.com/MC

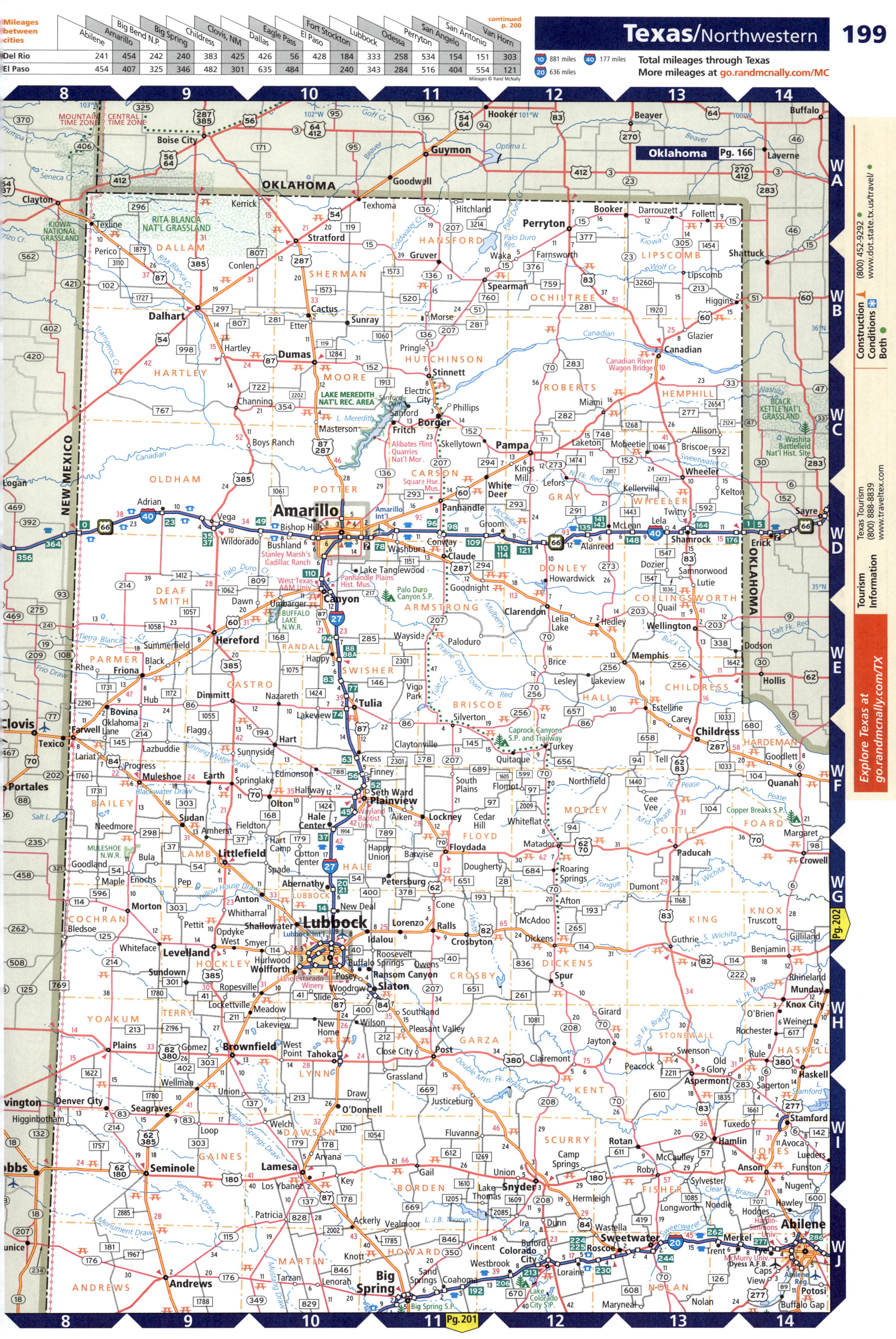

Index of cities Pg. 260 | Map legend Pg. 7

Mileages between cities

| | Abilene | Amarillo | Big Bend N.P. | Big Spring | Childress | Clovis, NM | Dallas | Eagle Pass | El Paso | Fort Stockton | Lubbock | Odessa | Perryton | San Angelo | San Antonio | Van Horn |
|---|---|---|---|---|---|---|---|---|---|---|---|---|---|---|---|---|
| Lubbock | 163 | 120 | 349 | 106 | 141 | 103 | 345 | 390 | 343 | 224 | | 138 | 240 | 194 | 390 | 302 |
| Odessa | 168 | 258 | 210 | 61 | 279 | 204 | 352 | 314 | 284 | 85 | 138 | | 377 | 132 | 352 | 164 |

Mileages © Rand McNally

Texas state facts

Nickname: The Lone Star State
Capital: Austin, EJ-6
Population: 24,326,974 (rank: 2nd)
Largest city: Houston, 2,242,193, EK-10
Land area: 261,797 sq. mi. (rank: 2nd)
Highest point: Guadalupe Peak, 8,749 ft., WK-4

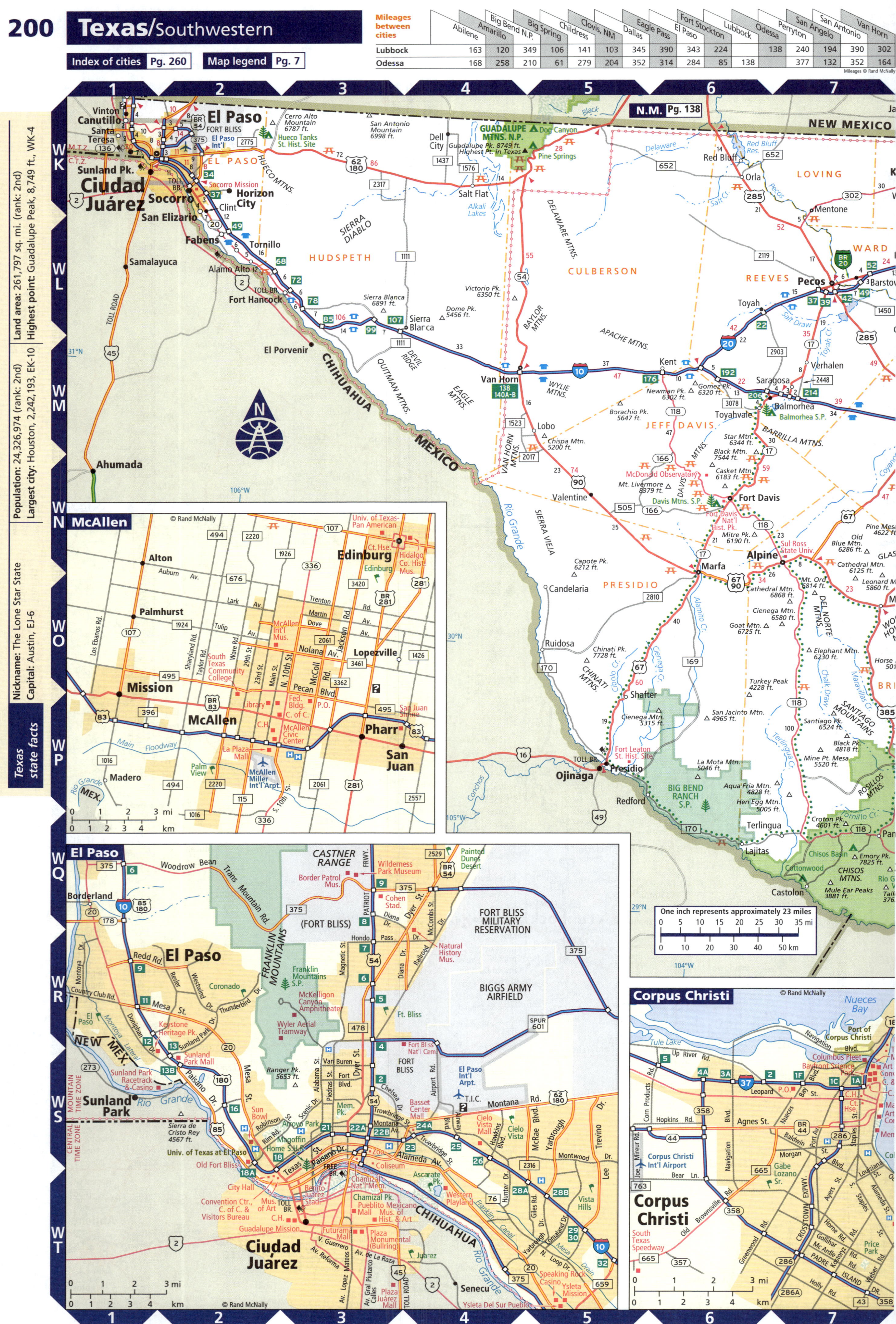

**Mileages between cities**

| | Abilene | Amarillo | Big Bend N.P. | Big Spring | Childress | Clovis, NM | Dallas | Eagle Pass | El Paso | Fort Stockton | Lubbock | Odessa | Perryton | San Angelo | San Antonio | Van Horn |
|---|---|---|---|---|---|---|---|---|---|---|---|---|---|---|---|---|
| San Angelo | 88 | 318 | 290 | 86 | 226 | 296 | 269 | 212 | 404 | 162 | 194 | 132 | 377 | | 213 | 282 |
| San Antonio | 250 | 510 | 404 | 299 | 408 | 493 | 276 | 143 | 554 | 315 | 390 | 352 | 556 | 213 | | 434 |



Total mileages through Texas: 10 – 881 miles; 20 – 636 miles; 40 – 177 miles

More mileages at go.randmcnally.com/MC

Tourism Information: Texas Tourism (800) 888-8839 www.traveltex.com

Construction Conditions Both: (800) 452-9292 www.dot.state.tx.us/travel/

Explore Texas at go.randmcnally.com/TX

Pg. 199
Pg. 202
Pg. 204

Map labels: Midland, Odessa, Monahans, Fort Stockton, Crane, McCamey, Rankin, Big Lake, Iraan, Ozona, Sonora, Eldorado, San Angelo, Ballinger, Eden, Menard, Junction, Rocksprings, Del Rio, Ciudad Acuña, Brackettville, Uvalde, Eagle Pass, Piedras Negras, Crystal City, Carrizo Springs, Sanderson, Dryden, Langtry, Comstock, Amistad Reservoir, Rio Grande, COAHUILA, MEXICO, NUEVO LEÓN, Austin, West Lake Hills, Rollingwood, Sunset Valley, Del Valle

Index of cities Pg. 260 | Map legend Pg. 7

**Mileages between cities**

| | Abilene | Austin | Beaumont | Brownsville | Dallas | Houston | Laredo | Lufkin | Paris | San Angelo | San Antonio | Shreveport, LA | Texarkana | Tyler | Waco | Wichita Falls |
|---|---|---|---|---|---|---|---|---|---|---|---|---|---|---|---|---|
| Abilene | | 221 | 449 | 524 | 179 | 377 | 396 | 363 | 285 | 88 | 250 | 368 | 358 | 280 | 183 | 151 |
| Austin | 221 | | 242 | 353 | 193 | 157 | 237 | 224 | 296 | 208 | 81 | 325 | 366 | 224 | 99 | 299 |



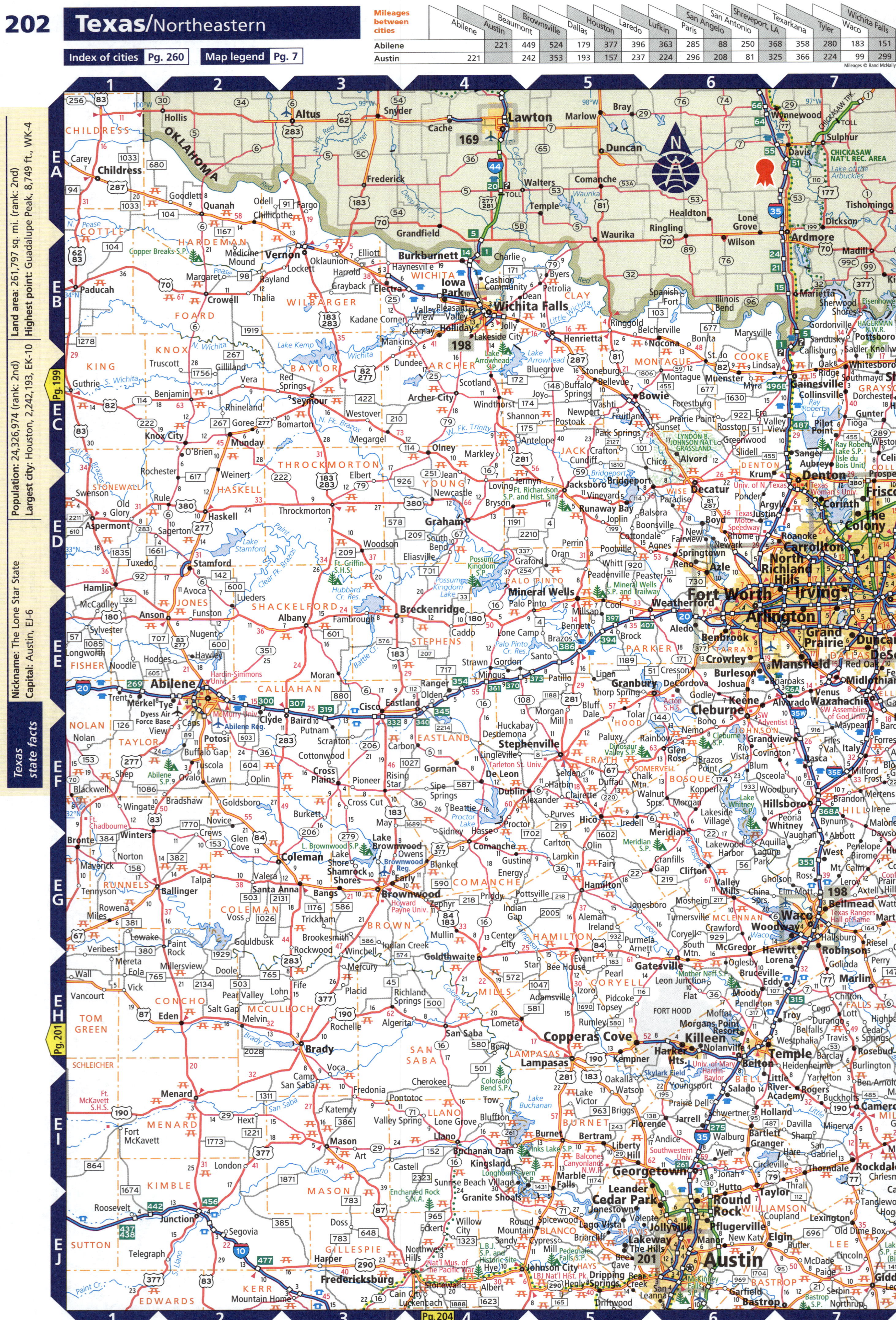

| Mileages between cities | Abilene | Austin | Beaumont | Brownsville | Dallas | Houston | Laredo | Lufkin | Paris | San Angelo | San Antonio | Shreveport, LA | Texarkana | Tyler | Waco | Wichita Falls |
|---|---|---|---|---|---|---|---|---|---|---|---|---|---|---|---|---|
| Brownsville | 524 | 353 | 439 | | 547 | 354 | 204 | 473 | 622 | 491 | 274 | 596 | 650 | 530 | 435 | 614 |
| Corpus Christi | 387 | 217 | 292 | 156 | 410 | 207 | 138 | 328 | 496 | 355 | 138 | 449 | 504 | 392 | 316 | 477 |



continued p. 204

Total mileages through Texas: 10 — 811 miles; 20 — 636 miles; 30 — 223 miles; 35 — 504 miles

More mileages at go.randmcnally.com/MC

Index of cities Pg. 260 | Map legend Pg. 7

Mileages between cities

| | Abilene | Austin | Beaumont | Brownsville | Dallas | Houston | Laredo | Lufkin | Paris | San Angelo | San Antonio | Shreveport, LA | Texarkana | Tyler | Waco | Wichita Falls |
|---|---|---|---|---|---|---|---|---|---|---|---|---|---|---|---|---|
| Dallas | 179 | 193 | 282 | 547 | | 228 | 428 | 183 | 106 | 269 | 276 | 187 | 177 | 100 | 96 | 139 |
| Houston | 377 | 157 | 85 | 354 | 228 | | 348 | 118 | 299 | 368 | 197 | 242 | 295 | 199 | 184 | 375 |



Texas state facts

Nickname: The Lone Star State
Capital: Austin, EI-6
Population: 24,326,974 (rank: 2nd)
Largest city: Houston, 2,242,193, EK-10
Land area: 261,797 sq. mi. (rank: 2nd)
Highest point: Guadalupe Peak, 8,749 ft., WK-4

| Mileages between cities | Abilene | Austin | Beaumont | Brownsville | Dallas | Houston | Laredo | Lufkin | Paris | San Angelo | San Antonio | Shreveport, LA | Texarkana | Tyler | Waco | Wichita Falls |
|---|---|---|---|---|---|---|---|---|---|---|---|---|---|---|---|---|
| San Antonio | 250 | 81 | 280 | 274 | 276 | 197 | 154 | 314 | 380 | 213 | | 406 | 451 | 309 | 180 | 341 |
| Shreveport, LA | 368 | 325 | 206 | 596 | 187 | 242 | 565 | 120 | 154 | 455 | 406 | | 72 | 98 | 226 | 324 |

Mileages © Rand McNally

10 811 miles, 20 636 miles, 30 223 miles, 35 504 miles

Total mileages through Texas
More mileages at go.randmcnally.com/MC

Index of cities Pg. 262 | Map legend Pg. 7

Mileages between cities

| | Blanding | Cedar City | Grand Jct., CO | Las Vegas, NV | Logan | Moab | Ogden | Page, AZ | Park City | Price | Provo | Richfield | St. George | Salt Lake City | Vernal | Wendover |
|---|---|---|---|---|---|---|---|---|---|---|---|---|---|---|---|---|
| Grand Junction, CO | 186 | 335 | | 506 | 363 | 112 | 319 | 380 | 286 | 164 | 240 | 224 | 389 | 283 | 140 | 401 |
| Logan | 388 | 330 | 363 | 499 | | 313 | 46 | 457 | 113 | 199 | 124 | 239 | 385 | 82 | 252 | 199 |

Mileages © Rand McNally

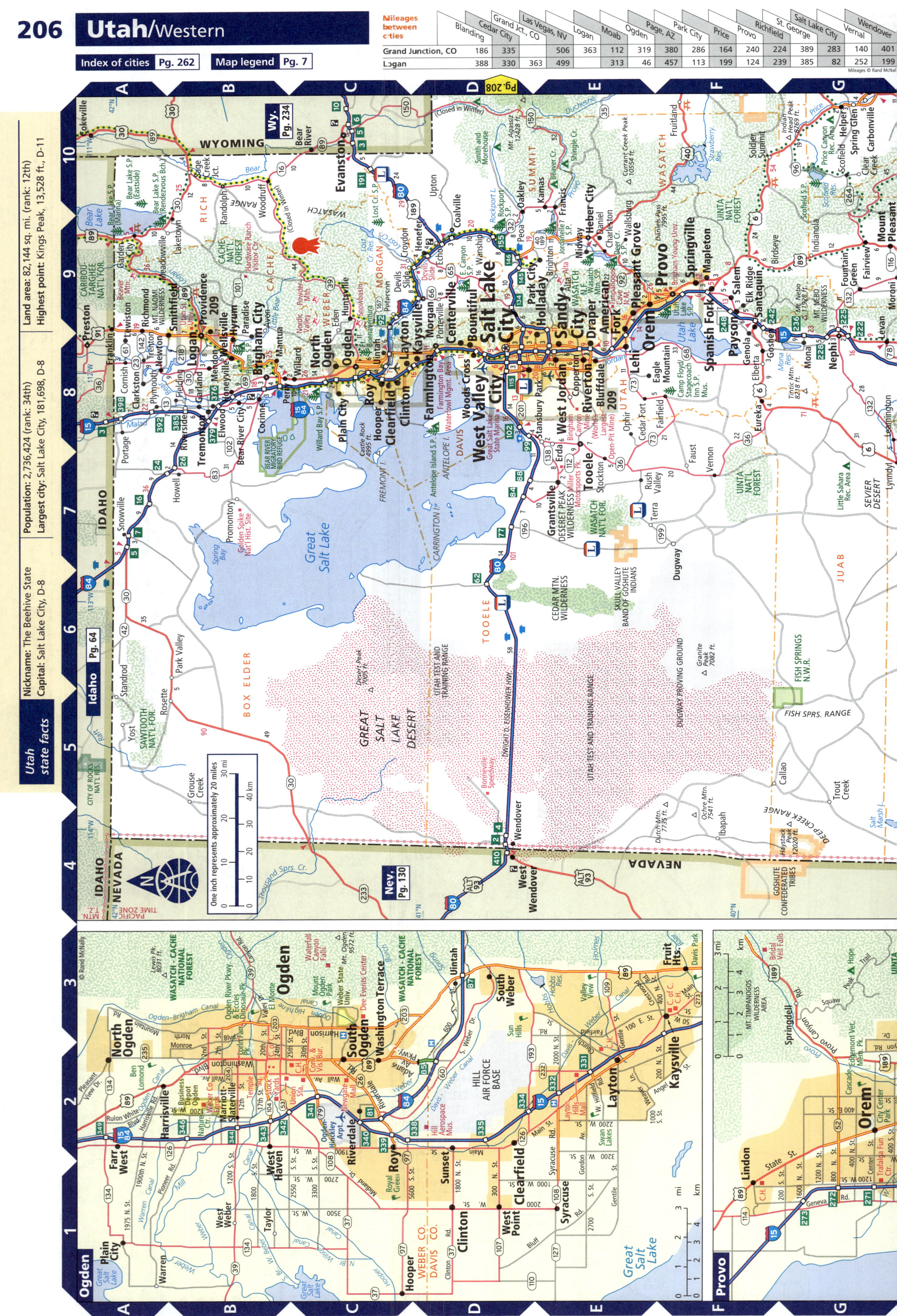

Mileages between cities

| | Blanding | Cedar City | Grand Jct., CO | Las Vegas, NV | Logan | Moab | Ogden | Page, AZ | Park City | Price | Provo | Richfield | St. George | Salt Lake City | Vernal | Wendover |
|---|---|---|---|---|---|---|---|---|---|---|---|---|---|---|---|---|
| Moab | 74 | 287 | 112 | 456 | 313 | | 269 | 268 | 238 | 115 | 190 | 174 | 341 | 234 | 207 | 352 |
| Richfield | 249 | 114 | 224 | 282 | 239 | 174 | 194 | 219 | 166 | 121 | 115 | | 169 | 159 | 232 | 270 |

continued p. 208



Total mileages through Utah

15: 401 miles · 70: 232 miles · 80: 196 miles · 84: 119 miles

More mileages at go.randmcnally.com/MC

Index of cities Pg. 262 | Map legend Pg. 7

**Mileages between cities**

| | Blanding | Cedar City | Grand Jct., CO | Las Vegas, NV | Logan | Moab | Ogden | Page, AZ | Park City | Price | Provo | Richfield | St. George | Salt Lake City | Vernal | Wendover |
|---|---|---|---|---|---|---|---|---|---|---|---|---|---|---|---|---|
| St. George | 415 | 55 | 389 | 117 | 385 | 341 | 341 | 154 | 308 | 286 | 261 | 169 | | 304 | 401 | 333 |
| Salt Lake City | 308 | 250 | 283 | 419 | 82 | 234 | 37 | 377 | 30 | 119 | 43 | 159 | 304 | | 172 | 121 |

Mileages between cities

| | Blanding | Cedar City | Grand Jct., CO | Las Vegas, NV | Logan | Moab | Ogden | Page, AZ | Park City | Price | Provo | Richfield | St. George | Salt Lake City | Vernal | Wendover |
|---|---|---|---|---|---|---|---|---|---|---|---|---|---|---|---|---|
| Vernal | 281 | 345 | 140 | 514 | 252 | 207 | 207 | 450 | 145 | 112 | 154 | 232 | 401 | 172 | | 291 |
| Wendover | 426 | 317 | 401 | 361 | 199 | 352 | 154 | 503 | 150 | 237 | 161 | 270 | 333 | 121 | 291 | |



15 401 miles · 80 196 miles · 70 232 miles · 84 119 miles

Total mileages through Utah
More mileages at go.randmcnally.com/MC

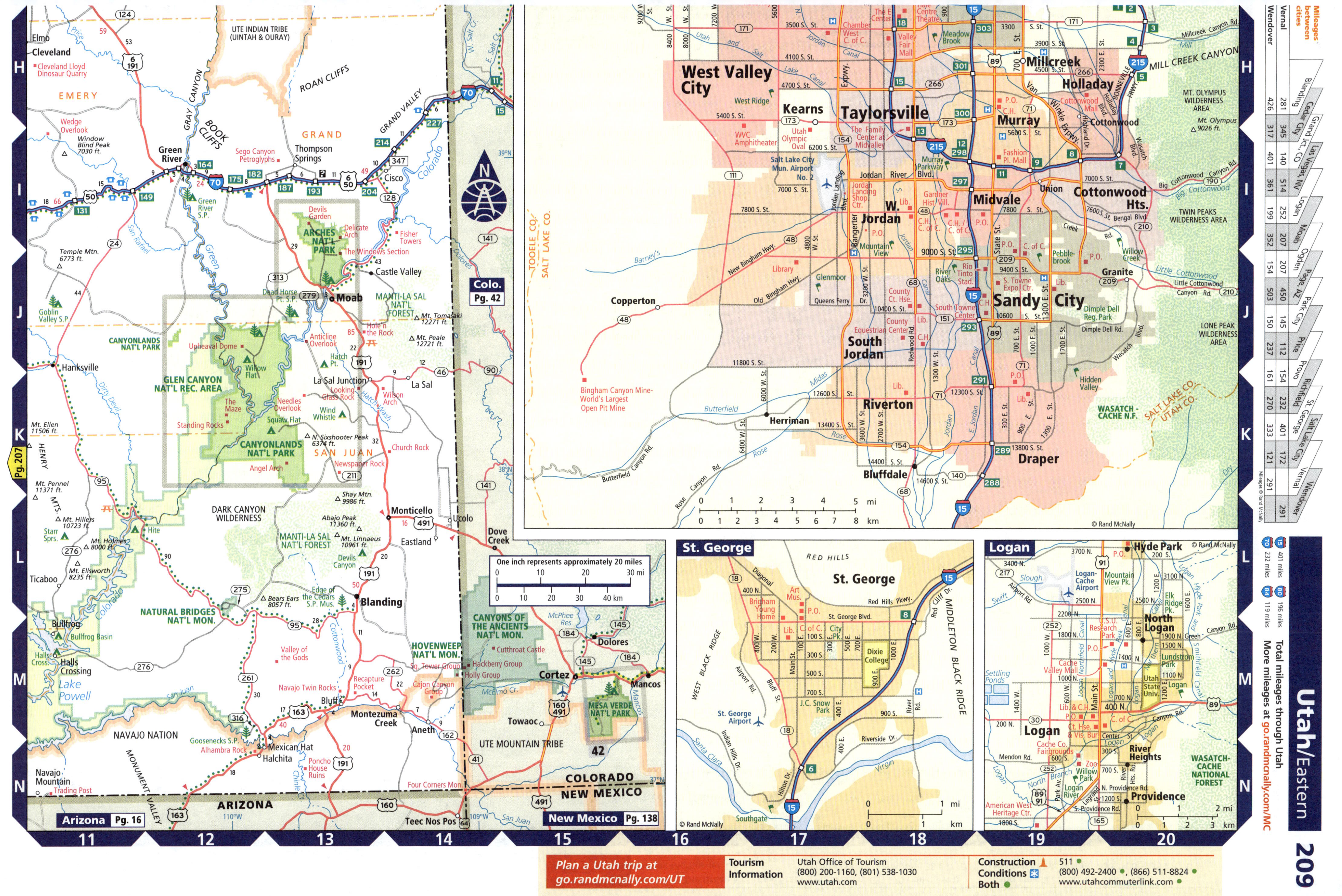


Tourism Information: Utah Office of Tourism (800) 200-1160, (801) 538-1030 www.utah.com

Construction · Conditions · Both: 511, (800) 492-2400, (866) 511-8824, www.utahcommuterlink.com

Plan a Utah trip at go.randmcnally.com/UT

Index of cities Pg. 262 | Map legend Pg. 7

**Mileages between cities**

| | Albany, NY | Brattleboro | Burlington | Montpelier | Newport | Rutland | St. Johnsbury | White River Jct. |
|---|---|---|---|---|---|---|---|---|
| Albany, NY | | 78 | 151 | 156 | 230 | 90 | 187 | 128 |
| Brattleboro | 78 | | 151 | 115 | 164 | 73 | 121 | 62 |
| Burlington | 151 | 151 | | 39 | 76 | 67 | 75 | 90 |
| Montpelier | 156 | 115 | 39 | | 78 | 66 | 37 | 54 |



**Vermont state facts**

Nickname: The Green Mountain State
Capital: Montpelier, E-5

Population: 621,270 (rank: 49th)
Largest city: Burlington, 38,897, D-2

Land area: 9,250 sq. mi. (rank: 43rd)
Highest point: Mount Mansfield, 4,393 ft., D-4

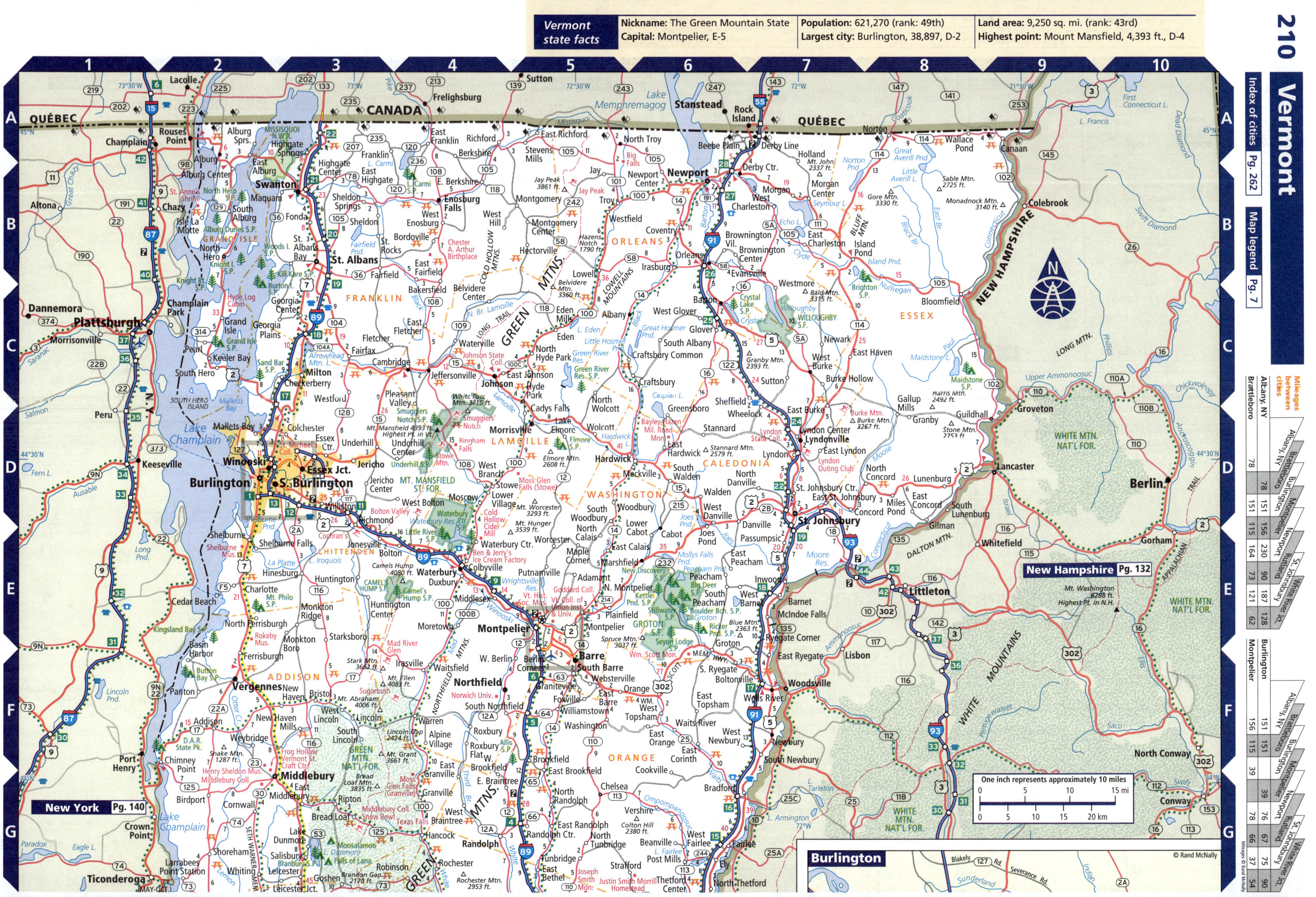

Mileages between cities

| | Albany, NY | Brattleboro | Burlington | Montpelier | Newport | Rutland | St. Johnsbury | White River Jct. |
|---|---|---|---|---|---|---|---|---|
| Newport | 230 | 164 | 76 | 78 | | 147 | 43 | 102 |
| Rutland | 90 | 73 | 67 | 66 | 147 | | 105 | 45 |
| St. Johnsbury | 187 | 121 | 75 | 37 | 43 | 105 | | 60 |
| White River Jct. | 128 | 62 | 90 | 54 | 102 | 45 | 60 | |



89 130 miles | 91 177 miles | 93 11 miles | 4 64 miles

Total mileages through Vermont
More mileages at go.randmcnally.com/MC

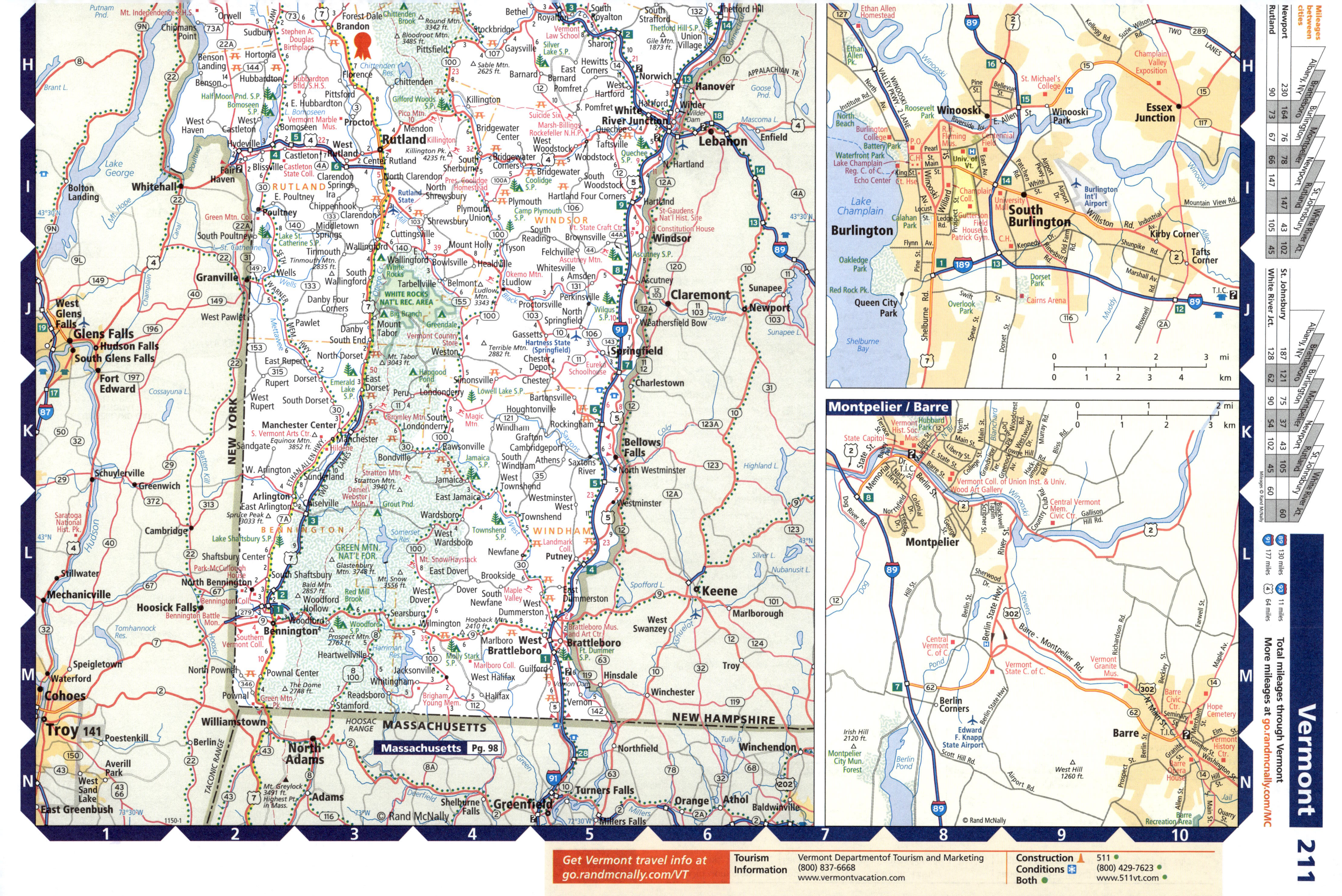

Index of cities Pg. 262 | Map legend Pg. 7

**Mileages between cities**

| | Bristol | Chincoteague | Danville | Emporia | Fredericksburg | Harrisonburg | Lynchburg | Manassas | Norfolk | Richmond | Roanoke | Virginia Beach | Washington, DC | Williamsburg | Winchester | Wytheville |
|---|---|---|---|---|---|---|---|---|---|---|---|---|---|---|---|---|
| Bristol | | 510 | 192 | 341 | 323 | 242 | 200 | 347 | 407 | 321 | 145 | 423 | 377 | 370 | 310 | 67 |
| Charlottesville | 253 | [illegible] | 131 | 136 | 66 | 61 | 65 | 81 | 157 | 71 | 117 | 174 | 116 | 121 | 128 | 183 |



**Virginia state facts**

**Nickname:** Old Dominion
**Capital:** Richmond, J-14
**Population:** 7,769,089 (rank: 12th)
**Largest city:** Virginia Beach, 433,746, L-18
**Land area:** 39,594 sq. mi. (rank: 37th)
**Highest point:** Mount Rogers, 5,729 ft., M-1

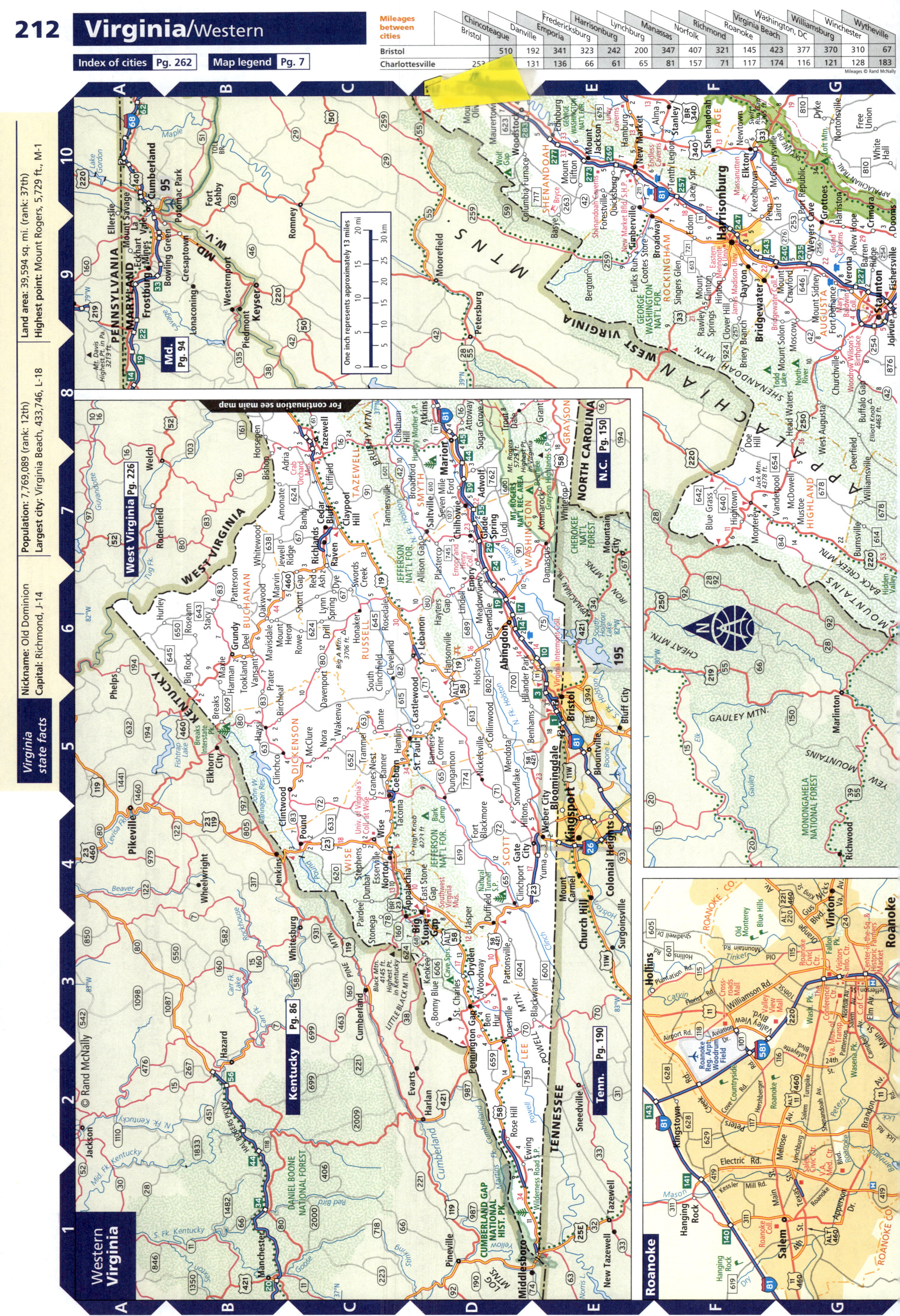

**Mileages between cities**

| | Bristol | Chincoteague | Danville | Emporia | Fredericksburg | Harrisonburg | Lynchburg | Manassas | Norfolk | Richmond | Roanoke | Virginia Beach | Washington, DC | Williamsburg | Winchester | Wytheville |
|---|---|---|---|---|---|---|---|---|---|---|---|---|---|---|---|---|
| Danville | 192 | 300 | | 115 | 197 | 163 | 68 | 215 | 191 | 144 | 89 | 206 | 247 | 199 | 230 | 124 |
| Norfolk | 407 | 104 | 191 | 78 | 139 | 216 | 189 | 177 | | 91 | 276 | 17 | 189 | 41 | 222 | 340 |

continued p. 214



**Total mileages through Virginia**

64: 298 miles; 81: 325 miles; 85: 69 miles; 95: 179 miles

More mileages at go.randmcnally.com/MC

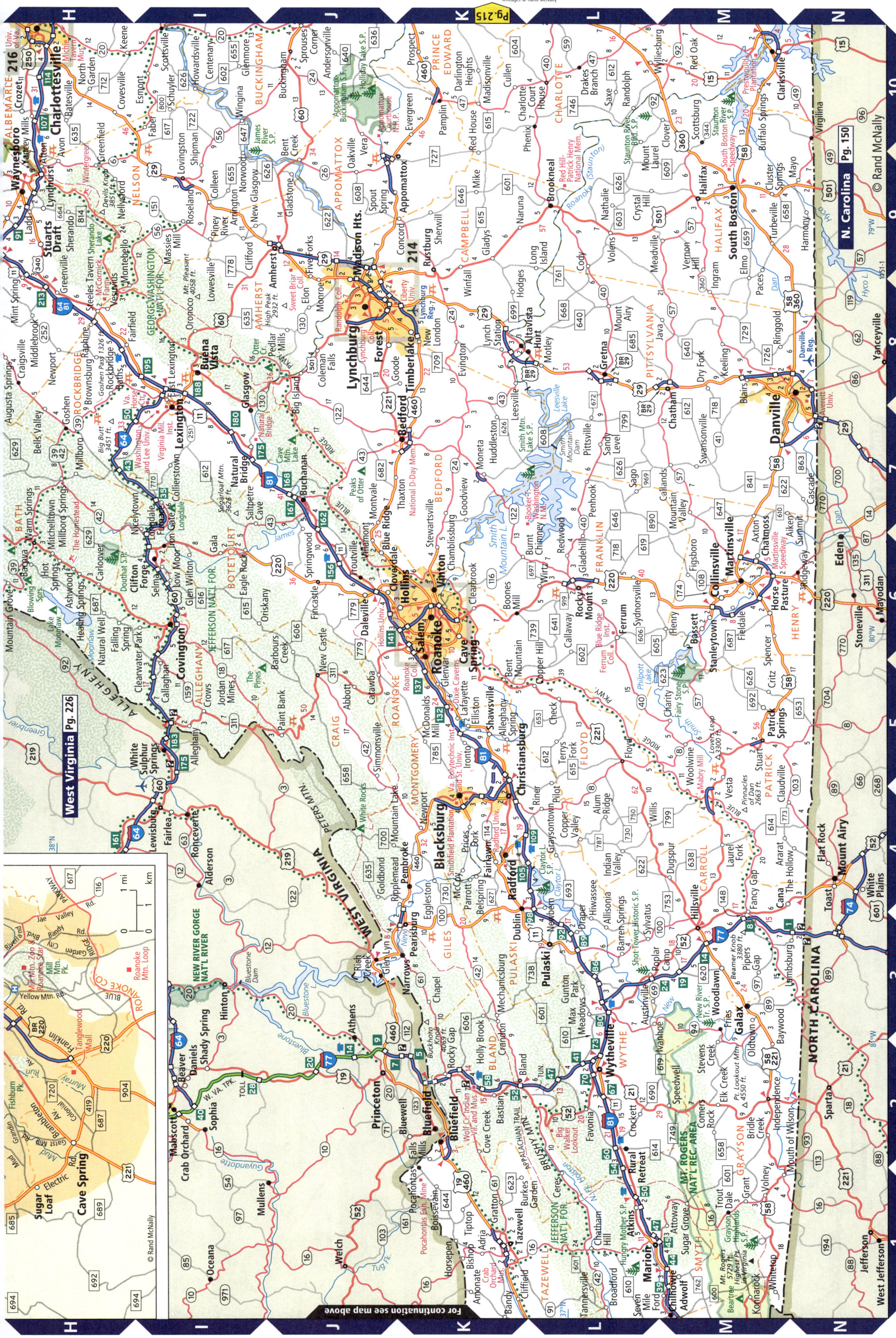

Pg. 215

Explore Virginia go.randmcnally.com/VA

Tourism Information: Virginia Tourism Corporation (800) 847-4882 www.virginia.org

Construction / Conditions / Both: 511 (800) 578-4111 (800) 367-7623 www.511virginia.org

Index of cities Pg. 262 | Map legend Pg. 7

Mileages between cities

| | Bristol | Chincoteague | Danville | Emporia | Fredericksburg | Harrisonburg | Lynchburg | Manassas | Norfolk | Richmond | Roanoke | Virginia Beach | Washington, DC | Williamsburg | Winchester | Wytheville |
|---|---|---|---|---|---|---|---|---|---|---|---|---|---|---|---|---|
| Richmond | 321 | 190 | 144 | 66 | 56 | 130 | 114 | 96 | 91 | | 187 | 105 | 107 | 50 | 135 | 253 |
| Roanoke | 145 | 378 | 89 | 176 | 192 | 111 | 53 | 214 | 276 | 187 | | 292 | 241 | 238 | 178 | 77 |



Virginia state facts

Nickname: Old Dominion
Capital: Richmond, J-14
Population: 7,769,089 (rank: 12th)
Largest city: Virginia Beach, 433,746, L-18
Land area: 39,594 sq. mi. (rank: 37th)
Highest point: Mount Rogers, 5,729 ft., M-1

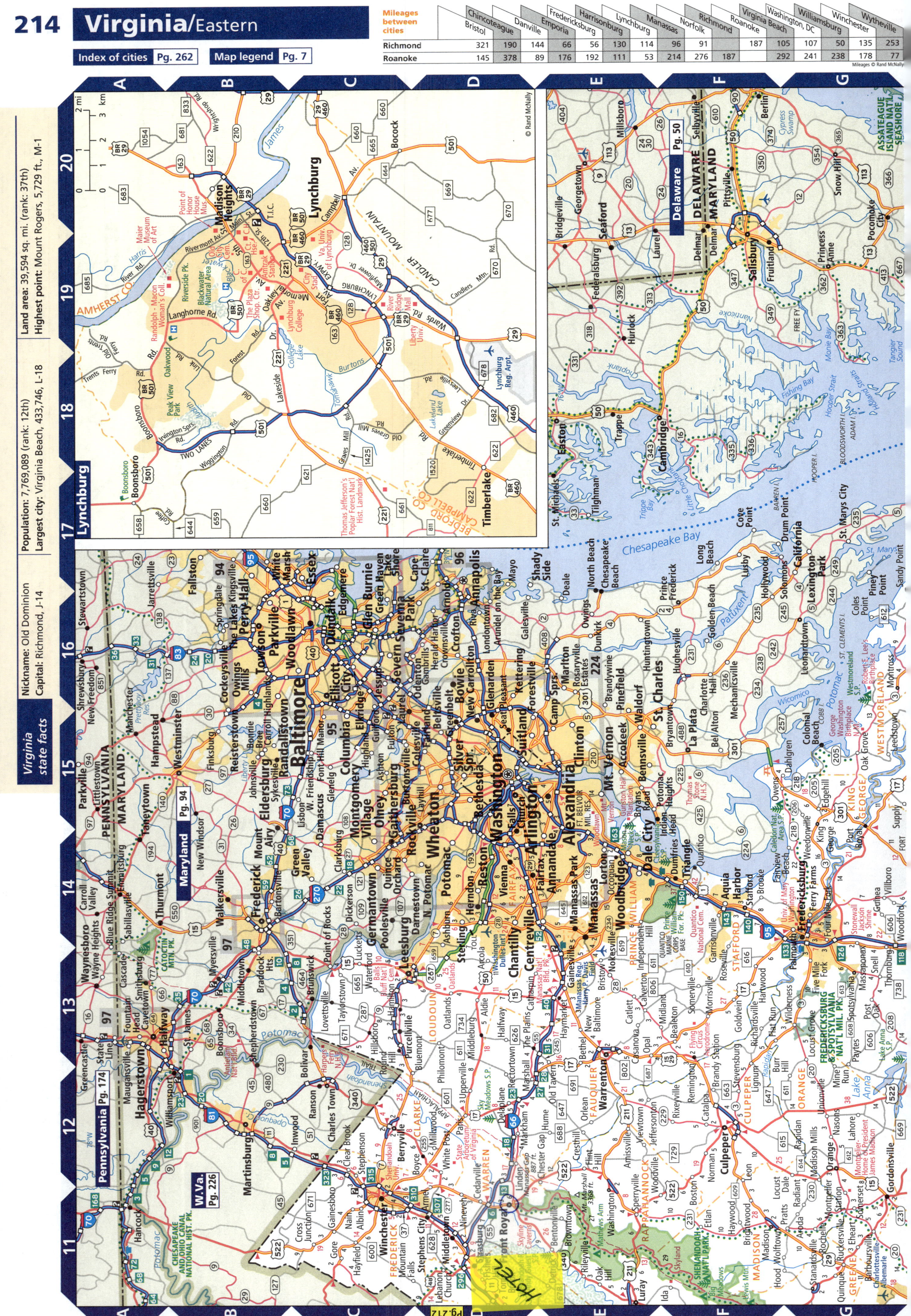

**Mileages between cities**

| | Bristol | Chincoteague | Danville | Emporia | Fredericksburg | Harrisonburg | Lynchburg | Manassas | Norfolk | Richmond | Roanoke | Virginia Beach | Washington, DC | Williamsburg | Winchester | Wytheville |
|---|---|---|---|---|---|---|---|---|---|---|---|---|---|---|---|---|
| Washington, DC | 377 | 168 | 247 | 174 | 53 | 132 | 182 | 32 | 189 | 107 | 241 | 205 | | 153 | 76 | 307 |
| Winchester | 310 | 244 | 230 | 200 | 83 | 68 | 164 | 54 | 222 | 135 | 178 | 236 | 76 | 181 | | 244 |



64 298 miles
81 325 miles
85 69 miles
95 179 miles

Total mileages through Virginia
More mileages at go.randmcnally.com/MC

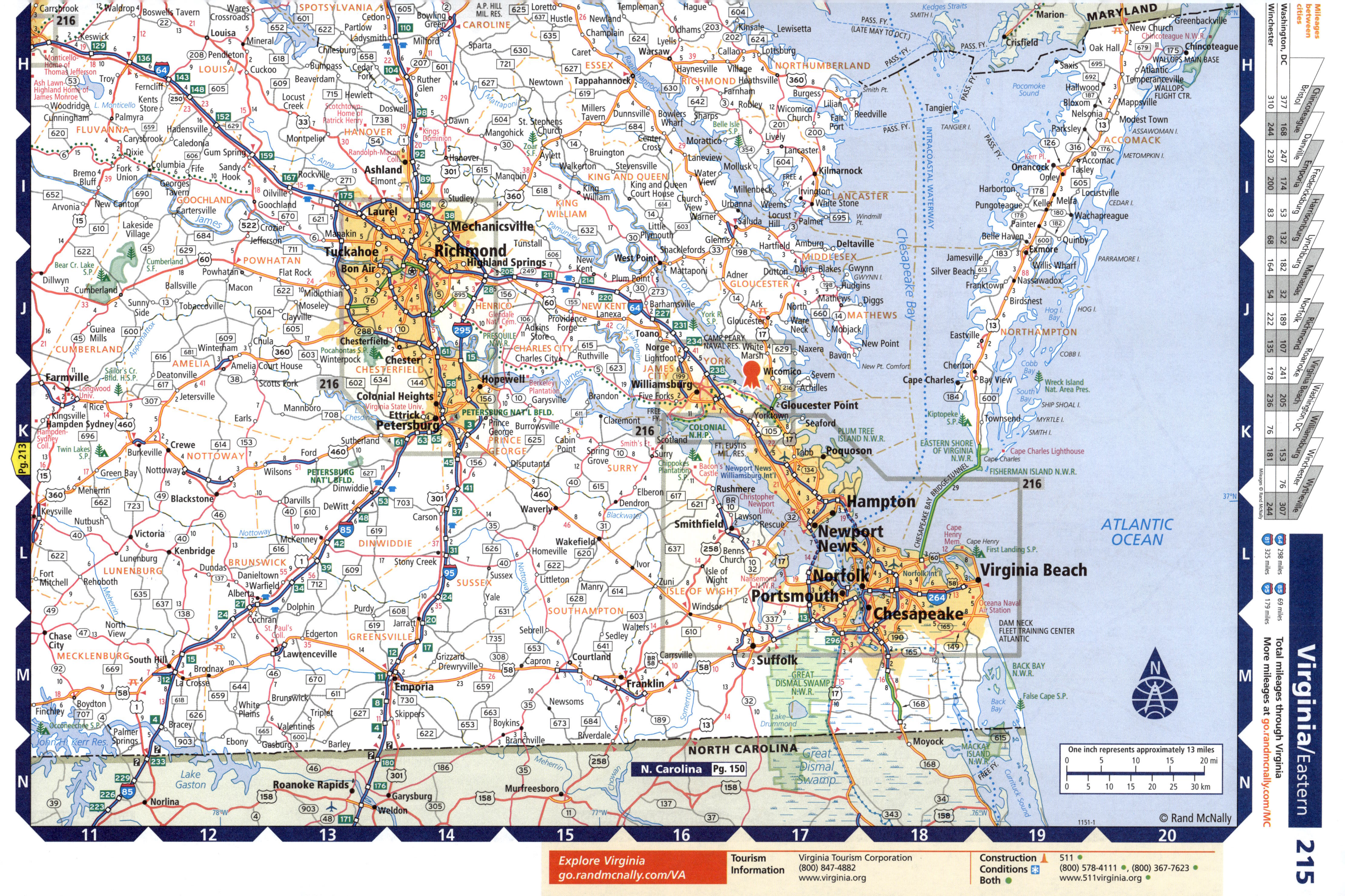

**City sights to see**

- Agecroft Hall and Gardens, Richmond, C-7
- Children's Museum of Virginia, Portsmouth, M-6
- Chrysler Museum of Art, Norfolk, L-6
- Colonial Williamsburg, Williamsburg, F-2
- Edgar Allen Poe Museum, Richmond, C-8
- First Landing State Park, Virginia Beach, L-9
- Hermitage Foundation Museum, Norfolk, K-5

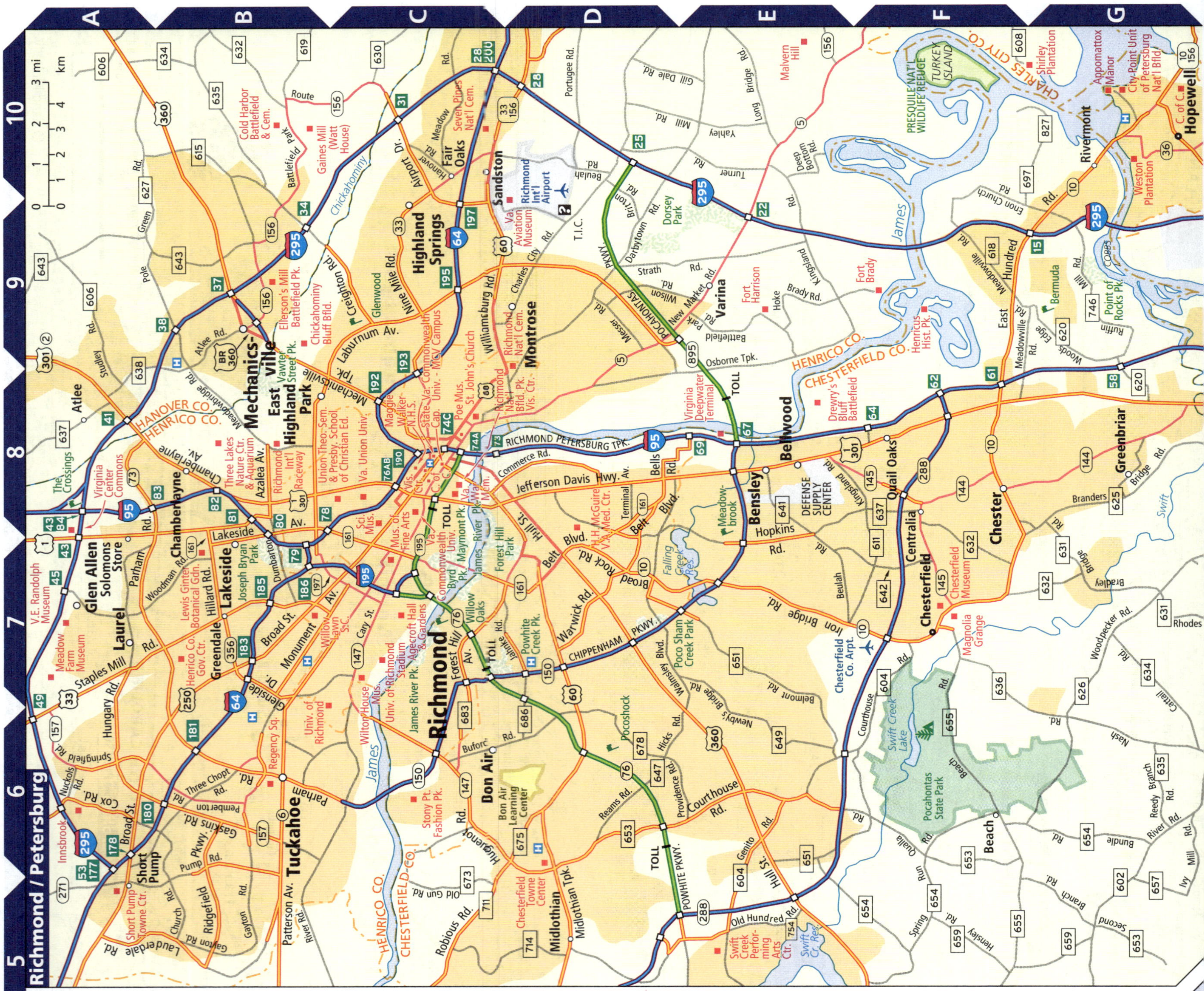

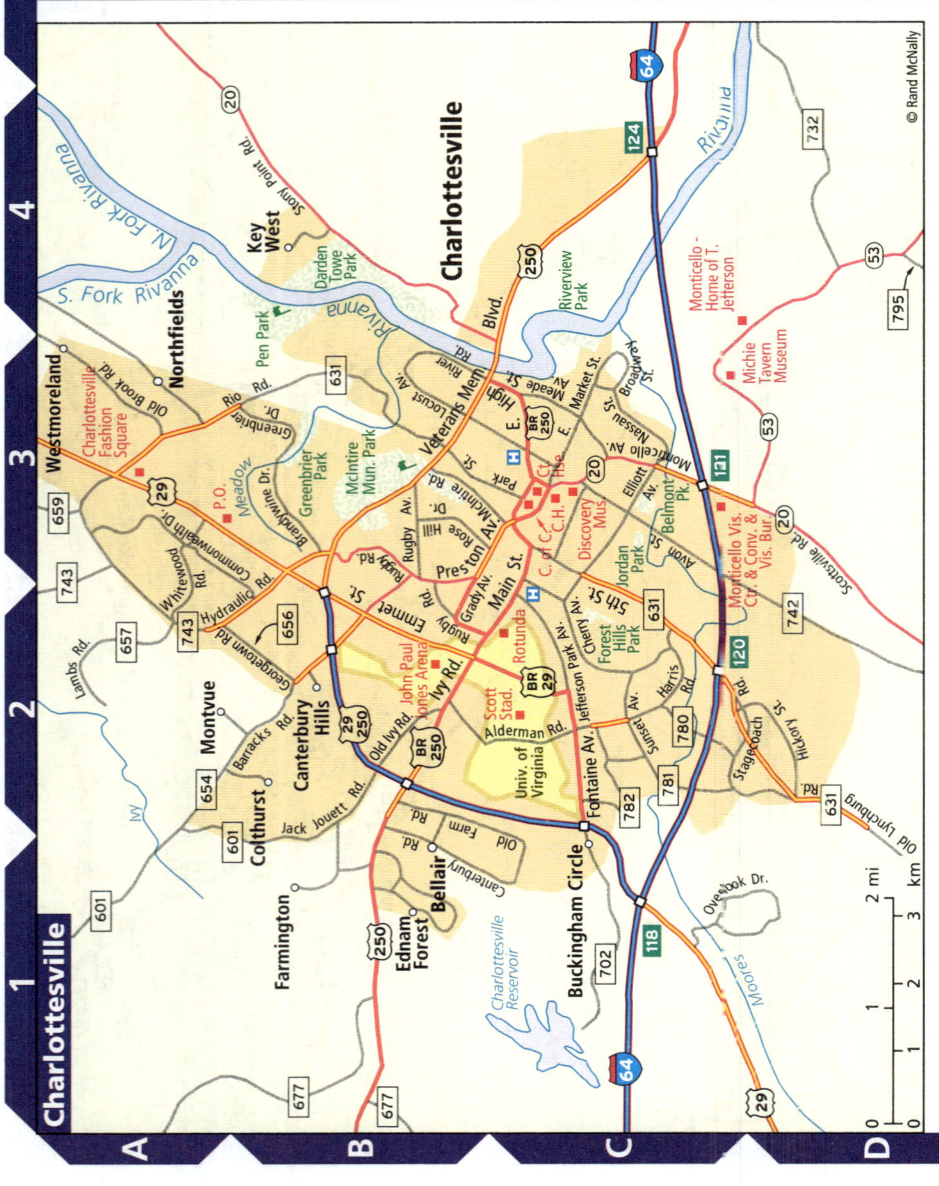

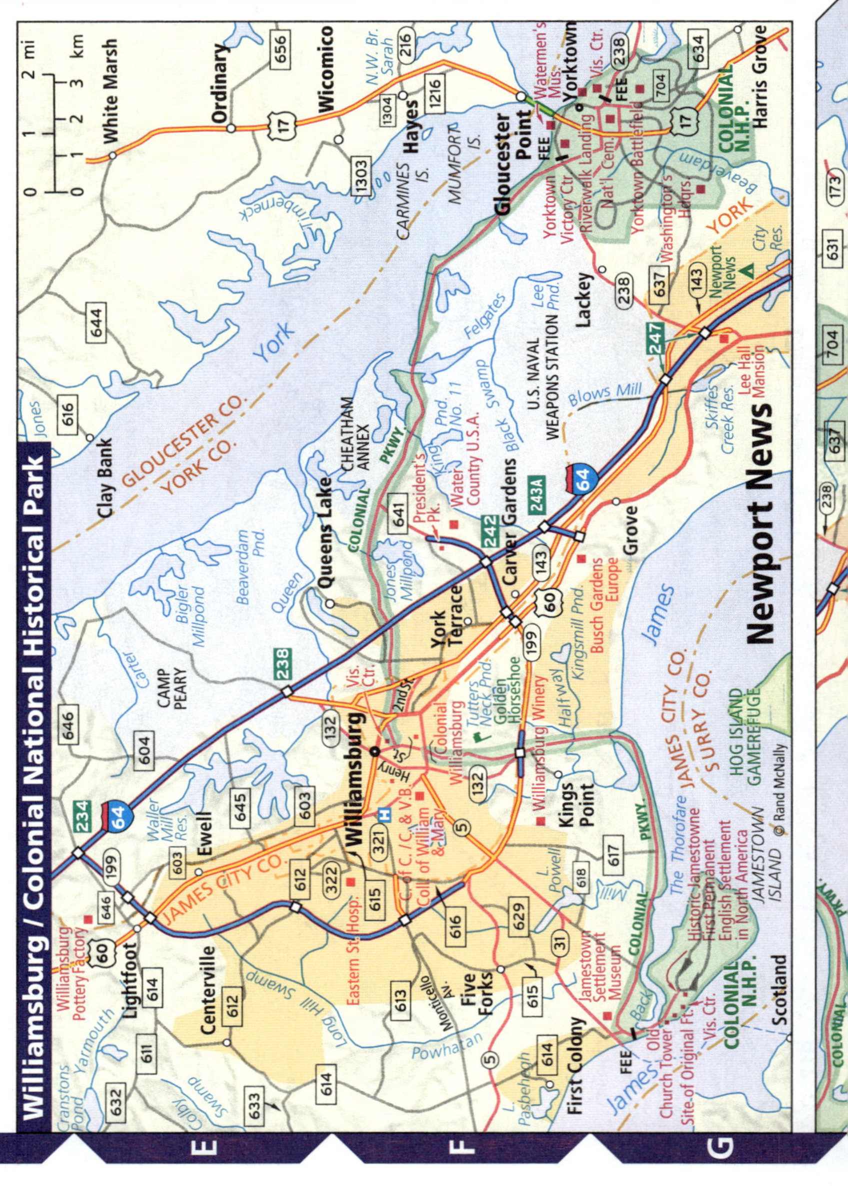

Richmond / Petersburg

Charlottesville

Williamsburg / Colonial National Historical Park

- Historic Jamestowne, Williamsburg, G-1
- Monticello, Charlottesville, D-4
- Nauticus, Norfolk, L-6
- Ocean Breeze Waterpark, Virginia Beach, M-10
- Old Cape Henry Lighthouse, Virginia Beach, K-9
- Three Lakes Nature Center and Aquarium, Richmond, B-8
- Virginia State Capitol, Richmond, C-8

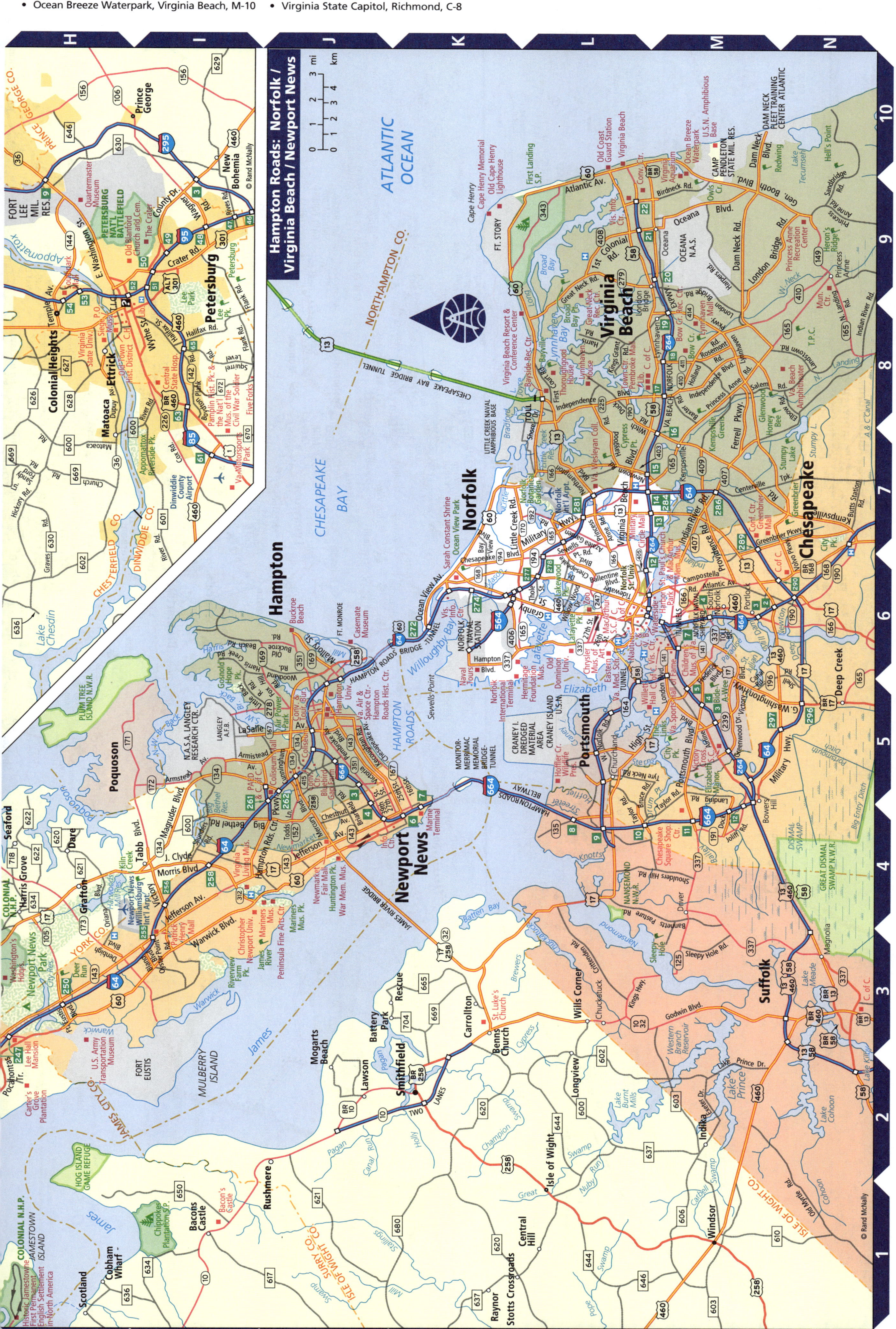

Index of cities Pg. 263 | Map legend Pg. 7

**Washington state facts**

Nickname: The Evergreen State
Capital: Olympia, H-6
Population: 6,549,224 (rank: 13th)
Largest city: Seattle, 598,541, F-7
Land area: 66,544 sq. mi. (rank: 20th)
Highest point: Mount Rainier, 14,411 ft., I-8

Mileages between cities

| | Aberdeen | Bellingham | Colville | Kennewick | Longview | Olympia | Omak | Port Angeles | Portland, OR | Seattle | Spokane | Tacoma | The Dalles, OR | Vancouver, BC | Wenatchee | Yakima |
|---|---|---|---|---|---|---|---|---|---|---|---|---|---|---|---|---|
| Bellingham | 198 | | 317 | 306 | 216 | 149 | 201 | 118 | 261 | 89 | 361 | 121 | 326 | 52 | 182 | 224 |
| Kennewick | 312 | 306 | 209 | | 254 | 263 | 189 | 340 | 213 | 223 | 138 | 235 | 130 | 359 | 132 | 82 |

Mileages © Rand McNally

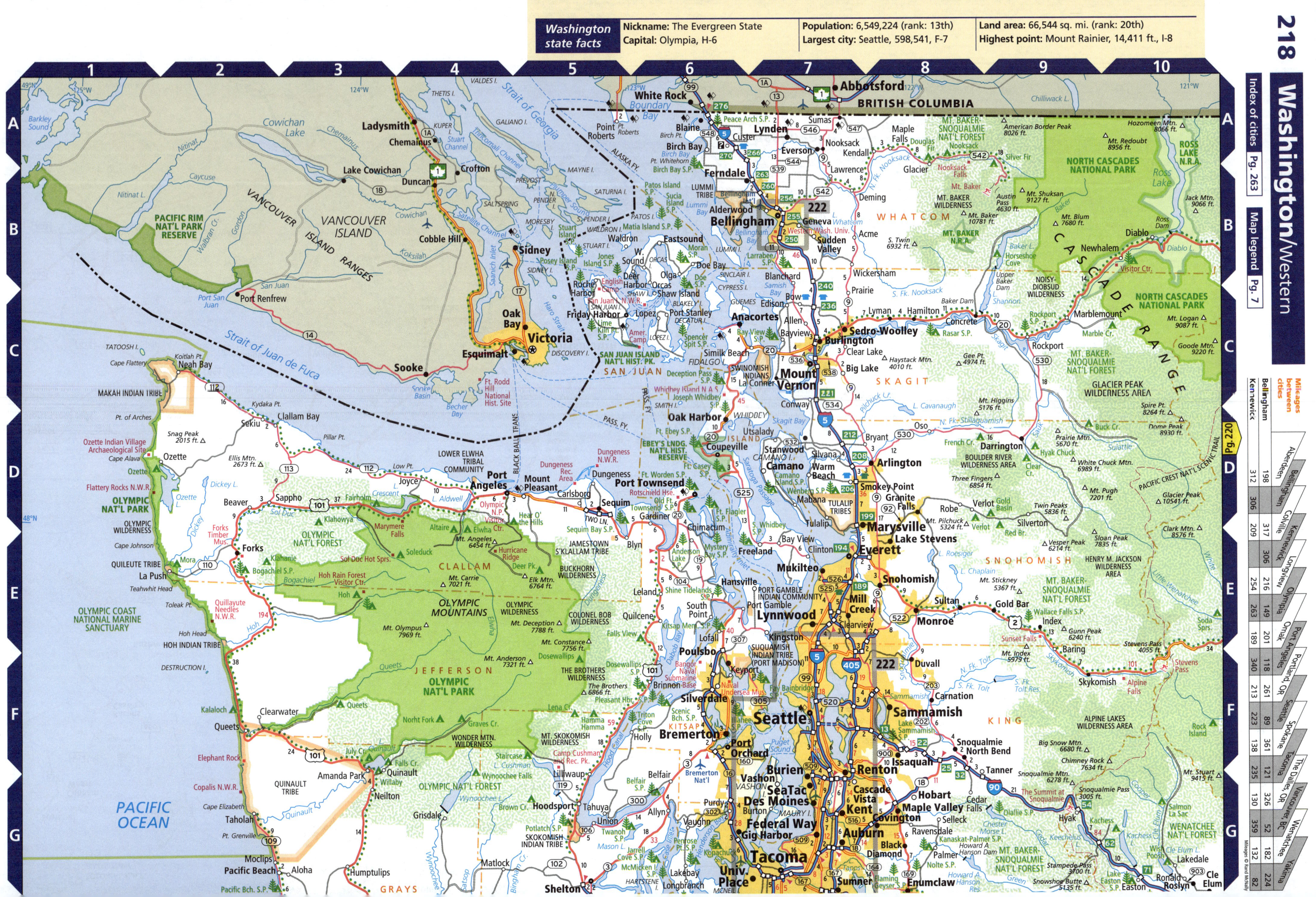

**Mileages between cities**

| | Aberdeen | Bellingham | Colville | Kennewick | Longview | Olympia | Omak | Port Angeles | Portland, OR | Seattle | Spokane | Tacoma | The Dalles, OR | Vancouver, BC | Wenatchee | Yakima |
|---|---|---|---|---|---|---|---|---|---|---|---|---|---|---|---|---|
| Lewiston, ID | 402 | 396 | 173 | 124 | 381 | 353 | 237 | 431 | 339 | 313 | 102 | 325 | 256 | 449 | 228 | 204 |
| Portland, OR | 141 | 261 | 422 | 213 | 48 | 113 | 377 | 228 | | 172 | 351 | 141 | 83 | 313 | 291 | 185 |

continued p. 220



5 277 miles | 90 297 miles
82 133 miles | 101 373 miles

Total mileages through Washington
More mileages at go.randmcnally.com/MC

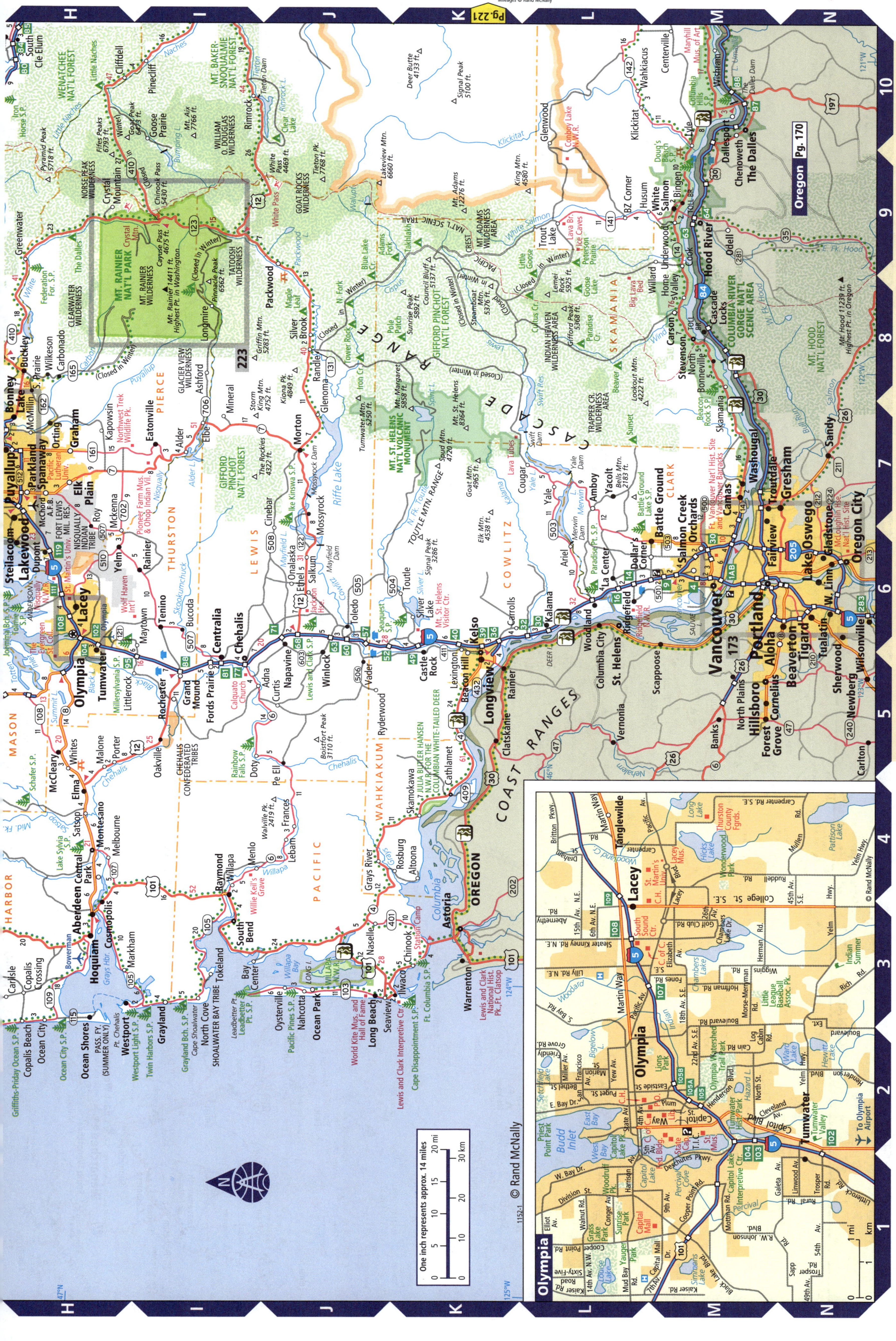

Pg. 221

Pg. 170

223

Get more Washington info at go.randmcnally.com/WA

Tourism Information
Washington State Tourism
(800) 544-1800
www.experiencewa.com

Construction
Conditions
Both
511
(800) 695-7623
www.wsdot.wa.gov/traffic/

One inch represents approx. 14 miles

Index of cities Pg. 263

Map legend Pg. 7

**Mileages between cities**

| | Aberdeen | Bellingham | Colville | Kennewick | Longview | Olympia | Omak | Port Angeles | Portland, OR | Seattle | Spokane | Tacoma | The Dalles, OR | Vancouver, BC | Wenatchee | Yakima |
|---|---|---|---|---|---|---|---|---|---|---|---|---|---|---|---|---|
| Seattle | 108 | 89 | 350 | 223 | 127 | 60 | 236 | 83 | 172 | | 278 | 32 | 249 | 141 | 148 | 141 |
| Spokane | 367 | 361 | 71 | 138 | 386 | 319 | 139 | 396 | 351 | 278 | | 291 | 268 | 413 | 169 | 201 |



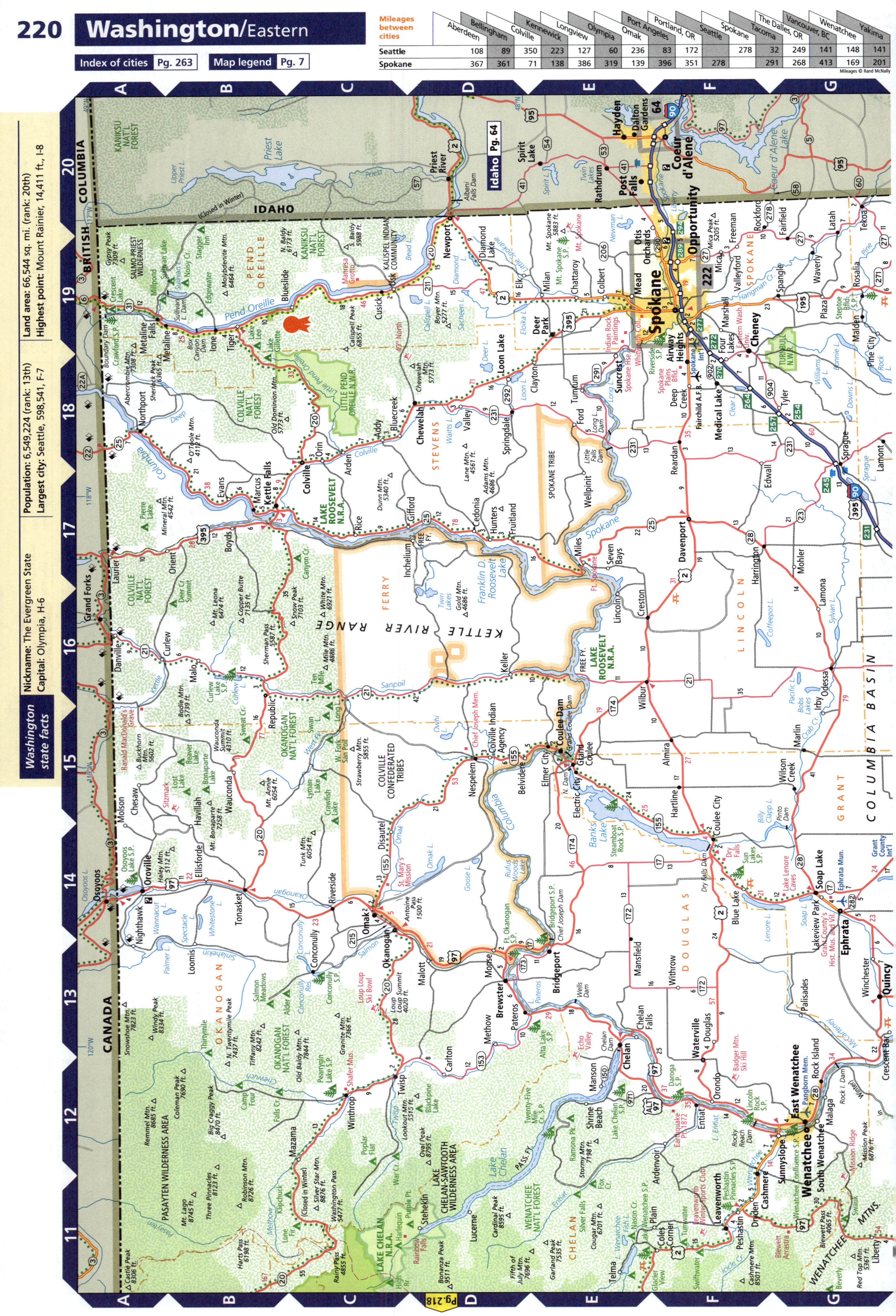

**Mileages between cities**

| | Aberdeen | Bellingham | Colville | Kennewick | Longview | Olympia | Omak | Port Angeles | Portland, OR | Seattle | Spokane | Tacoma | The Dalles, OR | Vancouver, BC | Wenatchee | Yakima |
|---|---|---|---|---|---|---|---|---|---|---|---|---|---|---|---|---|
| Tacoma | 77 | 121 | 362 | 235 | 96 | 28 | 248 | 106 | 141 | 32 | 291 | | 217 | 174 | 160 | 153 |
| Yakima | 230 | 224 | 272 | 82 | 166 | 181 | 192 | 259 | 185 | 141 | 201 | 153 | 102 | 276 | 106 | |



5 277 miles · 90 297 miles · 82 133 miles · 101 373 miles

Total mileages through Washington
More mileages at go.randmcnally.com/MC

Yakima

Oregon Pg. 170

Pg. 219

One inch represents approx. 14 miles



Get more Washington info at go.randmcnally.com/WA

Tourism Information: Washington State Tourism (800) 544-1800 www.experiencewa.com

Construction / Conditions / Both: 511 (800) 695-7623 www.wsdot.wa.gov/traffic/

**City sights to see**

- Experience Music Project, Seattle, H-1
- Frye Art Museum, Seattle, J-3
- Klondike Gold Rush National Historical Park, Seattle, K-2
- Logmire Museum, Logmire, N-1
- Museum of Glass, Tacoma, L-6
- Nordic Heritage Museum, Seattle, C-7
- Pacific Science Center, Seattle, H-1

- Pike Place Market, Seattle, J-2
- Point Defiance Zoo & Aquarium, Tacoma, K-5
- Seattle Aquarium, Seattle, J-1
- Space Needle, Seattle, H-1
- Washington State History Mus., Tacoma, L-6
- Whatcom Museum of History and Art, Bellingham, E-2
- Woodland Park Zoo, Seattle, C-7

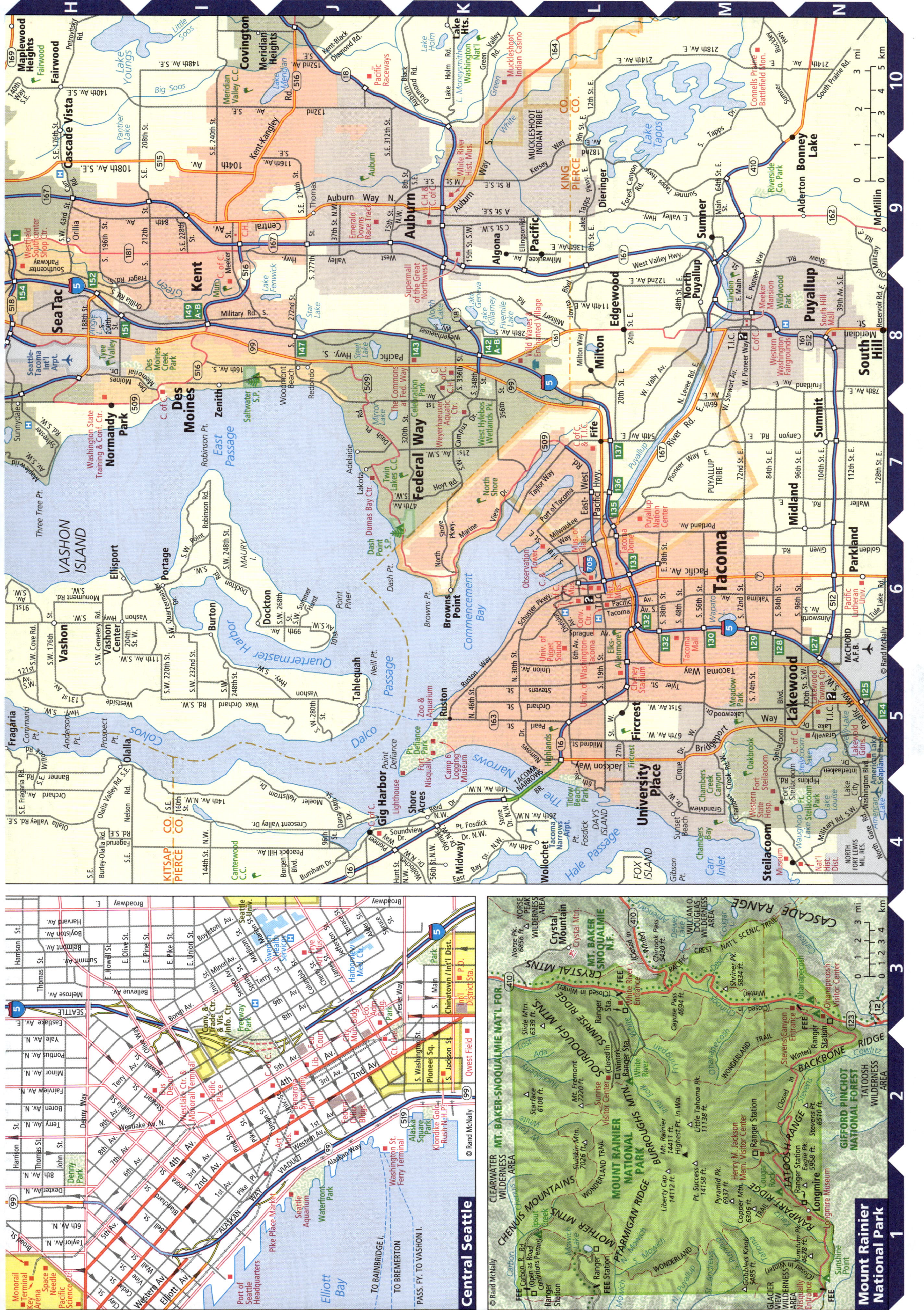

Index of cities Pg. 239 | Map legend Pg. 7

**City sights to see**

- Arlington National Cemetery, Arlington, Va., G-6
- Frederick Douglass National Historic Site, G-7
- Freedom Park, Arlington, Va., G-5
- John F. Kennedy Ctr. for the Performing Arts, K-3
- National Arboretum, F-7
- National Mall, M-7
- National Postal Museum, K-9

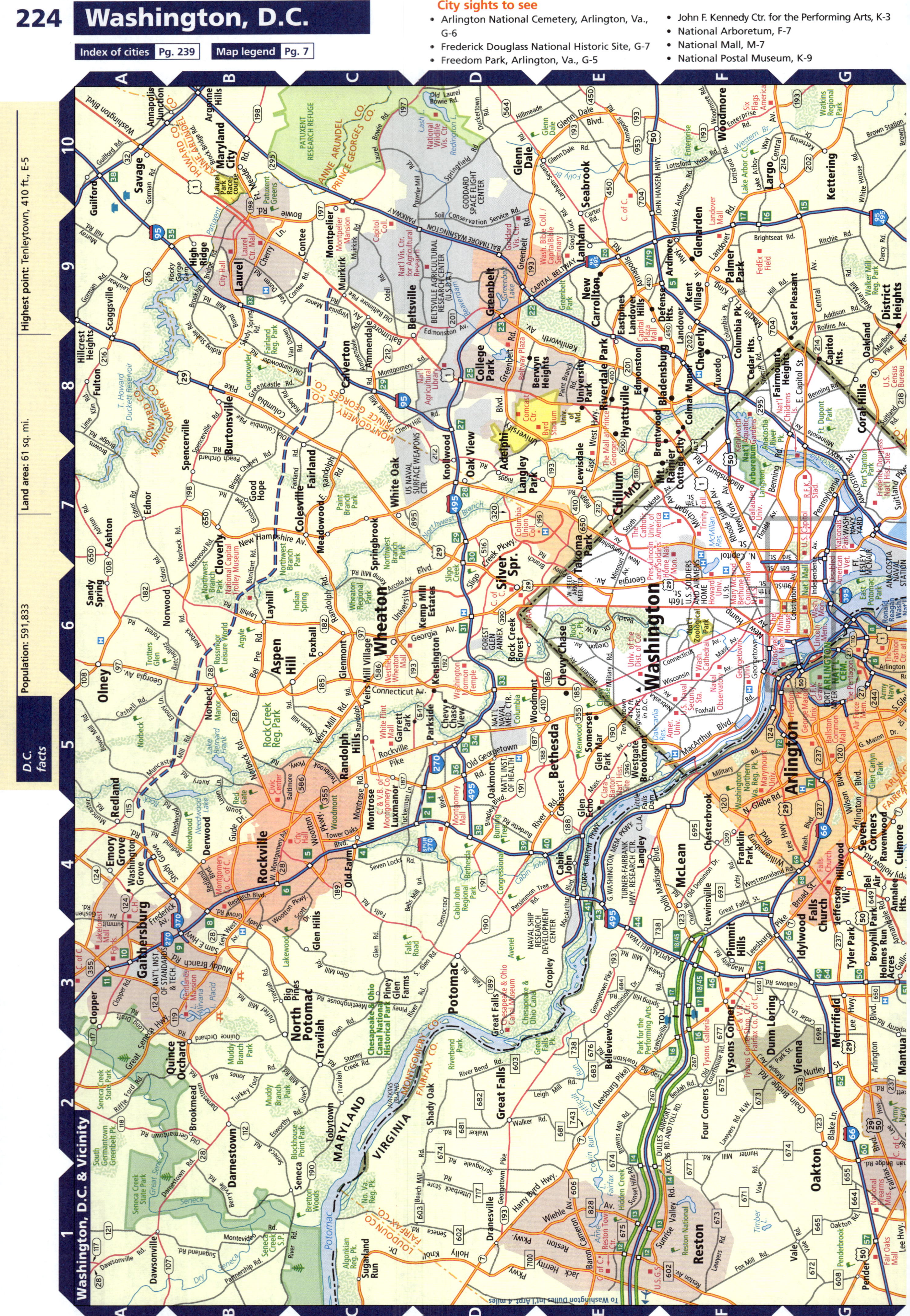

- National Zoological Park, F-6
- Patuxent Research Refuge National Wildlife Visitor Center, Laurel, Md., D-10
- The Pentagon, Arlington, Va., G-6
- Supreme Court of the United States, M-10
- United States Botanic Garden, M-8
- Wolf Trap National Park for the Performing Arts, Vienna, Va., E-2

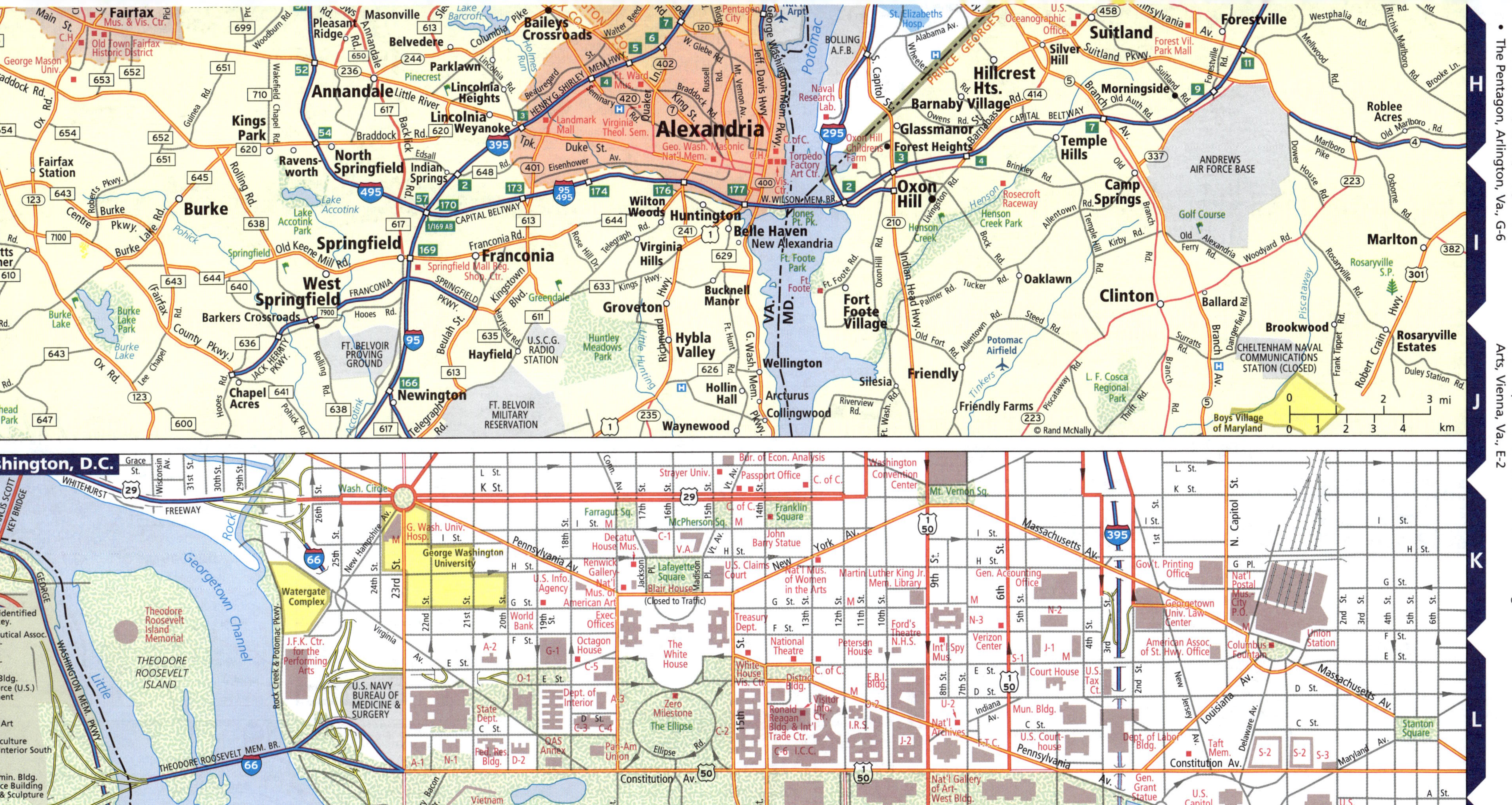

Index of cities Pg. 263 | Map legend Pg. 7

**Mileages between cities**

| | Bluefield | Charleston | Clarksburg | Cumberland, MD | Martinsburg | Petersburg | Wheeling | Wh. Sulphur Sprs. |
|---|---|---|---|---|---|---|---|---|
| Beckley | 50 | 59 | 136 | 239 | 267 | 184 | 236 | 59 |
| Charleston | 106 | | 123 | 225 | 304 | 193 | 177 | 120 |

| | Bluefield | Charleston | Clarksburg | Cumberland, MD | Martinsburg | Petersburg | Wheeling | Wh. Sulphur Sprs. |
|---|---|---|---|---|---|---|---|---|
| Cumberland, MD | 288 | 225 | 109 | | 79 | 66 | 155 | 194 |
| Huntington | 158 | 51 | 174 | 276 | 355 | 244 | 228 | 172 |

Mileages © Rand McNally

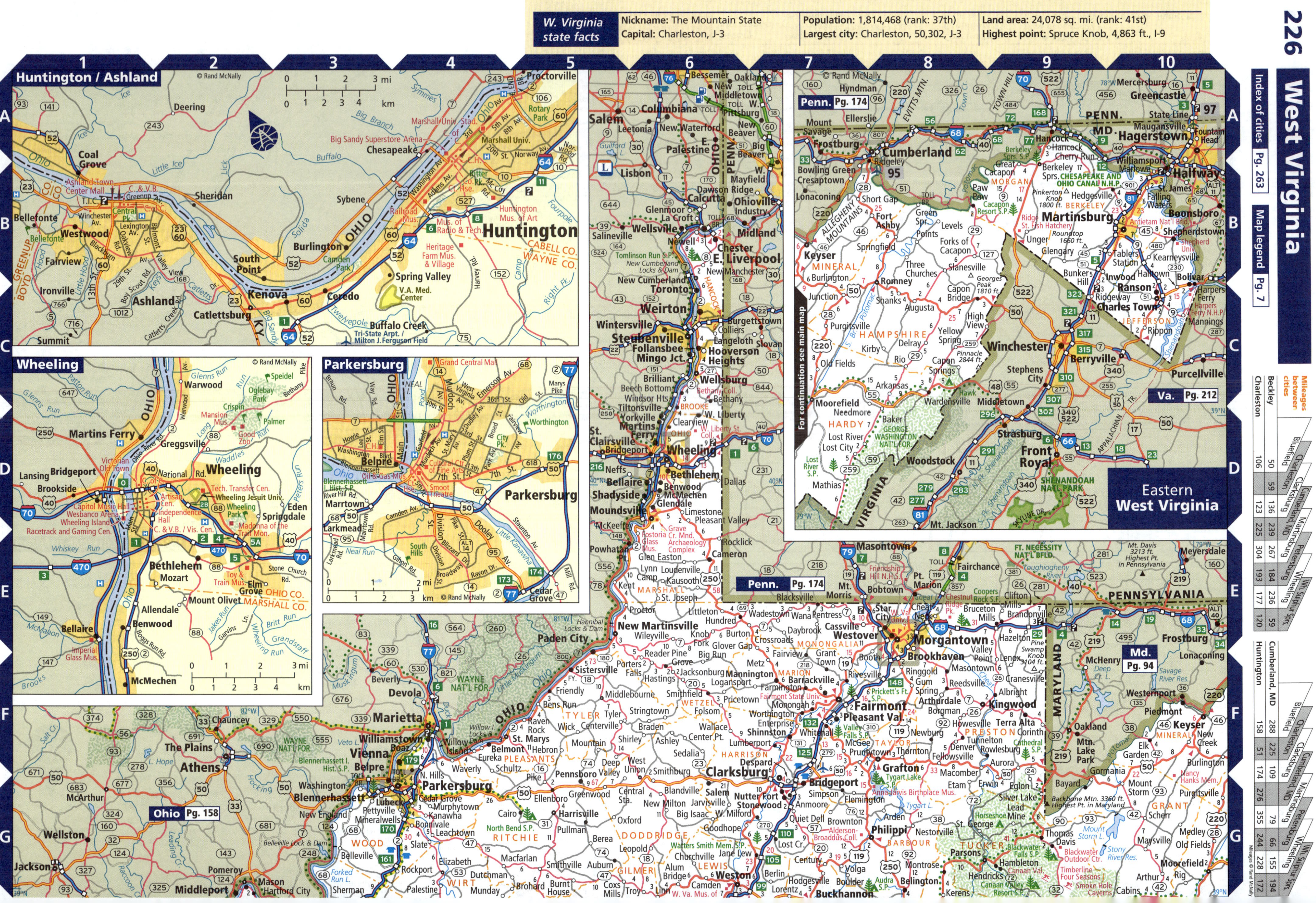

**Mileages between cities**

| | Bluefield | Charleston | Clarksburg | Cumberland, MD | Martinsburg | Petersburg | Wheeling | Wh. Sulphur Sprs. |
|---|---|---|---|---|---|---|---|---|
| Morgantown | 218 | 154 | 38 | 73 | 151 | 103 | 78 | 187 |
| Parkersburg | 183 | 76 | 72 | 181 | 259 | 172 | 104 | 198 |
| Wheeling | 283 | 177 | 114 | 155 | 225 | 179 | | 262 |
| White Sulphur Sprs. | 79 | 120 | 155 | 194 | 208 | 125 | 262 | |

Mileages © Rand McNally

**Total mileages through West Virginia**

- 64: 189 miles
- 70: 14 miles
- 77: 187 miles
- 79: 161 miles

More mileages at go.randmcnally.com/MC

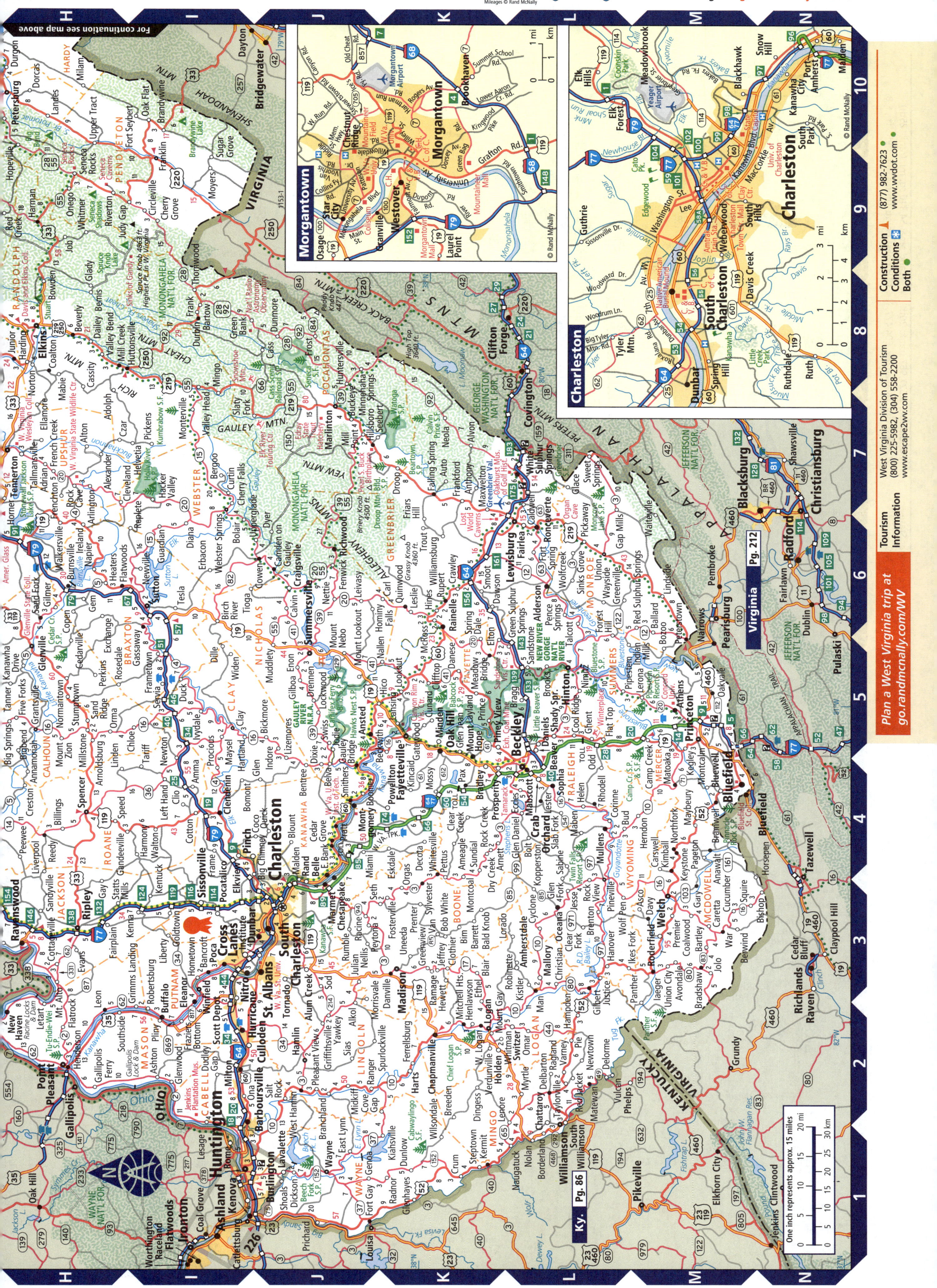

For continuation see map above

Morgantown

Charleston

**Tourism Information**
West Virginia Division of Tourism
(800) 225-5982, (304) 558-2200
www.escape2wv.com

Construction
Conditions
Both
(877) 982-7623
www.wvdot.com

Plan a West Virginia trip at go.randmcnally.com/WV

One inch represents approx. 15 miles

Index of cities Pg. 264 | Map legend Pg. 7

| Mileages between cities | Beloit | Chicago, IL | Dubuque, IA | Eau Claire | Green Bay | Hayward | La Crosse | Madison | Milwaukee | Oshkosh | Rhinelander | Sheboygan | Sturgeon Bay | Superior | Wausau | Wisconsin Dells |
|---|---|---|---|---|---|---|---|---|---|---|---|---|---|---|---|---|
| Chicago, IL | 96 | | 177 | 315 | 206 | 420 | 281 | 146 | 90 | 175 | 338 | 145 | 245 | 462 | 281 | 195 |
| Eau Claire | 223 | 315 | 192 | | 192 | 106 | 86 | 177 | 243 | 181 | 155 | 228 | 237 | 149 | 98 | 124 |

Mileages © Rand McNally

Mileages between cities

| | Beloit | Chicago, IL | Dubuque, IA | Eau Claire | Green Bay | Hayward | La Crosse | Madison | Milwaukee | Oshkosh | Rhinelander | Sheboygan | Sturgeon Bay | Superior | Wausau | Wisconsin Dells |
|---|---|---|---|---|---|---|---|---|---|---|---|---|---|---|---|---|
| Green Bay | 184 | 206 | 233 | 192 | | 283 | 203 | 138 | 116 | 52 | 136 | 64 | 44 | 326 | 96 | 132 |
| La Crosse | 188 | 281 | 119 | 86 | 203 | 190 | | 143 | 209 | 153 | 214 | 195 | 248 | 233 | 170 | 90 |

continued p. 230



39 182 miles
43 192 miles
90 189 miles
94 341 miles

Total mileages through Wisconsin
More mileages at go.randmcnally.com/MC

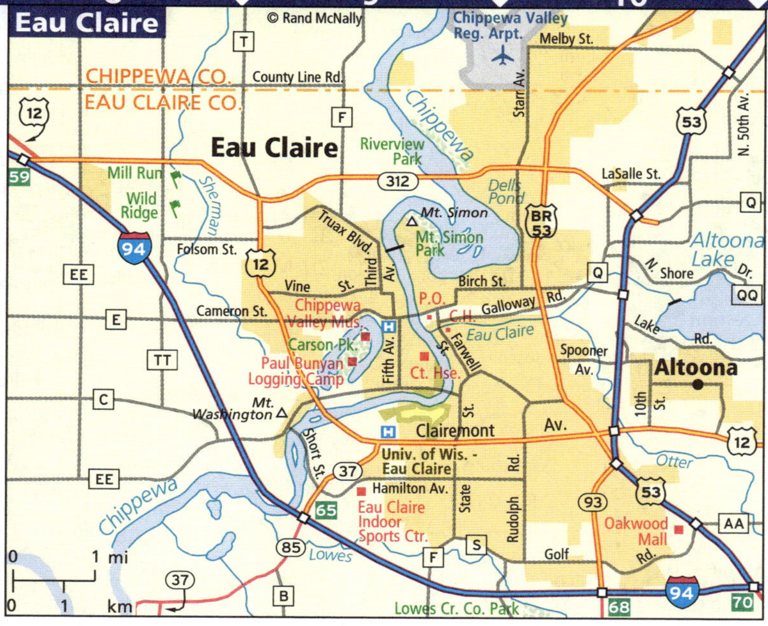

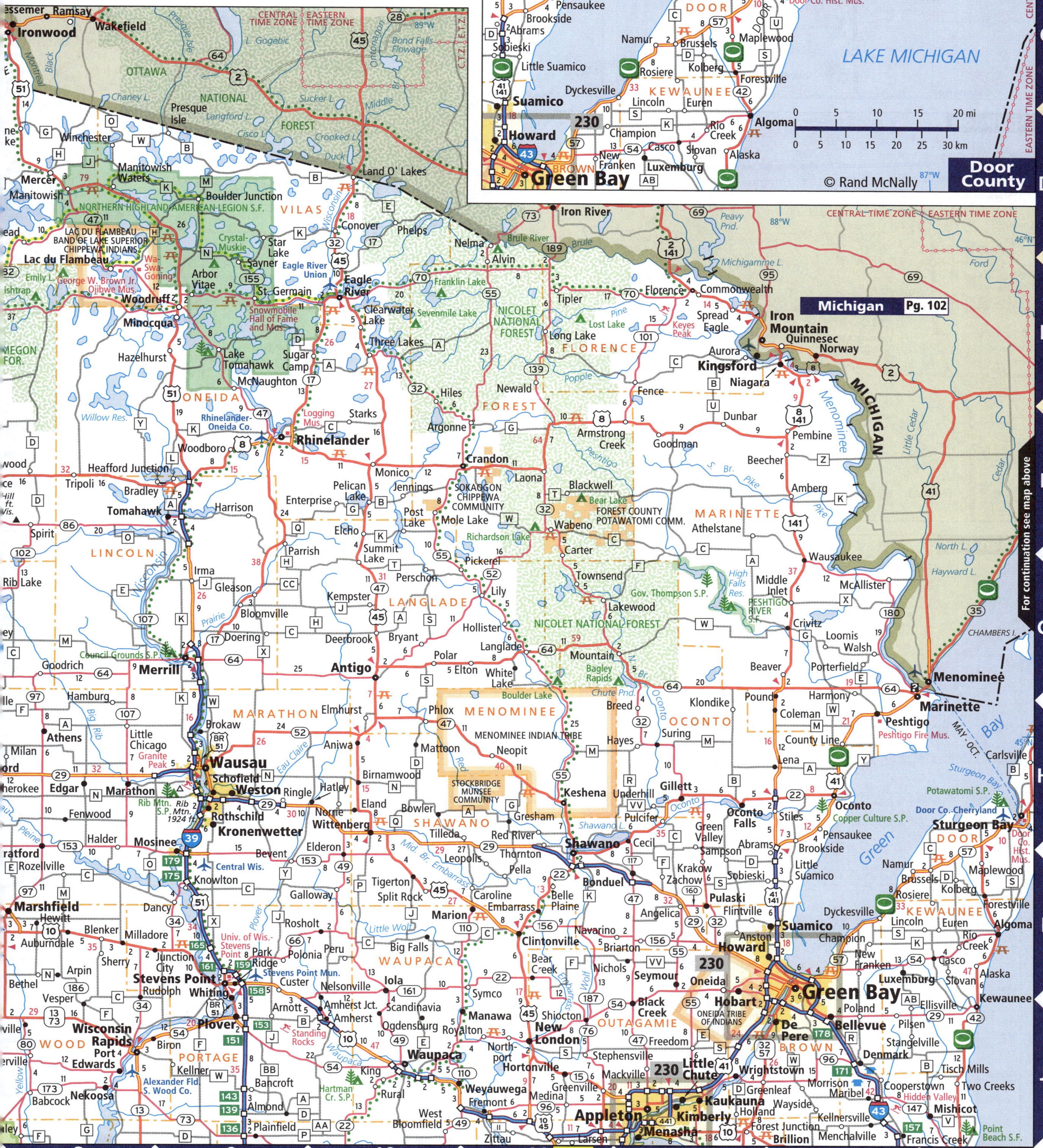

Eau Claire



For continuation see main map

Mich. Pg. 102

Door County



Michigan Pg. 102

For continuation see map above

Pg. 231

Construction 511 (866) 511-9472 www.511wi.gov
Conditions
Both

Tourism Information
Wisconsin Department of Tourism
(800) 432-8747, (608) 266-2161
www.travelwisconsin.com

Explore Wisconsin at go.randmcnally.com/WI

Index of cities Pg. 264 | Map legend Pg. 7

**Mileages between cities**

| | Beloit | Chicago, IL | Dubuque, IA | Eau Claire | Green Bay | Hayward | La Crosse | Madison | Milwaukee | Oshkosh | Rhinelander | Sheboygan | Sturgeon Bay | Superior | Wausau | Wisconsin Dells |
|---|---|---|---|---|---|---|---|---|---|---|---|---|---|---|---|---|
| Madison | 54 | 146 | 93 | 177 | 138 | 282 | 143 | | 78 | 87 | 200 | 117 | 185 | 325 | 143 | 57 |
| Milwaukee | 74 | 90 | 171 | 243 | 116 | 348 | 209 | 78 | | 86 | 244 | 54 | 155 | 390 | 187 | 123 |

Mileages © Rand McNally

| Mileages between cities | Beloit | Chicago, IL | Dubuque, IA | Eau Claire | Green Bay | Hayward | La Crosse | Madison | Milwaukee | Oshkosh | Rhinelander | Sheboygan | Sturgeon Bay | Superior | Wausau | Wisconsin Dells |
|---|---|---|---|---|---|---|---|---|---|---|---|---|---|---|---|---|
| Superior | 370 | 462 | 339 | 149 | 326 | 70 | 233 | 325 | 390 | 332 | 182 | 388 | 370 | | 232 | 271 |
| Wausau | 189 | 281 | 239 | 98 | 96 | 189 | 170 | 143 | 187 | 103 | 59 | 158 | 141 | 232 | | 112 |

Mileages © Rand McNally

39 182 miles · 43 192 miles · 90 189 miles · 94 341 miles

Total mileages through Wisconsin
More mileages at go.randmcnally.com/MC

## City sights to see

- Angel Museum, Beloit, N-6
- Betty Brinn Children's Museum, Milwaukee, L-3
- Golden Rondelle Theatre, Racine, J-10
- Harley Davidson Museum, Milwaukee, M-2
- Henry Maier Festival Park, Milwaukee, M-3
- J.M. Kohler Arts Center, Sheboygan, F-10

- Kenosha History Center, Kenosha, L-10
- Miller Brewery, Milwaukee, E-5
- Milwaukee Art Museum & War Memorial, Milwaukee, L-3
- Milwaukee Public Museum, Milwaukee, L-2
- Mitchell Park Horticultural Conservatory, Milwaukee, F-5
- Petit National Ice Center, Milwaukee, F-4



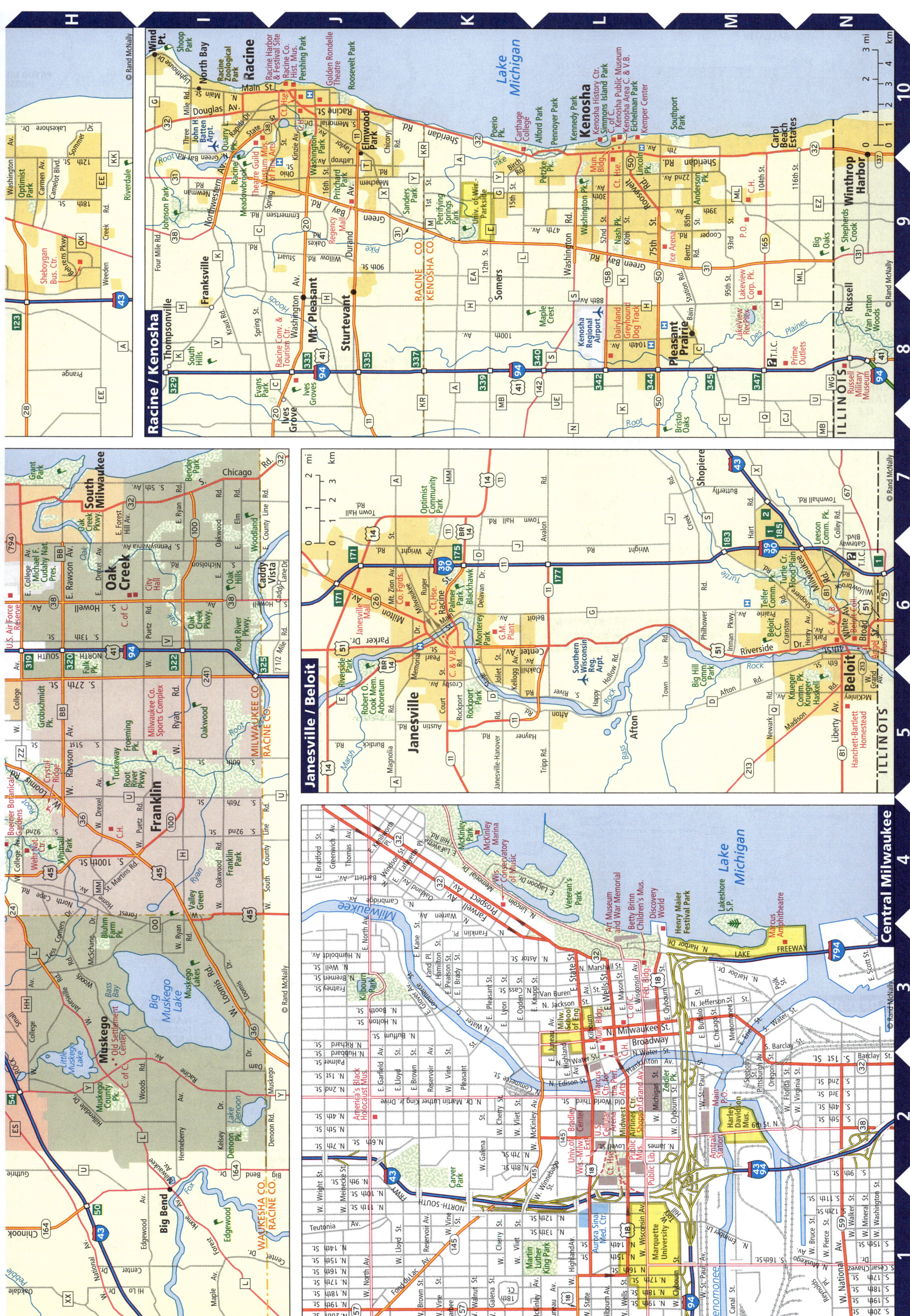

Index of cities Pg. 264 | Map legend Pg. 7

**Mileages between cities**

| | Casper | Cheyenne | Cody | Evanston | Gillette | Laramie | Sheridan | Spearfish, SD |
|---|---|---|---|---|---|---|---|---|
| Casper | | 178 | 213 | 325 | 126 | 147 | 148 | 219 |
| Cheyenne | 178 | | 392 | 357 | 244 | 49 | 324 | 290 |
| Cody | 213 | 392 | | 376 | 250 | 363 | 148 | 344 |
| Jackson | 283 | 432 | 177 | 190 | 411 | 383 | 325 | 504 |

Mileages by Rand McNally

**Wyoming state facts**

Nickname: The Equality State
Capital: Cheyenne, H-13
Population: 532,668 (rank: 50th)
Largest city: Cheyenne, 56,915, H-13
Land area: 97,100 sq. mi. (rank: 9th)
Highest point: Gannett Peak, 13,804 ft., E-6

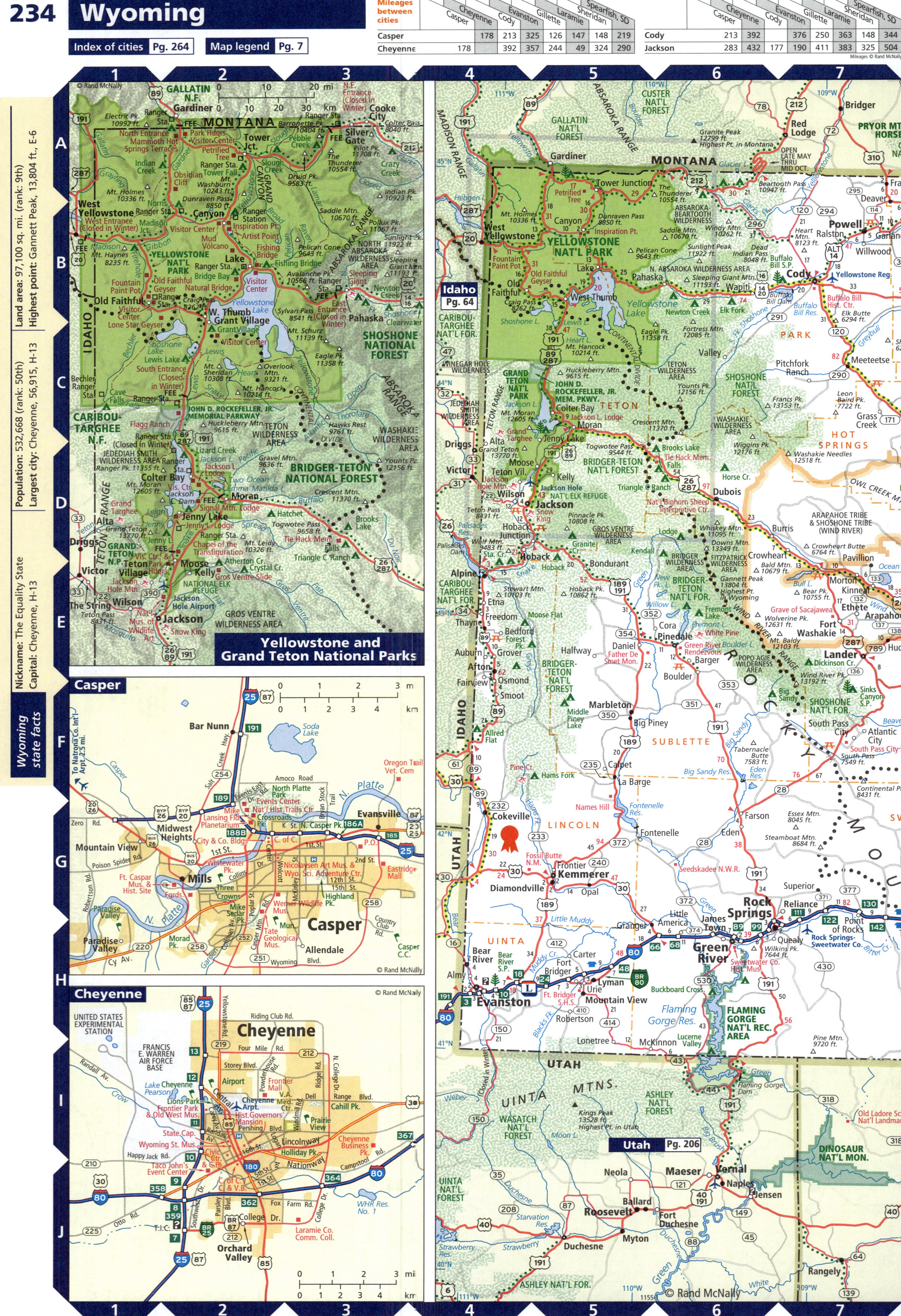

Mileages between cities

| | Casper | Cheyenne | Cody | Evanston | Gillette | Laramie | Sheridan | Spearfish, SD |
|---|---|---|---|---|---|---|---|---|
| Riverton | 119 | 272 | 138 | 238 | 248 | 222 | 213 | 341 |
| Rock Springs | 225 | 257 | 278 | 100 | 351 | 207 | 373 | 444 |
| Sheridan | 148 | 324 | 148 | 473 | 103 | 294 | | 196 |
| Spearfish, SD | 219 | 290 | 344 | 544 | 93 | 296 | 196 | |

Mileages © Rand McNally

Total mileages through Wyoming: 25 301 miles; 90 209 miles; 80 403 miles; 20 505 miles

More mileages at go.randmcnally.com/MC

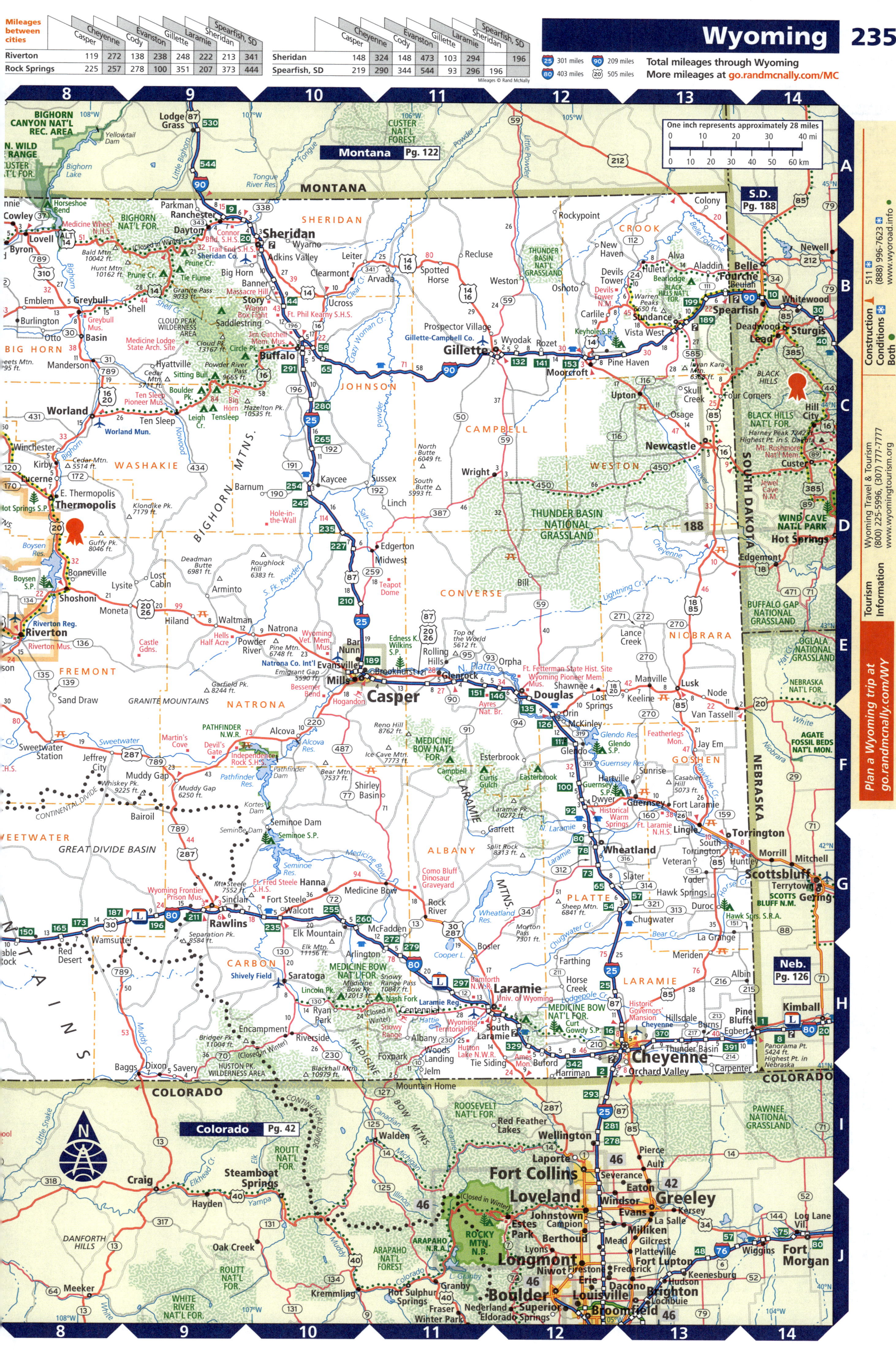

# Index

## United States Counties, Cities and Towns

2000 Census populations or latest available estimates

### Alabama

Page locator
Map keys A - J, Atlas pages 10 - 11
Map keys K - T, Atlas pages 12 - 13
* City keyed to p. 60



### Alaska

Page locator
Map keys A - J, Atlas pages 14 - 15



### Arizona

Page locator
Map keys A - J, Atlas pages 16 - 17
Map keys K - T, Atlas pages 18 - 19
* City keyed to pp. 20 - 21



*, †, ‡, §, ¥ See explanation under state title in this index.

County names are listed in capital letters and in boldface type.

## Arkansas

Page locator

| Map keys | Atlas pages |
|---|---|
| 1 - 10 | 22 - 23 |
| 11 - 20 | 24 - 25 |



## California

Page locator

| Map keys | Atlas pages |
|---|---|
| NA 1 - NN 10 | 26 - 27 |
| NA 11 - NN 20 | 28 - 29 |
| SA 1 - SN 10 | 34 - 35 |
| SA 11 - SN 20 | 36 - 37 |

* City keyed to p. 19
† City keyed to pp. 30 - 31
‡ City keyed to pp. 32 - 33
§ City keyed to pp. 38 - 39
◊ City keyed to pp. 40 - 41



*, †, ‡, §, ◊ See explanation under state title in this index.

County names are listed in capital letters and in boldface type.

# 238 California - Colorado

## California (continued)

Gazelle, 136 ... NC-5
Genesee, 150 ... NJ-8
Georgetown, 962 ... NJ-8
Gerber, 1100 ... NC-5
Geyserville, 1000 ... NJ-4
Gilroy, 49934 ... SL-8
Glamis ... SM-18
Glen Avon, 14853 ... F-16
Glen Ellen, 992 ... NK-5
Glencoe, 400 ... NJ-8
Glendale, 197176 ... SJ-11
Glendora, 49410 ... *D-12
Glenhaven, 360 ... NJ-4
**GLENN CO.,** 26453 ... **NH-5**
Glenn, 110 ... NH-5
Glennville, 210 ... SE-10
Glenview, 160 ... NJ-5
Goffs ... SG-18
Gold Run, 200 ... NJ-8
Goleta, 55204 ... SI-7
Gonzales, 8537 ... SC-4
Goodyears Bar, 100 ... NH-8
Gorman, 50 ... SH-10
Goshen, 2394 ... SD-8
Granada Hills ... *C-12
Grand Terrace, 12204 ... §E-18
Grangeville, 300 ... SD-8
Graniteville ... NI-9
Grass Valley, 12232 ... NI-8
Graton, 1815 ... NK-4
Grayson, 1077 ... NM-7
Greeley Hill, 200 ... NM-10
Green Brae, 3400 ... ND-12
Greenacres, ... SL-5
Greenfield, 15222 ... SD-4
Greenview, 200 ... NB-8
Greenville, 1160 ... NF-8
Greenwood, 600 ... NJ-8
Grenada, 351 ... NB-5
Gridley, 6478 ... NH-7
Grimes, 370 ... NJ-6
Grizzly Flat, 400 ... NJ-9
Groveland, 500 ... NM-10
Grover Beach, 13137 ... SG-5
Guadalupe, 6593 ... SG-6
Gualala, 1500 ... NJ-3
Guasti ... §E-15
Guerneville, 2441 ... NK-4
Guernsey, 238 ... SD-8
Gustine, 5094 ... SA-7
Hacienda Hts., 53122 ... §F-10
Half Moon Bay, 12449 ... NM-5
Hamburg, 140 ... NB-5
Hamilton City, 1903 ... NG-6
Hanford, 50103 ... SD-7
Happy Camp, 1200 ... NB-3
Harbor City, ... *F-11
Hardwick, 140 ... SC-8
Harmony ... SF-4
Harris ... ND-7
Hat Creek, 200 ... NE-7
Hathaway Pines, 350 ... NJ-9
Havasu Lake, 410 ... *F-10
Havilah ... SF-10
Hawaiian Gardens, 15229 ... NH-5
Hawkinsville, 45 ... NB-5
Hawthorne, 84305 ... §F-6
Hayfork, 2315 ... NE-4
Hayward, 142061 ... NM-6
Healdsburg, 10771 ... NJ-4
Heber, 2988 ... SM-17
Helendale, 700 ... NE-14
Helm ... SC-7
Hemet, 70991 ... SK-14
Henderson Vil., 200 ... NL-7
Herald, 500 ... NJ-8
Hercules, 24484 ... NC-15
Hermosa Beach, 19350 ... §F-6
Hesperia, 85883 ... SI-13
Hi Vista ... SH-12
Hickman, 457 ... NM-7
Hidden Hills, 2003 ... *D-2
Hidden Valley, 1400 ... NJ-6
Highgrove, 3445 ... *F-18
Highland, 51096 ... §J-13
Hillsborough, 10844 ... NL-10
Hilt, 30 ... NA-5
Hinkley, 1000 ... SG-13
Hobart Mills ... NH-9
Hollister, 34877 ... SB-4
Hollywood ... §F-6
Hollywood by the Sea ... *C-8
Holt ... NL-7
Holtville, 5412 ... SM-18
Home Gardens, 9461 ... §K-11
Homeland Valley, 3500 ... ND-11
Homewood, 350 ... NI-10
Honcut, 150 ... NI-7
Honeydew, 80 ... NF-2
Hoopa, 1200 ... NC-3
Hope Ranch, 1600 ... §B-3
Hopeton ... NL-8
Hopland, 630 ... NI-4
Hornbrook, 286 ... NA-5
Horntown ... NL-5
Horse Creek, 200 ... NB-6
Hughson, 6312 ... NM-8
**HUMBOLDT CO.,** 126518 ... **NE-2**
Humphreys Station ... SB-8
Huntington Beach, 192620 ... SK-11
Huntington Harbor ... SK-11
Huntington Park, 60898 ... §F-7
Hyampom, 150 ... NE-3
Idria ... SC-5
Ignacio ... NB-11
Igo ... NE-5
**IMPERIAL CO.,** 142361 ... **SM-18**
Imperial, 13648 ... SM-17
Imperial Beach, 26543 ... SN-15
Incline, 60 ... NM-1
Independence, 574 ... SB-11
Indian Wells, 5177 ... SK-16
Indian Hts., 2800 ... §J-3
Industry, 913 ... §F-11
Inglewood, 112714 ... §F-6
Ingot ... NE-6
Inverness, 1421 ... NK-4
**INYO CO.,** 17945 ... **SC-13**
Inyokern, 984 ... SF-12
Ione ... NJ-8
Ione ... NJ-8
Iowa Hill ... NI-8
Irvine, 207500 ... SL-12
Irwin, 600 ... NN-8
Irwindale, 1439 ... *D-11
Isla Vista, 21069 ... SI-7
Island Mtn. ... NG-3
Isleton, 838 ... NL-7
Ivanhoe, 4474 ... SD-9
Ivanpah ... SF-18
Jackson, 4320 ... NK-9
Jamesburg, 600 ... SC-4
Jamestown, 3017 ... NL-9
Jamul, 5920 ... SN-14
Janesville, 1000 ... NF-9
Johannesburg, 176 ... SF-12
Johnsondale ... SL-5
Johnsonville, 400 ... NF-9
Johnsville, 71 ... NG-8
Jolon ... SD-4
Joshua Tree, 4207 ... SJ-16
Julian, 1621 ... SM-15
Junction City, 600 ... NE-4
June Lake, 600 ... NM-10
Keddie, 96 ... NG-8
Keeler, 66 ... SC-12
Keene, 339 ... SG-10
Kelsey, 250 ... NJ-8
Kelseyville, 2928 ... NJ-4
Kelso ... SG-18
Kensington, 4936 ... ND-14
Kentfield, 6351 ... ND-11
Kenwood, 900 ... NK-5
Kerman, 12737 ... SC-7
**KERN CO.,** 661645 ... **SF-10**
Kernville, 1736 ... SC-11
Kettleman City, 1499 ... SE-7
Keyes, 4575 ... NM-8
King City, 11627 ... SD-5
Kings Beach, 4037 ... NI-10
Kingsburg, 11064 ... SC-8
**KINGS CO.,** 129461 ... **SD-7**
Kirkville ... NL-6
Kirkwood, 100 ... NJ-9
Kit Carson ... NJ-10
Klamath, 865 ... NB-2
Klamath Glen, 290 ... NB-2
Klamath River ... NB-3
Kneeland, 30 ... ND-2
Knights Ferry, 700 ... NM-9
Knights Landing ... NJ-6
Kramer Junction, 200 ... SG-13
Kyburz, 150 ... NJ-9
La Barr Meadows, 500 ... NI-8
La Cañada Flintridge, 20671 ... *C-12
La Crescenta, 18532 ... *C-11
La Grange, 340 ... NM-9
La Habra, 59155 ... §J-11
La Honda, 1500 ... NM-5
La Jolla ... SN-14
La Mesa, 54573 ... SN-14
La Mirada, 49809 ... §G-10
La Palma, 15603 ... §H-10
La Porte, 43 ... NH-8
La Puente, 40642 ... §F-10
La Quinta, 43865 ... SK-16
Ladera, 1430 ... NM-5
Ladera Hts., 6568 ... §F-6
Lafayette, 25013 ... ND-15
Laguna Beach, 23995 ... §L-12
Laguna Hills, 31818 ... SL-12
Laguna Niguel, 64469 ... SL-12
Laguna Woods, 18170 ... *G-13
Lake Alpine, 100 ... SB-9
Lake Arrowhead, 8934 ... §B-20
Lake City, 100 ... NB-9
Lake Elsinore, 50952 ... SK-13
Lake Forest, 100 ... NB-9
Lake Forest, 75566 ... SK-12
Lake Hughes, 300 ... SH-11
Lake Isabella, 3315 ... SF-11
Lake of the Pines, 3717 ... NI-8
Lake View Terrace ... §B-5
Lake Wildwood, 4868 ... NI-8
Lakehead, 450 ... NF-5
Lakeland Vil., 5626 ... *K-18
Lakeport, 4943 ... NJ-4
Lakeshore ... NG-5
Lakeside, 19560 ... SM-14
Lakeview, 1619 ... SJ-14
Lakewood, 80048 ... §G-8
Lamont, 13296 ... SF-10
Lancaster, 145469 ... SH-11
Landers, 2300 ... SJ-16
**LASSEN CO.,** 33828 ... **NE-9**
Lathrop, 11917 ... NL-7
Lawndale, 31346 ... §H-5
Laytonville, 1301 ... NG-3
Le Grand, 1760 ... SA-7
Lebec, 1285 ... SH-10
Lee Vining, 300 ... NL-12
Leggett, 300 ... NG-3
Lemon Cove, 298 ... SC-9
Lemon Cove ... SN-1
Lemon Grove, 24947 ... §J-5
Lemon Hts., 2800 ... §J-13
Lemoore, 23873 ... SD-8
Lennox, 22950 ... §G-6
Leona Valley, 500 ... NL-6
Lewiston, 1305 ... NE-4
Liberty Farms ... NK-7
Libfarm ... NK-7
Likely, 120 ... NB-9
Lincoln, 43602 ... NJ-7
Lincoln Acres, 1650 ... NK-4
Lincoln Hts., 4276 ... *F-5
Linda, 13474 ... NI-7
Linden, 1103 ... NK-8
Lindsay, 10571 ... SD-9
Litchfield ... NE-9
Little Lake, 50 ... NE-2
Little Valley ... NE-7
Littlerock, 1402 ... SJ-11
Live Oak, 8245 ... NH-7
Live Oak Springs, 240 ... SN-16
Livermore, 80188 ... NM-6
Livingston, 13058 ... NM-8
Loch Lomond, 300 ... NJ-4
Lockeford, 3179 ... NL-8
Lockwood, 350 ... NL-9
Lodgepole ... SL-2
Lodi, 61301 ... NL-7
Lodoga ... NH-5
Lokern ... SF-7
Loleta, 600 ... ND-2
Loma Linda, 21601 ... SJ-13
Loma Rica, 2075 ... NI-7
Lomita, 20707 ... NH-10
Lompoc, 41099 ... SH-6
London, 1848 ... SC-8
Lone Pine, 1655 ... SC-11
Long Barn, 250 ... NK-10
Long Beach, 463789 ... SK-11
Longvale, 15 ... NH-3
Lookout, 150 ... NC-8
Loomis, 6795 ... NJ-7
**LOS ANGELES CO.,** 9519338 ... **SJ-11**
Los Alamitos, 11656 ... §H-9
Los Alamos, 1372 ... SH-6
Los Altos, 28349 ... NN-6
Los Altos Hills, 8592 ... NL-16
Los Angeles, 3803869 ... SA-13
Los Banos, 34968 ... SB-5
Los Gatos, 29324 ... SA-3
Los Molinos, 1952 ... NG-6
Los Nietos, 8200 ... *F-9
Los Olivos, 850 ... SH-7
Los Oros, 5000 ... SF-5
Los Ranchos, 425 ... NC-11
Los Serranos, 8000 ... *F-14
Lost Hills, 1938 ... SF-8
Lost Lake, 110 ... SJ-10
Lotus, 700 ... NJ-8
Lower Lake, 1755 ... NJ-5
Loyalton, 771 ... NH-10
Lucas Valley, 4000 ... NL-5
Lucerne, 3000 ... NJ-5
Lucerne Valley, 2100 ... SH-14
Ludlow ... SG-15
Lynwood, 70025 ... §G-7
Lyoth ... NL-17
Lytle Creek, 500 ... §E-15
Macdoel, 140 ... NB-6
Madeline ... NC-9
**MADERA CO.,** 123109 ... **SB-7**
Madera, 56100 ... SB-7
Magalia, 10956 ... NG-7
Malaga, 745 ... SC-7
Malibu, 13009 ... SJ-10
Mammoth Lakes, 7402 ... NM-11
Manhattan Beach, 36665 ... §G-5
Manteca, 65028 ... NM-8
Manton, 300 ... NE-7
Maricopa, 1133 ... SG-9
**MARIN CO.,** 247289 ... **NL-4**
Marin City, 2500 ... *G-1
Marina, 17943 ... SC-3
Marinwood, 2300 ... NL-11
**MARIPOSA CO.,** 17130 ... **SA-7**
Mariposa, 1373 ... NN-10
Markleeville, 197 ... NJ-11
Martell ... NJ-8
Martinez, 35145 ... NL-6
Marysville, 12000 ... NI-7
Massack ... NG-8
Maxwell, 850 ... NJ-5
Mayten, 30 ... NB-5
McArthur, 365 ... NE-8
McCloud, 1343 ... NC-6
McFarland, 12098 ... SE-9
McKinleyville, 13599 ... ND-2
McKittrick, 160 ... SF-8
Mead Valley, 4000 ... §J-19
Meadow Vista, 3096 ... NJ-8
Mecca, 5402 ... SL-16
Meeks Bay, 17 ... NJ-10
Mendocino, 955 ... NI-3
**MENDOCINO CO.,** 86265 ... **NH-3**
Mendota, 10339 ... SB-6
Menifee, 10339 ... SK-13
Menlo Park, 30785 ... NM-5
Mentone ... NH-9
**MERCED CO.,** 210554 ... **SA-6**
Merced, 78958 ... SA-7
Meridian, 300 ... NJ-6
Meyers, 2500 ... NJ-10
Middletown, 1100 ... NJ-5
Midway City, 8485 ... §H-10
Milford, 250 ... NF-9
Mill Creek, 17 ... NF-8
Mill Valley, 13268 ... NL-5
Millbrae, 20800 ... NL-13
Milpitas, 67503 ... NN-6
Milton, 90 ... NL-8
Mineral, 143 ... NF-7
Mineral Hts., 2800 ... §J-3
Mira Loma, 17617 ... SJ-13
Mira Mesa ... §J-12
Mira Monte, 7177 ... SI-9
Miracle Hot Springs ... SF-10
Miranda, 250 ... NF-3
Miramonte ... §G-12
Mission Hills ... §B-1
Mission Viejo, 94003 ... SL-12
Mi-Wuk Vil., 1485 ... NL-10
**MODOC CO.,** 9449 ... **NC-9**
Modesto, 202967 ... NM-8
Mojave, 3836 ... SG-11
Mokelumne Hill, 800 ... NK-9
**MONO CO.,** 12853 ... **NL-12**
Mono Hot Springs ... SA-9
Monolith ... NK-12
Monrovia, 37651 ... *D-11
Montague, 1454 ... NB-5
Montalvin Manor ... ND-14
Montara, 2900 ... NC-14
Montara, 43591 ... *K-15
Montclair, 36530 ... §J-14
Monte Nido, 850 ... §D-2
Monte Rio, 1104 ... NK-4
Monte Sereno, 3602 ... NN-18
Montebello, 61906 ... §F-9
Montecito, 10000 ... SJ-8
Monterey, 27763 ... SC-3
**MONTEREY CO.,** 401762 ... **SD-4**
Monterey Park, 61234 ... §E-8
Montgomery Creek, 100 ... NE-6
Montrose, 27763 ... SC-3
Moorpark, 36372 ... SJ-10
Moraga, 17050 ... NE-15
Morena Vil., 190871 ... *F-4
Morgan Hill, 38096 ... SA-3
Mormon Bar, 130 ... NN-10
Morongo Valley, 3923995 ... SJ-15
Morro Bay, 10333 ... SF-5
Moss Beach, 1953 ... NM-4
Mott ... NB-5
Mountain Center, 100 ... SK-15
Mountain Meadows ... NG-9
Mountain Ranch, 1557 ... NK-9
Mountain View, 71348 ... NL-17
Mugginsville, 60 ... NB-5
Murphys, 2213 ... NK-10
Murrieta, 97874 ... SJ-13
Muscoy, 10644 ... §E-16
Myers Flat, 150 ... NF-3
Napa, 74587 ... NK-6
**NAPA CO.,** 124279 ... **NK-5**
Napa Junction, 10 ... NL-6
National City, 54869 ... SN-14
Navarro, 250 ... NI-3
Needles, 5348 ... SH-19
Nelson, 150 ... NG-7
**NEVADA CO.,** 99032 ... **NI-9**
Nevada City, 2929 ... NI-8
New Cuyama, 360 ... SG-8
Newark, 41781 ... NM-6
Newberry Springs, 2948 ... SG-15
Newcastle, 2200 ... NJ-8
Newell, 480 ... ND-9
Newman, 10226 ... SA-5
Newport Beach, 81000 ... SK-12
Nice, 2509 ... NJ-4
Nicolaus, 240 ... NJ-7
Niland, 1143 ... SL-17
Nipomo, 15826 ... SG-5
Nipton, 25 ... SF-18
Niles ... NM-6
No. Highlands, 44187 ... *H-7
No. Hollywood ... *C-6
No. Richmond, 3096 ... ND-14
No. San Juan, 100 ... NI-8
Nord, 80 ... NG-6
Noreen ... NE-6
Norco, 26659 ... SJ-13
Norden, 100 ... NI-9
Norris Cyn. ... SL-5
North Fork, 1196 ... SA-8
Northridge ... *C-4
Norton AFB ... §H-17
Norwalk, 105549 ... §G-8
Novato, 51904 ... NL-5
Nubieber, 60 ... ND-8
Nuevo, 4135 ... *F-20
20824 ... **NF-9**
Oakhurst, 2829 ... NN-10
Oakdale, 18990 ... NL-8
Oakland, 404115 ... NM-5
Oakley, 32035 ... NL-6
Oakville, 220 ... NK-5
O'Brien, 30 ... ND-5
Occidental, 1272 ... NK-4
Ocean Beach ... *I-1
Ocean View, 100 ... NH-4
Oceano, 7260 ... SG-6
Oceanside, 169684 ... SL-13
Ocotillo, 296 ... SN-16
Ocotillo Wells, 200 ... SL-16
Oildale, 27885 ... SF-9
Ojai, 7775 ... SI-9
Olancha, 134 ... SC-12
Old River, 200 ... SG-9
Old Station, 108 ... NE-7
Olinda, 100 ... SK-11
Olinda ... §G-12
Olivehurst, 11061 ... NI-7
Olympic Valley ... NI-9
Omo Ranch ... NJ-9
O'Neals ... SA-8
Ono, 60 ... NE-5
Ontario, 171691 ... SJ-12
Onyx, 170 ... SF-11
**ORANGE CO.,** 2846289 ... **SK-12**
Orange, 136392 ... §I-12
Orange Cove, 10641 ... SB-9
Orange Pk. Acres, 6300 ... §I-13
Orangevale, 26705 ... *K-6
Orcutt ... SH-6
Ordbend, 80 ... NH-6
Oregon House, 300 ... NI-8
Orick, 400 ... NC-2
Orinda, 18445 ... NE-16
Orland, 7174 ... NG-6
Orleans, 375 ... NC-3
Oro Fino, 30 ... NB-5
Oro Grande, 310 ... SI-13
Orosi, 8668 ... SC-9
Oroville, 14684 ... NH-7
Otay ... NN-16
Outingdale, 400 ... NK-9
Oxnard, 185717 ... SJ-9
Pacheco, 3562 ... NC-17
Pacific Hts. ... ND-6
Pacific Grove, 14601 ... SC-3
Pacific Palisades ... §E-3
Pacifica, 37739 ... NM-5
Paicines ... SB-4
Pajaro, 3420 ... SB-3
Pala, 1050 ... SK-14
Palermo, 5720 ... NH-7
Palm Desert, 49436 ... SK-15
Palm City ... §M-4
Palm Springs, 47952 ... SK-15
Palmdale, 143197 ... SH-11
Palo Alto, 59395 ... NN-5
Palo Cedro, 1247 ... NE-6
Palo Verde, 236 ... SK-19
Palomar ... NJ-15
Palomar Pk., 520 ... NJ-15
Palos Verdes Estates, 13585 ... §H-5
Paradise, 26640 ... NG-7
Paramount, 55236 ... §H-8
Parker Dam, 30 ... SI-19
Parkfield ... SE-5
Parkway, 14280 ... *L-7
Parlier, 13275 ... SC-8
Pasadena, 143080 ... SJ-11
Paskenta, 100 ... NG-5
Paso Robles, 28715 ... SE-5
Patrick Creek ... NA-3
Patterson, 19570 ... NM-7
Pauma Valley, 800 ... SL-14
Paxton, 21 ... NG-8
Paynes Creek, 100 ... NF-7
Paynesville ... NJ-11
Peanut ... NE-4
Pearblossom, 700 ... SI-11
Peardale, 700 ... NI-8
Pearsonville, 24 ... SE-12
Pebble Beach, 2200 ... SC-3
Pedley, 11708 ... *F-17
Penngrove, 1100 ... NK-5
Penryn, 1200 ... NJ-8
Peppermint Hill, 250 ... NH-5
Perris, 55643 ... SK-13
Pescadero, 300 ... NM-5
Petaluma, 54666 ... NL-4
Peters, 200 ... NL-8
Petrolia, 30 ... NF-2
Phelan, 2500 ... SI-12
Phillipsville, 170 ... NF-3
Philo, 250 ... NI-4
PiŃeon Hills, 5000 ... SI-12
Pico Rivera, 63138 ... §F-9
Piedmont, 10481 ... NE-15
Piercy, 400 ... NG-3
Pine Grove, 820 ... NK-9
Pine Valley, 1501 ... SM-15
Pinecrest ... NJ-10
Pinedale, 7260 ... SB-8
Pinole, 19039 ... NC-14
Pioneer, 900 ... NJ-9
Pioneer Pt., 1003 ... NJ-11
Piru, 1196 ... SI-10
Pismo Beach, 8573 ... SG-5
Pittsburg, 64148 ... NL-6
Pixley, 2586 ... SE-9
Placentia, 49692 ... §I-12
**PLACER CO.,** 248399 ... **NI-9**
Placerville, 9994 ... NJ-8
Plainview, 1000 ... SD-9
Planada, 4369 ... SA-6
Plaster City, 60 ... SM-17
Platina ... NE-5
Playa Del Rey, 10000 ... *E-4
Pleasant Grove, 300 ... NJ-7
Pleasant Hill, 32862 ... NL-6
Pleasanton, 68603 ... NM-6
**PLUMAS CO.,** 20824 ... **NF-9**
Plymouth, 1033 ... NJ-8
Pt. Arena, 460 ... NJ-3
Pt. Reyes Station, 818 ... NL-4
Pollock Pines, 4291 ... NJ-9
Pomona, 152699 ... §J-14
Pondosa ... NC-7
Pope Valley, 250 ... NJ-5
Port Costa, 800 ... NL-14
Port Hueneme, 21478 ... SJ-9
Porter Ranch ... NG-9
Porterville, 51830 ... SD-9
Portola, 2086 ... NG-9
Portola Valley, 4525 ... *K-15
Potrero, 560 ... SN-15
Potter Valley, 1025 ... NH-4
Poway, 48858 ... SM-14
Princeton, 300 ... NH-6
Proberta, 350 ... NF-6
Prunedale, 16514 ... SB-3
Pulga ... NG-7
Pumpkin Center, 520 ... SE-9
Quartz Hill, 9890 ... SH-11
Quincy, 1879 ... NG-8
Quincy Jct. ... NG-8
Rackerby ... NI-7
Rail Road Flat, 549 ... NK-9
Rainbow, 2025 ... SL-14
Rainbow Glen ... SL-14
Raisin City, 115 ... SC-7
Ramona, 15691 ... SM-14
Ranchita, 300 ... NL-11
Rancho Cordova, 55060 ... NK-7
Rancho Cucamonga, 171176 ... §J-14
Rancho Mirage, 16714 ... SK-15
Rancho Murieta, 4193 ... NK-8
Rancho Palos Verdes, 41106 ... §I-6
Rancho Rinconada ... NM-7
Rancho Santa Fe, 3252 ... SM-14
Rancho Santa Margarita, 43591 ... *K-15
Randsburg, 77 ... SF-12
Raymond, 200 ... SA-8
Red Bluff, 14025 ... NF-6
Red Mtn., 130 ... SF-12
Red Mtn., 400 ... NC-8
Redding, 90000 ... NE-6
Redlands, 69600 ... SJ-13
Redondo Beach, 66882 ... §H-6
Redway, 1200 ... NF-3
Redwood City, 76815 ... NM-5
Redwood Valley, 1900 ... NI-4
Reedley, 24194 ... SC-8
Represa ... NJ-8
Requa ... NB-2
Rescue, 1200 ... NJ-8
Reseda ... *D-4
Rialto, 98700 ... SJ-13
Richardson Springs, 100 ... NG-7
Richfield, 450 ... NG-6
Richgrove, 2723 ... SE-9
Richmond, 102285 ... ND-13
Richvale, 300 ... NH-6
Ridgecrest, 25538 ... SF-11
Rimforest, 200 ... §E-19
Rio Dell, 3182 ... NE-3
Rio Linda, 10486 ... *J-7
Rio Oso, 260 ... NJ-7
Ripley, 760 ... SL-19
Ripon, 14473 ... NM-8
Ripperdan, 350 ... SB-7
River Pines, 200 ... NJ-9
Riverbank, 20606 ... NM-8
**RIVERSIDE CO.,** 1545387 ... **SK-15**
Riverside, 295357 ... SJ-13
Robbins, 300 ... NJ-6
Robles Del Rio, 500 ... SC-3
Rocklin, 52831 ... NJ-7
Rodeo, 8717 ... NC-14
Rohnert Pk., 40490 ... NK-4
Rolinda ... SB-7
Rolling Hills, 1908 ... §H-6
Rolling Hills Estates, 7860 ... §H-6
Romoland, 2764 ... SK-13
Roosevelt Beach, 10 ... NM-4
Rosamond, 14389 ... SH-11
Roseland, 6369 ... NK-4
Rosemead, 54412 ... §E-8
Rosemont, 8200 ... *L-7
Rosemont, 22904 ... *L-7
Roseville, 112260 ... NJ-7
Ross, 2278 ... NC-11
Rosser, 10238 ... §I-5
Rough And Ready, 900 ... NI-8
Round Mtn., 200 ... NE-7
Rovana, 450 ... SC-11
Rowland Hts., 48553 ... §F-11
Rubidoux, 29187 ... §I-18
Rumsey, 200 ... NJ-6
Running Springs, 4863 ... §F-20
Rutherford, 500 ... NK-5
**SACRAMENTO CO.,** 1423499 ... **NK-8**
Sacramento, 463794 ... NJ-7
St. Helena, 5817 ... NJ-5
Salida, 12560 ... NM-8
Salinas, 143640 ... SC-3
Salton Sea Beach, 392 ... SL-16
Salton City, 978 ... SM-17
Salyer, 175 ... ND-3
San Andreas, 2615 ... NK-9
San Anselmo, 11986 ... NL-11
San Antonio Hts., 2500 ... SI-12
Smartville, 90 ... NI-7
Smith River, 950 ... NA-1
Snelling, 300 ... NL-8
**SOLANO CO.,** 334542 ... **NK-6**
Solana Beach, 12625 ... SM-14
Soledad, 5148 ... SC-4
Solemint ... SI-11
Solvang, 5332 ... SH-7
Somes Bar, 150 ... NC-3
Somis, 1400 ... SI-9
**SONOMA CO.,** 2813833 ... **SM-15**
San Diego, 1279329 ... SN-14
San Dimas, 33043 ... *D-12
San Fernando, 23933 ... SI-11
San Francisco, 808976 ... NL-5
**SAN FRANCISCO CO.,** 308976 ... **NE-13**
S. Dos Palos, 1385 ... SA-6
S. El Monte, 21397 ... *E-10
S. Fontana, 52810 ... §J-15
S. Fork, 98 ... NG-4
S. Gate, 96640 ... §F-8
S. Laguna, 3500 ... SL-12
S. Lake Tahoe, 23333 ... NJ-10
S. Los Angeles, 54 ... §E-7
S. Pasadena, 24446 ... *D-8
S. S. San Gabriel, 7595 ... *E-9
S. San Francisco, 62502 ... NM-5
S. San Jose, 948279 ... NN-6
S. Whittier, 57156 ... *F-9
Sonora, 4559 ... NL-9
**SAN JOAQUIN CO.,** 563598 ... **NM-8**
San Joaquin, 4005 ... SC-7
San Juan, 20 ... SH-7
San Juan Bautista, 1700 ... SB-3
San Juan Capistrano, 34793 ... SL-12
San Leandro, 77880 ... NM-6
San Lorenzo, 21898 ... NN-16
**SAN LUIS OBISPO CO.,** 246681 ... **SG-6**
San Luis Obispo, 43636 ... SG-6
San Lucas, 269 ... SC-5
San Marcos, 79114 ... SL-14
San Martin, 4230 ... SA-4
**SAN MATEO CO.,** 707161 ... **SA-2**
San Mateo, 94197 ... NM-5
San Miguel, 1427 ... SE-5
San Pablo, 30729 ... NL-6
San Pedro ... §I-6
San Quentin, 100 ... ND-12
San Rafael, 55602 ... NL-5
San Ramon, 49161 ... NM-6
San Simeon, 800 ... SE-4
Sand City, 261 ... SC-3
Sanger, 25447 ... SC-8
Santa, 300 ... SG-10
Santa Ana, 339130 ... SK-12
**SANTA BARBARA CO.,** 1682585 ... **SH-8**
Santa Barbara, 86093 ... SI-8
Santa Clara, 108738 ... SA-3
**SANTA CLARA CO.,** 1682585 ... **SA-4**
Santa Cruz, 56124 ... SB-3
**SANTA CRUZ CO.,** 255602 ... **SB-3**
Santa Fe Springs, 17159 ... §G-9
Santa Margarita, 1259 ... SG-6
Santa Maria, 86356 ... SG-6
Santa Nella Vil., 87664 ... SJ-11
Santa Paula, 28618 ... SJ-9
Santa Rita Pk., 105 ... SB-5
Santa Rosa, 155796 ... NK-4
Santa Susana, 10000 ... *B-3
Santa Venetia, 4298 ... NL-5
Santa Ynez, 4584 ... SH-7
Saratoga, 29926 ... NN-17
Saticoy ... SJ-9
Sattley, 100 ... NH-9
Sausalito, 7158 ... NC-12
Sawyers Bar, 10 ... NC-4
Scott Bar, 150 ... NB-5
Scotts Valley, 11128 ... SB-3
Sea Ranch, 250 ... NJ-3
Seal Beach, 25070 ... §H-8
Searles, 250 ... SE-13
Seaside, 31970 ... SC-3
Sebastopol, 7718 ... NK-4
Sedco Hills, 3078 ... §L-19
Seeley, 1624 ... SM-17
Seiad Valley, 425 ... NB-4
Selma, 22897 ... SC-8
Sequoia ... SC-9
Sherwood Forest, 2600 ... §B-6
Shaffer, 15000 ... SF-9
Shandon, 986 ... SF-6
Shasta ... NE-5
**SHASTA CO.,** 163256 ... **NE-6**
Shasta, 950 ... NE-5
Shasta Lake, 10192 ... NE-5
Shaver Lake, 400 ... SA-9
Sheep Ranch, 70 ... NK-9
Shelter Cove, 500 ... NG-2
Sheridan, 1100 ... NJ-7
Sherman Oaks ... *C-6
Shingle Springs, 2643 ... NJ-8
Shingletown, 2222 ... NE-6
Shively, 110 ... NE-3
Shore Acres, 4000 ... NB-18
Shoshone, 52 ... SE-15
**SIERRA CO.,** 3555 ... **NH-9**
Sierra City, 250 ... NH-9
Sierra Madre, 10912 ... *D-9
Sierraville, 150 ... NH-9
Silver City ... SM-2
Silverado, 800 ... *J-13
Silverthrone, 12000 ... *C-19
Simi Valley, 105302 ... SI-10
**SISKIYOU CO.,** 44301 ... **NC-5**
Sisquoc, 250 ... NH-7
Sleepy Hollow ... NL-11
Skyforest, 100 ... §E-20
San Jose, 948279 ... §C-12
Vineberg ... NJ-10
Visalia, 121040 ... SD-9
Vista, 91144 ... SL-13
Volta, 275 ... SA-5
Volcano, 400 ... NK-9
Vorden ... NK-7
Walker, 530 ... NK-11
Wallace, 220 ... NL-8
Walnut, 30741 ... §F-11
Walnut Creek, 64296 ... NL-6
Walnut Grove, 669 ... NL-7
Walnut Pk., 16180 ... §F-7
Warner Springs, 100 ... SL-15
Wasco, 24628 ... SF-9
Washington, 200 ... NI-9
Waterford, 7936 ... NL-8
Waterloo ... NL-8
Watsonville, 50442 ... SB-3
Watts ... §F-7
Wawona, 200 ... NN-10
Weaverville, 3554 ... NE-4
Weed, 3024 ... NC-5
Weimar, 850 ... NJ-8
Weitchpec, 30 ... NC-3
Weldon, 2287 ... SF-11
Weott, 300 ... NF-3
W. Covina, 105790 ... §F-11
W. Hollywood, 36055 ... *D-6
W. Point, 746 ... NK-9
W. Sacramento, 47511 ... NK-6
Westhaven, 370 ... NC-2
Westlake Vil., 8469 ... SJ-10
Westley, 747 ... NM-8
Westminster, 88975 ... §H-10
Westmorland, 2197 ... SM-17
Westport, 160 ... NI-3
Westwood, 1998 ... NF-8
Wheatland, 3591 ... NI-7
Whiskeytown, 100 ... NE-5
Whispering Pines ... NJ-5
White River ... SD-10
Whitehorn, 300 ... NF-2
Whitethorn, 300 ... NF-3
Whitmore, 100 ... NE-7
Whittier, 82267 ... §F-10
Wilbur Springs ... NI-5
Wildomar, 14064 ... SK-13
Williams, 4792 ... NI-6
Willits, 4972 ... NI-4
Willow Creek, 1743 ... ND-3
Willow Ranch ... NB-10
Willows, 5882 ... NH-6
Wilmington ... §I-7
Wilseyville, 350 ... NK-9
Wilton, 4551 ... NK-8
Winchester, 295 ... SK-14
Windsor, 25362 ... NK-4
Winter Gardens, 20631 ... SM-14
Winterhaven, 379 ... SM-20
Winters, 6977 ... NK-6
Winton, 8832 ... NM-8
Wofford Hts., 2276 ... SL-11
Woodacre, 1393 ... NL-4
Woodbridge, 4107 ... NL-8
Woodcrest, 8342 ... §K-19
Woodfords, 100 ... NJ-10
Woodlake, 7414 ... SC-9
Woodland, 54567 ... NJ-6
Woodside, 5579 ... NL-15
Woodville, 1678 ... SD-9
Woody ... NF-10
Wrightwood, 3837 ... §C-12
Wyandotte, 100 ... NH-7
Yankee Hill ... SB-7
Yermo, 1623 ... SG-14
Yettem, 212 ... SC-9
**YOLO CO.,** 168660 ... **NJ-6**
Yolo, 450 ... NJ-6
Yorba Linda, 65717 ... §I-12
Yorkville, 130 ... NJ-4
Yosemite Forks ... NN-10
Yosemite Vil., 1500 ... NM-10
Yountville, 2268 ... NK-5
Yreka, 7309 ... NB-5
**YUBA CO.,** 60219 ... **NI-7**
Yucaipa, 49750 ... SJ-14
Yucca Valley, 20375 ... SJ-15
Zamora, 100 ... NJ-6
Zenia, 39 ... NF-3

## Colorado

Page locator — Map keys 1–10 Atlas pages 42–43, 11–20 pages 44–45
*City keyed to pp. 46–47

**ADAMS CO.,** 363857 ... **E-15**
Agate, 80 ... F-16
Aguilar, 570 ... L-14
Akron, 1711 ... D-17
Allenspark, 496 ... D-12
Allison, 110 ... N-7
Alma, 172 ... G-11
Almont, 70 ... J-9
Antonito, 788 ... N-11
Arapahoe, 100 ... H-20
Arapaho Park, 42 ... §L-7
**ARAPAHOE CO.,** 487967 ... **F-14**
**ARCHULETA CO.,** 9898 ... **M-8**
Arlington, 40 ... H-18
Arriba, 209 ... F-17
Arroya, 10 ... H-18
Arvada, 103761 ... E-13
Aspen, 5902 ... G-9
Atwood, 195 ... C-17
Ault, 1442 ... C-14
Aurora, 319057 ... F-14
Avondale, 754 ... J-15
Bailey, 150 ... F-12
Barnesville ... C-14
Barr Lake, 100 ... *F-9
Basalt, 3244 ... F-8
Bayfield, 2024 ... N-7
Bedrock ... J-4
Beecher Island, 200 ... F-20
Beulah, 280 ... J-13
**BENT CO.,** 5998 ... **K-18**
Berthoud, 5429 ... D-13
Bethune, 209 ... G-19
El Jebel, 4488 ... F-8
Black Forest, 13247 ... H-14
Black Hawk, 105 ... E-12
Blakeland, 100 ... §L-7
Blanca, 342 ... M-12
Blende, 1500 ... J-14
Bloom ... J-15
Blue Mtn., 15 ... E-4
Boncarbo, 30 ... L-13
Bonanza, 14 ... I-10
Bond, 100 ... E-9
Boone, 358 ... J-15
Booneville, 350 ... J-15
**BOULDER CO.,** 301494 ... **D-12**
Boulder, 94171 ... D-13
Bow Mar, 811 ... §K-5
Bowie, 40 ... H-7
Boyero ... F-18
Branson, 77 ... N-16
Breckenridge, 3420 ... F-11
Briggsdale, 110 ... C-15
Brighton, 31380 ... E-14
Bristol, 130 ... I-20
Broadmoor ... H-13
Brookvale, 280 ... F-12
Brush, 5227 ... D-17
Buckingham, 30 ... C-16
Buena Vista, 2134 ... H-11
Buffalo Creek, 300 ... F-12
Burlington, 3875 ... F-20
Burns ... E-9
Byers, 1200 ... F-15
Caddoa ... K-19
Cahone, 500 ... M-5
Calhan, 894 ... G-14
Campion, 1832 ... D-13
Campo, 117 ... N-18
Canon City, 15889 ... I-13
Capulin, 278 ... N-11
Carbondale, 6124 ... F-8
Carr, 60 ... B-13
Cascade, 1479 ... H-13
Castle Rock, 44369 ... F-14
Cattle Creek, 220 ... F-7
Cedaredge, 2252 ... H-7
Centennial, 99680 ... *L-13
Center, 2389 ... L-11
Central City, 566 ... E-12
**CHAFEE CO.,** 16242 ... **H-11**
Chama, 45 ... M-13
Cheney Center, 130 ... §L-13
Cherry Hills Vil., 6355 ... §J-7
**CHEYENNE CO.,** 2200 ... **H-19**
Cheyenne Wells, 915 ... H-19
Chimney Rock, 48 ... N-8
Chipita Park, 465 ... H-13
Chivington, 40 ... H-19
Chromo, 5 ... N-8
Cimarron, 20 ... J-8
Clark, 80 ... C-9
Clarkville ... E-20
**CLEAR CREEK CO.,** 9322 ... **F-12**
Clifton, 17345 ... G-5
Coal Creek, 355 ... J-13
Coaldale, 15 ... J-12
Coalmont, 50 ... C-9
Cokedale, 134 ... M-14
Collbran, 426 ... G-6
Colona, 70 ... J-7
Colorado City, 2018 ... K-14
Colorado Springs Estates, 691 ... *L-9
Columbine, 24728 ... §L-5
Columbine Valley, 1320 ... §K-5
Commerce City, 42472 ... *H-7
Como ... G-11
Conejos, 50 ... N-11
**CONEJOS CO.,** 8400 ... **M-10**
Conifer, 200 ... F-13
Cope, 100 ... E-18
**COSTILLA CO.,** 3663 ... **M-12**
Cotopaxi ... J-12
Cowdrey, 10 ... B-10
Craig, 9241 ... D-7
Creede, 290 ... K-9
Crested Butte, 1651 ... H-8
Crestone, 115 ... K-11
Cripple Creek, 1013 ... H-13
Crook, 124 ... B-18
**CROWLEY CO.,** 5518 ... **I-16**
Crowley, 165 ... J-16
Crystal, 160 ... H-8
**CUSTER CO.,** 3503 ... **J-12**
Dacono, 4052 ... D-13
Dailey ... C-18
De Boque, 522 ... F-6
Deckers ... G-13
Deer Trail, 577 ... F-15
Del Norte, 1603 ... L-10
Delagua ... M-14
**DELTA CO.,** 27834 ... **H-6**
Delta, 9020 ... H-6
**DENVER CO.,** 598707 ... ***H-9**
Dillon, 808 ... F-11
Dinosaur, 335 ... D-4
Divide, 140 ... H-13
**DOLORES CO.,** 1844 ... **K-5**
Dolores, 920 ... L-5
Dotsero ... F-8
**DOUGLAS CO.,** 175766 ... **G-13**
Dove Creek, 745 ... K-4
Doyleville ... I-9
Drake ... D-12
Dumont ... F-12
Dupont, 3650 ... *G-7
Durango, 16416 ... M-6
Durango West, 1050 ... M-6
Eads, 606 ... §19
**EAGLE CO.,** 41659 ... **E-9**
Eagle, 5897 ... F-9
E. Point, 2 ... G-2
Eastlake ... *F-7
Eaton, 4213 ... C-14
Eckley, 265 ... D-19
Edgewater, 5136 ... *J-6
Edwards, 8266 ... F-9
**EL PASO CO.,** 516929 ... **H-14**
**ELBERT CO.,** 19872 ... **F-15**
Elbert, 120 ... G-14
Eldora ... D-12
Eldorado Springs, 557 ... E-13
Elizabeth, 1434 ... F-14
Elk Springs, 25 ... D-5
Elliot, 300 ... K-16
Empire, 324 ... E-12
Englewood, 32669 ... *L-13
Erie, 6432 ... D-13
Estabrook ... §K-3
Estella ... H-13
Evans, 18842 ... C-14
Evergreen, 9216 ... *L-2
Fairplay, 658 ... F-11
Federal Hts., 11732 ... *G-6
Fernclift, 70 ... D-12
Firestone, 8171 ... D-13
Flagler, 567 ... F-18
Fleming, 438 ... B-18
Florence, 3623 ... I-13
Florissant, 119 ... H-12
Ft. Collins, 136509 ... C-13
Ft. Garland, 432 ... L-12
Ft. Logan ... *J-5
Ft. Lupton, 7533 ... D-14
Ft. Lyon ... K-18
Ft. Morgan, 10529 ... D-16
Fountain, 19669 ... H-14
Fowler, 1094 ... J-15
Foxton ... G-13
Franktown, 300 ... F-14
Fraser, 957 ... D-11
Frederick, 8277 ... D-13
**FREMONT CO.,** 47436 ... **I-12**
Frisco, 2696 ... E-11
Fruita, 7418 ... G-5
Fruitvale, 6936 ... G-5
Galatea ... H-18
Garcia, 20 ... N-12
Garden City, 260 ... C-14
Garden, 837 ... I-12
**GARFIELD CO.,** 54791 ... **F-6**
Garnett ... §K-7
Gateway, 50 ... H-4
Genoa, 136 ... G-18
Georgetown, 1027 ... E-12
Gilcrest, 1134 ... D-14
Gill, 235 ... C-14
Gilman ... F-10
**GILPIN CO.,** 4757 ... **E-12**
Glade Park, 60 ... H-4
Glendale, 4798 ... *J-7
Gleneagle ... H-14
Glenwood Springs, 9093 ... F-7
Golden, 18026 ... E-13
Goldfield, 53 ... H-13
Goodrich, 35 ... D-16
Gould ... C-10
Granada, 580 ... K-19
Granby, 1775 ... D-11
**GRAND CO.,** 13212 ... **D-10**
Grand Junction, 49688 ... G-5
Grand Lake, 565 ... D-11
Grand Mesa, 60 ... G-6
Grand Valley, 1216 ... F-6
Granite ... H-10
Grant, 80 ... F-12
Greeley, 91492 ... C-14
Green Mtn. Falls, 655 ... H-13
Greenwood Vil., 13925 ... *L-8
Greystone ... D-5
Grover, 144 ... B-15
Guffey, 25 ... H-12
Gulnare, 100 ... L-14
**GUNNISON CO.,** 14947 ... **H-7**
Gunnison, 5461 ... I-9
Gypsum, 6052 ... F-9
Hahns Peak, 40 ... B-9
Hamilton, 30 ... C-7
Hartman, 102 ... J-20
Hartsel, 80 ... H-11
Hasty, 200 ... J-18
Haswell, 70 ... J-17
Haxtun, 964 ... C-19
Hayden, 1579 ... C-8
Heaney, 55 ... C-10
Henderson, 600 ... *F-8
Hereford, 60 ... B-15
Hermosa, 260 ... L-6
Hesperus, 200 ... N-6
Hiawatha, 10 ... C-5
Highland Pk., 300 ... §C-12
**HINSDALE CO.,** 790 ... **K-7**
Hoehne, 95 ... M-15
Holly, 961 ... J-20
Holyoke, 2223 ... C-19
Homelake, 260 ... L-11
Hooper, 124 ... K-11
Hot Sulphur Springs, 512 ... D-11
Hotchkiss, 1084 ... H-7
Howard, 200 ... J-12
Hoyt, 30 ... E-15
**HUERFANO CO.,** 7862 ... **L-13**
Hugo, 742 ... G-17
Hygiene, 400 ... D-13
Idaho Springs, 1736 ... E-12
Idalia, 98 ... E-20
Ignacio, 767 ... M-7
Iliff, 216 ... B-18
Indian Hills, 1197 ... *L-2
Iowa ... G-13
**JACKSON CO.,** 1577 ... **B-10**
Jamestown, 213 ... D-12
Jansen, 210 ... M-14
Jaroso, 60 ... N-12
Jefferson, 60 ... F-11
Joes, 90 ... F-19
Johnstown, 9029 ... C-13
Julesburg, 1252 ... B-20
Karval, 60 ... H-17
Kassler ... *M-5
Keenesburg, 1177 ... D-14
Kelim, 40 ... F-19
Keota ... B-15
Kersey, 1460 ... C-14
Kim, 65 ... M-17
**KIOWA CO.,** 1622 ... **J-18**
Kiowa, 722 ... F-14
**KIT CARSON CO.,** 8011 ... **F-19**
Kit Carson, 202 ... H-18
Knob Hill ... *H-2
Kremmling, 1528 ... D-10
La Garita ... K-10
La Jara, 883 ... M-11
La Junta, 7568 ... J-16
**LA PLATA CO.,** 43941 ... **L-6**
La Salle, 1986 ... D-14
La Valley, 160 ... M-12
La Veta, 832 ... L-13
Lafayette, 25065 ... D-13
Laird ... E-19
Lake City, 375 ... K-7
Lake George, 385 ... G-12
**LAKE CO.,** 7812 ... **G-10**
Lakewood, 140989 ... *J-5
Lamar, 8843 ... J-19
**LARIMER CO.,** 291494 ... **C-12**
Larkspur, 234 ... F-13
**LAS ANIMAS CO.,** 15207 ... **M-15**
Las Animas, 2481 ... K-17
Las Mesitas, 60 ... N-10
Last Chance, 20 ... E-17
Lawson, 320 ... E-12
Lay ... D-7
**LINCOLN CO.,** 6087 ... **H-17**
Lincoln Park, 3904 ... I-13
Limon, 2047 ... G-16
Lindon ... E-17
Little Thompson ... *A-4
Littleton, 40844 ... E-13
Livermore, 60 ... C-12
Lochbuie, 4808 ... D-14
Log Lane Vil., 873 ... D-16
**LOGAN CO.,** 20504 ... **B-17**
Loma, 250 ... G-4
Lone Tree, 9348 ... *L-8
Longmont, 85150 ... D-13
Louisville, 19133 ... D-13
Louviers, 257 ... F-13
Loveland, 65287 ... C-13
Lucerne, 1100 ... C-14
Lyons, 2035 ... D-13
Mack, 240 ... G-4
Maher, 100 ... H-6
Manassa, 995 ... N-11
Mancos, 1086 ... M-5
Manitou Springs, 5161 ... H-13
Manzanola, 472 ... J-15
Marble, 65 ... G-8
Marvel, 30 ... M-6
Masonic Pk., 2 ... L-9
Maybell, 72 ... D-6
Mayfield ... I-13
Maysville ... I-11
McClave, 190 ... K-18
McCoy, 95 ... E-9
Mead, 3405 ... D-13
Meeker, 2387 ... E-6
Meredith, 25 ... F-8
Merino, 223 ... C-17
**MESA CO.,** 116255 ... **H-5**
Mesa, 780 ... G-6
Mesa Verde ... M-5
Milliken, 5610 ... C-14
Milner ... C-8
**MINERAL CO.,** 831 ... **L-8**
Minturn, 1185 ... F-10
Model, 110 ... M-15
**MOFFAT CO.,** 13184 ... **C-6**
Moffat, 126 ... K-11
Molina, 90 ... G-6
Monte Vista, 4009 ... L-10
**MONTEZUMA CO.,** 23830 ... **M-4**
Montezuma, 43 ... E-11
**MONTROSE CO.,** 33432 ... **J-6**
Montrose, 17989 ... I-6
Monument, 2588 ... G-13
**MORGAN CO.,** 27171 ... **D-16**
Mosca, 130 ... L-11
Mt. Crested Butte, 848 ... H-9
Mt. Princeton Hot Springs, 50 ... H-11
Mountain View, 520 ... *H-6
Nathrop ... H-11
Naturita, 693 ... J-5
Nederland, 1369 ... E-12
New Castle, 3796 ... F-7
Ninaview ... L-17
Niwot, 3650 ... D-13
Northglenn, 33697 ... *F-7
Norwood, 462 ... J-5
Nucla, 732 ... J-5
Nunn, 541 ... B-14
Oak Creek, 813 ... D-8
Ohio, 40 ... I-9
Olathe, 1742 ... I-6
Olney Springs, 344 ... J-16
Ophir, 132 ... K-6
Orchard, 120 ... D-15
Orchard City, 3203 ... H-6
Ordway, 1098 ... J-16
Ortiz ... N-11
**OTERO CO.,** 20311 ... **K-16**
Ouray, 470 ... J-7
**OURAY CO.,** 3742 ... **K-6**
Ovid, 291 ... B-20
Oxford, 90 ... M-7
Padroni, 50 ... B-18
Pagoda Junction, 25 ... D-8
Pagosa Junction ... N-8
Pagosa Springs, 1744 ... M-8
Palisade, 2840 ... G-5
Palmer Lake, 2326 ... G-13
Pandora ... K-6
Paoli, 37 ... B-19
Paonia, 1633 ... H-7
Paradox, 70 ... J-4
Paradise ... F-9
Park Center ... I-13
**PARK CO.,** 14523 ... **G-11**
Parker, 43767 ... F-14
Parlin, 50 ... I-9
Parshall, 130 ... D-11
Peckham, 135 ... D-14
Peetz, 244 ... B-18
Penrose, 4070 ... I-13
Peyton, 250 ... H-14
**PHILLIPS CO.,** 4480 ... **C-19**
Phippsburg, 220 ... D-9
Pierce, 884 ... C-14
Pine, 330 ... F-13
Pine Grove, 50 ... *F-2
Pine Junction, 950 ... F-12
Pinecliff ... E-12
Pitkin, 76 ... I-9
**PITKIN CO.,** 14872 ... **G-9**
Placerville, 150 ... J-6
Platoro, 10 ... M-10
Platteville, 2485 ... D-14
Pleasant View, 80 ... L-4
Poncha Springs, 456 ... I-11
Portland, 13 ... I-13
Powder Wash, 55 ... B-6
Powderhorn, 20 ... J-9
**POWERS CO.,** 6620 ... **K-19**
Pritchett, 135 ... M-18
Pryor ... J-14
**PUEBLO CO.,** 146155 ... **J-14**
Pueblo, 104951 ... J-14
Pueblo West, 21728 ... J-14
Punkin Center ... H-16
Radium ... E-10
Rand ... C-10
Rangely, 2138 ... D-5
Raymond, 60 ... D-12
Red Cliff, 323 ... F-10
Red Feather Lakes, 300 ... C-12
Redlands, 8043 ... G-5
Redmesa, 60 ... N-6
Redstone, 99 ... G-8
Redvale, 220 ... J-6
Ridgway, 900 ... J-7
**RIO BLANCO CO.,** 6666 ... **E-6**
**RIO GRANDE CO.,** 12413 ... **L-11**
Rockvale, 405 ... I-13
Rocky Ford, 3950 ... J-16
Roggen, 60 ... D-15
Rollinsville, 190 ... E-12
Romeo, 342 ... N-15
Rosedale ... *M-5
Roswell ... I-13
**ROUTT CO.,** 19690 ... **C-8**
Royal Gorge ... I-13
Rush ... H-16
Rustic ... C-12
Rye, 201 ... K-13
**SAGUACHE CO.,** 5917 ... **K-9**
Saguache, 581 ... K-11
St. Charles Mesa ... J-14
St. Elmo ... I-10
Salida, 5396 ... I-11
Salt Creek, 602 ... J-15
San Acacio, 70 ... N-12
San Isabel, 50 ... K-13
**SAN JUAN CO.,** 558 ... **L-7**

*, †, ‡, § See explanation under state title in this index. County names are listed in capital letters and in boldface type.

## Connecticut

Page locator

| Map keys | Atlas pages |
|---|---|
| A - J | 48 - 49 |



## Delaware

Page locator

| Map keys | Atlas pages |
|---|---|
| A - N | 50 - 51 |



## District of Columbia

Page locator

| Map keys | Atlas pages |
|---|---|
| A - N | 224 - 225 |



## Florida

Page locator

| Map keys | Atlas pages |
|---|---|
| A - J | 52 - 53 |
| K - T | 54 - 55 |

* City keyed to p. 51
† City keyed to pp. 56 - 57



*, †, ‡, §, ◊ See explanation under state title in this index.

County names are listed in capital letters and in boldface type.

## Georgia

Page locator

| Map keys | Atlas pages |
|---|---|
| A - J | 58 - 59 |
| K - T | 60 - 61 |

* City keyed to p. 62 - 63



## Hawaii

Page locator

| Map keys | Atlas pages |
|---|---|
| F - N | 62 - 63 |



*, †, ‡, §, ◊ See explanation under state title in this index.

County names are listed in capital letters and in boldface type.

## Idaho

Page locator

| Map keys | Atlas pages |
|---|---|
| A - N | 64 - 65 |



## Illinois

Page locator

| Map keys | Atlas pages |
|---|---|
| A - J | 66 - 67 |
| K - T | 68 - 69 |

* City keyed to pp. 70 - 71
† City keyed to pp. 72 - 73
‡ City keyed to p. 120 - 121
§ City keyed to p. 233



*, †, ‡, §, ◊ See explanation under state title in this index.

County names are listed in capital letters and in boldface type.

## Indiana

Page locator

| Map keys | Atlas pages |
|---|---|
| A - J | 74 - 75 |
| K - T | 76 - 77 |

* City keyed to pp. 72 - 73



*, †, ‡, §, ◊ See explanation under state title in this index.

County names are listed in capital letters and in boldface type.

## Iowa

Page locator

| Map keys | Atlas pages |
|---|---|
| 1 - 10 | 78 - 79 |
| 11 - 20 | 80 - 81 |

* City keyed to p. 68
† City keyed to p. 128



*, †, ‡, §, ◊ See explanation under state title in this index.

County names are listed in capital letters and in boldface type.

## Kansas

Page locator

| Map keys | Atlas pages |
|---|---|
| 1 - 10 | 82 - 83 |
| 11 - 20 | 84 - 85 |

* City keyed to p. 117



## Kentucky

Page locator

| Map keys | Atlas pages |
|---|---|
| 1 - 10 | 86 - 87 |
| 11 - 20 | 88 - 89 |

* City keyed to p. 163
† City keyed to p. 226



*, †, ‡, §, ◊ See explanation under state title in this index.

County names are listed in capital letters and in boldface type

## Louisiana

Page locator

| Map keys | Atlas pages |
|---|---|
| A - J | 90 - 91 |



*, †, ‡, §, ◊ See explanation under state title in this index.

County / Parish names are listed in capital letters and in boldface type.

## Maine

Page locator

| Map keys | Atlas pages |
|---|---|
| A - J | 92 - 93 |

Abbot Vil., 150 C-5
Acton, 40 I-1
Addison, 250 E-9
Albion, 225 F-5
Alexander, 50 D-9
Alfred, 600 I-2
Allagash, 100 B-12
Allens Mills, 200 E-3
Alna, 130 G-4
Amherst, 130 E-7
Andover, 600 E-2
**ANDROSCOGGIN CO., 103793 G-3**
Anson, 818 E-4
**AROOSTOOK CO., 73938 B-8**
Ashland, 750 B-13
Athens, 275 D-4
Atlantic, 60 G-7
Auburn, 23177 G-3
Augusta, 18282 F-4
Aurora, 60 E-7
Bailey Island, 650 H-3
Bancroft, 30 B-8
Bangor, 31756 E-6
Bar Harbor, 2680 F-8
Bar Mills, 275 I-2
Baring, 125 D-10
Bass Harbor, 500 G-7
Bath, 8885 H-4
Bay Pt. H-4
Beals, 400 F-9
Beans Corner E-3
Belfast, 6721 F-6
Belgrade, 200 F-4
Belgrade Lakes, 300 F-4
Belmont Corner, 25 F-6
Benedicta, 60 A-7
Benton, 120 E-5
Bernard, 200 J-11
Berwick, 1993 J-1
Bethel, 1200 F-2
Biddeford, 21435 I-2
Bingham, 856 D-4
Blaine, 500 C-14
Blaisdell Corners, 200 I-1
Blanchard C-4
Blue Hill, 800 F-7
Blue Hill Falls, 100 F-7
Bolsters Mills, 90 G-2
Boothbay, 450 H-4
Boothbay Harbor, 1237 H-4
Bowdoinham, 400 G-4
Bowerbank, 70 C-5
Bradford, 80 D-6
Bremen, 100 G-5
Brewer, 9035 E-6
Bridgewater, 325 C-14
Bridgton, 2359 G-2
Brighton, 40 D-4
Bristol, 300 H-5
Brooklin, 225 G-7
Brooks, 400 F-6
Brooksville, 120 F-7
Brookton, 170 B-9
Brownville, 350 C-6
Brownville Junction, 800 C-6
Brunswick, 14816 H-4
Bryant Pond, 500 F-2
Buckfield, 400 F-3
Bucks Harbor, 120 E-10
Bucksport, 2970 F-6
Burlington, 140 C-7
Burnham, 200 E-5
Byron, 75 E-2
Calais, 3184 C-10
Cambridge, 125 D-5
Camden, 3934 G-6
Canaan, 400 E-5
Canton, 400 F-3
Cape Elizabeth, 8854 I-3
Cape Neddick, 2997 J-2
Cape Porpoise, 500 I-2
Caratunk, 70 C-4
Cardville, 150 D-7
Caribou, 8093 B-14
Carmel, 550 E-6
Carrabassett, 225 D-3
Carthage E-2
Casco, 450 G-2
Castine, 800 F-6
Center Lovell, 200 G-1
Charleston, 275 D-6
Chebeague Island, 333 H-3
Cherryfield, 350 E-8
Chester, 100 C-7
Chesterville, 110 F-3
Chesuncook A-5
China, 450 F-5
Chisholm, 1399 F-3
Clark Island, 120 G-5
Clayton Lake, 20 B-12
Clifton, 150 E-7
Clinton, 1305 E-5
Coburn Gore, 15 C-2
Columbia Falls, 250 E-9
Coopers Mills, 250 G-5
Corea, 300 F-8
Corinna, 1100 D-5
Cornish, 600 H-2
Cornville E-4
Costigan, 60 D-7
Cranberry Isles, 90 J-12
Crescent Lake, 200 G-2
Crouseville, 175 B-14
**CUMBERLAND CO., 265612 H-2**
Cumberland Center, 2596 H-3
Cushing, 50 G-5
Cutler, 200 E-10
Damariscotta, 1100 G-5
Danforth, 500 B-8
Deblois, 30 E-8
Dedham, 125 E-7
Deer Isle, 400 G-7
Denmark, 125 G-2
Dennysville, 150 D-10
Derby C-6
Detroit, 175 E-5
Dexter, 2201 D-5
Dickey A-12
Dixfield, 1137 F-2
Dixmont, 80 E-5
Dixmont Center E-6
Dover-Foxcroft, 2592 D-5
Dresden Mills, 140 G-4
Dry Mills, 350 H-3
Dyer Brook, 60 A-8
Eagle Lake, 600 B-13
E. Andover, 150 E-2
E. Blue Hill, 150 F-7
E. Brownfield, 75 H-1
E. Corinth, 400 D-6
E. Dixfield, 120 E-3
E. Eddington, 250 E-7
E. Exeter D-6
E. Hampden, 1200 J-6
E. Hiram, 250 H-2
E. Holden, 175 E-7
E. Lamoine, 50 F-7
E. Livermore, 60 F-3
E. Machias, 300 E-10
E. Millinocket, 1701 B-7
E. Newport, 120 E-5
E. Peru, 100 F-3
E. Sebago, 325 H-2
E. Sumner, 100 F-3
E. Vassalboro, 300 F-4
E. Waterboro, 250 I-2
E. Wilton, 350 E-3
E. Winn C-7
Eastbrook, 125 E-8
Easton, 300 B-14
Eastport, 1536 D-10
Eaton, 40 B-9
Eden, 75 G-12
Edgecomb, 110 H-4
Eggemoggin F-6
Eliot, 90 J-2
Ellsworth, 7103 F-7
Emery Mills, 200 I-1
Enfield, 350 C-7
Estcourt Station, 25 A-12
Etna, 140 E-6
Eustis, 75 C-2
Exeter Corners, 80 D-6
Fairbanks, 180 E-3
Fairfield, 2569 E-4
Falmouth, 7610 H-3
Falmouth Foreside, 1964 H-10
Farmingdale, 1935 G-4
Farmington, 4098 E-3
Farmington Falls, 300 E-3
Fayette, 100 F-3
Forest City, 45 B-9
Ft. Fairfield, 1600 B-14
Ft. Kent, 1978 A-13
Frankfort, 180 E-6
**FRANKLIN CO., 29467 D-3**
Franklin, 300 E-8
Freedom, 150 F-5
Freeport, 1813 H-3
Frenchboro, 40 G-7
Frenchville, 140 A-13
Friendship, 600 G-5
Frye, 60 E-2
Fryeburg, 1549 G-1
Fryeburg Center G-1
Gardiner, 6100 G-4
Garland, 150 D-5
Georgetown, 150 H-4
Gilead, 50 F-1
Glenburn Center, 60 D-6
Goodwins Mills, 50 I-2
Gorham, 4164 H-2
Gouldsboro, 200 F-8
Grand Isle, 375 A-14
Grand Lake Stream, 140 C-9
Gray, 1100 H-3
Great Pond, 50 D-8
Greeleys Landing C-5
Green Lake E-7
Greenbush D-7
Greenbush, 80 D-7
Greene, 500 G-3
Greenfield, 40 D-7
Greenville, 1319 C-4
Greenville Junction, 450 C-4
Greenwood F-2
Grindstone, 50 B-7
Grove, 40 D-9
Guerette, 45 A-13
Guilford, 945 D-5
Hall Quarry, 100 H-12
Hallowell, 2437 F-4
Hampden, 4126 E-6
Hampden Highlands, 800 E-6
**HANCOCK CO., 51791 D-8**
Hancock, 400 F-8
Hancock Pt., 40 F-8
Hanover, 175 F-2
Harborside, 100 F-6
Harmony, 275 D-5
Harrington, 300 E-9
Harrison, 600 G-2
Hartford, 60 F-3
Hartland, 872 D-5
Haynesville, 70 A-8
Hebron, 300 G-3
Highland Lake H-8
Hinckley, 120 E-4
Hiram, 350 H-2
Hodgdon, 350 A-8
Holden, 125 E-7
Hollis Center, 250 I-2
Hope, 100 G-5
Houlton, 5270 C-14
Howland, 1210 C-7
Hudson, 150 D-6
Hulls Cove, 250 F-8
Indian River, 40 E-9
Island Falls, 600 A-8
Isle au Haut, 75 G-7
Islesboro, 140 F-6
Islesford, 80 I-13
Jackman, 800 B-3
Jackson, 100 E-6
Jacksonville, 150 E-10
Jay, 600 F-3
Jefferson, 275 G-5
Jonesboro, 220 E-9
Jonesport, 1100 F-9
Keegan, 200 A-14
Kenduskeag, 350 D-6
Kennebago Lake Camps D-2
**KENNEBEC CO., 117114 F-4**
Kennebunk, 4804 I-2
Kennebunkport, 1376 I-2
Kents Hill, 200 F-3
Kingfield, 850 D-3
Kingman, 140 B-8
Kittery, 4884 J-2
Kittery Pt., 1135 J-2
Knights Landing, 20 C-6
Knowles Corner C-13
**KNOX CO., 39618 G-5**
Kokadjo, 10 B-5
Lagrange, 120 D-6
Lake Moxie C-4
Lakewood, 80 E-4
Lambert Lake, 130 B-9
Lebanon, 90 I-1
Lee, 250 C-7
Leeds, 80 F-3
Levant, 250 E-6
Lewiston, 35131 G-3
Liberty, 180 F-5
Limerick, 400 H-2
Limestone, 1453 B-14
Limington, 220 H-2
**LINCOLN CO., 33616 G-5**
Lincoln, 2933 C-7
Lincolnville, 300 F-6
Linneus, 100 A-8
Lisbon, 1400 G-3
Lisbon Falls, 4420 G-3
Litchfield Corners, 200 G-4
Little Deer Isle, 150 F-7
Littleton, 200 C-14
Livermore, 200 F-3
Livermore Falls, 1626 F-3
Locke Mills, 200 F-2
Lookout I-14
Lovell, 225 G-1
Lubec, 800 E-11
Ludlow, 40 C-14
Lynchville G-2
Machias, 1376 E-9
Machiasport, 250 E-10
Macwahoc, 90 B-8
Madawaska A-13
Madawaska, 3326 B-14
Madison, 2733 E-4
Madrid E-2
Manchester, 650 F-4
Manset, 200 I-12
Mapleton, 500 B-14
Maplewood, 30 H-1
Mars Hill, 1200 C-14
Martin G-5
Martinsville, 100 H-5
Matinicus, 50 H-6
Mattawamkeag, 650 B-7
Mechanic Falls, 2450 G-3
Meddybemps, 70 D-10
Medway, 900 B-7
Mexico, 1946 F-2
Milbridge, 600 F-8
Milford, 2197 D-7
Millinocket, 5190 B-7
Milo, 1898 C-6
Minturn, 90 G-7
Molunkus, 35 B-7
Monarda A-7
Monhegan, 75 H-5
Monmouth, 600 G-3
Monroe, 140 E-6
Monson, 500 C-5
Monticello, 400 C-14
Moose River, 200 B-3
Moosehead, 15 B-4
Morrill, 150 F-5
Moscow, 400 D-4
Mt. Vernon, 200 F-4
Naples, 600 G-2
New Harbor, 500 H-5
New Portland, 200 D-3
New Sharon, 300 E-3
New Sweden, 60 B-14
New Vineyard, 220 E-3
Newagen, 125 H-4
Newburgh Vil., 120 E-6
Newcastle, 600 G-5
Newfield, 225 H-1
Newport, 1754 E-5
Newry, 150 F-2
Nobleboro, 150 G-5
Norridgewock, 1557 E-4
N. Amity A-9
N. Anson, 750 E-4
N. Berwick, 1580 J-2
N. Bradford, 50 D-6
N. Bridgton, 500 G-2
N. E. Carry A-4
N. Fryeburg, 70 G-1
N. Haven, 275 G-6
N. Jay, 200 F-3
N. Lovell G-1
N. Monmouth, 200 F-3
N. New Portland, 225 D-3
N. Newcastle G-4
N. Palermo F-5
N. Parsonsfield, 50 H-1
N. Penobscot F-7
N. Searsmont F-5
N. Sebago, 275 H-2
N. Sedgwick F-7
N. Shapleigh, 125 I-1
N. Turner, 350 F-3
N. Vassalboro, 900 F-4
N. Waldoboro, 175 G-5
N. Waterboro, 200 I-2
N. Waterford, 300 G-2
N. Whitefield, 60 G-4
N. Windham, 4568 H-2
Northeast Harbor, 650 F-7
Northport, 90 F-6
Norway, 2623 G-2
Oak Hill, 700 I-3
Oakland, 2758 F-4
Oceanville, 70 G-7
Ogunquit, 974 J-2
Olamon, 220 D-7
Old Orchard Beach, 8856 I-3
Old Town, 7709 D-7
Oquossoc, 150 D-2
Orient, 30 A-9
Orland, 300 F-6
Orono, 8253 D-7
Orrington, 400 E-6
Otter Creek, 250 I-13
Owls Head, 350 G-6
**OXFORD CO., 54755 F-2**
Oxford, 1300 G-2
Palermo, 120 F-5
Palmyra, 160 E-5
Parkman, 80 D-5
Passadumkeag, 150 C-7
Patten, 1000 A-7
Peaks Island J-10
Pembroke, 275 D-10
**PENOBSCOT CO., 144919 A-7**
Penobscot, 100 F-7
Perham, 150 B-13
Perry, 90 D-10
Phillips, 750 E-3
Phippsburg, 150 H-4
Pine Pt., 200 I-3
**PISCATAQUIS CO., 17235 A-5**
Pittsfield, 3217 E-5
Pittston, 150 G-4
Pittston Farm A-4
Pleasant Hill, 900 H-9
Pleasant Pond, 15 C-4
Plymouth, 200 E-5
Poland, 350 G-3
Popham Beach, 40 H-4
Port Clyde, 600 H-5
Portage, 350 B-13
Porter, 200 H-1
Portland, 62561 H-3
Pownal Center, 50 H-3
Presque Isle, 9045 B-14
Prides Corner H-8
Princeton, 750 C-9
Prospect, 100 F-6
Prospect Harbor, 400 F-8
Prouts Neck, 50 I-3
Purgatory, 300 G-4
Randolph, 1911 G-4
Rangeley, 700 D-2
Raymond, 600 H-2
Readfield, 400 F-4
Richmond, 1864 G-4
Ripley, 110 D-5
Robbinston, 200 D-10
Rockland, 7436 G-6
Rockport, 1200 G-6
Rockwood, 300 B-4
Rome, 200 E-4
Roque Bluffs, 100 E-9
Round Pond, 400 H-5
Roxbury, 180 E-2
Rumford, 4795 F-2
Sabattus, 1300 G-3
Saco, 18125 I-2
**SAGADAHOC CO., 35214 G-4**
St. Agatha, 500 A-13
St. Albans, 250 D-5
St. Francis, 300 A-13
St. George, 275 G-5
St. John, 125 A-13
Salem, 50 D-3
Salsbury Cove, 250 G-12
Sanford, 10133 I-2
Sangerville, 650 D-5
Scarborough, 3867 I-3
Seal Cove, 100 F-7
Seal Harbor, 250 I-13
Searsmont, 175 F-5
Searsport, 1102 F-6
Seawall, 50 J-12
Sebago Lake, 600 H-2
Sebasco Estates, 200 H-4
Sebec, 100 C-6
Sebec Lake, 5 C-5
Seboeis, 30 C-6
Seboomook A-4
Sedgwick, 150 F-7
Shapleigh, 175 I-1
Sherman Mills, 450 A-7
Sherman Station, 325 A-7
Shin Pond C-13
Shirley Mills, 150 C-4
Sidney F-4
Skowhegan, 6696 E-4
Small Pt. Beach H-4
Smithfield, 225 E-4
Smyrna Mills, 250 C-14
Solon, 450 D-4
**SOMERSET CO., 50888 B-3**
Somesville, 250 H-12
S. Addison, 100 F-9
S. Arm, 15 E-2
S. Berwick, 3000 J-2
S. Brewer J-6
S. Bristol, 500 H-5
S. China, 500 F-5
S. Eliot, 3445 J-2
S. Gouldsboro, 175 G-14
S. Harpswell, 225 H-3
S. Lagrange, 60 D-6
S. Lebanon, 225 J-1
S. Liberty F-5
S. Lincoln, 180 C-7
S. Lubec, 120 E-10
S. Orrington, 300 E-6
S. Paris, 2237 G-2
S. Portland, 23803 I-3
S. Sanford, 4173 I-2
S. Surry F-7
S. Waterford, 175 G-2
S. Windham, 400 H-2
S. Woodstock, 65 F-2
Southwest Harbor, 1200 F-7
Springfield, 120 C-8
Springvale, 3488 I-2
Spruce Head, 220 G-6
Stacyville, 35 A-7
Standish, 500 H-2
Starboard, 50 E-10
Starks, 135 E-4
Steep Falls, 400 H-2
Stetson, 150 D-6
Steuben, 150 F-8
Stockton Springs, 325 F-6
Stonington, 800 G-7
Stow G-1
Stratton, 400 D-3
Strong, 700 E-3
Sullivan, 175 F-8
Sunshine, 60 G-7
Surry, 225 F-7
Swanville, 150 F-6
Temple, 225 E-3
Tenants Harbor, 400 G-5
The Forks, 10 C-4
Thomaston, 2714 G-5
Thorndike, 140 E-5
Topsfield, 125 C-9
Topsham, 6271 H-4
Town Hill, 60 G-12
Tremont, 80 J-11
Trenton, 170 F-7
Troy, 60 E-5
Turner, 500 G-3
Union, 600 G-5
Unity, 486 E-5
Upper Gloucester, 300 G-3
Upton, 50 E-1
Van Buren, 2369 A-14
Vanceboro, 190 B-9
Vassalboro, 200 F-4
Veazie, 1633 E-7
Vienna, 70 F-4
Vinalhaven, 800 G-6
Waite, 35 C-9
**WALDO CO., 36280 F-5**
Waldoboro, 1291 G-5
Walnut Hill, 250 H-3
Waltham, 75 E-7
Warren, 400 G-5
Washburn, 1200 B-14
**WASHINGTON CO., 33941 E-9**
Washington, 240 F-5
Waterboro, 750 I-2
Waterford Flat, 120 G-2
Waterville, 16016 F-4
Wayne, 250 F-3
Weld, 200 E-3
Wellington, 45 D-4
Wells, 1200 J-2
Wells Beach, 500 J-2
Wesley, 50 D-9
W. Baldwin, 150 H-2
W. Bethel, 175 F-1
W. Bowdoin, 10 G-3
W. Buxton, 300 H-2
W. Enfield, 600 C-7
W. Forks, 30 C-3
W. Jonesport F-9
W. Kennebunk, 809 I-2
W. Levant D-6
W. Mills, 100 E-3
W. Minot, 125 G-3
W. Newfield, 100 H-1
W. Paris, 550 F-2
W. Peru, 120 F-2
W. Poland, 325 G-2
W. Rockport, 250 G-6
W. Scarborough, 900 I-3
W. Sumner, 25 F-2
W. Tremont, 80 G-7
W. Trenton G-11
Westbrook, 16534 H-3
Weston, 90 B-8
Whitefield, 100 G-4
Whiting, 110 E-10
Whitneyville, 150 E-9
Willimantic, 40 C-5
Wilsons Mills, 30 D-1
Wilton, 2290 E-3
Windsor, 120 F-4
Winn, 200 C-7
Winslow, 7743 F-4
Winter Harbor, 900 F-8
Winterport, 1307 E-6
Winthrop, 2893 F-4
Woodland, 1044 D-10
Woolwich, 600 H-4
Wytopitlock, 150 B-8
Yarmouth, 3560 H-3
**YORK CO., 186742 I-1**
York Beach, 900 J-2
York Harbor, 3321 J-2
York Vil., 1000 J-2

## Maryland

Page locator

| Map keys | Atlas pages |
|---|---|
| 1 - 10 | 94 - 95 |
| 11 - 20 | 96 - 97 |

* City keyed to pp. 224 - 225

~ Independent city not included in any county

Abell, 160 I-13
Aberdeen, 13993 B-15
Abingdon, 2000 C-15
Accident, 330 B-2
Accokeek, 7349 G-12
Adamstown, 400 C-10
Adelina, 150 H-13
Adelphi, 14998 *D-7
Airey, 30 H-16
Alesia, 230 A-12
**ALLEGANY CO., 74930 B-5**
Allen, 520 I-18
Allens Fresh, 100 H-12
Alpha, 100 C-12
Altamont C-2
American Corner F-17
Ammendale *C-9
Andersontown, 30 F-17
Andrews, 130 I-16
Annapolis, 36524 E-14
Annapolis Junction, 800 *B-10
**ANNE ARUNDEL CO., 489656 F-13**
Antietam, 50 C-8
Appleton Acres, 200 A-17
Aquasco, 180 H-13
Arbutus, 20116 K-3
Ardmore *F-9
Argonne Hills *B-10
Arnold, 23422 E-14
Arundel Gardens, 1250 L-5
Arundel on the Bay, 1100 F-14
Arundel Vil., 2700 L-5
Ashton, 1500 D-12
Aspen Hill, 50228 *C-6
Athol, 50 H-17
Avenue, 450 I-13
Avilton A-3
Baden, 270 G-13
Baldwin, 420 B-14
Ballard *I-9
**BALTIMORE CO., 754292 B-13**
Baltimore, 636919 ~C-13
Baltimore Corner E-17
Baltimore Highlands, 6700 K-4
Barclay, 146 D-17
Barnaby Vil., 5300 *H-7
Barnesville, 197 D-10
Barrelville, 300 A-4
Barstow, 200 H-13
Barton, 452 B-3
Bartonsville, 1650 C-10
Bay City E-15
Bay Ridge, 2200 C-20
Bay View, 300 B-16
Baynesville G-6
Bayside Beach, 950 D-14
Beachville, 300 J-14
Beallsville, 400 D-10
Beantown G-12
Beaver Creek, 350 B-3
Beaver Dam, 50 K-13
Bel Air, 9826 B-15
Bel Alton, 1000 H-12
Belhaven, 500 N-3
Bellevue, 330 G-15
Belltown F-1
Beltsville, 15690 E-12
Bembe Beach B-20
Benedict, 600 H-13
Benevola B-9
Bennsville, 7325 G-12
Benson B-14
Bentley Springs, 130 A-13
Berlin, 4056 I-20
Berwyn Hts., 2932 *D-8
Bestgate B-19
Bethesda, 55277 E-11
Bethlehem, 150 G-16
Betterton, 354 C-16
Beverly Beach, 170 F-14
Big Pines, 500 *B-3
Big Pool, 300 B-7
Big Spring, 30 B-8
Birdsville, 250 F-14
Bishop, 40 H-20
Bishops Head, 200 I-16
Bishopville, 280 H-20
Bittinger, 70 B-2
Bivalve, 350 I-17
Black Horse, 130 B-14
Bladensburg, 7563 *F-8
Bloomington, 350 B-3
Bloomsbury K-2
Bonnie Brae, 1250 C-12
Boonsboro, 3399 B-9
Boonsboro E-17
Boring, 300 B-13
Bowens, 50 H-13
Bowie, 52544 E-13
Bowleys Quarters, 6314 C-14
Bowling Green, 1800 E-4
Boxiron, 110 J-19
Boyds, 900 D-10
Bozman, 440 F-15
Braddock Hts., 4627 C-10
Bradshaw, 500 C-14
Brandywine, 1410 G-12
Breezy Pt., 230 G-14
Brentwood, 2795 *F-8
Bridgeport, 860 M-17
Bridgetown, 80 E-17
Brighton, 110 D-12
Brighton F-3
Brightwood Acres N-17
Brinklow, 300 D-12
Bristol, 30 F-13
Broad Run C-9
Brookeville, 131 D-11
Brooklandville, 1700 F-4
Brooklyn Pk., 10938 E-5
Brookmead, 260 *B-2
Brookmont, 3202 *E-5
Brookview, 62 H-17
Brookwood, 2200 *J-10
Broomes Island, 550 H-14
Browns Woods Villa, 400 A-20
Brownsville, 210 C-9
Brownsville, 100 E-16
Bruceville, 60 G-16
Brunswick, 5228 C-9
Bryans Road, 4912 G-11
Bryantown E-15
Bryantown, 1000 H-12
Buckeystown, 350 C-10
Bucktown, 30 H-16
Budds Creek I-13
Burkittsville, 186 C-9
Burrisville, 100 E-16
Burrsville, 130 F-17
Burtonsville, 7305 E-12
Bush, 50 B-15
Bushwood, 400 I-13
Butler, 350 B-13
Butlertown, 160 D-16
Byrdtown, 110 K-17
Cabin John, 1734 *E-4
California, 9300 I-14
Callaway, 470 I-14
**CALVERT CO., 74563 H-13**
Calvert, 150 A-16
Calvert Beach, 600 H-14
Calverton, 12610 *C-8
Cambridge, 11752 H-16
Camp Springs, 17968 F-12
Canton J-5
Cape Isle of Wight, 750 J-20
Cape St. Claire, 8022 E-14
Capitol Hts., 4096 *G-8
Cardiff, 300 A-15
Carmichael, 30 E-16
Carney, 28264 G-7
**CAROLINE CO., 29772 F-17**
**CARROLL CO., 150897 B-12**
Carroll Highlands, 2000 C-12
Carrollton B-12
Carsins Run, 100 B-15
Cascade, 1000 A-10
Castleton, 200 A-15
Catonsville, 39820 J-2
Catonsville Manor, 2600 J-3
Cavetown, 1486 B-9
Cayots I-17
Cearfoss, 300 A-8
**CECIL CO., 85951 B-17**
Cecilton, 493 C-17
Cedar Beach, 500 D-14
Cedar Grove, 500 D-11
Cedar Hts., 1150 *F-8
Cedar Lawn, 400 M-16
Cedarville, 300 G-13
Centreville, 3458 E-16
Ceresville, 70 C-20
Champ, 100 J-17
Chance, 377 J-16
Chaney G-13
Chaneyville, 90 G-13
Chaptico, 850 I-13
**CHARLES CO., 120546 H-12**
Charlestown, 1103 B-16
Charlotte Hall, 1214 H-13
Chase, 920 C-14
Chattolanee G-2
Cheltenham, 950 G-13
Cherry Hill, 500 A-17
Chesaco Pk., 450 I-7
Chesapeake Beach, 3399 G-14
Chesapeake City, 830 B-17
Chesapeake Haven, 250 C-16
Chesapeake Hts., 550 I-18
Chester, 3723 E-15
Chestertown, 4899 D-16
Chesterville, 100 D-16
Chestnut Hill Estates J-1
Chestnut Ridge, 230 E-2
Cheverly, 6372 *F-8
Chevy Chase, 9381 *E-5
Chevy Chase View, 901 *D-5
Chewsville, 293 B-9
Chicamuxen, 100 H-11
Childs, 370 A-17
Chillum, 34252 *E-7
Choptank, 100 G-16
Christs Rock, 120 H-16
Church Creek, 78 H-16
Church Hill, 670 D-16
Churchton F-14
Churchville, 600 B-15
Claggettsville, 400 C-11
Claiborne, 170 F-15
Clarksburg, 1834 D-11
Clarksville, 830 D-12
Clarysville, 70 A-4
Clear Spring, 461 A-8
Clements, 270 I-13
Cliffs City, 60 D-16
Clinton, 26064 F-12
Clopper, 1000 *A-3
Clover Hill, 3260 L-19
Cloverly, 7835 *B-7
Cobb Island, 900 I-12
Cockeysville, 19388 B-13
Cohasset *E-4
Cokesbury, 110 J-18
Coleman, 130 C-16
Colesville, 19810 E-12
College Pk., 26925 *D-8
Colmar Manor, 1256 *F-8
Colonial Pk. N-17
Colora, 120 A-16
Coltons Pt., 500 J-13
Columbia, 88254 D-12
Columbia Pk., 1000 *F-9
Compton, 250 I-13
Comus D-10
Conowingo, 150 A-15
Contee *C-9
Cooksville, 500 C-12
Coral Hills, 10720 *G-8
Cordova, 592 F-16
Cornersville, 100 G-15
Corriganville, 930 A-4
Cottage City, 1122 *F-8
Cove Pt., 1200 I-14
Coxs Corner I-17
Craigtown, 200 B-16
Cranberry, 200 B-12
Crapo, 130 I-16
Creagerstown, 150 B-10
Crellin, 200 C-1
Cresaptown, 1850 B-4
Crestwood, 200 I-18
Crisfield, 2749 K-17
Crocheron, 120 I-16
Crofton, 20091 E-13
Croom, 120 G-13
Cropley, 1700 *E-3
Crosby, 50 D-15
Crownsville, 1670 E-13
Crumpton, 550 D-16
Crystal Beach, 250 C-16
Cub Hill G-7
Cumberland, 20495 A-4
Damascus, 11430 D-11
Dameron, 350 J-15
Dames Quarter, 188 J-17
Daniel, 100 C-12
Danville, 270 B-3
Dares Beach, 500 H-14
Dargan, 100 C-9
Darlington, 800 B-15
Darnestown, 6378 E-10
Davidsonville, 780 F-13
Dawson, 120 B-3
Dawsonville, 120 D-10
Dayton, 800 D-12
Deal Island, 578 J-16
Deale, 4796 F-14
Deer Pk., 382 C-2
Defense Hts., 2250 *E-9
Delight F-1
Delmar, 3356 H-18
Denton, 4022 F-17
Dentsville, 50 H-12
Derwood, 2000 *B-4
Detour, 520 B-11
Dickerson, 1000 D-10
District Hts., 6034 *G-8
Dominion, 470 E-15
Doncaster, 200 H-11
**DORCHESTER CO., 30674 H-16**
Dorrs Corner, 170 E-13
Dorsey, 1350 M-2
Doubs, 160 C-10
Dowell, 300 I-14
Downsville, 380 B-8
Drayden, 110 J-14
Drum Pt., 1600 I-14
Drury, 300 F-13
Drybranch A-14
Dublin, 500 A-15
Dumbarton, 1200 G-3
Dundalk, 62306 D-14
Dunkirk, 2363 G-13
Dynard, 50 I-13
Eagle Harbor, 56 H-13
Earleville, 500 C-16
E. New Market, 266 G-17
Easton, 14677 F-16
Eastpines, 1250 *E-8
Eastpoint J-7
Eastview, 330 M-18
Eden, 793 I-18
Edesville, 160 D-15
Edgemere, 9248 D-14
Edgewater, 1400 C-18
Edgewood, 23378 C-15
Edmonston, 1329 *E-8
Ednor *A-7
Eldersburg, 27741 C-12
Eldorado, 57 G-17
Elioak, 290 D-12
Elk Mills, 700 A-17
Elk Neck, 130 B-16
Elkridge, 22042 D-13
Elkton, 14842 B-17
Ellerslie, 1400 A-4
Ellerton B-9
Ellicott City, 56397 D-13
Elliott, 200 I-16
Ellwood, 130 G-17
Elmwood, 1650 H-7
Emmitsburg, 2365 A-10
Emmorton, 1400 B-15
Emory Grove, 5300 *A-4
Epping Forest A-19
Essex, 39078 C-14
Etchison, 200 D-11
Evergreen Pk., 70 D-14
Evitts Creek L-10
Ewell, 250 K-16
Ewingville, 120 D-16
Fair Haven, 500 G-14
Fair Hill, 100 A-17
Fairbank G-15
Fairland, 21738 E-12
Fairlee, 250 D-15
Fairmount, 537 J-17
Fairmount Hts., 1495 *F-8
Fairplay, 440 B-8
Fairview, 680 M-7
Fallston, 8427 B-14
Farmington, 180 A-16
Faulkner, 450 H-12
Feagaville, 220 C-10
Federalsburg, 2625 G-17
Fells Pt. J-5
Ferndale, 16056 L-4
Fiddlersburg, 180 M-17
Finksburg, 1000 B-12
Finzel A-3
Fishing Creek, 700 I-15
Flintstone, 500 A-5
Flohrville, 800 C-12
Florence, 100 C-11
Font Hill, 890 K-1
Font Hill Manor, 2500 D-12
Forest Hts., 2554 *H-7
Forest Hill, 900 B-14
Forestville, 12712 F-12
Fork, 500 C-14
Ft. Foote Vil., 5000 *I-6
Ft. Howard, 890 D-14
Fountain Green, 1400 B-15
Fountain Head, 1350 A-9
Fountaindale C-10
Fowblesburg, 250 B-12
Foxhall *C-6
**FREDERICK CO., 195277 B-10**
Frederick, 59213 C-10
Freeland, 600 A-13
Friendly, 10938 *J-7
Friendly Farms *J-7
Friendship, 170 G-14
Friendsville, 503 A-1
Frizzelburg, 280 B-11
Frostburg, 7719 A-3
Fruitland, 4431 I-18
Fullerton, 2400 H-7
Fulton, 1150 *A-8
Funkstown, 937 B-9
Furnace Branch M-5
Gaither, 300 C-12
Gaithersburg, 58744 D-11
Galena, 504 C-17
Galestown, 97 H-17
Galesville, 1000 F-14
Gamber, 250 C-12
Gambrills, 1500 E-13
Gapland, 240 C-9
Garland, 1800 L-4
**GARRETT CO., 29846 B-2**
Garrett Pk., 955 *D-5
Garrison, 7969 G-2
George Island Landing J-19
Georgetown, 100 C-17
Germantown, 55419 D-11
Gibson Island, 400 E-14
Gingerville Manor Estates B-18
Girdletree, 117 J-19
Glassmanor, 8800 *H-7
Glen Arm, 730 C-14
Glen Burnie, 38922 D-13
Glen Echo, 255 *E-4
Glen Hills, 1400 *C-3
Glen Mar Pk. *E-5
Glenarden, 6306 F-12
Glencoe, 220 B-13
Glenelg, 1000 D-12
Glenmar M-10
Glenmont *C-6
Glenn Dale, 12609 *E-10
Glenwood, 200 D-12
Glyndon, 900 B-13
Golden Beach, 2669 H-13
Golden Hill, 40 I-15
Goldsboro, 216 E-17
Golts, 150 C-17
Good Hope *B-7
Goodwill, 50 J-18
Gorman, 150 C-1
Gortner, 230 C-1
Goshen, 530 D-11
Governors Run, 30 H-14
Graceham, 120 B-10
Granite, 950 C-12
Grantsville, 649 A-2
Grasonville, 2193 E-15
Gratitude D-15
Great Mills, 700 I-14
Green Haven, 17415 D-14
Green Hill, 70 I-17
Green Valley, 12262 C-10
Greenbelt, 21098 E-12
Greenberry Hills, 1400 N-16
Greensboro, 1998 E-17
Greentree, 3400 N-4
Guilford, 1000 D-12
Hack Pt., 500 C-17
Hagerstown, 39728 B-9
Halethorpe K-3
Halfway, 10065 B-8
Hampden I-4
Hampstead, 5467 B-12
Hampton, 5004 F-5
Hancock, 1743 A-7
Hanesville, 30 D-16
Hanover, 210 L-3
Harewood Pk., 1100 C-15
**HARFORD CO., 218590 B-14**
Harmans, 400 M-3
Harmony, 150 F-17
Harney, 240 A-11
Harrisonville, 270 G-1
Harundale M-5
Harwood, 450 F-13
Harwood Pk. D-13
Havre de Grace, 13085 B-16
Hebbville, 2500 H-2
Hebron, 1044 H-18
Helen, 70 I-13
Henderson, 122 E-17
Herald Harbor, 2313 E-14
Hereford, 700 B-13
Hernwood Hts., 960 H-1
Hickory, 610 B-15
Hidden Pt., 70 A-20
High Pt., 3700 M-6
High Ridge, 2500 *B-9
Highfield, 210 A-10
Highland, 1700 D-12
Highland Beach, 110 F-14
Hill Top, 50 H-11
Hillcrest Hts., 380 *A-9
Hillcrest Hts., 16359 *H-7
Hillsboro, 157 F-16
Hillsmere Shores, 2977 C-20
Hobbs, 60 F-17
Hollywood, 1000 I-14
Homewood K-9
Honga, 100 I-15
Hoopersville, 250 I-15
Hopewell K-17
**HOWARD CO., 247842 D-12**
Hoyes B-1
Hudson, 250 G-15
Hughesville, 1537 H-13
Huntingtown, 2436 G-13
Hurlock, 2048 G-17
Hutton, 110 C-1
Huyett, 100 A-8
Hyattstown, 120 D-10
Hyattsville, 15376 *E-8
Hydes, 250 B-14
Ijamsville, 400 C-10
Ilchester, 890 K-2
Indian Head, 3758 G-11
Indian Springs, 50 B-7
Ingleside, 350 E-17
Ironshire I-20
Ironsides, 180 H-11
Island Creek, 100 H-14
Issue, 430 I-12
Jacksonville, 600 B-14
Jacobsville N-7
Jarrettsville, 2756 B-14
Jefferson, 300 C-9
Jefferson Hts., 800 M-17
Jessup, 7865 D-13
Johnsville, 550 B-11
Johnsville, 1200 C-12
Jonestown, 250 G-17
Joppatowne, 11391 C-15
Keedysville, 862 B-9
Kemp Mill Estates, 3050 *D-6
Kemptown C-11
Kennedyville, 200 C-16
Kensington, 1946 *D-5
**KENT CO., 19197 D-16**
Kent Vil., 2700 *F-9
Kentmore Pk., 250 C-16
Kenwood, 4100 I-7
Kettering, 11008 F-13
Keymar, 520 B-11
Keysers Ridge, 30 A-2
Kingston, 50 J-17
Kingstown, 1644 D-16
Kingsville, 4214 C-14
Kitzmiller, 278 C-2
Knollwood, 1250 *D-7
Knoxville, 340 C-9
La Plata, 8879 H-12
La Vale, 4613 K-9
Ladiesburg, 230 B-11
Lake Shore, 13065 E-14
Lakesville, 30 I-16
Landover, 640 *F-8
Landover Hills, 1514 *E-9
Langley Pk., 16214 *E-7
Lanham, 5200 *E-9
Lansdowne, 8700 K-4
Lappans, 30 B-9
Largo, 8408 *F-10
Laurel, 22329 E-12
Lawsonia K-17
Layhill, 1600 *B-6
Laytonsville, 353 D-11
Le Gore, 110 B-10
Leitersburg, 523 A-9
Leonardtown, 2214 I-13
Level, 400 B-15
Lewisdale, 3100 *E-7
Lewistown, 270 B-10
Lexington Pk., 11021 I-14
Liberty Grove, 90 A-16
Liberty Manor, 920 H-1
Libertytown, 750 B-11
Libertytown I-19
Lime Kiln C-10
Lineboro, 350 A-12
Linhigh, 2100 H-7
Linkwood, 240 H-16
Linthicum, 7539 L-4
Linwood, 130 B-11
Lipins Corner, 80 N-5
Lisbon, 1150 C-11
Little Orleans, 200 B-6
Lloyds, 30 G-15
Loch Lynn Hts., 435 C-1
Loch Raven Vil. G-6
Lochearn, 25269 H-3
Locust Grove, 30 C-16
Lonaconing, 1125 B-3
Londontown, 4000 F-14
Long Beach, 1100 H-14
Long Green, 500 B-14
Longwoods, 50 F-16
Loretta Hts. B-19
Lothian, 300 F-13
Louisville, 200 C-12
Love Pt., 290 E-15
Loveville, 640 I-13
Lower Marlboro, 170 G-13
Luke, 73 B-3
Lusby, 1666 H-14
Lutherville, 4000 F-5
Luxmanor, 2500 *C-5
Lynch, 200 C-16
Lynne Acres, 6700 H-2
Lyons Creek, 50 G-13
Maddox, 230 I-13
Madison, 160 H-15
Madonna, 200 B-14
Magnolia C-15
Malcolm, 50 G-13
Manchester, 3546 A-12
Manokin, 250 J-17
Manor B-14
Manor View E-7
Marbury H-11
Mardela Springs, 357 H-17
Margate, 1500 M-5
Marion, 1050 K-17
Marley, 1500 M-5
Marlton, 7798 F-13
Marriottsville, 110 C-12
Marshall Hall, 230 G-11
Marston, 50 B-11
Martinsburg D-10
Marydel, 143 D-17
Maryland City, 6814 *B-10
Maryland Line, 400 A-13
Mason Springs, 30 H-11
Massey, 150 C-17
Matthews, 30 F-16
Maugansville, 2295 A-8
Mayo, 3153 F-14
Mays Chapel, 11427 F-4
McAlpine, 4400 K-1
McCoole, 350 C-3
McDaniel, 170 F-15
McHenry, 1000 B-1
Meadowcliff F-6
Meadowood, 1100 *C-7
Mechanicsville, 1000 H-13
Melitota, 70 D-15
Melrose, 200 A-12
Middle River, 23958 I-8
Middleburg, 310 B-11
Middletown, 100 A-13
Middletown, 2859 C-9
Midland, 440 B-3
Midlothian, 450 B-3
Milford, 690 H-2
Millers, 180 A-12
Millersville, 950 E-13
Millington, 445 D-17
Monie, 110 J-17
Monkton, 800 B-13
Monrovia C-11
Montevideo, 230 M-2
**MONTGOMERY CO., 873341 D-11**
Montgomery Vil., 38051 D-11
Montpelier, 3600 *C-9
Montrose *C-5
Morgantown, 200 I-12
Morganza, 150 I-13
Morningside, 1305 *H-9
Moscow, 230 B-3
Motters A-10
Mt. Aetna, 838 B-9
Mt. Airy, 8761 C-11
Mt. Harmony G-13
Mt. Hebron, 2000 I-1
Mt. Lena, 501 B-9
Mt. Pleasant, 330 C-10
Mt. Pleasant Beach, 2100 M-6
Mt. Rainier, 8322 *F-7
Mt. Savage, 1300 A-4
Mt. Vernon, 761 I-17
Mt. Victoria, 150 I-12
Mt. Zion, 80 E-17
Mountain Lake Pk., 2106 C-1
Muirkirk, 650 *C-9
Myersville, 1511 B-9
Nanjemoy, 240 H-11
Nanticoke, 580 I-17
Neavitt, 300 G-15
New Carrollton, 12460 E-12
New Germany, 30 B-2
New London, 170 C-11
New Market, 465 C-11
New Midway, 380 B-10
New Windsor, 1352 B-11
Newark, 339 I-19
Newburg, 720 I-12
Newcomb, 250 G-15
Newport, 80 H-12
Newtown, 80 C-15
Newtown, 330 H-12
Nikep, 270 B-3
Norbeck *B-5
Normans, 100 E-15
Norrisville, 250 A-14
N. Beach, 1869 G-14
North East, 2862 B-16
N. Point J-8
N. Point Village J-8
N. Potomac, 23044 E-11
Norwood, 250 *B-6
Notch Cliff F-7
Nottingham, 240 H-8
Oak Court B-18
Oak Pk., 30 C-1
Oak Ridge, 580 N-16
Oak View, 3400 *D-7
Oakland, 1844 C-1
Oakland C-12
Oakland, 2500 *G-8
Oaklawn, 1600 *I-8
Oakley, 50 I-13
Oakmont *D-5
Oakville I-13
Oakwood A-15
Ocean City, 7042 I-20
Ocean Pines, 10496 I-20
Odenton, 20534 E-13
Oella J-2
Old Farm, 3450 *C-4
Oldtown, 600 B-5
Olivet, 500 I-14
Olney, 31438 D-11
Oraville, 50 H-13
Orchard Beach, 2200 M-6
Orchard Hills, 1250 L-16
Oriole, 100 J-17
Overlea, 12148 H-7
Owings, 1325 G-14
Owings Mills, 20193 C-13
Oxford, 706 G-15
Oxon Hill-Glassmanor, 35355 *I-7
Palmer Pk. *F-9
Paradise Beach, 350 M-8
Paramount, 1450 L-17
Park Hall, 700 J-14
Parkside, 4650 *D-5
Parkton, 500 B-13
Parkville, 31118 C-14
Parole, 14031 B-19
Parsonsburg, 680 I-19
Patapsco, 200 B-12
Pen Mar, 210 A-10
Pendennis Mt., 350 A-20
Perry Hall, 28705 C-14
Perryman, 2461 C-15
Perryville, 3802 B-16
Pfeiffer Corners L-1
Phoenix, 600 B-13
Pikesville, 29123 H-3
Pinefield, 8800 G-12
Pinehurst on the Bay, 570 D-14
Pinesburg, 250 B-8
Piney Glen Farms, 470 *C-3
Piney Grove A-6
Piney Pt., 1500 J-14
Pinto, 300 B-4
Pioneer City N-3
Piscataway G-12
Pisgah, 300 H-11
Pittsville, 1182 I-19
Pleasant Hill, 250 A-17
Pleasant Hills, 2851 B-14
Plum Pt., 200 G-14
Pocomoke City, 3879 J-18
Pt. Lookout, 70 K-15
Pt. of Rocks, 1200 D-9
Pt. Pleasant, 1000 M-5
Pomfret, 620 H-12
Pomona, 60 D-16
Pomonkey, 500 G-11
Poolesville, 5674 D-10
Popes Creek, 80 I-12
Poplar Grove, 550 B-15
Poplar Hill G-13
Poplar Springs, 220 C-11
Port Deposit, 701 B-16
Port Herman, 270 B-17
Port Republic, 200 H-14
Port Tobacco, 19 H-12
Pot Spring F-5
Potomac, 44822 E-11
Potomac Hts., 1154 G-11
Powellville, 170 I-19
Preston, 672 G-17
Price, 180 E-16
Prince Frederick, 1432 H-14
**PRINCE GEORGES CO., 801515 G-12**
Princess Anne, 3015 J-18
Principio Furnace, 130 B-16
Providence, 1400 F-6
Public Landing, 230 J-19
Pumphrey, 5317 L-4
Putnam, 300 B-14
Putty Hill H-7
Pylesville, 350 A-14
Quaker Neck Landing, 30 D-16
Quantico, 300 I-17
Queen Anne, 166 F-16
**QUEEN ANNES CO., 40563 E-16**
Queenstown, 635 E-15
Quince Orchard, 5500 E-11
Randallstown, 30870 C-13
Randolph Hills, 6150 *C-5
Rawlings, 850 B-4
Rayville, 100 B-13
Redhouse, 30 C-1
Redland, 16998 *A-4
Rehobeth, 50 K-18
Reids Grove, 50 H-17
Reisterstown, 22438 C-12
Relay K-3
Reliance, 30 G-17
Rhodes Pt., 150 K-16
Rhodesdale, 50 H-17
Riderwood G-4
Ridge, 450 J-15
Ridgely, 1517 E-17
Ridgeway, 500 N-4
Ringgold, 160 A-9
Ripley, 110 H-11
Rippling Ridge, 2100 L-5
Rising Sun, 1809 A-16
Rison, 450 H-11
Riva, 3966 E-14
Riverdale Pk., 6418 *E-8
Riviera Beach, 12695 M-7
Roberts, 50 D-16
Roblee Acres *H-10
Rock Creek Forest *D-6
Rock Hall, 1486 D-15
Rock Hill Beach, 1350 M-7
Rock Pt., 280 I-12
Rockdale, 6700 H-2
Rockville, 60734 E-11
Rocky Ridge, 900 B-10
Rodgers Forge G-5
Rohrersville, 170 C-9
Romancoke, 470 F-15
Rosaryville Estates, 1250 G-13
Rose Haven, 550 G-14
Rosedale, 19199 I-7
Rosemont, 310 C-9
Rossville, 11515 H-7
Royal Oak, 550 G-15
Royal Oak, 50 I-17
Rumbley, 200 J-17
Ruthsburg, 30 E-16
Ruxton G-4
Sabillasville, 1100 A-10
St. Augustine B-17
St. Charles, 33379 G-12
St. Clement Shores, 400 I-13
St. George Island, 250 J-14
St. Inigoes, 500 J-14
St. James, 1657 B-8
St. Leonard, 536 H-14
St. Margarets, 350 A-20
St. Martin I-20
**ST. MARYS CO., 86211 I-13**
St. Marys City, 3300 J-14
St. Michaels, 1068 F-15
Salem, 60 H-17
Salisbury, 28200 I-18
Sanders Pk., 450 M-7
Sandgates, 450 H-13
Sandy Bottom D-15
Sandy Hook, 120 C-9
Sandy Spring, 1200 *A-6
Sandyville, 650 B-12
Sang Run, 180 B-1
Sassafras, 70 C-17
Savage, 8000 *A-10
Scaggsville, 270 N-9
Scientists Cliffs, 750 H-14
Scotland, 350 J-15
Seabrook, 8000 *E-9
Seat Pleasant, 4829 F-12
Secretary, 487 G-16
Security, 270 M-17
Selbysport, 50 A-1
Seneca E-10
Severn, 35076 D-13
Severn Grove A-19
Severna Pk., 28507 E-14
Shady Side, 5559 F-14
Shallmar, 150 C-2
Shane A-14
Sharpsburg, 661 C-8
Sharptown, 620 H-17
Shawsville, 80 B-14
Shelltown, 100 K-18
Sherwood, 250 F-15
Sherwood Forest, 630 E-14
Shiloh, 130 I-12
Shipley L-4
Shookstown, 550 M-18
Showell, 400 H-20
Silesia, 240 G-12
Silver Hill, 1850 *H-8
Silver Run, 180 A-12
Silver Sands, 350 M-6
Silver Spring, 76540 E-12
Simpsonville N-9
Skidmore, 450 E-14
Smithsburg, 2908 A-9
Smithville, 50 F-17
Smithville, 100 H-15
Snow Hill, 2321 J-19
Solomons, 1536 I-14
**SOMERSET CO., 24747 J-17**
Somerset, 1169 *E-5
S. River Pk., 550 F-14
Sparks B-13
Sparrows Pt. L-8
Spence J-19
Spencerville, 2850 *B-7
Spring Gap, 450 B-4
Springbrook, 1100 *C-7
Springdale, 3850 B-13
Starkey Corner D-16
Starr, 30 E-16
Stemmers Run I-8
Stepney, 350 B-15
Stevenson, 750 G-3
Stevensville, 5880 E-15
Still Pond, 400 C-16
Stockton, 143 J-19
Stonecrest Hill, 1250 K-1
Street, 900 A-14
Sudlersville, 392 D-17
Suitland, 26750 F-12
Summit Pk., 2900 G-3
Sunderland, 600 G-14
Sunnybrook, 980 B-14
Sunshine, 100 D-11
Swan Pt., 280 D-14
Swanton, 300 B-2
Sweet Air Manor, 450 B-14
Sykesville, 4419 C-12
Takoma Pk., 17701 *E-7
**TALBOT CO., 33812 F-16**
Tall Timbers, 400 J-14
Taneytown, 5422 A-11
Tanyard G-16
Taylors Island, 400 H-15
Taylorsville, 150 C-11
Taylorville, 30 I-20
Temple Hills, 7792 *H-8
Templeville, 82 D-17
The Lakes, 9000 C-13
Thomas H-15
Thurmont, 6037 B-10
Tilghman, 1000 G-15
Tilghmanton, 170 B-8
Timber Ridge C-11
Timberview L-2
Timonium, 11000 F-5
Tobytown, 210 *C-2
Toddville, 250 I-16
Tolchester Beach, 150 D-15
Tompkinsville, 150 I-12
Towson, 51793 C-13
Tracys Landing, 440 F-14
Trappe, 1143 G-16
Travilah, 7442 *C-2
Tunis Mills, 70 F-15
Tuscarora, 150 D-10
Tuxedo *F-8
Tyaskin, 350 I-17
Tylerton, 150 K-16
Union Bridge, 1077 B-11
Union Mills, 150 A-12
Uniontown, 350 B-11
Unionville, 220 B-11
Unionville, 80 F-16
Unionville, 230 J-18
Unity, 80 D-11
University Pk., 2282 *E-8
Upper Crossroads, 300 B-14
Upper Fairmount, 450 J-17
Upper Falls, 450 C-14
Upper Hill, 190 J-17
Upper Marlboro, 657 F-13
Upperco, 700 B-12
Urbana, 330 C-10
Vale Summit, 200 B-3
Valley Lee, 500 J-14
Van Bibber, 1000 C-15
Van Lear Manor N-15
Venice on the Bay, 300 M-8
Venton, 150 J-17
Vienna, 319 H-17
Viers Mill Vil. *C-5
Villa Nova, 940 H-2
Waldorf, 22312 G-12
Walkersville, 5606 B-10
Wallville H-14
Walston, 50 I-18
Warfieldsburg B-12
Warwick, 850 C-17
**WASHINGTON CO., 131923 A-7**
Washington Grove, 566 *A-4
Waterloo M-1
Waterview I-17
Webster Vil., 380 B-15
Welcome, 600 H-11
Wenona, 390 J-16
W. Denton, 150 F-17
W. Edmondale, 3450 J-3
W. Friendship, 1200 C-12
W. Ocean City, 3311 I-20
W. River F-14
Westernport, 1945 B-3
Westgate *E-5
Westminster, 17689 B-12
Westover, 800 J-18
Westwood G-13
Wetipquin, 80 I-17
Weverton, 140 C-9
Whaleyville, 124 I-19
Wheaton, 134800 E-12
White Hall, 700 B-13
White Marsh, 8485 C-14
White Oak, 20973 *C-7
White Plains G-12
White Pt. Beach, 450 J-13
Whiteford, 750 A-15
Whitehaven, 80 I-17
Whitehouse B-13
Whiteleysburg E-17
Whiton, 60 I-19
**WICOMICO CO., 84644 H-18**
Widgeon I-17
Wild Rose Shores, 300 C-19
Willards, 967 I-19
Williamsburg, 100 G-17
Williamsport, 2278 B-8
Williston, 110 F-17
Willow Beach Colony, 230 G-14
Wilson, 50 A-8
Wilton Farm Acres, 1900 I-1
Winchester on the Severn, 700 A-19
Winfield, 200 C-11
Wingate, 230 I-16
Wiseburg B-13
Wittman, 600 F-15
Wolfsville, 150 B-9
Wood Pt. M-16
Woodbine, 500 C-11
Woodensburg, 150 B-12
Woodfield D-11
Woodland, 80 B-3
Woodlawn, 150 B-16
Woodlawn, 36079 C-13
Woodlawn Hts., 2400 L-4
Woodmont *D-5
Woodmoor, 8400 H-2
Woodmore, 6077 *F-10
Woodsboro, 914 B-10
Woodstock, 850 C-12
Woodville C-11
Woolford, 290 H-15
**WORCESTER CO., 46543 I-19**
Worthington, 2350 K-1
Worton, 500 D-16
Wye Mills, 400 E-16
Yellow Springs, 450 B-10

## Massachusetts

Page locator

| Map keys | Atlas pages |
|---|---|
| 1 - 10 | 98 - 99 |
| 11 - 20 | 100 - 101 |

* City keyed to p. 185

Abington, 14605 G-15
Acoaxet, 220 L-14
Acton, 2300 D-12
Acushnet, 3220 J-15
Adams, 5784 C-3
Adams Shore M-8
Adamsville C-5
Agawam, 28091 G-5
Aldenville L-12
Alford, 160 F-1
Allerton L-10
Amesbury, 12327 A-15
Amherst, 16100 E-6
Andover, 7900 C-14
Arlington, 42389 E-14
Ashburnham, 1500 C-9
Ashby, 700 C-10
Ashfield, 1050 D-4
Ashland, 12066 F-12
Ashley Falls, 850 G-1
Ashley Hts., 110 I-15
Assinippi, 1400 G-15
Assonet, 1300 I-14
Athol, 8370 C-8
Attleboro, 42833 H-13
Auburn, 15005 F-10
Avon, 4558 G-14
Ayer, 2960 D-11
Baldwinville, 1852 C-8
Ballardvale, 1400 C-13
Bancroft E-3
**BARNSTABLE CO., 222230 J-17**
Barnstable, 46184 J-18
Barre, 1150 E-8
Barre Plains, 400 E-8
Barrowsville, 250 H-13
Bass River J-19
Bayview, 200 K-15
Beach Pt., 50 G-19
Becket, 600 E-3
Becket Center, 80 F-3
Bedford, 12996 D-13
Beechwood, 500 F-15
Belchertown, 2626 F-7
Bellingham, 4497 G-12
Belmont, 24194 E-13
Berkley, 200 I-14
**BERKSHIRE CO., 134953 E-2**
Berkshire, 200 D-2
Berlin, 700 E-11
Bernardston, 900 C-6
Beverly, 39343 D-15
Billerica, 6850 D-13
Blackstone, 5000 H-11
Blandford, 950 F-4
Bliss Corner, 5466 F-20
Bolton, 550 E-11
Bondsville, 1876 F-7
Boston, 609023 E-14
Bourne, 1443 J-17
Bournedale, 150 I-17
Boxboro, 500 D-11
Boxford, 2340 C-14
Boylston Center, 950 E-11
Bradstreet, 200 E-6
Braintree, 33698 F-14
Braleys, 350 J-15
Brant Rock, 1800 G-16
Brayton Pt. F-18
Brewster, 2212 J-20
Bridgewater, 6664 H-15
Briggsville B-3
Brighton K-6
Brightwood M-11
Brimfield, 800 G-8
**BRISTOL CO., 534678 J-14**
Brockton, 93007 G-14
Brookfield, 1200 F-8
Brookline, 57107 E-14
Bryantville, 1800 G-15
Buckland, 200 C-4
Buena Vista Shores, 700 I-15
Burlington, 22876 D-13
Buzzards Bay, 3549 J-17
Byfield, 1300 B-15

*, †, ‡, §, ◊ See explanation under state title in this index. County / Parish names are listed in capital letters and in boldface type.

Cambridge, 105596 ... E-14
Canton, 18530 ... G-14
Carlisle, 800 ... D-12
Carver, 1500 ... I-16
Cataumet, 1500 ... J-17
Cedarville, 450 ... I-17
Central Vil., 400 ... K-14
Chaffin, 4000 ... E-10
Charlemont, 800 ... C-4
Charlton, 550 ... G-9
Charlton City, 1400 ... G-9
Charlton Depot, 400 ... G-9
Chartley, 750 ... H-13
Chatham, 1667 ... J-20
Chelmsford, 34200 ... C-12
Chelsea, 41577 ... E-14
Cherry Valley, 1200 ... F-10
Cheshire, 1200 ... D-2
Cheshire Harbor, 60 ... C-2
Chester, 800 ... F-3
Chesterfield, 700 ... E-4
Chicopee, 54941 ... G-5
Chicopee Falls ... L-12
Chilmark, 200 ... M-16
City Mills, 250 ... G-12
Clayton ... G-2
Clifton ... I-10
Cliftondale ... I-8
Clinton, 7884 ... E-11
Cochesett, 300 ... H-14
Cochituate, 6768 ... E-12
Cohasset, 6000 ... F-16
College Hill ... C-19
Collinsville ... C-12
Colrain, 500 ... C-5
Cominsville ... D-17
Concord, 4700 ... D-12
Congamond, 900 ... H-4
Conway, 800 ... D-5
Cordaville, 2515 ... F-11
Coury Hts. ... E-20
Crescent Mills ... F-4
Cummington, 300 ... D-4
Cushman, 400 ... E-6
Cuttyhunk, 70 ... L-15
Dalton, 7155 ... D-2
Danvers, 25212 ... C-15
Dartmouth, 5000 ... F-20
Dawson, 150 ... E-10
Dedham, 23464 ... F-13
Deerfield, 700 ... D-5
Dennis, 2798 ... J-19
Dennis Port, 3612 ... J-19
Dighton, 1100 ... I-14
Dodge, 200 ... G-9
Dorchester Center ... L-7
Dorothy Pond, 1700 ... C-20
Douglas, 400 ... G-10
Dover, 2216 ... F-13
Dracut, 25594 ... C-13
Drury, 150 ... C-3
Dudley, 3700 ... G-9
**DUKES CO., 14987 ... L-16**
Dunstable, 900 ... C-12
Duxbury, 1426 ... G-16
Dwight, 200 ... E-6
E. Acton, 950 ... D-12
E. Arlington ... J-6
E. Billerica, 3850 ... C-13
E. Boston ... K-8
E. Braintree ... N-8
E. Brewster, 800 ... I-20
E. Bridgewater, 3400 ... H-15
E. Brimfield, 150 ... G-8
E. Brookfield, 1410 ... F-9
E. Charlemont, 200 ... C-4
E. Deerfield ... D-6
E. Dennis, 3299 ... J-19
E. Douglas, 2319 ... G-10
E. Falmouth, 6615 ... K-17
E. Foxboro, 1000 ... G-13
E. Freetown, 850 ... I-15
E. Harwich, 4744 ... J-20
E. Holliston ... N-2
E. Junction ... *J-10
E. Leverett, 170 ... E-6
E. Longmeadow, 13367 ... G-6
E. Lynn ... I-9
E. Mansfield, 350 ... H-13
E. Marion, 370 ... J-16
E. Millbury, 1000 ... F-10
E. Northfield ... B-6
E. Orleans, 1200 ... I-20
E. Otis, 200 ... F-3
E. Pembroke ... G-16
E. Pepperell, 2034 ... C-11
E. Princeton, 200 ... D-10
E. Sandwich, 3720 ... J-18
E. Springfield ... L-12
E. Sudbury, 800 ... E-12
E. Templeton, 1350 ... D-9
E. Walpole, 3800 ... G-13
E. Wareham, 1500 ... J-16
E. Weymouth ... N-9
E. Windsor, 100 ... D-3
Eastham, 1250 ... I-20
Easthampton, 16195 ... F-5
Easton, 530 ... H-14
Eastondale, 750 ... H-14
Edgartown, 1500 ... L-17
Edgemere ... C-20
Egypt, 900 ... F-16
Ellisville, 120 ... I-17
Elmwood, 500 ... H-15
Elmwood ... K-11
Endicott ... M-6
Erving, 420 ... C-7
**ESSEX CO., 723419 ... C-14**
Essex, 1426 ... C-16
Everett, 37353 ... E-14
Fairhaven, 16132 ... J-15
Fairlawn ... B-20
Fairview ... K-12
Fall River, 90931 ... J-14
Falmouth, 4115 ... K-17
Farley, 50 ... C-6
Fayville, 1000 ... F-12
Felchville ... L-3
Fisherville, 2750 ... G-11
Fiskdale, 2156 ... G-8
Fitchburg, 40239 ... C-10
Five Corners, 110 ... H-14
Flint ... F-18
Florida, 50 ... C-3
Forest Pk. ... M-12
Forestdale, 3992 ... J-17
Forge Vil., 850 ... C-12
Foxboro, 4220 ... G-13
Foxvale, 500 ... G-13
Framingham, 66910 ... F-12
Framingham Center ... L-2
**FRANKLIN CO., 71535 ... D-5**
Franklin, 32148 ... G-12
Furnace ... E-8
Gardner, 20682 ... C-9
Gay Head, 130 ... M-16
Georgetown, 2200 ... B-14
Gilbertville, 1050 ... E-8
Gill, 150 ... C-6
Gleasondale, 250 ... E-11
Glen Grove, 270 ... G-9
Glendale, 400 ... F-1
Glenridge, 350 ... M-4
Gloucester, 30243 ... C-16
Goshen, 600 ... D-4
Goss Hts., 100 ... F-4
Grafton, 1650 ... F-11
Granby, 1344 ... F-6
Graniteville, 1200 ... C-12
Granville, 450 ... G-4
Gray Gables, 600 ... J-17
Great Barrington, 2459 ... F-1
Great Brook Valley ... A-20
Green Harbor, 1500 ... G-16
Greendale ... A-19
Greenfield, 13716 ... C-5
Greenwood ... I-7
Griswoldville, 380 ... C-5
Groton, 1113 ... C-11
Groveland, 3800 ... B-14
Hadley, 970 ... E-6
Halifax, 980 ... H-15
Hamilton, 1000 ... C-15
**HAMPDEN CO., 456228 ... G-4**
Hampden, 900 ... G-7
**HAMPSHIRE CO., 152251 ... E-4**
Hancock, 200 ... D-1
Hanover, 2500 ... G-15
Hanover Center, 1000 ... G-15
Hanson, 2044 ... G-15
Harding, 850 ... F-13
Hardwick, 530 ... E-8
Hartsville, 125 ... G-2
Harvard, 1200 ... D-11
Harwich, 2000 ... J-20
Harwich Port, 1809 ... J-20
Hatchville, 330 ... K-17
Hatfield, 1298 ... E-6
Hathorne, 230 ... C-15
Haverhill, 61275 ... B-14
Hawley, 60 ... D-4
Hayden Row, 200 ... N-1
Haydenville, 550 ... E-5
Head of Westport, 650 ... K-14
Heath, 150 ... C-4
Heaven Hts., 330 ... I-15
Highland Lake, 200 ... G-13
Highland Pk. ... K-11
Highlands ... K-11
Hingham, 5352 ... F-15
Hingham Center ... M-10
Hinsdale, 1050 ... D-3
Hixville, 150 ... J-14
Hodges Vil., 100 ... G-9
Holbrook, 10785 ... G-14
Holden, 4200 ... E-10
Holland, 1444 ... G-8
Holliston, 12926 ... F-12
Holly Woods, 70 ... J-16
Holyoke, 39947 ... F-5
Hoosac Tunnel ... C-3
Hopedale, 4158 ... G-11
Hopkinton, 2628 ... F-12
Horseneck Beach ... L-14
Hortonville, 260 ... I-13
Houghs Neck ... M-9
Houghtonville ... B-3
Housatonic, 1335 ... F-1
Hubbardston, 800 ... D-9
Hudson, 14388 ... E-11
Hull, 11050 ... E-15
Humarock, 500 ... G-16
Huntington, 1200 ... F-4
Hyde Pk. ... M-6
Idlewood ... F-20
Indian Mound Beach ... J-17
Indian Orchard ... L-13
Interlaken, 300 ... E-1
Ipswich, 4161 ... C-15
Island Creek, 160 ... H-16
Islington, 4800 ... F-13
Jamesville ... C-18
Jefferson, 1000 ... E-9
Katama, 400 ... M-17
Kent Pk., 550 ... G-16
Kingston, 5380 ... H-16
Knightville ... F-4
Lake Mattawa, 150 ... D-7
Lake Pleasant, 300 ... D-6
Lake View ... B-20
Lakeside, 600 ... J-14
Lakeville, 150 ... I-15
Lambs Grove, 150 ... F-9
Lancaster, 900 ... D-11
Lane Vil., 130 ... C-9
Lanesborough, 1000 ... D-2
Lawrence, 70014 ... B-13
Lee, 2021 ... E-2
Leicester, 3200 ... F-9
Leino Pk., 230 ... D-9
Lenox, 1667 ... E-2
Lenox Dale, 600 ... E-2
Leominster, 41055 ... D-10
Leverett, 500 ... D-6
Lexington, 30355 ... D-13
Leyden, 605 ... B-6
Lincoln, 2850 ... E-13
Linwood, 750 ... G-11
Lithia, 100 ... D-4
Little Nahant ... J-9
Little Neck, 50 ... C-16
Littleton Common, 2816 ... D-12
Lobsterville, 50 ... M-16
Locks Vil., 200 ... D-6
Long Plain, 300 ... J-15
Longmeadow, 15633 ... G-6
Lovell Corners ... N-9
Lowell, 103615 ... C-13
Ludlow ... C-18
Ludlow, 18820 ... G-6
Lunenburg, 1695 ... C-10
Lynn, 86957 ... D-15
Lynnfield, 11542 ... D-14
Lyonsville, 654 ... C-5
Madaket, 350 ... M-19
Malden, 55597 ... E-14
Manchaug, 800 ... G-10
Manchester, 5286 ... C-16
Manomet, 1600 ... H-17
Mansfield, 5100 ... H-13
Maplewood ... C-20
Marblehead, 20377 ... D-15
Marion, 1400 ... J-16
Marlboro, 220 ... B-15
Marlborough, 37932 ... E-11
Marshfield, 4246 ... G-16
Marshfield Hills, 2369 ... G-16
Mashpee, 800 ... J-17
Matfield, 700 ... H-14
Mattapoisett, 1660 ... J-16
Maynard, 10433 ... E-12
Medfield, 6670 ... F-13
Medford, 55573 ... E-14
Medway, 4000 ... G-12
Melrose, 26708 ... D-14
Melrose Highlands ... I-7
Mendon, 1000 ... G-11
Menemsha, 250 ... M-16
Merrimac, 2200 ... B-14
Merrimacport, 250 ... B-14
Methuen, 44055 ... B-13
Middleboro, 5580 ... I-15
Middlefield, 350 ... E-3
**MIDDLESEX CO., 1465396 ... C-11**
Middleton, 4921 ... C-14
Mile Oak Center, 2050 ... G-6
Milford, 24230 ... G-12
Mill River, 400 ... G-2
Millbury, 5000 ... F-10
Millers Falls, 1072 ... C-6
Millis, 3800 ... G-12
Millville, 2236 ... H-11
Milton, 26062 ... F-14
Milton Center ... M-7
Milton Vil. ... M-7
Monomoy, 250 ... M-20
Monponsett, 730 ... H-15
Monroe Bridge, 90 ... B-4
Monson Center, 2103 ... G-7
Montague, 830 ... D-6
Monterey, 600 ... F-2
Montgomery, 325 ... F-4
Montville, 90 ... G-2
Monument Beach, 2438 ... J-17
Moores Corner, 100 ... D-6
Morningdale, 1150 ... E-10
Morseville ... M-3
Mt. Hermon, 650 ... C-6
Myricks, 60 ... I-14
Mystic Grove, 150 ... F-9
Nabnasset, 3600 ... C-12
Nahant, 3632 ... E-15
**NANTUCKET CO., 9520 ... M-20**
Nantucket, 3830 ... M-20
Nashaquitsa, 150 ... M-16
Natick, 32200 ... F-12
Needham, 28911 ... F-13
Needham Hts. ... L-5
New Ashford, 80 ... C-2
New Bedford, 91365 ... J-15
New Boston, 110 ... G-3
New Braintree, 500 ... E-8
New Lenox, 800 ... E-2
New Marlborough, 185 ... G-2
New Salem, 450 ... D-7
Newburyport, 17542 ... B-15
Newton, 82139 ... E-13
Newton Center ... L-5
Newton Highlands ... L-5
Newton Lower Falls ... L-4
Nipmuck Pond ... G-11
Nonquitt, 200 ... K-15
**NORFOLK CO., 650308 ... F-13**
Norfolk, 550 ... G-13
N. Acton, 900 ... D-12
N. Adams, 13711 ... C-3
N. Amherst, 6019 ... E-6
N. Andover, 22792 ... B-14
N. Ashburnham, 800 ... C-9
N. Attleboro, 11900 ... H-13
N. Billerica, 5500 ... C-13
N. Blandford ... F-3
N. Brookfield, 2527 ... F-8
N. Carver, 600 ... H-16
N. Chatham, 950 ... J-20
N. Cohasset, 500 ... F-15
N. Dighton, 1250 ... I-14
N. Eastham, 1915 ... I-20
N. Easton, 4500 ... G-14
N. Egremont, 300 ... F-1
N. Falmouth, 3355 ... J-17
N. Grafton, 3150 ... F-10
N. Hadley, 350 ... E-6
N. Hanover, 980 ... G-15
N. Harwich, 500 ... J-19
N. Hatfield, 470 ... E-5
N. Lancaster ... D-11
N. Leverett, 200 ... D-6
N. Marshfield, 450 ... G-16
N. Middleboro, 250 ... H-15
N. New Salem, 50 ... D-7
N. Orange, 80 ... C-7
N. Otis ... F-3
N. Oxford, 1200 ... G-10
N. Pembroke, 2913 ... G-16
N. Quincy ... M-8
N. Reading, 12002 ... C-14
N. Rehoboth ... I-13
N. Saugus ... I-8
N. Scituate, 5065 ... F-16
N. Sudbury, 2600 ... E-12
N. Swansea, 900 ... J-13
N. Tewksbury, 1200 ... C-13
N. Tisbury, 200 ... L-16
N. Truro, 950 ... G-20
N. Uxbridge, 1800 ... G-11
N. Weymouth ... M-9
N. Woburn ... H-6
Northampton, 28379 ... E-5
Northborough, 6257 ... E-11
Northbridge, 3600 ... G-11
Northbridge Center, 220 ... G-11
Northfield, 1141 ... C-6
Norton Center, 2618 ... H-13
Norton Grove, 2100 ... H-13
Norwell, 1200 ... G-16
Norwood, 28587 ... F-13
Nutting Lake, 3200 ... D-13
Oak Bluffs, 2000 ... L-17
Oak Island ... J-8
Oakdale, 600 ... E-10
Oakdale ... K-11
Oakdale ... M-6
Oakham, 600 ... E-9
Ocean Bluff, 2500 ... G-16
Ocean Grove, 3012 ... J-13
Ocean Hts. ... L-17
Oldham Vil., 900 ... G-15
Onset, 1292 ... J-16
Orange, 3945 ... C-7
Orleans, 1716 ... I-20
Otis, 650 ... F-3
Otter River, 630 ... C-9
Oxford, 5899 ... G-10
Packard Hts., 50 ... C-8
Pakachoag ... C-19
Palmer, 3900 ... G-7
Parkwood Beach, 250 ... J-16
Paxton, 1600 ... E-9
Peabody, 51331 ... D-15
Pelham, 500 ... E-7
Pembroke, 2100 ... G-16
Pepperell, 2517 ... C-11
Peru, 400 ... D-3
Petersham, 700 ... D-8
Phillipston, 200 ... D-8
Phillipston Four Corners, 50 ... D-8
Pigeon Cove, 1800 ... C-16
Pilgrim Hts. ... G-19
Pine Lake, 750 ... E-12
Pine Pt. ... M-12
Pine Rest, 750 ... E-12
Pinehurst, 6941 ... D-13
Pittsfield, 42652 ... D-2
Plainfield, 150 ... D-4
Plainville, 6871 ... H-13
Pleasant Lake, 450 ... J-20
**PLYMOUTH CO., 472822 ... H-15**
Plymouth, 7658 ... H-16
Plympton, 500 ... H-15
Pocasset, 2671 ... J-17
Pt. Independence, 870 ... J-16
Polpis, 100 ... M-20
Pondville ... D-19
Popponesset Beach ... K-18
Princeton, 1100 ... D-10
Priscilla Beach, 450 ... H-17
Provincetown ... G-19
Provincetown, 3192 ... G-19
Quaise, 40 ... M-20
Quidnet, 100 ... M-20
Quincy, 92339 ... F-14
Quinsigamond Vil. ... C-19
Quissett, 200 ... K-16
Randolph, 30963 ... G-14
Raynham, 2100 ... H-14
Raynham Center, 3633 ... H-14
Reading, 23708 ... D-14
Rehoboth, 400 ... I-13
Revere, 60204 ... E-14
Rexhame, 600 ... G-16
Rice Square ... C-19
Richmond, 250 ... E-1
Richmond Furnace ... E-1
Risingdale, 100 ... F-1
River Pines, 3600 ... D-13
Riverdale, 110 ... G-11
Riverdale ... M-6
Riverside, 350 ... C-6
Rochdale, 1150 ... F-9
Rochester, 600 ... J-15
Rock, 600 ... I-15
Rock Harbor ... I-20
Rockland, 16123 ... G-15
Rockport, 5606 ... C-16
Rockville, 300 ... G-12
Roosterville, 50 ... G-3
Rowe, 80 ... C-4
Rowley, 1434 ... B-15
Roxbury ... L-7
Royalston, 370 ... C-8
Russell, 900 ... F-4
Russells Mills, 160 ... K-14
Rutland, 2205 ... E-9
Sagamore, 3544 ... J-17
Sagamore Beach, 1100 ... I-17
Sagamore Highlands, 100 ... I-17
Salem, 41256 ... D-15
Salisbury, 4484 ... A-15
Salisbury Beach ... A-15
Salisbury Hts. ... A-18
Salisbury Plains, 600 ... A-15
Sand Hill, 1800 ... F-16
Sandisfield, 225 ... G-2
Sandwich, 3058 ... J-17
Sandy Beach, 60 ... E-9
Saugus, 26078 ... D-14
Savoy, 150 ... D-3
Saxonville ... L-2
Scituate, 5069 ... F-16
Sea View, 100 ... G-16
Seekonk, 13046 ... I-13
Segreganset, 400 ... I-14
Sharon, 5941 ... G-13
Shattuckville, 170 ... C-5
Shawsheen Vil., 2600 ... C-14
Sheffield, 1150 ... G-1
Shelburne, 150 ... C-5
Shelburne Falls, 1951 ... C-5
Sheldonville, 450 ... H-12
Sherborn, 1400 ... F-12
Sherwood Forest ... E-20
Sherwood Forest, 400 ... F-3
Shirley, 1427 ... D-11
Shirley Center, 220 ... D-11
Shore Acres, 1200 ... F-16
Shore Acres, 50 ... K-15
Shrewsbury, 31600 ... F-10
Shutesbury, 700 ... D-7
Siasconset, 600 ... M-20
Silver Beach, 1300 ... J-16
Silver Lake, 2900 ... C-13
Silver Lake, 220 ... H-16
Sippewisset, 1500 ... K-17
Sixteen Acres ... M-13
Smith Mills, 4432 ... J-15
Somerset, 18234 ... J-14
Somerville, 75662 ... E-14
S. Acton, 3200 ... D-12
S. Amherst, 5039 ... E-6
S. Ashburnham, 1013 ... C-9
S. Ashfield, 70 ... D-5
S. Athol ... D-7
S. Barre, 550 ... E-8
S. Berlin, 300 ... E-11
S. Boston ... L-7
S. Byfield, 50 ... B-15
S. Carver ... I-16
S. Chatham, 950 ... J-20
S. Dartmouth, 9800 ... K-15
S. Deerfield, 1868 ... D-5
S. Dennis, 3679 ... J-19
S. Duxbury, 3062 ... H-16
S. Easton, 1500 ... G-14
S. Egremont, 600 ... G-1
S. Georgetown, 150 ... B-14
S. Hadley, 5500 ... F-6
S. Hadley Falls, 5100 ... F-6
S. Hamilton, 2800 ... C-15
S. Hanover, 1000 ... G-15
S. Harwich, 800 ... J-20
S. Hingham, 4000 ... F-15
S. Lancaster, 1742 ... D-11
S. Lee, 500 ... F-2
S. Lynnfield ... H-8
S. Mashpee ... K-17
S. Middleboro, 480 ... I-15
S. Natick ... M-4
S. Orleans, 750 ... J-20
S. Peabody ... H-9
S. Quincy ... M-8
S. Royalston, 450 ... C-8
S. Sandisfield ... G-2
S. Sutton, 175 ... G-10
S. Swansea, 1000 ... J-13
S. Walpole, 1300 ... G-13
S. Wellfleet, 600 ... H-20
S. Westport, 180 ... K-14
S. Williamstown ... C-2
S. Worthington ... E-4
S. Yarmouth, 11603 ... J-19
Southampton, 550 ... F-5
Southborough, 1500 ... F-11
Southbridge, 12878 ... G-9
Southfield, 200 ... G-2
Southwick, 1250 ... G-5
Spencer, 6032 ... F-9
Springdale ... K-11
Springfield, 150640 ... G-6
Squantum ... L-8
State Line, 80 ... E-1
Sterling, 1350 ... E-10
Sterling Junction, 600 ... E-10
Still River, 300 ... D-11
Stockbridge, 1100 ... F-1
Stoneham, 22219 ... D-14
Stoneville ... C-18
Stoughton ... G-14
Stow, 1250 ... E-12
Sturbridge, 2047 ... G-8
Sudbury, 1950 ... E-12
Sudbury Center ... E-12
**SUFFOLK CO., 689807 ... F-14**
Sunderland ... C-19
Sunderland, 1100 ... D-6
Surfside, 150 ... M-20
Sutton, 600 ... G-10
Swampscott, 14412 ... D-15
Swansea, 900 ... J-13
Swansea Center, 300 ... J-13
Swift River, 250 ... D-4
Tafts Corner, 50 ... F-9
Tatnuck ... B-18
Taunton, 55702 ... I-14
Teaticket, 1907 ... K-17
Templeton, 1000 ... D-8
Tewksbury, 11000 ... C-13
Texas, 430 ... G-9
Thorndike, 1150 ... F-7
Three Rivers, 2939 ... F-7
Tinkertown, 100 ... H-16
Tolland Center, 200 ... G-3
Tonset, 500 ... I-20
Topsfield, 2826 ... C-15
Touisset, 1500 ... J-13
Town Crest Vil., 120 ... C-2
Townsend, 1043 ... C-10
Townsend Harbor, 900 ... C-11
Truro, 550 ... H-20
Tully, 200 ... C-7
Turkey Hill Shores, 170 ... E-9
Turners Falls, 4441 ... C-6
Tyngsboro, 250 ... C-12
Tyringham, 200 ... F-2
Upton, 1200 ... G-11
Uxbridge, 3500 ... G-11
Vallersville, 150 ... I-17
Van Deusenville, 100 ... F-1
Vineyard Haven, 2048 ... L-17
Waban ... L-5
Wakeby, 100 ... J-17
Wakefield, 24804 ... D-14
Wales, 800 ... G-8
Walpole, 5867 ... G-13
Waltham, 60236 ... E-13
Wamesit, 2700 ... C-13
Wapping, 110 ... D-5
Waquoit Vil., 500 ... K-17
Ware, 6174 ... F-7
Wareham Center, 2874 ... J-16
Warren, 1452 ... F-8
Warrentown, 370 ... H-15
Warwick, 250 ... C-7
Washington, 100 ... E-3
Watertown, 32365 ... E-13
Waterville, 800 ... C-8
Waterville, 200 ... H-15
Wauwinet, 80 ... M-20
Wayland, 2500 ... E-12
Webster, 11600 ... G-10
Wellesley, 26613 ... F-13
Wellesley Farms ... L-4
Wellesley Fells ... L-4
Wellesley Hills ... L-4
Wellfleet, 1200 ... H-20
Wellington ... J-7
Wellville, 110 ... C-9
Wendell, 400 ... D-7
Wendell Depot, 90 ... C-7
Wenham, 4212 ... C-15
W. Acton, 5200 ... D-12
W. Andover, 2000 ... C-13
W. Becket, 50 ... F-2
W. Berlin, 100 ... E-11
W. Boxford, 800 ... B-14
W. Boylston, 3300 ... E-10
W. Bridgewater, 2100 ... H-14
W. Brookfield, 1610 ... F-8
W. Chatham, 1446 ... J-20
W. Chesterfield, 200 ... E-4
W. Chop ... L-17
W. Concord, 5632 ... D-12
W. Cummington, 150 ... D-3
W. Deerfield ... D-5
W. Dennis, 2570 ... J-19
W. Dudley, 300 ... G-9
W. Falmouth, 1867 ... K-17
W. Foxboro, 1100 ... G-13
W. Granville, 175 ... G-4
W. Groton, 900 ... C-11
W. Hanover, 1700 ... G-15
W. Harwich, 1500 ... J-19
W. Hatfield, 350 ... E-5
W. Hawley ... C-4
W. Hingham ... M-10
W. Lynn ... I-8
W. Mansfield, 650 ... H-13
W. Medford ... J-6
W. Medway, 2000 ... G-12
W. Millbury, 300 ... G-10
W. Natick ... M-3
W. Newbury, 1100 ... B-14
W. Otis ... F-2
W. Pelham, 550 ... E-6
W. Roxbury ... M-6
W. Springfield, 27459 ... G-5
W. Springfield, 27899 ... M-11
W. Sterling, 100 ... D-10
W. Stockbridge, 850 ... E-1
W. Stockbridge Center, 120 ... E-1
W. Sutton, 200 ... G-10
W. Tatnuck ... B-18
W. Tisbury, 600 ... L-16
W. Townsend, 700 ... C-10
W. Upton, 1100 ... G-11
W. Walpole, 150 ... G-13
W. Wareham, 1908 ... I-16
W. Warren, 1300 ... F-8
W. Wrentham, 150 ... H-12
W. Yarmouth, 6460 ... J-19
Westborough, 3983 ... F-11
Westfield, 40608 ... G-5
Westford, 1400 ... C-12
Westhampton, 300 ... F-5
Westminster, 1100 ... D-9
Weston, 10200 ... E-13
Westport Factory, 250 ... K-14
Westport Pt., 600 ... K-14
Westwood, 6500 ... F-13
Weymouth, 53988 ... F-15
Weymouth Hts. ... N-9
Whalom, 1400 ... D-10
Whately, 630 ... D-5
Wheelockville, 500 ... G-11
Wheelwright, 500 ... E-8
White Horse Beach, 1250 ... H-17
White Island Shores, 2133 ... I-17
Whitinsville, 6340 ... G-11
Whitman, 13240 ... G-15
Wilbraham, 3544 ... G-6
Wilkinsonville, 900 ... F-10
Williamsburg, 1200 ... E-5
Williamstown, 4754 ... B-2
Williamsville, 100 ... D-8
Williamsville, 220 ... F-1
Willimansett ... K-11
Wilmington, 21363 ... D-13
Winchendon, 4246 ... C-9
Winchendon Springs, 500 ... C-9
Winchester, 20810 ... D-14
Windsor, 200 ... D-3
Winnecunnet, 450 ... H-14
Winthrop, 18303 ... E-14
Woburn, 36871 ... D-14
Wollaston ... M-8
Woodland Pk. ... C-18
Woods Hole, 925 ... K-16
Woodville, 250 ... F-11
**WORCESTER CO., 750963 ... D-8**
Worcester, 175011 ... F-10
Woronoco, 300 ... G-4
Worthington Center, 125 ... E-4
Worthington Corners, 550 ... E-4
Wrentham, 2250 ... G-13
Wyoming ... I-7
Yarmouth, 1300 ... J-19
Yarmouth Port, 5395 ... J-19

## Michigan

Page locator

| Map keys | Atlas pages |
|---|---|
| A - J | 102 - 103 |
| K - T | 104 - 105 |

* City keyed to pp. 106 - 107

Acme, 500 ... J-6
Ada, 2300 ... P-6
Adair, 130 ... P-13
Adamsville, 190 ... T-5
Addison, 586 ... S-9
Adrian, 21391 ... S-10
Advance ... H-7
Aetna ... N-5
Afton, 150 ... H-9
Agnew, 180 ... P-5
Ahmeek, 145 ... A-13
Akron, 432 ... N-11
Alabaster, 110 ... L-11
Alamo, 140 ... R-6
Alanson, 783 ... G-8
Alaska, 280 ... P-6
Alba, 370 ... I-7
Alberta, 30 ... B-13
Albion, 9103 ... R-8
Albright Shores, 150 ... M-9
**ALCONA CO., 11719 ... J-11**
Alden, 450 ... I-7
**ALGER CO., 9862 ... D-5**
Alger, 180 ... L-10
Algonac, 4540 ... Q-14
**ALLEGAN CO., 105665 ... Q-5**
Allegan, 4816 ... Q-5
Allen, 212 ... S-8
Allen Pk., 26014 ... *L-5
Allendale, 11555 ... P-5
Allenton, 360 ... P-13
Allenville ... F-8
Alma, 9223 ... N-8
Almont, 2744 ... P-12
Aloha, 100 ... G-9
**ALPENA CO., 31314 ... I-11**
Alpena, 10465 ... I-11
Alpha, 175 ... D-13
Alpine, 250 ... O-6
Alto, 300 ... P-7
Altona, 30 ... N-7
Alverno ... G-9
Amasa, 400 ... C-13
Amble, 90 ... N-7
Anchorville ... Q-13
Ann Arbor, 114386 ... R-11
**ANTRIM CO., 23110 ... I-7**
Antrim, 230 ... I-7
Applegate, 268 ... N-13
Arbutus Beach, 250 ... I-8
Arcadia, 320 ... K-4
**ARENAC CO., 17269 ... L-10**
Argentine, 2285 ... P-10
Argyle, 160 ... N-13
Armada, 1657 ... P-13
Arnold, 50 ... E-1
Ashley, 507 ... O-9
Ashton, 110 ... L-6
Assinins ... B-13
Assyria, 50 ... R-7
Athens, 1034 ... S-7
Atlanta, 757 ... I-9
Atlas, 290 ... P-11
Attica, 420 ... O-12
Au Gres, 926 ... L-11
Au Sable, 1533 ... K-12
Au Train, 490 ... D-3
Auburn, 2031 ... N-10
Auburn Hills, 20931 ... *F-5
Augusta, 857 ... R-7
Aurelius, 270 ... Q-9
Austin, 70 ... E-1
Averill, 400 ... M-9
Avoca, 270 ... O-13
Azalia, 120 ... S-11
Bach, 30 ... M-12
Backus Beach, 730 ... J-11
Bad Axe, 3069 ... M-12
Bailey, 330 ... O-6
Baldwin, 1171 ... M-5
Bancroft, 587 ... P-10
Banfield, 40 ... R-7
Bangor, 1840 ... R-5
Bannister, 250 ... O-9
Bar Lake, 280 ... K-4
**BARAGA CO., 8746 ... B-13**
Baraga, 1185 ... B-13
Barbeau, 100 ... E-9
Bark River, 280 ... G-2
Baroda, 868 ... S-4
Barron Lake, 1570 ... T-5
**BARRY CO., 56755 ... Q-7**
Barryton, 370 ... M-7
Barton City, 350 ... J-11
Barton Hills, 325 ... *A-9
Barton Lake, 420 ... S-6
Batavia, 120 ... S-7
Bates, 30 ... J-6
Bath, 730 ... P-9
Battle Creek, 52053 ... R-7
Bauer, 190 ... P-5
**BAY CO., 110157 ... M-10**
Bay City, 33874 ... N-10
Bay Mills, 170 ... D-8
Bay Pk. ... M-11
Bay Port, 600 ... M-11
Bay View ... H-8
Bayshore, 260 ... H-7
Beadle Lake, 1550 ... R-7
Beal City, 345 ... M-8
Bear Lake, 288 ... K-5
Beaverton, 1060 ... M-9
Bedford, 260 ... R-7
Beecher, 12793 ... *A-6
Beechwood, 30 ... C-12
Belding, 5691 ... O-7
Bell Oak, 30 ... Q-10
Bellaire, 1123 ... I-7
Belleville, 3556 ... R-11
Bellevue, 1357 ... R-8
Belmont, 730 ... P-6
Bendon, 50 ... J-5
Bennington, 50 ... P-9
Bentheim, 100 ... Q-5
Bentley, 100 ... L-10
Benton Harbor, 10814 ... S-4
Benton Hts., 5458 ... S-4
**BENZIE CO., 15998 ... J-5**
Benzonia, 473 ... J-5
Bergland, 670 ... B-11
Berkley, 14787 ... *H-5
**BERRIEN CO., 162453 ... S-4**
Berrien Center, 250 ... S-4
Berrien Springs, 1913 ... S-4
Bertrand, 5500 ... T-4
Berville, 310 ... P-13
Bessemer, 1846 ... C-10
Bete Grise ... A-13
Beulah, 385 ... J-5
Beverly Hills, 9865 ... *H-5
Big Bay, 265 ... B-14
Big Rapids, 10257 ... M-6
Big Rock ... I-9
Bingham Farms, 989 ... *H-5
Birch Beach, 360 ... O-14
Birch Creek ... H-1
Birch Run, 1640 ... O-11
Birmingham, 18932 ... *G-5
Bitely, 200 ... M-5
Black River, 200 ... I-12
Blaine ... O-14
Blanchard, 450 ... N-7
Blaney Pk., 30 ... E-5
Bliss, 50 ... G-8
Blissfield, 3171 ... T-10
Bloomfield, 43000 ... *G-4
Bloomfield Hills, 3772 ... *G-5
Bloomingdale, 494 ... R-5
Boon, 110 ... K-6
Borculo, 400 ... P-5
Bostwick Lake, 800 ... O-6
Boyne City, 3142 ... H-7
Boyne Falls, 330 ... H-8
Bradley, 170 ... Q-6
Brampton, 260 ... F-2
**BRANCH CO., 45787 ... T-7**
Branch, 140 ... M-5
Brant, 80 ... O-9
Bravo, 40 ... Q-5
Breckenridge, 1284 ... N-9
Breedsville, 219 ... R-5
Brethren, 240 ... K-5
Brevort, 130 ... E-7
Bridgeport, 7849 ... N-10
Bridgeton, 30 ... O-5
Bridgewater, 140 ... R-10
Bridgman, 2400 ... S-4
Brighton, 7204 ... Q-11
Brimley, 560 ... D-8
Brinton, 70 ... M-7
Britton, 650 ... S-11
Brockway, 30 ... O-13
Brohman, 200 ... M-5
Bronson, 2274 ... T-7
Brookfield, 50 ... R-8
Brooklyn, 1340 ... S-9
Brown City, 1252 ... O-13
Brownlee Pk., 2588 ... R-7
Brownstown, 23000 ... *M-5
Bruce Crossing, 300 ... C-12
Brunswick, 50 ... N-5
Brutus, 160 ... G-8
Buchanan, 4354 ... T-4
Buckley, 550 ... K-6
Bullock Creek, 750 ... N-9
Bunker Hill, 130 ... Q-9
Burdickville, 30 ... I-5
Burlington, 382 ... S-7
Burnips, 190 ... Q-5
Burnside, 50 ... O-12
Burr Oak, 750 ... T-7
Burt Lake, 120 ... G-8
Burton, 29845 ... P-11
Burton, 30 ... P-9
Butler, 30 ... S-8
Butternut, 130 ... O-8
Byron, 562 ... P-10
Byron Center, 3777 ... P-6
Cadillac, 10245 ... K-6
Cadmus, 100 ... S-10
Calderwood ... C-12
Caledonia, 1354 ... P-6
**CALHOUN CO., 137985 ... R-7**
California, 50 ... T-8
Calumet, 800 ... A-13
Calvin Center, 70 ... T-5
Cambria, 280 ... T-8
Camden, 524 ... T-8
Cannonsburg, 240 ... P-6
Canton, 76366 ... *J-2
Capac, 2063 ... P-13
Caribou Lake, 30 ... E-10
Carland, 80 ... O-9
Carleton, 2572 ... S-12
Carlshend, 120 ... D-2
Carlton Center, 70 ... Q-7
Carney, 212 ... G-1
Caro, 4020 ... N-11
Carp Lake, 350 ... G-8
Carrollton, 6602 ... N-10
Carson City, 1177 ... O-8
Carsonville, 470 ... N-13
Cascade, 2300 ... P-6
Caseville, 835 ... L-12
Casnovia, 311 ... O-6
Caspian, 882 ... D-12
**CASS CO., 51104 ... S-5**
Cass City, 2493 ... N-12
Cassopolis, 1920 ... S-5
Castle Pk., 50 ... Q-5
Cathro, 290 ... H-11
Cedar, 230 ... I-6
Cedar Lake, 290 ... N-8
Cedar River, 40 ... H-2
Cedar Springs, 3250 ... O-6
Cedarville, 1070 ... E-9
Cement City, 419 ... S-9
Center Line, 8135 ... *H-7
Central ... A-13
Central Lake, 965 ... I-7
Centreville, 1521 ... S-6
Ceresco, 230 ... R-7
Champion, 450 ... C-13
Channing, 550 ... D-13
Chapin, 60 ... O-9
**CHARLEVOIX CO., 26090 ... H-7**
Charlevoix, 2656 ... H-7
Charlotte, 8984 ... Q-8
Chase, 250 ... M-6
Chassell, 710 ... A-13
Chatham, 211 ... D-3
**CHEBOYGAN CO., 26448 ... H-9**
Cheboygan, 4942 ... G-9
Chelsea, 5005 ... R-10
Chesaning, 2345 ... O-10
Chester, 50 ... Q-8
Chesterfield, 37500 ... *F-9
**CHIPPEWA CO., 38543 ... D-8**
Chippewa Lake, 160 ... M-7
Christmas, 200 ... D-3
Clam River ... I-7
**CLARE CO., 31252 ... M-8**
Clare, 3093 ... M-8
Clarion, 120 ... H-8
Clarkston, 913 ... Q-11
Clarksville, 310 ... P-7
Clawson, 12157 ... *G-6
Clayton, 303 ... T-10
Clear Lake, 230 ... K-9
Clifford, 310 ... O-12
Climax, 750 ... R-7
**CLINTON CO., 64753 ... P-8**
Clinton, 95648 ... *G-8
Clinton, 2391 ... S-10
Clio, 2489 ... O-11
Cloverdale, 250 ... Q-7
Cloverville, 1950 ... O-5
Clyde, 210 ... *E-1
Coats Grove ... Q-7
Coe, 40 ... N-8
Cohoctah, 110 ... P-10
Cohoctah Center, 40 ... P-10
Coldwater, 10514 ... S-8
Coleman, 1229 ... M-9
Colling ... N-11
Coloma, 1459 ... S-4
Colon, 1166 ... S-7
Columbiaville, 786 ... O-12
Commerce, 5000 ... Q-11
Comstock, 5600 ... P-3
Comstock Pk., 10674 ... P-6
Concord, 1072 ... R-8
Conklin, 290 ... O-5
Constantine, 2103 ... T-6
Conway, 350 ... G-8
Cooks, 160 ... F-4
Cooper Center, 250 ... R-6
Coopersville, 4138 ... P-5
Copemish, 214 ... K-5
Copper City, 193 ... A-13
Copper Harbor, 70 ... A-14
Coral, 230 ... N-7
Corinne ... E-6
Cornell, 80 ... F-2
Corunna, 3285 ... P-10
Cottage Pk., 50 ... D-9
Covert, 640 ... R-4
Covington, 300 ... C-13
**CRAWFORD CO., 14273 ... J-8**
Crooked Lake, 230 ... R-7
Cross Vil., 150 ... G-7
Croswell, 2457 ... O-13
Crump, 60 ... M-10
Crystal, 680 ... O-8
Crystal Falls, 1594 ... D-13
Crystal Valley, 110 ... M-4
Cumber ... M-12
Curran, 300 ... J-10
Curtis, 620 ... E-6
Curtisville, 50 ... J-10
Custer, 315 ... M-4
Cutlerville, 15114 ... *D-2
Dafter, 160 ... D-9
Daggett, 253 ... G-1
Dalton, 320 ... O-4
Dansville, 436 ... Q-9
Darragh ... J-7
Davis, 160 ... *E-8
Davisburg, 450 ... *D-2
Davison, 5136 ... P-11
Dayton, 150 ... T-4
De Tour Vil., 405 ... E-10
Dearborn, 86477 ... R-12
Dearborn Hts., 51972 ... *J-4
Decatur, 1775 ... S-5
Decker, 100 ... N-12
Deckerville, 890 ... N-13
Deer Pk., 20 ... C-6
Deerfield, 930 ... S-11
Deerton, 230 ... D-3
Deford, 200 ... N-12
Delano ... L-11
**DELTA CO., 38520 ... F-3**
Delta Mills, 330 ... A-1
Delton, 500 ... Q-7
Denton, 330 ... *L-2
Detroit, 912062 ... R-12
Detroit Beach, 2289 ... S-12
Devereaux, 60 ... R-8
DeWitt, 4428 ... P-9
Dexter, 3272 ... R-10
**DICKINSON CO., 27472 ... C-13**
Dighton, 80 ... L-7
Dimondale, 1319 ... Q-8
Diorite, 200 ... C-14
Disco ... *E-7
Dixboro, 500 ... R-11
Dollar Bay, 1050 ... A-13
Dollarville, 90 ... D-6
Donken, 30 ... B-12
Dorr, 1600 ... Q-6
Doster, 40 ... R-6
Douglas, 1174 ... Q-5
Dowagiac, 5703 ... S-5
Dowling, 150 ... Q-7
Drayton Plains ... *E-4
Drummond Island, 300 ... E-10
Dryden, 777 ... P-12
Dublin, 50 ... L-5
Duck Lake, 500 ... R-8
Duel ... M-10
Duffield, 40 ... P-10
Dundee, 4253 ... S-11
Durand, 3740 ... P-10
Dutton, 630 ... *D-3
Eagle, 120 ... P-8
Eagle Harbor, 70 ... A-13
Eagle Lake, 1160 ... T-5
Eagle River, 50 ... A-13
E. Grand Rapids, 10467 ... *B-3
E. Jordan, 2234 ... H-7
E. Lake, 563 ... L-4
E. Lansing, 45857 ... Q-9
E. Leroy, 190 ... S-7
E. Saugatuck, 100 ... Q-5
E. Tawas, 2723 ... K-11
Eastmanville, 60 ... P-5
Easton, 30 ... O-10
Eastpointe, 32497 ... Q-13
Eastport, 160 ... I-7
Eastwood, 6265 ... P-3
**EATON CO., 103655 ... Q-8**
Eaton Rapids, 5246 ... Q-8
Eau Claire, 623 ... S-4
Eben Junction, 410 ... D-3
Eckerman, 250 ... D-7
Eckford, 70 ... R-8
Ecorse, 9858 ... *L-6
Eden, 300 ... Q-9
Edenville, 410 ... M-9
Edmore, 1222 ... N-7
Edwardsburg, 1108 ... T-5
Elba, 200 ... P-11
Elberta, 437 ... J-4
Elk Rapids, 1672 ... I-6
Elkton, 758 ... M-12
Ellsworth, 455 ... H-7
Elm Hall, 200 ... N-8
Elmdale, 30 ... P-7
Elmer ... N-13
Elmira, 230 ... I-8
Elo, 30 ... B-12
Elsie, 979 ... O-9
Elwell, 270 ... N-8
**EMMET CO., 31437 ... G-7**
Emmett, 243 ... P-13
Empire, 367 ... J-5
Engadine, 480 ... E-6
Ensign, 40 ... F-3
Ensley Center, 30 ... O-6
Entrican, 80 ... O-7
Epoufette, 70 ... E-7
Epsilon, 50 ... H-8
Erie, 690 ... T-11
Escanaba, 12214 ... F-2
Essexville, 3493 ... N-10
Estey, 100 ... M-9
Estral Beach, 464 ... S-12
Eureka, 210 ... O-9
Evart, 1667 ... M-7
Evergreen Shores, 180 ... F-8
Ewen, 470 ... C-11
Fair Haven ... Q-13
Fair Plain, 7828 ... S-4
Fairfax, 40 ... S-7
Fairfield, 240 ... T-10
Fairgrove, 591 ... N-11
Fairport, 100 ... G-3
Fairview, 570 ... J-10
Falmouth, 280 ... K-7
Fargo, 90 ... O-13
Farmington, 9839 ... *H-3
Farmington Hills, 78522 ... Q-12
Farrandville, 820 ... O-11
Farwell, 807 ... M-8
Fawn River, 30 ... T-7
Felch, 300 ... D-14
Felch Mtn., 40 ... D-14
Fennville, 1393 ... Q-5
Fenton, 11678 ... P-11
Fenwick, 150 ... O-7
Ferndale, 21112 ... *I-6
Ferry, 200 ... N-4
Ferrysburg, 3053 ... O-4
Fibre, 30 ... E-8
Fife Lake, 461 ... J-7
Filer City, 280 ... L-4
Filion, 200 ... M-12
Fillmore, 50 ... Q-5
Findley, 40 ... S-7
Fitchburg, 80 ... R-9
Flat Rock, 8771 ... S-12
Flint, 112900 ... P-11
Flowerfield, 130 ... S-6
Flushing, 7815 ... O-10
Ford River, 280 ... G-2
Forest Hills, 20942 ... *B-3
Forest Lake, 60 ... D-3
Forester, 80 ... N-13
Forestville, 119 ... M-13
Foster City, 350 ... F-1
Fostoria, 370 ... O-12
Fountain, 174 ... L-5
Fowler, 1054 ... P-8
Fowlerville, 3101 ... Q-10
Frankenmuth, 4652 ... O-11
Frankentrost, 50 ... N-11
Frankfort, 1451 ... J-4
Franklin, 2914 ... *H-4
Fraser, 14902 ... *G-8
Frederic, 570 ... J-8
Free Soil, 176 ... L-4
Freeland, 5147 ... N-10
Freeport, 436 ... Q-7
Fremont, 4195 ... N-5
Frontier, 210 ... T-9
Fruitport, 1080 ... O-5
Fulton, 400 ... S-7
Gaastra, 300 ... D-12
Gagetown, 366 ... M-12
Gaines, 349 ... P-10
Galesburg, 1933 ... R-7
Galien, 554 ... T-4
Ganges, 80 ... Q-5
Garden, 223 ... F-4
Garden City, 26743 ... *J-4
Garden Corners, 110 ... F-4
Garnet, 90 ... E-7
Gay, 60 ... A-13
Gaylord, 3593 ... I-8
**GENESEE CO., 436141 ... O-10**
Genesee, 1500 ... O-11
Gera, 20 ... N-11
Germfask, 250 ... E-5
Gibbs City ... C-12
Gibraltar, 4826 ... S-12
Gilchrist, 20 ... E-6
Gilford, 230 ... N-11
Gingellville, 320 ... *E-4
Girard, 200 ... S-8
Gladstone, 5067 ... F-3
**GLADWIN CO., 26023 ... L-9**
Gladwin, 2906 ... L-9
Glen Arbor, 280 ... I-5
Glen Haven, 60 ... I-5
Glendale ... R-5
Glendora, 40 ... T-4
Glenn, 270 ... Q-4
Glennie, 400 ... J-11
Gobles, 782 ... R-5
Goetzville, 100 ... E-10
**GOGEBIC CO., 17370 ... C-11**
Gomins, 170 ... J-10
Good Hart, 60 ... G-7
Goodells, 340 ... P-13
Goodison, 190 ... *E-6
Goodrich, 1619 ... P-11
Gould City, 200 ... E-6
Gowen, 350 ... O-7
Grand Beach, 240 ... T-3
Grand Blanc, 7551 ... P-11
Grand Haven, 10608 ... P-4
Grand Junction, 230 ... R-5
Grand Ledge, 7669 ... Q-8
Grand Marais, 330 ... C-5
Grand Rapids, 193396 ... P-6
**GRAND TRAVERSE CO., 77654 ... J-6**
Grandville, 16718 ... *C-1
Grant, 856 ... O-6
Grass Lake, 1170 ... R-10
**GRATIOT CO., 42285 ... O-8**
Grattan, 130 ... O-7
Grawn, 300 ... J-6
Grayling, 1835 ... J-8
Green Lake, 400 ... Q-6
Greenbush, 250 ... J-12
Greenland, 340 ... B-12
Greenville, 8154 ... O-7
Gregory, 420 ... R-10
Greilickville, 1415 ... J-6
Gresham ... Q-8
Grind Stone City, 70 ... L-13
Gros Cap, 70 ... F-8
Grosse Ile, 10894 ... *M-6
Grosse Pt., 4996 ... *J-8
Grosse Pt. Farms, 8614 ... *I-9
Grosse Pt. Park, 10956 ... *J-8
Grosse Pt. Shores, 2509 ... *I-9
Grosse Pt. Woods, 15154 ... *I-9
Gulliver, 100 ... F-5
Gwinn, 1965 ... E-2
Hadley, 580 ... P-12
Hagensville ... H-10
Hale, 400 ... K-10
Hamburg, 1050 ... R-11
Hamilton, 1170 ... Q-5
Hamtramck, 20512 ... R-12
Hancock, 4158 ... A-12
Hannah ... J-6
Hanover, 413 ... S-9
Harbert, 600 ... T-3
Harbor Beach, 1617 ... M-13
Harbor Springs, 1547 ... G-7
Hardwood, 200 ... F-1
Harper Woods, 12584 ... *I-8
Harrietta, 169 ... K-6
Harris, 30 ... G-2
Harrisburg ... O-5
Harrison, 24461 ... *G-10
Harrison, 1993 ... L-8
Harrisville, 491 ... J-12
Hart, 1915 ... M-4
Hartford, 2453 ... R-5
Hartland, 570 ... Q-11
Harvey, 1321 ... D-2
Haslett, 11292 ... Q-9
Hastings, 6858 ... Q-7
Hawkins, 20 ... M-6
Hawks, 70 ... H-10
Hazel Pk., 17955 ... *I-7
Hell, 40 ... R-10
Hemlock, 1585 ... N-9
Henderson, 200 ... O-10
Hermansville, 550 ... G-1
Herron, 100 ... I-11
Hersey, 363 ... M-6
Hesperia, 949 ... N-5
Hessel, 270 ... E-9
Hickory Corners, 250 ... R-7
Higgins Lake, 460 ... K-8
Highland, 800 ... Q-11
Highland Pk., 14153 ... *I-6
Hilliards, 30 ... Q-6
Hillman, 673 ... I-10
**HILLSDALE CO., 46527 ... T-8**
Hillsdale, 7762 ... S-9
Hodunk ... S-7
Holland, 34076 ... P-5
Holly, 6273 ... P-11
Holt, 11315 ... Q-9
Holton, 580 ... N-5
Homer, 1747 ... S-8
Homestead, 60 ... D-9
Honor, 287 ... J-5
Hope, 100 ... M-9
Hopkins, 552 ... Q-6
Hopkinsburg, 50 ... Q-6
Horton, 380 ... S-9
Horton Bay ... H-7
**HOUGHTON CO., 36016 ... B-12**
Houghton, 6878 ... A-12
Houghton Lake, 3749 ... K-8
Houghton Pt., 120 ... K-8
Howard City, 1577 ... N-6
Howardsville, 40 ... S-6
Howell, 9694 ... Q-10
Hoxeyville, 60 ... L-6
Hubbard Lake, 210 ... I-11
Hubbardston, 387 ... O-8
Hubbell, 1105 ... A-13
Hudson, 2313 ... T-9
Hudsonville, 6997 ... P-5
Hulbert, 350 ... D-7
Huntington Woods, 5811 ... *H-6
**HURON CO., 36079 ... M-12**
Huron Beach, 200 ... G-10
Huron Mtn., 30 ... B-14
Huronia Hts., 460 ... O-14
Ida, 1020 ... S-11
Idlewild, 270 ... M-6
Imlay City, 3687 ... P-12
Ina ... L-7
Indian Lake, 640 ... S-5
Indian River, 2008 ... G-8
Ingalls, 160 ... H-1
**INGHAM CO., 279320 ... Q-9**
Inkster, 26621 ... *K-4
Interlochen, 300 ... J-6
**IONIA CO., 61518 ... P-8**
Ionia, 12589 ... P-7
**IOSCO CO., 27339 ... K-11**
**IRON CO., 13138 ... C-12**
Iron Mtn., 7763 ... D-13
Iron River, 3386 ... D-12
Irons, 50 ... L-5
Ironton, 110 ... H-7
Ironwood, 5368 ... C-10
Irving, 120 ... Q-7
**ISABELLA CO., 63351 ... N-8**
Ishpeming, 6480 ... D-1
Ithaca, 3017 ... O-8
**JACKSON CO., 158422 ... R-9**
Jackson, 33518 ... R-9
Jamestown, 350 ... P-5
Jasper, 330 ... T-10
Jeddo, 110 ... O-14
Jenison, 17211 ... P-6
Jennings, 70 ... K-7
Jerome, 150 ... S-9
Johannesburg, 250 ... I-9
Jones, 290 ... S-6
Jonesville, 2228 ... S-8
Joppa ... R-7
Juddville, 30 ... O-10
Juniata, 100 ... N-11
**KALAMAZOO CO., 238603 ... S-6**
Kalamazoo, 72179 ... R-6
Kalamo, 80 ... Q-8
Kaleva, 464 ... K-5
**KALKASKA CO., 16571 ... J-7**
Kalkaska, 2172 ... J-7
Karlin, 100 ... J-6
Kawkawlin, 450 ... M-10
Keego Harbor, 2838 ... *F-4
Keeler, 180 ... S-5
Kendall, 250 ... R-6
**KENT CO., 574335 ... O-6**
Kent City, 1088 ... O-6
Kenton, 250 ... C-12
Kentwood, 47419 ... P-6
Kewadin, 570 ... I-7
**KEWEENAW CO., 2301 ... A-13**
Kilmanagh, 30 ... M-12
Kinde, 472 ... L-12
Kinderhook, 150 ... T-8
Kings Mill, 50 ... O-12
Kingsford, 5297 ... D-13
Kingsley, 1499 ... J-6
Kingston, 423 ... N-12
Kinneville, 30 ... R-9
Kinross, 50 ... E-8
Kipling, 310 ... F-3
Kiva ... E-2
Klinger Lake, 1000 ... T-6
Koss, 30 ... H-1
La Grange, 120 ... S-5
La Salle, 250 ... S-12
Lac La Belle, 30 ... A-13
Lacey, 80 ... Q-7
Lachine, 200 ... I-11
Lacota, 250 ... R-5
Laingsburg, 1248 ... P-9
**LAKE CO., 11333 ... L-6**
Lake, 550 ... M-7
Lake Angelus, 313 ... *E-4
Lake Ann, 266 ... J-5
Lake City, 919 ... K-7
Lake George, 430 ... L-8
Lake Leelanau, 370 ... I-6
Lake Linden, 1050 ... A-13
Lake Michigan Beach, 1509 ... R-4
Lake Nepessing, 690 ... P-12
Lake Odessa, 2215 ... P-7
Lake Orion, 2713 ... P-12
Lakeland, 1050 ... Q-10
Lakeport, 350 ... O-14
Lakeside, 1100 ... T-3
Lakeview, 1091 ... N-7
Lakeville, 340 ... P-12
Lakewood, 1560 ... P-3
Lakewood Club, 1353 ... N-4
Lamb, 140 ... P-13
Lambertville, 9299 ... T-11
Lamont, 300 ... P-5
Langston, 150 ... O-7
L'Anse, 1882 ... B-13
Lansing, 113968 ... Q-9
**LAPEER CO., 87904 ... O-12**
Lapeer, 9018 ... O-12
Laporte, 150 ... N-10
LaSalle ... *L-6
Lathrup Vil., 4060 ... *H-5
Laurium, 2003 ... A-13
Lawrence, 994 ... R-5
Lawton, 1800 ... S-5
Layton Corners, 30 ... O-10
Le Roy, 256 ... L-6
Leaton, 30 ... M-8
**LEELANAU CO., 21119 ... I-6**
Leland, 680 ... I-6
**LENAWEE CO., 98890 ... S-10**
Lennon, 491 ... P-10
Leonard, 331 ... P-12
Leoni, 90 ... R-9
Leonidas, 220 ... S-7
Leslie, 2272 ... R-9
Level Pk., 3490 ... R-7
Levering, 500 ... G-8
Lewiston, 990 ... I-9
Lexington, 1046 ... O-14
Lexington Hts., 520 ... O-14
Liberty, 110 ... S-9
Lilley ... M-5
Limestone, 60 ... E-3
Lincoln, 346 ... J-11
Lincoln Pk., 35317 ... *L-5
Linden, 3358 ... P-11
Linwood, 500 ... M-10
Litchfield, 1392 ... S-8
Little Lake, 800 ... E-2
**LIVINGSTON CO., 156951 ... Q-10**
Livonia, 91220 ... R-12
Long Lake, 170 ... K-10
Long Rapids, 50 ... H-10
Loomis, 100 ... M-8
Loretto, 400 ... D-14
Lost Lake Woods, 339 ... J-11
Lovells, 130 ... J-9
Lowell, 4153 ... P-7
Lucas, 150 ... L-7
**LUCE CO., 7024 ... D-6**
Ludington, 8324 ... M-4
Lum, 270 ... O-12
Luna Pier, 1520 ... T-12
Lupton, 250 ... K-10
Luther, 322 ... L-6
Luzerne, 610 ... J-9
Lyons, 713 ... P-8
**MACKINAC CO., 11943 ... E-7**
Mackinac Island, 462 ... F-8
Mackinaw City, 840 ... F-8
**MACOMB CO., 788149 ... P-13**
Macomb, 160 ... *E-8
Macon, 100 ... S-10
Madison Hts., 29531 ... *H-6
Mancelona, 1355 ... I-7
Manchester, 2222 ... S-10
**MANISTEE CO., 24527 ... K-5**
Manistee, 6110 ... L-4
Manistique, 3085 ... F-4
Manitou Beach, 1400 ... S-9
Manton, 1167 ... K-6
Maple City, 240 ... I-5
Maple Hill ... N-6
Maple Rapids, 598 ... O-8
Maple Ridge, 30 ... L-10
Mapleton, 100 ... I-6
Marcellus, 1040 ... S-6
Marenisco, 700 ... C-11
Marine City, 4377 ... P-14
Marion, 802 ... L-7
Marlette, 1984 ... N-12
Marne, 490 ... P-5
**MARQUETTE CO., 64634 ... E-1**
Marquette, 20916 ... D-2
Marshall, 7121 ... R-8
Martin, 407 ... Q-6
Marysville, 9943 ... P-14
**MASON CO., 28274 ... L-4**
Mason, 8099 ... Q-9
Masonville ... F-3
Mass City, 430 ... B-12
Matherton, 140 ... O-8
Mattawan, 2795 ... R-6
Maybee, 520 ... S-11
Mayfield, 170 ... J-6
Mayville, 988 ... N-12
McBain, 583 ... L-7
McBride, 227 ... O-7
McFarland, 70 ... E-2
McGregor, 60 ... N-13
McKinley, 220 ... J-10
McMillan, 200 ... D-6
Meade, 160 ... *E-9
Mears, 360 ... M-4
Meauwataka ... K-6
**MECOSTA CO., 40553 ... N-7**
Mecosta, 433 ... N-7
Medina, 70 ... T-9
Melita, 30 ... L-10
Melstrand, 60 ... D-4
Melvin, 150 ... O-13
Melvindale, 9812 ... *K-5
Memphis, 1110 ... P-13
Mendon, 909 ... S-6
**MENOMINEE CO., 25326 ... G-1**
Menominee, 8324 ... I-1
Meredith, 200 ... L-8
Merrill, 719 ... N-9
Merritt, 120 ... K-8
Merriweather, 130 ... B-11
Mesick, 451 ... K-6
Metamora, 483 ... P-12
Metz, 90 ... H-10
Michiana, 193 ... T-3
Michigamme, 287 ... C-13
Michigan Center, 4641 ... R-9
Middleton, 500 ... O-8
Middleville, 2697 ... Q-6
**MIDLAND CO., 82874 ... N-9**
Midland, 40917 ... N-9
Midland Pk., 450 ... R-7
Mikado, 150 ... J-11
Milan, 5657 ... S-11
Milford, 6482 ... Q-11
Millbrook, 110 ... N-7
Millburg, 180 ... S-4
Millersburg, 242 ... H-10
Millett, 770 ... Q-8
Millington, 1070 ... O-11
Millville ... Q-10
Milwood ... P-2
Minden City, 226 ... M-13
Mio, 2016 ... J-10
**MISSAUKEE CO., 14478 ... K-7**
Mohawk, 600 ... A-13
Moline, 750 ... Q-6
**MONROE CO., 145945 ... T-11**
Monroe, 21374 ... S-12
Montague, 2294 ... N-4
**MONTCALM CO., 61266 ... O-7**
Monterey Center ... Q-5
Montgomery, 367 ... T-8
**MONTMORENCY CO., 10315 ... I-9**
Montrose, 1480 ... O-10
Moore Pk., 130 ... S-6
Moorestown, 100 ... K-7
Mooreville, 70 ... S-11
Moran, 200 ... F-8
Morenci, 2259 ... T-10
Morley, 481 ... N-6
Morrice, 872 ... P-10
Morseville, 30 ... O-10
Moscow, 190 ... S-9
Mottville, 200 ... T-6
Mt. Clemens, 16781 ... Q-13
Mt. Forest, 30 ... M-10
Mt. Morris, 3168 ... O-11
Mt. Pleasant, 26675 ... N-8
Muir, 623 ... P-8
Mullet Lake, 250 ... G-9
Mulliken, 550 ... Q-8
Munger, 200 ... N-11
Munising, 2311 ... D-3
Munith, 350 ... R-9
Munson, 50 ... T-9
**MUSKEGON CO., 170200 ... O-5**
Muskegon, 39401 ... O-4
Muskegon Hts., 11623 ... O-4
Nadeau, 260 ... G-1
Nahma, 210 ... F-3
Napoleon, 1254 ... S-9
Nashville, 1655 ... Q-7
National City, 80 ... K-11
National Mine, 490 ... D-1
Naubinway, 350 ... E-6
Needmore ... Q-8
Negaunee, 4456 ... D-1
Nelson ... O-9
Nessen City, 40 ... K-5
Nestoria ... C-13
New Baltimore, 11346 ... Q-13
New Boston, 1300 ... R-12
New Buffalo, 2423 ... T-3
New Era, 456 ... N-4
New Greenleaf, 30 ... M-12
New Haven, 5132 ... P-13
New Holland, 130 ... P-5
New Hudson, 110 ... *H-1
New Lothrop, 578 ... O-10
New Richmond, 180 ... Q-5
New Swanzy, 570 ... E-2
New Troy, 500 ... T-4
**NEWAYGO CO., 47874 ... N-5**
Newaygo, 1626 ... N-6
Newberry, 1517 ... D-6
Newport, 1070 ... S-12
Niles, 11258 ... T-4
Nirvana, 50 ... M-6
Nisula, 100 ... B-12
N. Adams, 485 ... S-9
N. Bradley, 70 ... M-9
N. Branch, 977 ... O-12
N. Epworth ... L-4
N. Lake, 120 ... D-1
N. Lake, 380 ... O-11
N. Lake, 270 ... R-5
N. Lakeport, 300 ... O-14
N. Manitou, 170 ... I-5
N. Morenci, 90 ... T-10
N. Muskegon, 3915 ... O-4
N. Star, 300 ... O-9
N. Street, 80 ... O-14
Northland, 60 ... E-1
Northport, 631 ... I-6
Northville, 5996 ... *I-2
Northwood, 1100 ... O-2
Norton Shores, 23307 ... O-4
Norwalk ... K-5
Norway, 2821 ... D-13
Norwood, 100 ... H-7
Nottawa, 270 ... S-6
Novi, 54054 ... Q-11
Nunica, 360 ... O-5
Oak Grove, 150 ... Q-10
Oak Hill, 920 ... L-4
Oak Pk., 30547 ... *H-5
Oakhurst, 50 ... M-11
**OAKLAND CO., 1194156 ... P-12**
Oakley, 315 ... O-10
Oakville, 50 ... S-11
**OCEANA CO., 26873 ... N-5**
Oden, 350 ... G-8
Ogden Center, 90 ... T-10
**OGEMAW CO., 21645 ... K-10**
Oil City ... N-9
Okemos, 22805 ... Q-9
Old Mission, 300 ... I-6
Olivet, 1813 ... R-8
Omena, 60 ... I-6
Omer, 297 ... L-10
Onaway, 902 ... H-9
Onekama, 597 ... K-5
Onondaga, 250 ... R-9
Onsted, 993 ... S-10
**ONTONAGON CO., 7818 ... B-11**
Ontonagon, 1509 ... B-11
Orangeville, 250 ... Q-6
Orchard Lake, 2209 ... *G-4
Orleans, 230 ... P-7
Ortonville, 1482 ... P-11
**OSCEOLA CO., 23197 ... L-7**
**OSCODA CO., 9418 ... J-9**
Oscoda, 992 ... K-12
Oshtemo, 2000 ... R-6
Osseo, 320 ... S-9
Ossineke, 1059 ... I-11
Otisville, 803 ... O-11
**OTSEGO CO., 23301 ... I-8**
Otsego, 3821 ... R-6
**OTTAWA CO., 238314 ... P-5**
Ottawa Lake, 190 ... T-11
Otter Lake, 411 ... O-11
Overisel, 160 ... Q-5
Ovid, 1399 ... P-9
Owendale, 265 ... M-12
Owosso, 14963 ... P-10
Oxford, 3526 ... P-12
Painesdale, 450 ... A-12
Palmer, 449 ... D-1
Palms, 30 ... M-13
Palmyra, 370 ... T-10
Palo, 250 ... O-8
Paradise, 340 ... C-7
Parchment, 1808 ... O-2
Paris, 230 ... M-6
Parisville, 50 ... M-13
Park Lake, 390 ... P-9
Parkdale, 550 ... K-4

***, †, ‡, §, ◊ See explanation under state title in this index.**

**County names are listed in capital letters and in boldface type.**

Parkers Corners, 40 Q-10
Parma, 846 R-9
Parshallville, 80 Q-11
Partello R-8
Paulding, 100 C-11
Paw Paw, 3222 R-5
Paw Paw Lake, 3944 R-4
Payment, 110 D-9
Paynesville, 30 C-12
Peacock, 60 L-5
Pearl, 60 Q-5
Pearl Beach, 3224 Q-14
Peck, 562 O-13
Pelkie, 160 B-12
Pellston, 772 G-8
Pennfield, 100 R-7
Pentwater, 942 M-4
Perkins, 400 F-2
Perrinton, 422 O-8
Perronville, 60 F-2
Perry, 2025 P-9
Peshawbestown, 90 I-6
Petersburg, 1120 S-11
Petoskey, 6028 H-7
Pewamo, 549 P-8
Phoenix, 20 A-13
Pickford, 450 E-9
Pierport, 40 K-4
Pierson, 181 O-6
Pigeon, 1066 M-12
Pinckney, 2423 Q-10
Pinconning, 1314 M-10
Pine Creek S-7
Pine Run, 660 O-11
Piney Woods, 500 L-8
Pinnebog, 60 L-12
Pittsburg P-9
Pittsford, 530 T-9
Plainfield, 40 Q-10
Plainwell, 3856 R-6
Pleasant Ridge, 2469 *H-6
Pleasant Valley N-9
Plymouth, 8487 *J-2
Pokagon, 140 S-5
Pompeii, 200 O-9
Pontiac, 66095 Q-12
Port Austin, 652 L-12
Port Hope, 273 L-13
Port Huron, 30869 P-14
Port Sanilac, 620 N-14
Port Sheldon, 180 P-5
Portage, 46133 R-6
Portland, 3687 P-8
Posen, 267 H-11
Poseyville, 100 N-9
Potterville, 2147 Q-8
Powers, 415 G-1
Prairieville, 180 Q-6
Prattville, 200 T-9
Prescott, 270 L-10
**PRESQUE ISLE CO., 14411 H-9**
Presque Isle, 120 H-11
Princeton, 180 E-1
Prudenville, 1737 K-8
Pulaski, 130 S-8
Pullman, 600 Q-5
Quanicassee, 130 N-11
Quincy, 1594 S-8
Quinnesec, 1187 D-13
Raco, 50 D-8
Ralph, 60 E-1
Ramsay, 1080 C-10
Rankin, 220 P-11
Ransom, 130 T-9
Rapid City, 420 J-7
Rapid River, 800 F-3
Ravenna, 1211 O-5
Ray, 60 T-8
Ray Center *D-9
Reading, 1071 T-8
Red Oak J-9
Redford, 51622 *J-4
Reed City, 2330 M-6
Reeman, 110 N-5
Reese, 1346 N-11
Remus, 480 N-7
Republic, 614 C-13
Rexton, 160 E-7
Rhodes, 130 M-9
Richland, 752 R-6
Richmond, 5606 P-13
Richmondville, 30 N-13
Richville, 370 N-11
Ridgeway, 200 S-10
Riga, 400 T-11
Riley Center, 60 P-13
River Rouge, 8449 *K-6
Riverdale, 400 N-8
Riverside, 290 S-4
Riverview, 11817 *M-6
Rives Junction, 450 R-9
Roberts Landing, 50 Q-14
Rochester, 11002 Q-12
Rochester Hills, 69014 Q-12
Rock, 440 E-2
Rockford, 5411 O-6
Rockland, 270 B-12
Rockwood, 3160 S-12
Rodney, 160 M-7
Rogers City, 3031 G-10
Rollin, 110 S-9
Romeo, 3737 P-12
Romulus, 22963 R-12
Roosevelt Pk., 3769 O-4
**ROSCOMMON CO., 25469 K-8**
Roscommon, 1059 K-8
Rose Center, 170 *E-1
Rose City, 685 K-10
Roseburg, 30 O-13
Rosebush, 341 M-8
Rosedale D-9
Roseville, 46782 Q-13
Rothbury, 433 N-4
Royal Oak, 57110 *H-6
Rudyard, 1100 E-8
Rumely, 90 D-3
Russellville, 220 O-11
Ruth, 230 M-13
**SAGINAW CO., 210039 O-10**
Saginaw, 55620 N-10
Sagola, 180 D-13
St. Charles, 2016 O-10
**ST. CLAIR CO., 164235 P-13**
St. Clair, 5763 P-14
St. Clair Shores, 60364 Q-13
St. Helen, 2993 K-9
St. Ignace, 2335 F-8
St. James, 400 F-6
St. Johns, 7278 P-9
**ST. JOSEPH CO., 62422 T-6**
St. Joseph, 8458 S-4
St. Louis, 6971 N-8
Salem, 350 *I-1
Saline, 8882 R-11
Samaria, 480 T-11
Sand Creek, 110 T-10
Sand Lake, 515 O-6
Sand River, 80 D-2
Sands, 60 D-2
Sandusky, 2609 N-13
Sanford, 910 M-9
**SANILAC CO., 44547 N-13**
Saranac, 1296 P-7
Saugatuck, 1009 Q-5
Sault Ste. Marie, 14087 D-9
Sawyer, 500 T-4
Schaffer, 150 F-2
**SCHOOLCRAFT CO., 8903 E-4**
Schoolcraft, 1512 S-6
Scofield S-11
Scotts, 350 R-7
Scottville, 1250 M-4
Sears, 90 M-7
Sebewaing, 1750 M-11
Selkirk, 40 K-10
Seney, 200 D-5
Seven Harbors, 4700 *E-1
Shabbona, 90 N-12
Shady Shores, 350 K-10
Shaftsburg, 270 P-9
Shelby, 65159 *E-7
Shelby, 1885 N-4
Shelbyville, 160 Q-6
Sheldon, 590 *K-2
Shepardsville, 70 P-9
Shepherd, 1353 N-8
Sheridan, 694 O-7
Sherman, 30 K-6
Sherman City, 30 M-7
Sherwood, 308 S-7
**SHIAWASSEE CO., 71687 P-10**
Shiawasseetown, 80 P-10
Shields, 6590 N-10
Shingleton, 320 D-4
Shoreham, 823 S-4
Sidnaw, 300 C-12
Sidney, 120 O-7
Silver City, 60 B-11
Silverwood, 170 N-12
Six Lakes, 420 N-7
Skandia, 300 D-2
Skanee, 100 B-13
Skeels L-8
Skidway Lake, 3147 L-10
Snover, 300 N-13
Sodus, 180 S-4
Somerset, 140 S-9
Somerset Center, 440 S-9
S. Boardman, 250 J-7
S. Branch, 150 K-10
S. Haven, 5138 R-4
S. Ionia, 100 P-7
S. Lyon, 10959 Q-11
S. Range, 688 A-12
S. Rockwood, 1654 S-12
Southfield, 75392 Q-12
Southgate, 27739 *M-5
Spalding, 590 G-1
Sparlingville P-14
Sparr, 30 I-8
Sparta, 4054 O-6
Spratt I-10
Spring Arbor, 2188 R-9
Spring Lake, 2420 O-5
Springfield, 5037 R-7
Springport, 653 R-8
Springville, 60 S-10
Spruce, 120 I-11
Stalwart, 60 E-9
Standish, 1981 L-10
Stanton, 1483 O-7
Stanwood, 198 N-6
Stephenson, 803 H-1
Sterling, 485 L-10
Sterling Hts., 127160 Q-12
Steuben, 70 E-4
Stevensville, 1131 S-4
Stockbridge, 1274 R-10
Stonington G-3
Stony Lake, 50 N-4
Stronach, 210 L-4
Strongs, 380 D-7
Sturgis, 10916 T-7
Sumner, 170 O-8
Sumnerville, 180 S-5
Sunfield, 583 Q-8
Sunrise Hts., 1350 R-7
Suttons Bay, 578 I-6
Swartz Creek, 5223 P-10
Sylvan Lake, 1639 *F-4
Tallmadge, 80 P-5
Tallman, 70 L-5
Tawas City, 1873 K-11
Taylor, 60619 R-12
Tekonsha, 679 S-8
Temperance, 7757 T-11
Temple, 300 L-7
Thomaston, 80 C-10
Thompson, 200 F-4
Thompsonville, 440 K-5
Thornville, 60 P-12
Three Lakes, 150 C-13
Three Oaks, 1695 T-4
Three Rivers, 7179 S-6
Tipton, 200 S-10
Topinabee, 400 G-8
Torch Lake, 80 I-7
Torch River, 410 I-7
Tower, 300 H-9
Traunik, 100 E-3
Traverse Bay, 40 A-13
Traverse City, 14398 J-6
Trenary, 300 E-3
Trenton, 17965 S-12
Trombly, 40 E-2
Trout Creek, 230 C-12
Trout Lake, 370 E-7
Trowbridge Pk., 2012 D-2
Troy, 80264 Q-12
Trufant, 500 O-7
Turk Lake, 350 O-7
Turner, 125 L-10
**TUSCOLA CO., 58266 N-11**
Tuscola, 100 O-11
Tustin, 226 L-6
Twin Lake, 1613 O-5
Twin Lakes, 150 B-12
Twining, 172 L-10
Tyre, 40 M-13
Ubly, 771 M-13
Unadilla, 100 R-10
Union, 200 T-5
Union City, 1717 S-7
Union Lake, 8500 *F-3
Union Pier, 1100 T-3
Unionville, 567 M-11
Utica, 4924 *F-7
Valley Center O-13
Valley Farms, 950 A-3
**VAN BUREN CO., 76263 R-5**
Vandalia, 392 S-5
Vanderbilt, 571 H-8
Vandercook Lake, 4809 R-9
Vassar, 2665 N-11
Vermontville, 785 Q-7
Vernon, 790 P-10
Verona, 80 M-13
Vestaburg, 420 N-8
Vickeryville, 100 O-8
Vicksburg, 2194 S-6
Vil. of Lake Isabella, 1127 N-8
Virginia Pk., 1800 Q-5
Vogel Center, 110 L-7
Volinia, 40 S-5
Vriesland, 150 P-5
Vulcan, 100 D-14
Wacousta, 310 P-8
Wadhams, 150 P-14
Wakefield, 1846 C-10
Wakelee, 80 S-5
Waldenburg, 110 *F-8
Waldron, 560 T-9
Walhalla, 380 M-5
Walker, 23722 *B-1
Walkerville, 253 M-5
Wallace, 150 H-1
Walled Lake, 6876 *G-2
Walloon Lake, 240 H-8
Waltz, 500 *N-3
Warren, 133939 Q-13
Washington, 1850 Q-12
**WASHTENAW CO., 322895 R-10**
Waterford, 71981 Q-12
Waterloo, 190 R-10
Waters, 150 I-8
Watersmeet, 800 C-11
Watertown, 100 N-13
Watervliet, 1737 R-4
Watrousville, 200 N-11
Watson, 30 Q-6
Watton, 150 C-12
Waucedah, 50 F-1
Wayland, 3809 Q-6
**WAYNE CO., 2061162 R-12**
Wayne, 17186 *K-3
Webberville, 1480 Q-10
Weidman, 879 M-8
Wells, 1000 F-2
Wellston, 320 L-5
Wequetonsing G-7
W. Bloomfield, 64000 *G-4
W. Branch, 1816 K-9
W. Olive, 200 P-5
W. Windsor Q-8
Westland, 78961 R-12
Weston, 300 T-10
Westphalia, 816 P-8
Westville, 50 O-7
Westwood, 9122 P-1
Wetmore, 480 D-4
**WEXFORD CO., 30484 K-6**
Wheeler, 320 N-9
White Cloud, 1390 N-6
White Lake, 3200 *E-2
White Pigeon, 1559 T-6
White Pine, 910 B-11
Whitefish Pt., 50 C-7
Whitehall, 2803 N-4
Whitmore Lake, 6574 R-11
Whittaker, 240 S-11
Whittemore, 440 K-10
Wildwood, 80 K-5
Willard M-10
Williamsburg, 400 J-6
Williamston, 3765 Q-9
Willis, 300 *M-1
Willow, 430 *N-3
Wilmot, 140 N-12
Wilson, 150 G-2
Winegars L-9
Winn, 450 N-8
Winona, 40 B-12
Wisner, 110 N-11
Witch Lake, 80 C-13
Wixom, 13425 Q-11
Wolf Lake, 4455 O-5
Wolverine, 335 H-8
Wolverine Lake, 4303 *G-2
Woodbury, 80 Q-7
Wooden Shoe Vil., 330 L-9
Woodhaven, 12634 *M-5
Woodland, 488 Q-7
Woodland Beach, 2179 S-12
Woodland Pk., 200 M-5
Woodville, 50 M-6
Wyandotte, 24739 R-12
Wyman, 50 N-7
Wyoming, 70462 P-6
Yale, 1956 O-13
Ypsilanti, 21464 R-11
Yuma, 80 K-6
Zeeland, 5448 P-5
Zilwaukee, 1651 N-10

## Minnesota

Page locator

| Map keys | Atlas pages |
|---|---|
| A - J | 108 - 109 |
| K - T | 110 - 111 |

* City keyed to pp. 112 - 113

Ada, 1462 H-2
Adams, 755 T-10
Adrian, 1216 T-3
Afton, 2839 P-10
Ah-gwah-ching, 50 I-6
**AITKIN CO., 15301 K-9**
Aitkin, 1948 K-8
Akeley, 412 I-6
Albany, 2113 M-6
Albert Lea, 17389 T-9
Alberta, 136 N-3
Albertville, 6360 O-8
Alborn, 200 I-11
Alden, 612 S-9
Aldrich, 47 K-6
Alexandria, 11237 M-5
Alida, 20 H-5
Alma City, 40 R-8
Almelund, 150 N-10
Almora, 50 K-5
Alpha, 116 T-6
Altura, 421 R-12
Alvarado, 341 E-1
Alvwood G-7
Amboy, 537 S-7
Amiret, 60 Q-4
Amor K-4
Andover, 30442 O-9
Andree, 30 M-9
Angle Inlet, 75 B-5
Angora, 100 G-11
Angus, 50 F-2
Annandale, 3100 O-7
**ANOKA CO., 298084 N-9**
Anoka, 17289 O-9
Apple Valley, 50004 P-9
Appleton, 2950 O-3
Arco, 85 Q-3
Arden Hills, 9608 *D-6
Argyle, 599 E-2
Arlington, 2028 Q-7
Arnesen C-5
Arnold, 3032 J-12
Ash Creek, 25 T-3
Ash Lake, 30 E-10
Ashby, 478 L-4
Askov, 356 L-10
Atwater, 1027 O-6
Audubon, 445 J-3
Aurora, 1746 G-11
Austin, 22800 T-10
Averill, 40 I-2
Avoca, 130 S-4
Avon, 1315 M-7
Babbitt, 1587 G-12
Backus, 297 J-7
Badger, 459 D-3
Bagley, 1288 G-5
Baker, 90 J-2
Balaton, 632 R-3
Bald Eagle, 1800 *C-9
Ball Club, 75 H-8
Barnesville, 2308 J-2
Barnum, 595 K-10
Barrett, 344 M-4
Barrows, 75 K-7
Barry, 22 N-2
Bassett, 50 H-12
Battle Lake, 772 K-4
Baudette, 925 D-6
Baxter, 8271 K-7
Bay Lake, 80 K-8
Bayport, 3291 O-10
Bayview, 40 L-9
Bear River G-10
Beardsley, 230 N-2
Beauford, 60 R-8
Beaulieu, 30 H-4
Beaver Bay, 170 H-13
Beaver Creek, 260 T-2
Bechyn, 30 P-5
Becida H-5
**BECKER CO., 30000 I-4**
Becker, 4176 N-8
Bejou, 93 H-3
Belgrade, 714 N-5
Bellaire, 1900 *D-9
Belle Plaine, 6648 Q-8
Belle Prairie, 500 L-7
Bellechester, 163 Q-11
Bellingham, 179 O-3
**BELTRAMI CO., 39650 E-5**
Beltrami, 92 G-2
Belview, 376 Q-5
Bemidji, 13541 H-6
Bena, 100 H-7
Benedict, 90 I-6
Benson, 3049 N-4
**BENTON CO., 34226 M-7**
Bergen, 35 S-5
Bernadotte, 25 Q-7
Beroun, 100 M-10
Bertha, 426 K-5
Bethel, 539 N-9
Big Bend City, 60 O-4
Big Falls, 236 E-8
Big Lake, 9795 N-8
**BIG STONE CO., 5820 N-2**
Bigelow, 221 T-4
Bigfork, 443 G-8
Bingham Lake, 169 S-5
Birch Beach C-6
Birchdale, 30 D-7
Birchwood, 901 *D-9
Bird Island, 1111 P-6
Biscay, 108 P-7
Biwabik, 974 G-11
Bixby, 100 S-10
Black Hammer, 50 T-13
Blackberry, 50 I-9
Blackduck, 761 G-6
Blaine, 55042 O-9
Blakeley, 100 Q-8
Blomkest, 171 P-5
Blooming Prairie, 1967 S-10
Bloomington, 81280 P-9
**BLUE EARTH CO., 55941 S-7**
Blue Earth, 3247 T-7
Bluffton, 209 K-5
Bock, 110 M-9
Bois Fort, 350 F-9
Bonanza Grove N-2
Bongards P-8
Borup, 83 I-2
Bovey, 665 H-9
Bowlus, 231 M-7
Bowstring, 25 G-8
Boy River, 35 I-7
Boyd, 188 P-3
Braham, 1646 M-9
Brainerd, 13691 K-7
Brandon, 439 L-4
Bratsberg S-13
Breckenridge, 3196 K-2
Breezy Pt., 1614 J-7
Brennyville, 35 M-8
Brevik, 30 I-7
Brewster, 483 S-4
Bricelyn, 331 T-8
Brimson, 40 H-12
Britt, 80 G-11
Brook Pk., 157 L-10
Brooklyn Center, 27402 *D-3
Brooklyn Pk., 71308 *C-3
Brooks, 133 G-3
Brookston, 90 J-11
Brooten, 616 N-5
Browerville, 691 L-6
**BROWN CO., 26911 R-6**
Browns Valley, 603 M-1
Brownsdale, 688 S-10
Brownsville, 518 S-14
Brownton, 784 P-7
Bruno, 106 K-11
Brunswick, 80 M-9
Buckman, 231 M-7
Buffalo, 14222 O-8
Buffalo Lake, 731 P-6
Buhl, 983 H-10
Burnsville, 59139 *J-4
Burr, 20 P-2
Burtrum, 132 M-6
Butler J-5
Butterfield, 503 S-6
Butternut, 30 R-7
Buyck, 25 F-11
Byron, 4550 R-11
Caledonia, 2821 T-13
Callaway, 210 I-4
Calumet, 355 H-9
Cambria, 80 R-7
Cambridge, 7692 N-9
Campbell, 209 L-2
Canby, 1710 P-3
Cannon City, 60 Q-9
Cannon Falls, 4034 Q-10
Canton, 323 T-12
Canyon, 70 I-11
Carlisle, 30 K-3
Carlos, 421 L-5
**CARLTON CO., 31671 J-10**
Carlton, 770 J-11
**CARVER CO., 70205 P-8**
Carver, 2915 P-8
**CASS CO., 27150 I-7**
Cass Lake, 828 H-6
Castle Danger, 35 I-13
Castle Rock, 150 Q-10
Cedar Mills, 53 P-7
Center City, 608 N-10
Centerville, 3737 *B-8
Ceylon, 372 T-6
Champlin, 23503 *A-2
Chandler, 239 S-3
Chanhassen, 23480 P-9
Chaska, 24116 P-9
Chatfield, 2518 S-12
Cherry, 100 H-10
Cherry Grove, 70 T-11
Chester, 200 R-11
Chickamaw Beach, 145 J-7
**CHIPPEWA CO., 13088 O-4**
**CHISAGO CO., 41101 N-10**
Chisago City, 4202 N-10
Chisholm, 4594 H-10
Chokio, 424 N-3
Circle Pines, 5395 *B-7
Clara City, 1310 P-5
Claremont, 594 R-10
Clarissa, 592 L-6
Clarkfield, 846 P-4
Clarks Grove, 704 S-9
**CLAY CO., 51229 I-2**
Clear Lake, 478 N-8
Clearbrook, 534 G-5
**CLEARWATER CO., 8423 G-5**
Clearwater, 1493 N-8
Clements, 169 Q-5
Clementson, 20 D-7
Cleveland, 697 Q-8
Climax, 235 G-2
Clinton, 401 N-2
Clinton Falls, 40 R-9
Clitherall, 113 K-4
Clontarf, 153 N-4
Cloquet, 11387 J-11
Clotho, 25 L-5
Cloverdale, 40 L-10
Cloverton, 35 L-11
Coates, 161 P-10
Cobden, 59 R-6
Cohasset, 2541 H-8
Cokato, 2711 O-7
Cold Spring, 3751 N-7
Coleraine, 1008 H-9
Collegeville, 200 N-7
Collis, 20 M-2
Cologne, 1416 P-8
Columbia Hts., 18210 *E-5
Comfrey, 344 R-6
Comstock, 117 J-2
Concord, 60 R-10
Conger, 124 T-9
**COOK CO., 5168 F-14**
Cook, 618 F-10
Coon Rapids, 61832 *A-4
Corcoran, 5630 O-9
Cormorant J-3
Corning S-10
Correll, 44 O-3
Corvuso, 25 P-6
Cosmos, 555 P-6
Cottage Grove, 33623 P-10
Cotton, 200 I-11
**COTTONWOOD CO., 12167 S-5**
Cottonwood, 1118 Q-4
Courtland, 617 R-7
Cove, 75 L-8
Craigville F-8
Crane Lake, 300 E-11
Croftville, 40 B-12
Cromwell, 203 J-10
Crookston, 7790 G-2
Crosby, 2243 K-8
Cross Lake, 2108 J-7
Crow River, 20 O-5
**CROW WING CO., 55099 K-7**
Crow Wing, 25 K-7
Crown, 50 N-9
Crystal, 21707 *E-3
Currie, 198 R-4
Cushing, 70 L-6
Cusson, 40 F-10
Cutler, 100 K-8
Cuyuna, 273 K-8
Cyrus, 267 M-4
**DAKOTA CO., 355904 Q-9**
Dakota, 320 S-14
Dalbo, 100 M-9
Dalton, 241 L-4
Danube, 478 P-5
Danvers, 91 N-4
Darfur, 121 R-6
Darwin, 304 O-7
Dassel, 1320 O-7
Dawson, 1357 P-3
Dayton, 4643 O-9
De Graff, 115 O-4
Debs, 25 G-5
Deephaven, 3775 *G-1
Deer Creek, 325 K-5
Deer River, 918 H-8
Deerwood, 573 K-8
Delano, 5309 O-8
Delavan, 205 S-8
Delft, 100 S-5
Delhi, 63 Q-5
Dellwood, 1044 *D-9
Denham, 40 K-10
Dennison, 164 Q-10
Dent, 184 K-4
Detroit Lakes, 8147 J-4
Dexter, 314 S-11
Dilworth, 3677 I-2
**DODGE CO., 17731 S-10**
Dodge Center, 2598 R-10
Donaldson, 35 D-2
Donnelly, 243 M-3
Dora Lake, 40 G-8
Doran, 51 L-2
Dorothy, 25 F-3
Dorset, 80 I-6
**DOUGLAS CO., 32821 L-4**
Douglas, 150 R-11
Dover, 637 R-12
Dovray, 57 R-4
Downer, 90 J-2
Dresbach, 150 S-14
Duelm, 70 N-8
Duluth, 84284 J-12
Duluth Hts. A-8
Dumont, 105 M-2
Dundas, 1080 Q-9
Dundee, 98 S-4
Dunnell, 175 T-6
Dunvilla, 100 J-3
Duquette, 70 K-11
Duxbury, 20 L-11
Eagan, 63985 *J-7
Eagle Bend, 567 L-5
Eagle Lake, 2285 R-8
E. Beaver Bay, 20 H-13
E. Bethel, 11895 N-9
E. Chain, 200 T-7
E. Grand Forks, 7852 F-1
E. Gull Lake, 1027 K-7
E. Lake, 30 K-9
E. Lake Francis Shores, 170 N-9
E. Union, 60 P-8
Easton, 188 S-8
Ebro, 70 H-4
Echo, 242 Q-5
Eden Prairie, 61195 *I-1
Eden Valley, 954 N-6
Edgerton, 988 S-3
Edgewood, 120 N-9
Edina, 45608 *H-3
Effie, 85 G-8
Eitzen, 228 T-13
Elba, 210 R-12
Elbow Lake, 1184 L-3
Eldred, 50 G-2
Elgin, 938 R-12
Elizabeth, 183 K-3
Elk River, 23194 N-9
Elko, 472 Q-9
Elkton, 147 T-11
Ellendale, 673 S-9
Ellsworth, 513 T-3
Elmdale, 98 M-7
Elmore, 656 T-7
Elrosa, 178 N-6
Ely, 3473 F-12
Elysian, 616 R-8
Embarrass, 100 G-11
Emily, 991 J-8
Emmons, 405 T-9
Englund D-2
Erdahl, 50 L-4
Erhard, 149 K-3
Ericsburg, 90 D-9
Erskine, 403 G-3
Esko, 550 J-11
Essig, 50 Q-6
Estes Brook, 35 M-8
Euclid, 150 F-2
Evan, 92 Q-6
Evansville, 568 L-4
Eveleth, 3582 H-11
Everdell K-2
Evergreen J-5
Excelsior, 2291 P-9
Eyota, 1765 R-12
Fairfax, 1208 Q-6
Fairhaven, 200 N-7
Fairmont, 10221 T-7
Falcon Hts., 5643 *F-6
**FARIBAULT CO., 16181 S-8**
Faribault, 22009 R-9
Farming, 50 N-6
Farmington, 19057 P-10
Farwell, 49 M-4
Federal Dam, 93 H-7
Felton, 201 I-2
Fergus Falls, 13635 K-3
Fernando, 25 P-7
Fertile, 839 G-3
Fifty Lakes, 421 J-7
**FILLMORE CO., 21122 T-12**
Fillmore, 100 S-11
Finland, 300 H-13
Finlayson, 323 L-10
Fisher, 403 G-2
Flensburg, 216 L-7
Flom, 90 I-3
Floodwood, 497 I-10
Florence, 56 R-3
Florenton, 100 G-11
Florian E-2
Foley, 2381 M-8
Forada, 193 M-5
Forbes, 110 H-11
Forest City, 80 O-7
Forest Lake, 17373 O-10
Foreston, 539 M-8
Ft. Ripley, 73 L-7
Fosston, 1470 G-4
Fountain, 340 S-12
Fox C-4
Fox Lake, 100 S-6
Foxhome, 126 K-3
Franklin, 465 Q-6
Frazee, 1375 J-4
**FREEBORN CO., 32584 T-9**
Freeborn, 286 S-9
Freeburg, 60 T-14
Freedhem, 50 L-7
Freeport, 469 M-6
Fremont S-12
French Lake, 30 O-7
French River, 400 I-12
Fridley, 25868 *D-5
Frontenac, 300 Q-11
Frost, 221 T-8
Fulda, 1196 S-4
Funkley, 16 G-7
Garden City, 200 R-7
Garfield, 242 L-4
Garrison, 215 K-8
Garvin, 141 R-4
Gary, 194 H-3
Gatzke, 50 E-4
Gaylord, 2123 Q-7
Gem Lake, 442 *E-8
Gemmell F-7
Geneva, 445 S-9
Genoa, 60 R-11
Genola, 80 L-7
Gentilly, 100 G-2
Georgetown, 128 I-2
Georgeville, 30 N-6
Gheen, 60 F-10
Ghent, 367 Q-3
Gibbon, 738 Q-6
Giese L-10
Gilbert, 1755 H-11
Gilfillan Q-5
Gilman, 202 M-8
Glen, 100 K-9
Glencoe, 5550 P-7
Glendorado, 20 N-8
Glenville, 678 T-9
Glenwood, 2525 M-5
Gluek O-4
Glyndon, 1155 I-2
Godahl, 20 R-6
Golden Valley, 20344 *F-3
Gonvick, 287 G-4
Good Thunder, 569 R-8
**GOODHUE CO., 44127 Q-10**
Goodhue, 989 Q-11
Goodland, 150 I-9
Goodridge, 100 F-4
Goodview, 3565 R-13
Gordonsville, 100 T-9
Graceton, 30 D-6
Graceville, 557 N-2
Granada, 294 S-7
Grand Falls E-8
Grand Marais, 1418 B-12
Grand Meadow, 916 S-11
Grand Portage, 200 A-13
Grand Rapids, 8743 H-9
Grandy, 150 M-9
Granger, 100 T-12
Granite Falls, 2908 P-4
**GRANT CO., 6289 M-3**
Grant, 4028 *D-10
Grasston, 114 M-10
Grattan, 40 G-7
Greaney, 25 F-10
Green Isle, 435 P-8
Green Valley, 100 Q-4
Greenbush, 739 D-3
Greenfield, 3110 O-8
Greenleaf, 40 O-7
Greenleafton, 80 T-12
Greenwald, 177 N-6
Grey Eagle, 326 M-6
Grogan, 30 R-6
Grove City, 581 O-6
Grove Lake, 30 M-5
Grygla, 224 E-4
Guckeen, 100 T-7
Gully, 102 G-4
Gutches Grove L-6
Guthrie, 60 H-6
Hackensack, 283 I-7
Hader, 30 Q-10
Hadley, 70 R-4
Hallock, 1020 C-2
Halma, 66 D-2
Halstad, 541 H-2
Ham Lake, 15020 O-9
Hamburg, 546 P-8
Hammond, 181 R-11
Hampton, 657 Q-10
Hancock, 685 N-4
Hanley Falls, 285 P-4
Hanover, 2423 O-8
Hanska, 409 R-7
Harding, 106 L-8
Hardwick, 205 S-3
Harmony, 1076 T-12
Harris, 1252 N-10
Hartland, 270 S-9
Hassman J-8
Hastings, 22126 P-10
Hasty, 40 N-8
Hatfield, 43 S-3
Hawick, 120 N-6
Hawley, 1874 I-3
Hay Creek, 60 Q-11
Hayfield, 1337 S-10
Haypoint I-8
Hayward, 234 T-9
Hazel Run, 55 P-4
Hector, 1072 P-6
Heidelberg, 129 Q-8
Henderson, 918 Q-8
Hendricks, 648 Q-2
Hendrum, 285 H-2
**HENNEPIN CO., 1116200 O-8**
Henning, 800 K-5
Henriette, 106 M-10
Herman, 422 M-3
Hermantown, 9361 J-11
Heron Lake, 733 S-5
Hewitt, 259 K-5
Hibbing, 16209 H-10
High Forest, 100 S-11
Highland S-12
Hill City, 440 I-9
Hillman, 32 L-8
Hills, 594 T-2
Hilltop, 683 *D-5
Hillview, 20 J-5
Hinckley, 1409 L-10
Hines, 60 G-6
Hitterdal, 199 I-3
Hoffman, 640 M-4
Hokah, 561 S-14
Holdingford, 723 M-7
Holland, 198 R-3
Hollandale, 274 S-9
Holloway, 96 O-3
Holmes City, 90 M-4
Holt, 84 E-3
Holyoke, 60 K-11
Homer, 150 R-13
Hope, 120 S-9
Hopkins, 16961 *G-2
**HOUSTON CO., 19718 S-13**
Houston, 956 S-13
Hovland, 250 B-13
Howard Lake, 1987 O-7
Hoyt Lakes, 1956 G-12
**HUBBARD CO., 18376 I-6**
Hubbard, 150 J-5
Hugo, 13139 *B-9
Humboldt, 52 C-1
Hunters Pk. A-9
Huntersville, 20 J-6
Huntley, 150 S-7
Hutchinson, 13913 P-7
Ihlen, 103 S-2
Imogene, 70 T-7
Independence, 3546 O-8
Inger, 80 G-8
International Falls, 5906 D-9
Inver Grove Hts., 33383 *J-8
Iona, 155 S-4
Iron Junction, 85 H-11
Ironton, 528 K-8
Isabella, 100 G-13
**ISANTI CO., 31287 M-9**
Isanti, 6123 N-9
Island Lake G-6
Island View, 150 D-9
Isle, 909 L-9
**ITASCA CO., 43992 H-9**
Ivanhoe, 589 Q-3
**JACKSON CO., 11268 T-5**
Jackson, 3305 T-5
Jacobson, 150 I-9
Janesville, 2242 R-8
Jasper, 564 S-2
Jeffers, 362 R-5
Jenkins, 403 J-7
Johnsburg, 25 T-10
Johnson, 28 N-3
Jordan, 5445 P-8
Judson, 150 R-7
Kabekona, 60 H-6
Kabetogama, 100 E-10
**KANABEC CO., 14996 M-9**
Kanaranzi, 60 T-3
**KANDIYOHI CO., 41203 O-5**
Kandiyohi, 530 O-6
Karlstad, 674 D-2
Kasota, 690 R-8
Kasson, 5564 R-10
Keewatin, 1070 H-10
Kelliher, 306 F-7
Kellogg, 472 Q-12
Kelsey, 30 I-11
Kennedy, 211 D-2
Kenneth, 59 S-3
Kensington, 267 M-4
Kent, 100 K-2
Kenyon, 1680 R-10
Kerkhoven, 674 O-5
Kerrick, 71 K-11
Kettle River, 171 K-10
Kiester, 477 T-8
Kilkenny, 154 R-9
Kimball, 704 N-7
Kinbrae, 21 S-4
Kingsdale L-11
Kingston, 118 O-7
Kinney, 190 H-10
**KITTSON CO., 5285 D-2**
Klossner, 50 Q-7
Knife River, 300 I-12
**KOOCHICHING CO., 14355 E-7**
Kragnes, 60 I-2
La Crescent, 4912 S-14
La Prairie, 593 H-9
La Salle, 83 R-6
**LAC QUI PARLE CO., 8067 P-3**
Lac qui Parle, 50 O-3
Lafayette, 515 Q-7
**LAKE CO., 11058 G-13**
Lake Benton, 633 R-3
Lake Bronson, 196 D-2
Lake Center, 25 J-3
Lake City, 5285 Q-12
Lake Crystal, 2598 R-7
Lake Elmo, 7484 O-10
Lake Fremont, 5313 N-9
Lake George, 200 I-6
Lake Henry, 86 N-6
Lake Hubert, 300 K-7
Lake Itasca, 50 H-5
Lake Lillian, 244 P-6
**LAKE OF THE WOODS CO., 4522 D-6**
Lake Pk., 812 I-3
Lake Shore, 1068 K-7
Lake Wilson, 232 R-3
Lakefield, 1614 S-5
Lakeland, 1791 P-10
Lakeville, 54542 P-9
Lamberton, 771 R-5
Lamoille, 50 R-13
Lancaster, 305 C-2
Landfall, 634 *G-9
Lanesboro, 728 S-12
Lansing, 250 S-10
Laporte, 148 I-6
Larsmont, 50 I-12
Lastrup, 87 L-8
Lauderdale, 2212 *F-6
Lawler, 60 K-9
Lawndale J-3
Lax Lake H-13
Le Center, 2286 Q-8
Le Roy, 906 T-11
**LE SUEUR CO., 25426 Q-8**
Le Sueur, 4252 Q-8
Leader, 50 K-6
Leaf Valley L-4
Leavenworth, 40 R-6
Lengby, 76 H-4
Lenora, 35 T-12
Leonard, 28 G-5
Leonidas, 55 H-11
Leota, 230 S-3
Lester Prairie, 1747 P-8
Lewiston, 1480 R-12
Lewisville, 253 S-7
Lexington, 1997 *B-6
Lilydale, 615 *H-7
**LINCOLN CO., 6429 Q-3**
Lincoln, 70 L-6
Linden Grove F-10
Lindstrom, 3995 N-10
Lino Lakes, 20174 O-10
Lismore, 224 S-3
Litchfield, 6560 O-6
Little Canada, 9570 *E-8
Little Falls, 8121 L-7
Little Marais, 50 H-14
Little Rock, 30 M-7
Little Sauk, 70 M-6
Littlefork, 680 E-9
Lockhart, 30 H-2
Loman, 100 D-8
London, 50 T-10
Long Beach, 300 M-5
Long Lake, 1795 O-9
Long Pt. C-6
Long Prairie, 2809 L-6
Long Siding, 100 M-9
Longville, 178 I-7
Lonsdale, 3041 Q-9
Louisburg, 23 O-3
Lowry, 257 M-4
Lucan, 200 Q-5
Lutsen, 260 B-11
Luverne, 4425 S-3
Luxemburg, 150 N-7
Lydia, 80 P-9
Lyle, 558 T-10
Lynd, 404 Q-3
**LYON CO., 25425 Q-4**
Mabel, 709 T-13
Madelia, 2144 R-7
Madison, 1563 O-3
Magnolia, 208 T-3
**MAHNOMEN CO., 5190 H-4**
Mahnomen, 1185 H-3
Mahtomedi, 8025 *D-10
Mahtowa, 150 J-11
Makinen, 50 H-11
Mallory, 25 F-2
Malmo, 50 K-9
Manannah, 50 O-6
Manchester, 75 S-9
Manhattan Beach, 52 J-7
Mankato, 36245 R-8
Mansfield, 25 T-9
Mantorville, 1160 R-10
Maple Bay, 60 G-3
Maple Grove, 61542 *B-1
Maple Hill, 50 B-12
Maple Island, 150 S-10
Maple Lake, 2035 O-8
Maple Plain, 1983 O-8
Mapleton, 1658 S-8
Mapleview, 175 S-10
Maplewood, 36087 *E-8
Marble, 672 H-9
Marcell, 100 G-8
Margie, 40 F-8
Marietta, 151 O-2
Marine on St. Croix, 630 O-10
Marion, 100 S-11
Markham, 50 H-11
Markville, 80 L-11
**MARSHALL CO., 10155 E-2**
Marshall, 12647 Q-4
**MARTIN CO., 21802 T-6**
Marty, 50 N-7
Matawan, 50 S-8
Max, 25 G-8
Mayer, 1560 P-8
Maynard, 354 P-4
Mazeppa, 771 R-11
McGrath, 63 L-9
McGregor, 378 J-9
McIntosh, 604 G-4
McKinley, 83 H-11
**MCLEOD CO., 34898 P-7**
Meadowlands, 106 I-10
Medford, 1222 R-9
Medicine Lake, 340 *F-2
Medina, 5076 O-9
**MEEKER CO., 22644 O-7**
Meire Grove, 144 M-6
Melby, 40 L-4
Melrose, 3145 M-6
Menahga, 1196 J-5
Mendota, 202 *H-6
Mendota Hts., 11594 *I-7
Mentor, 139 G-3
Meriden, 150 R-9
Merrifield, 400 K-7
Middle River, 298 E-3
Midway, 25 J-5
Miesville, 137 Q-10
Milaca, 3003 M-8
Milan, 310 O-3
**MILLE LACS CO., 22330 L-9**
Millerville, 111 L-4
Millville, 170 R-11
Milroy, 246 Q-4
Miltona, 370 L-5
Minneapolis, 382605 O-9
Minneiska, 108 R-12
Minneota, 1366 Q-3
Minnesota City, 217 R-13
Minnesota Lake, 638 S-8
Minnetonka, 50081 *G-1
Minnetonka Mills *G-2
Mizpah, 71 F-7
Moland R-10
Money Creek, 80 S-13
Montevideo, 5233 O-4
Montgomery, 3266 Q-9
Monticello, 11780 N-8
Montrose, 2789 O-8
Moorhead, 36012 I-2
Moose Lake, 2386 K-10
Mora, 3416 M-9
Morgan, 811 Q-6
Morrill, 70 M-8
Morris, 4996 N-3
**MORRISON CO., 31712 L-8**
Morristown, 1039 R-9
Morton, 402 Q-5
Motley, 639 K-6
Mound, 9546 P-8
Mounds View, 12005 *C-6
Mountain Iron, 2921 H-11
Mountain Lake, 1928 S-6
**MOWER CO., 38603 T-10**
Murdock, 272 O-5
Murphy City, 30 G-13
**MURRAY CO., 9165 R-3**
Myrtle, 59 T-10
Nashua, 60 L-3
Nashwauk, 915 H-9
Nassau, 74 O-2
Naytahwaush, 583 H-4
Nebish, 30 G-6
Nelson, 164 M-5
Nerstrand, 226 Q-10
Nett Lake, 250 F-10
Nevis, 389 I-6
New Auburn, 502 P-7
New Brighton, 21563 *D-5
New Germany, 371 P-8
New Hope, 20428 *D-2
New London, 1319 N-6
New Market, 332 Q-9
New Munich, 344 M-6
New Prague, 6712 Q-9
New Richland, 1156 S-9
New Trier, 114 Q-10
New Ulm, 13048 Q-7
New York Mills, 1203 K-5
Newburg, 40 T-13
Newfolden, 340 E-3
Newport, 3484 *I-9
Nickerson, 40 K-11
**NICOLLET CO., 29771 Q-7**
Nicollet, 1074 R-7
Nicols *J-5
Nielsville, 83 G-2
Nimrod, 72 J-6
Nisswa, 2110 K-7
**NOBLES CO., 20832 S-3**
Nodine, 50 S-13
Norcross, 56 M-3
**NORMAN CO., 7442 H-2**
Norseland, 30 Q-7
N. Branch, 10417 N-10
N. Mankato, 12530 R-8
N. Oaks, 4689 *D-7
N. Prairie, 50 M-7
N. St. Paul, 11247 *F-9
Northcote C-1
Northfield, 19633 Q-10
Northome, 207 F-7
Northrop, 237 S-7
Norwood Young America, 3669 P-8
Noyes, 25 C-1
Oak Center, 40 Q-11
Oak Grove, 8245 N-9
Oak Pk., 130 M-8
Oak Pk. Hts., 4680 O-10
Oakdale, 27034 *F-9
Oakland, 130 S-10
Odessa, 103 N-3
Odin, 115 S-6
Ogema, 142 I-4
Ogilvie, 467 M-9
Okabena, 169 S-5
Oklee, 358 F-4
Old Frontenac, 250 Q-11
Olivia, 2362 P-5
**OLMSTED CO., 124277 R-11**
Onamia, 907 L-8
Onigum, 60 I-6
Opole, 90 M-7
Orleans, 10 C-2
Ormsby, 136 S-6
Orono, 7732 O-8
Oronoco, 949 R-11
Orr, 242 F-10
Orrock, 50 N-8
Ortonville, 1974 N-2
Osage, 300 I-5
Osakis, 1572 M-5
Oslo, 319 E-1
Oslo S-10
Osseo, 2551 *C-2
Ostrander, 199 T-11
Otisco, 100 S-9
Otsego, 13874 O-9
Ottawa, 100 Q-8
**OTTER TAIL CO., 57159 L-4**
Ottertail, 533 K-4
Outing, 250 J-8
Owatonna, 24855 R-9
Oylen, 25 J-6
Padua, 25 M-5
Palisade, 108 J-9
Palmers I-12
Palo H-11
Park Rapids, 3648 I-5
Parkers Prairie, 1008 L-5
Paynesville, 2262 N-6
Pease, 198 M-8
Pelican Rapids, 2311 J-3
Pelland D-9
Pemberton, 257 R-8
Pengilly, 850 H-9
**PENNINGTON CO., 13584 F-3**
Pennington, 90 H-7
Pennock, 491 O-5
Pequot Lakes, 1925 J-7
Perham, 2730 J-4
Perley, 107 H-2
Petersburg, 35 T-6
Peterson, 254 S-13
Pierz, 1371 L-7
Pillager, 492 K-7
Pillsbury, 45 L-6
Pilot Grove, 25 T-7
**PINE CO., 26530 L-10**
Pine Center, 80 L-8
Pine City, 3285 M-10
Pine Island, 3326 R-11
Pine River, 877 J-7
Pine Springs, 392 *E-10
Pinecreek C-4
Pinewood, 60 G-5
**PIPESTONE CO., 9895 R-2**
Pipestone, 4131 R-2
Pitt, 25 D-6
Plainview, 3225 R-12
Plato, 317 P-8
Pleasant Grove, 70 S-11
Plummer, 255 F-3
Plymouth, 71486 *E-1
Pt. Douglas, 65 P-10
**POLK CO., 31369 G-3**
Ponemah, 874 F-6
Ponsford, 200 I-5
**POPE CO., 11236 N-4**
Popple Creek M-8
Porter, 163 P-3
Potsdam, 30 R-11
Pratt, 40 R-10
Preston, 1324 S-12
Priam, 30 O-5
Princeton, 4763 N-9
Prinsburg, 449 P-5
Prior Lake, 23836 P-9
Proctor, 2839 J-11
Prosper, 50 T-12
Puposky, 35 G-6
Quamba, 107 M-9
Racine, 510 S-11
Radium, 30 E-2
Ramey, 50 M-8
**RAMSEY CO., 511035 O-10**
Ramsey, 23734 O-9
Randall, 633 L-7
Randolph, 351 Q-10
Ranier, 168 D-9
Rapidan, 150 R-8
Ray, 80 E-9
Raymond, 763 O-5
Reading, 150 S-4
Reads Landing, 250 Q-12
**RED LAKE CO., 4299 F-3**
Red Lake, 1430 F-6
Red Lake Falls, 1503 F-3
Red Wing, 15687 Q-11
Redby, 957 F-6
**REDWOOD CO., 16815 Q-5**
Redwood Falls, 5077 Q-5
Regal, 38 N-6
Remer, 357 I-8
Reno, 40 T-14
**RENVILLE CO., 17154 P-6**
Renville, 1207 P-5
Revere, 90 R-5
**RICE CO., 56665 Q-9**
Rice, 1454 M-7
Riceford, 20 T-13
Richfield, 32728 *I-4
Richmond, 1268 N-7
Richville, 123 K-4
Richwood, 100 I-4
Ridgeway, 40 S-13
Rindal, 25 H-3
Riverton, 106 K-8
Robbin, 30 D-1
Robbinsdale, 13518 *E-3
Rochert, 100 J-4
Rochester, 100413 R-11
**ROCK CO., 9721 S-3**
Rock Creek, 1502 M-10
Rock Dell, 40 S-11
Rockford, 4051 O-8
Rockville, 1253 N-7
Rogers, 7536 O-9
Rollag, 30 J-3
Rollingstone, 641 R-13
Ronneby, 32 M-8
Roosevelt, 161 C-5
Roscoe, 108 N-6
Rose City, 40 L-5
Rose Creek, 342 T-10
**ROSEAU CO., 16338 D-4**
Roseau, 2758 C-4
Roseland, 250 P-5
Rosemount, 21197 P-10
Rosen, 40 O-2
Rosendale O-6
Roseville, 32815 *E-6
Rosewood, 25 E-3
Ross, 20 C-4
Rothsay, 478 K-3
Round Lake, 403 T-4
Roy Lake, 100 H-4
Royalton, 976 M-7
Rush City, 2981 M-10
Rush Pt., 60 M-10
Rushford, 1727 S-13
Rushford Vil., 771 S-13
Rushmore, 367 T-4
Russell, 335 Q-3
Rustad, 40 J-2
Ruthton, 259 R-3
Rutledge, 195 K-10
Sabin, 441 J-2
Sacred Heart, 492 P-5
Saginaw, 100 J-11
St. Anthony, 8341 *E-5
St. Anthony, 79 M-6
St. Augusta, 3376 N-7
St. Bonifacius, 2364 P-8
St. Charles, 3577 S-12
St. Clair, 780 R-8
St. Cloud, 66948 N-7
St. Francis, 7261 N-9
St. George, 60 Q-6
St. Hilaire, 279 F-3
St. James, 4281 S-6
St. Joseph, 6114 N-7
St. Leo, 91 P-3
**ST. LOUIS CO., 200528 G-11**
St. Louis Pk., 45465 *G-3
St. Martin, 283 N-6
St. Michael, 15278 O-8
St. Nicholas, 50 N-7
St. Paul, 279590 O-10
St. Paul Pk., 5239 *I-9
St. Peter, 10971 Q-8
St. Rosa, 46 M-6
St. Stephen, 803 M-7
St. Thomas, 40 Q-8
St. Vincent, 97 C-1
St. Wendel, 50 M-7
Salem Corners, 30 S-11
Salol, 80 C-4
Sanborn, 385 R-5
Sandstone, 2792 L-10
Santiago, 75 N-8
Sargeant, 71 S-10
Sartell, 13915 M-7
Sauk Centre, 3945 M-6
Sauk Rapids, 12013 N-7
Saum, 40 F-6
Sawyer, 200 J-11
Scandia, 3933 O-10
Scanlon, 843 J-11
Schroeder, 350 G-14
**SCOTT CO., 89498 Q-9**
Seaforth, 71 Q-5
Searles, 150 R-7
Sebeka, 650 J-5
Sedan, 60 N-5
Shafer, 378 N-10
Shakopee, 33455 P-9
Sheldon, 50 S-13
Shelly, 239 H-2
Sherack F-2
Sherburn, 1019 T-6
**SHERBURNE CO., 64417 N-8**
Sheshebee, 100 J-9
Shevlin, 155 G-5
Shieldsville Q-9
Shooks, 30 F-7
Shoreview, 25111 *D-7
Shorewood, 7426 P-9
**SIBLEY CO., 15356 Q-7**
Side Lake, 180 G-10
Silica, 50 H-10
Silver Bay, 1840 H-13
Silver Creek, 170 N-8
Silver Lake, 799 P-7
Silverdale F-10
Simpson, 90 S-11
Sioux Valley, 30 T-5
Slayton, 1835 R-4
Sleepy Eye, 3423 R-6
Smiths Mill, 50 R-8
Snellman, 20 I-5
Sobieski, 181 L-7
Sogn, 30 Q-10
Solway, 73 H-5
Soudan, 900 G-11
S. Haven, 195 N-7
S. St. Paul, 19552 *H-8
Spafford T-5
Spicer, 1087 O-6
Spring Grove, 1239 T-13
Spring Hill, 49 N-6
Spring Lake, 50 G-8
Spring Lake Pk., 6458 *C-5
Spring Valley, 2355 S-11
Springfield, 2148 R-5
Squaw Lake, 94 G-7
Stacy, 1383 N-10
Stanchfield, 200 M-9
Stanton, 50 Q-10
Staples, 3019 K-6
Starbuck, 1240 M-4
Stark, 75 N-10
**STEARNS CO., 133166 N-6**
**STEELE CO., 33680 S-9**
Steen, 175 T-3
Stephen, 639 E-2
**STEVENS CO., 10053 M-3**
Stewart, 536 P-7
Stewartville, 5553 S-11
Stillwater, 17896 O-10
Stockholm, 25 O-7
Stockton, 819 R-13
Storden, 248 R-5
Strandquist, 83 D-3
Strathcona, 28 D-3
Sturgeon, 30 G-10
Sturgeon Lake, 372 K-10
Sunburg, 101 N-5
Sunfish Lake, 523 *I-7
Svea, 100 O-5
Sveadahl R-6
Swan River, 50 I-9
Swanville, 338 M-6
Swatara, 50 I-8
**SWIFT CO., 11956 O-4**
Swift, 30 C-5
Swift Falls, 70 N-5
Syre I-3
Tabor, 40 F-2
Taconite, 293 H-9
Talmoon, 60 G-8
Tamarack, 54 J-10
Taopi, 94 T-11
Taunton, 195 Q-3
Taylors Falls, 1031 N-11
Tenney, 5 L-2
Tenstrike, 207 G-6
Terrace, 40 N-5
Terrebonne, 25 F-3
Theilman, 70 R-12
Thief River Falls, 8472 F-3
Tintah, 68 L-3
**TODD CO., 24426 L-6**
Tofte, 250 C-11
Togo G-9
Toivola, 30 I-10
Tower, 477 G-11
Tracy, 2065 R-4
Trail, 60 G-4
**TRAVERSE CO., 4134 M-2**
Trimont, 686 S-6
Trommald, 129 K-8
Trosky, 110 S-3
Troy S-12
Truman, 1172 S-7
Turtle River, 80 G-6
Twig, 200 I-11
Twin Lakes, 161 T-9
Twin Valley, 783 H-3
Two Harbors, 3287 I-13
Tyler, 1122 R-3
Ulen, 560 I-3
Underwood, 337 K-4
Union Hill Q-8
Upsala, 396 M-6
Urbank, 58 L-4
Utica, 251 S-12
Vadnais Hts., 12506 *D-8
Vasa, 70 Q-10
Verdi, 70 R-2
Vergas, 293 J-4
Vermillion, 425 P-10
Verndale, 529 K-5
Vernon Center, 330 S-7
Vesta, 309 Q-4
Victoria, 6643 P-8
Viking, 87 E-3
Villard, 220 M-5
Vining, 66 K-4
Viola, 90 R-12
Virginia, 8480 G-11
**WABASHA CO., 21610 R-12**
Wabasha, 2520 Q-12
Wabasso, 639 Q-5
Waconia, 9934 P-8
Wacouta, 80 Q-11
**WADENA CO., 13713 J-5**
Wadena, 3980 K-5
Wahkon, 333 L-9
Waite Pk., 6796 N-7
Waldorf, 232 S-8
Walker, 1289 I-6
Walnut Grove, 536 R-4
Walters, 77 T-8
Waltham, 183 S-10
Wanamingo, 1044 R-10
Wanda, 95 Q-5
Wannaska, 90 D-4
Warba, 176 I-9
Ward Springs, 60 M-6
Warman, 30 L-9
Warren, 1556 E-2
Warroad, 1656 C-5
Warsaw, 250 R-9
**WASECA CO., 19526 S-9**
Waseca, 9438 R-9
**WASHINGTON CO., 201130 O-10**
Wasioja, 70 R-10
Waskish, 75 F-7
Waterford, 75 Q-10
Watertown, 4118 O-8
Waterville, 1893 R-9
Watkins, 927 N-7
**WATONWAN CO., 11876 S-6**
Watson, 192 O-4
Waubun, 394 I-4
Waverly, 1076 O-8
Wawina, 25 I-10
Wayzata, 3884 *F-1
Wealthwood, 30 K-8
Weaver, 50 R-12
Webster, 150 Q-9
Wegdahl, 100 P-4
Welch, 80 Q-10
Welcome, 678 T-6
Wells, 2346 S-8
Wendell, 170 L-3
W. Albion O-7
W. Concord, 806 R-10
W. St. Paul, 18935 *H-7
W. Union, 89 M-5
Westbrook, 678 R-4
Westbury I-4
Westport, 66 M-5
Whalan, 60 S-12
Wheaton, 1422 M-2
Wheeler's Pt., 220 C-6
Whipholt, 130 I-7
White Bear Beach, 500 *C-9
White Bear Lake, 24095 O-10
White Earth, 424 I-4
White Rock, 50 Q-10
Wilbert, 40 T-6
Wilder, 66 S-5
**WILKIN CO., 7138 K-2**
Willernie, 519 *D-10
Williams, 178 D-6
Willmar, 17774 O-5
Willow River, 391 K-10
Wilmont, 319 S-4
Wilno, 35 Q-3
Wilton, 195 H-6
Windom, 4197 S-5
Winger, 196 G-3
Winnebago, 1335 S-7
**WINONA CO., 49985 R-12**
Winona, 26785 R-13
Winsted, 2418 P-8
Winthrop, 1256 Q-7
Winton, 187 F-12
Wirock, 25 S-4
Wirt, 30 G-8
Witoka, 90 S-13
Wolf Lake, 38 J-5
Wolverton, 103 J-2
Wood Lake, 381 P-4
Woodbury, 56128 *H-10
Woodland A-9
Woodland, 469 *G-1
Woodland L-9
Woodstock, 124 R-3
Worthington, 11036 T-4
Wrenshall, 386 J-11
**WRIGHT CO., 89986 O-8**
Wright, 91 J-10
Wrightstown, 30 K-5
Wykoff, 433 S-12
Wyoming, 3803 N-10
**YELLOW MEDICINE CO., 11080 P-3**
Zemple, 74 H-8
Zerkel H-5
Zim, 100 H-11
Zumbro Falls, 172 R-11
Zumbrota, 3074 R-11

## Mississippi

Page locator

| Map keys | Atlas pages |
|---|---|
| A - N | 114 - 115 |

Abbeville, 449 C-7
Aberdeen, 6089 D-9
Ackerman, 1524 F-8
Acona, 50 F-6
**ADAMS CO., 34340 J-3**
Agricola, 200 L-9
**ALCORN CO., 34558 B-9**
Alcorn, 200 I-4
Algoma, 549 C-8
Alligator, 196 D-5
**AMITE CO., 13599 K-4**
Amory, 7262 D-9
Anding, 60 G-5
Anguilla, 763 F-4
Ansley, 270 M-7
Arcola, 484 F-4
Arkabutla, 380 B-6
Arm, 80 J-6
Arnold Line, 60 E-1
Artesia, 473 E-9
Ashland, 564 B-8
Askew, 150 B-6
**ATTALA CO., 19661 F-7**
Auburn J-5
Austin, 70 B-5
Avalon, 100 E-6
Avera, 60 K-9
Avon, 150 F-4
Bailey, 150 H-9
Baldwyn, 3362 B-9
Banner, 100 D-7
Barton, 60 A-7
Bassfield, 274 J-7
Batesville, 7880 C-6
Battles J-9
Baxterville, 320 K-7
Bay St. Louis, 8052 M-8
Bay Springs, 2198 I-8
Beatty F-7
Beaumont, 973 K-8
Beauregard, 278 J-5
Beauvoir M-3
Becker, 260 D-9
Belen, 250 C-5
Bellefontaine, 250 E-8
Belmont, 1930 B-10
Belzoni, 2432 F-5
Benoit, 556 E-4
**BENTON CO., 8026 B-8**
Benton, 400 G-6
Bentonia, 482 G-5
Benwood, 60 D-7
Berclair E-5
Bethany, 30 C-9
Bethlehem, 120 B-8
Beulah, 435 D-4
Bexley K-9
Big Creek, 121 D-7
Big Pt., 115 L-9
Bigbee Valley, 80 F-10
Biggersville, 300 B-9
Biloxi, 45670 M-9
Black Hawk, 100 E-6
Blackwater, 80 G-9

## Missouri

Page locator

| Map keys | Atlas pages |
|---|---|
| 1 - 10 | 116 - 117 |
| 11 - 20 | 118 - 119 |

* City keyed to pp. 120 - 121
~ Independent city not included in any county



*, †, ‡, §, ◊ See explanation under state title in this index. County names are listed in capital letters and in boldface type.

## Montana

Page locator

| Map keys | Atlas pages |
|---|---|
| 1 - 10 | 122 - 123 |
| 11 - 20 | 124 - 125 |



## Nebraska

Page locator

| Map keys | Atlas pages |
|---|---|
| 1 - 10 | 126 - 127 |
| 11-20 | 128 - 129 |



*, †, ‡, §, ◊ See explanation under state title in this index.

County names are listed in capital letters and in boldface type.

## Nevada

Page locator
Map keys A - N — Atlas pages 130 - 131
* City keyed to p. 30
† City keyed to pp. 132 - 133
~ Independent city not included in any county



## New Hampshire

Page locator
Map keys A - N — Atlas pages 132 - 133



## New Jersey

Page locator
Map keys A - J — Atlas pages 134 - 135
Map keys K - T — Atlas pages 136 - 137
* City keyed to p. 50
‡ City keyed to pp. 146 - 147
§ City keyed to p. 148
◊ City keyed to p. 182



## New Mexico

Page locator
Map keys A - N — Atlas pages 138 - 139



*, †, ‡, §, ◊ See explanation under state title in this index.

County names are listed in capital letters and in boldface type.

## New York

Page locator

| Map keys | Atlas pages |
|---|---|
| SA1 - SJ14 | 140 - 141 |
| NA1 - NN10 | 142 - 143 |
| NA11 - NN20 | 144 - 145 |

* City keyed to pp. 146 - 147
† City keyed to pp. 148 - 149

## North Carolina

Page locator

| Map keys | Atlas pages |
|---|---|
| 1 - 10 | 150 - 151 |
| 11 - 20 | 152 - 153 |

* City keyed to pp. 154 - 155

## North Dakota

Page locator

| Map keys | Atlas pages |
|---|---|
| A - J | 156 - 157 |



## Ohio

Page locator

| Map keys | Atlas pages |
|---|---|
| NA1 - NN10 | 158 - 159 |
| NA11 - NN20 | 160 - 161 |
| SA1 - SN10 | 162 - 163 |
| SA11 - SN20 | 164 - 165 |

* City keyed to p. 226



*, †, ‡, §, ◊ See explanation under state title in this index.

County names are listed in capital letters and in boldface type.

Darbyville, 298 ... SC-9
**DARKE CO., 53309 ... SB-2**
Darlington, 35 ... SB-14
Darrtown, 400 ... SD-2
Dart, 30 ... SD-17
Darwin ... SF-14
Davisville, 45 ... SG-11
Dawn, 170 ... NM-2
Dayton, 154200 ... SC-4
De Graff, 1148 ... NM-5
Deavertown, 200 ... SC-14
Decatur, 250 ... SH-6
Deer Pk., 5751 ... SK-4
Deerfield, 520 ... NH-18
Deering, 90 ... SJ-11
Deersville, 82 ... NM-17
**DEFIANCE CO., 39500 ... NG-2**
Defiance, 16063 ... NG-3
DeForest, 1150 ... NB-12
**DELAWARE CO., 109989 ... NM-10**
Delaware, 33719 ... NM-9
Delhi Hills, 20000 ... SM-2
Delightful, 40 ... NG-18
Delisle, 50 ... SA-2
Dellroy, 285 ... NK-17
Delmont ... SC-11
Delphi, 100 ... NH-11
Delphos, 6730 ... NI-3
Delta, 2900 ... NE-4
Denmark ... NK-9
Dennison, 2878 ... NL-16
Dent, 7612 ... SL-1
Denver, 25 ... SF-9
Derby, 450 ... SC-8
Derwent, 250 ... SB-16
Deshler, 1797 ... NG-5
Deunquat, 50 ... NI-8
Devola, 2771 ... SD-16
Deweyville, 30 ... NH-5
Dexter, 70 ... SG-13
Dexter City, 160 ... SC-16
Dialton, 60 ... SA-5
Diamond, 200 ... NH-18
Dilles Bottom, 130 ... SB-19
Dillonvale, 721 ... NM-19
Dixie Hts. ... SK-8
Dixon, 70 ... NI-1
Doanville, 160 ... SE-13
Dobbston ... SJ-12
Dodds, 80 ... SD-4
Dodgeville, 110 ... NE-19
Dodson, 25 ... SB-3
Dodsonville, 70 ... SF-5
Dola, 200 ... NJ-6
Donnelsville, 282 ... SB-5
Dorcas ... SG-14
Dorset, 200 ... ND-19
Dover, 12466 ... NK-16
Doylestown, 2898 ... NI-15
Drakes, 70 ... SD-13
Drakesburg, 55 ... NG-17
Dresden, 1415 ... NN-13
Drexel, 2057 ... SC-3
Dry Run, 750 ... SI-9
Dublin, 38536 ... NN-8
Duffy, 230 ... SC-19
Dumontville ... SC-11
Dunbridge, 250 ... NF-6
Duncan Falls, 600 ... SB-14
Dundas, 220 ... SF-12
Dundee, 250 ... NK-15
Dunkinsville, 50 ... SH-7
Dunkirk, 940 ... NJ-6
Dunlap, 250 ... SF-2
Dupont, 258 ... NH-3
Durbin, 25 ... NB-15
Duvall, 50 ... SC-9
Dyesville ... SG-13
Eagle City, 70 ... SB-5
Eagleport, 35 ... SC-14
Eagleville, 35 ... NG-6
E. Akron ... NC-6
E. Alliance, 1175 ... NI-17
E. Batavia Hts., 150 ... SG-4
E. Cadiz ... NM-18
E. Canton, 1602 ... NJ-17
E. Claridon, 260 ... NE-17
E. Cleveland, 24507 ... NE-15
E. Danville, 100 ... SG-6
E. Fairfield, 100 ... NJ-19
E. Fultonham, 460 ... SB-13
E. Greenville, 180 ... NJ-15
E. Liberty, 500 ... NM-6
E. Liverpool, 11995 ... NK-20
E. Monroe, 120 ... SE-7
E. Norwalk, 80 ... NG-11
E. Palestine, 4690 ... NI-20
E. Richland, 300 ... SA-18
E. Rochester, 400 ... NJ-18
E. Sparta, 775 ... NJ-16
E. Springfield, 350 ... NL-19
E. Townsend, 80 ... NG-11
E. Trumbull, 80 ... NF-18
E. Union, 30 ... SC-16
Eastlake, 19520 ... NE-15
Easton, 200 ... NI-15
Eastview ... SL-9
Eaton, 8005 ... SC-1
Eaton Estates, 1409 ... NG-13
Eber, 30 ... SD-7
Eckmansville ... SH-6
Eden Pk., 430 ... SN-9
Edenton, 300 ... SF-4
Edgerton, 1961 ... NF-1
Edgewater Pk., 700 ... SB-10
Edgewood, 4762 ... NC-19
Edinburg, 250 ... NH-17
Edison, 451 ... NK-10
Edon, 340 ... NE-1
Elba, 90 ... SD-16
Eldorado, 502 ... SB-2
Elery, 25 ... NG-4
Elgin, 48 ... NJ-2
Elida, 1899 ... NJ-4
Elizabethtown, 500 ... SF-1
Elkton, 80 ... NI-19
Ellerton ... SN-6
Elliott Crossroads ... SD-14
Ellisonville, 50 ... SI-11
Ellsberry, 60 ... SH-6
Ellsworth, 260 ... NH-18
Elm Grove, 35 ... SG-8
Elmore, 1365 ... NF-8
Elmwood Place, 2481 ... SL-3
Elroy, 35 ... NM-1
Elyria, 54979 ... NF-13
Empire, 276 ... NK-20
Englewood, 12686 ... SB-3
Enon, 2542 ... SB-5
Enterprise, 170 ... SD-12
Epworth Hts., 150 ... SK-6
Era, 130 ... SC-8
**ERIE CO., 79551 ... NF-11**
Eris, 25 ... NN-5
Erlin, 80 ... NF-9
Espyville, 80 ... NK-8
Essex, 90 ... NL-7
Etna, 500 ... SB-11
Euclid, 47415 ... NE-15
Eureka, 100 ... SI-13
Evansport, 300 ... NF-3
Evansville, 200 ... NC-12
Evendale, 2955 ... SK-4
Everett, 25 ... NG-15
Ewing, 30 ... SE-12
Ewington, 65 ... SG-12
Excello, 50 ... SD-3
Fairborn, 32366 ... SB-5
Fairfax, 40 ... SG-7
Fairfax, 1855 ... SL-4
**FAIRFIELD CO., 122759 ... SB-11**
Fairfield, 50 ... NL-19
Fairfield, 42384 ... SE-2
Fairhaven, 200 ... SC-1
Fairhope, 1720 ... NJ-17
Fairlawn, 7030 ... NH-15
Fairpoint, 500 ... NN-18
Fairport Harbor, 3211 ... ND-17
Fairview, 80 ... SB-16
Fairview, 50 ... SF-6
Fairview Lanes, 1015 ... NF-10
Fairview Pk., 15762 ... NF-14
Fallsburg, 80 ... NM-12
Farmdale, 200 ... NF-20
Farmer, 120 ... NF-2
Farmers, 45 ... SE-6
Farmerstown, 330 ... NL-15
Farmersville, 937 ... SC-3
Fawcett ... SH-7
**FAYETTE CO., 28433 ... SD-8**
Fayette, 1287 ... NE-3
Fayetteville, 367 ... SF-5
Feesburg, 500 ... SH-5
Felicity, 930 ... SH-4
Fernald, 25 ... SJ-1
Fernell Hts., 330 ... SM-6
Fincastle, 85 ... SG-6
Findlay, 36987 ... NH-6
Finneytown, 13492 ... SK-3
Firestone Pk. ... NC-5
Fitchville, 350 ... NH-11
Five Points, 120 ... SC-8
Five Points, 2191 ... SD-4
Fivemile ... SG-5
Flat Rock, 300 ... NG-9
Fleming, 40 ... SE-16
Fletcher, 504 ... NN-4
Florence, 200 ... NG-12
Florence, 150 ... NN-19
Florence, 100 ... SC-16
Florida, 236 ... NG-4
Flushing, 864 ... NN-18
Fly, 80 ... SD-18
Footville ... NE-18
Foraker, 130 ... NJ-6
Forest, 1449 ... NJ-7
Forest Hill ... NB-6
Forest Hills, 200 ... NA-15
Forest Pk., 18514 ... SF-2
Forestville, 10978 ... SG-3
Ft. Jefferson, 80 ... SA-2
Ft. Jennings, 419 ... NI-3
Ft. Loramie, 1463 ... NL-3
Ft. McKinley, 3989 ... SC-4
Ft. Recovery, 1332 ... NL-1
Ft. Seneca, 230 ... NG-8
Ft. Shawnee, 3704 ... NJ-4
Foster ... SE-3
Fosterville ... ND-13
Fostoria, 13000 ... NH-7
Fowler, 250 ... NG-19
Fowlers Mill, 320 ... NE-17
Fox, 25 ... SD-9
Frankfort, 1043 ... SE-8
**FRANKLIN CO., 1068978 ... SB-8**
Franklin, 12941 ... SD-3
Franklin Furnace, 1537 ... SI-10
Franklin Square, 240 ... NI-19
Frazeysburg, 1291 ... NN-13
Frederick, 150 ... SB-3
Fredericksburg, 475 ... NJ-14
Fredericksdale ... SC-16
Fredericktown, 200 ... NJ-20
Fredericktown, 2484 ... NL-11
Freeburg, 90 ... NI-17
Freedom, 90 ... NG-17
Freeport, 394 ... NM-17
Fremont, 16655 ... NG-8
Frenchtown, 60 ... NM-2
Fresno, 180 ... NL-15
Friendship, 500 ... SI-9
Frost, 25 ... SF-15
Fruit Hill, 3945 ... SM-5
Fruitdale ... [illegible]
Fryburg, 150 ... NL-4
Fryburg ... NI-14
Fulda, 40 ... SC-16
**FULTON CO., 42084 ... NE-3**
Fulton, 270 ... NL-10
Fultonham, 156 ... SB-13
Funk, 110 ... NJ-13
Gahanna, 33892 ... SA-10
Galena, 548 ... NM-9
Galion, 10737 ... NI-10
**GALLIA CO., 31069 ... SH-12**
Gallia ... SH-11
Gallipolis, 4190 ... SH-13
Galloway, 250 ... SB-9
Gambier, 2061 ... NL-12
Ganges, 120 ... NI-11
Gano, 200 ... SF-3
Garden Acres, 250 ... NB-17
Garden City, 300 ... SH-14
Garfield, 80 ... NI-18
Garfield Hts., 27734 ... NF-15
Garrettsville, 2206 ... NG-17
Gates Mills, 2276 ... NE-16
**GEAUGA CO., 90895 ... NE-17**
Geneva, 6291 ... ND-18
Geneva-on-the-Lake, 1486 ... NC-18
Genoa, 2279 ... NF-7
Georges Run, 400 ... NM-20
Georgesville, 160 ... SB-8
Georgetown, 150 ... NM-18
Georgetown, 3430 ... SH-5
Gephart ... [illegible]
Gerald, 50 ... NF-4
Germano, 150 ... NL-18
Germantown, 35 ... SD-17
Germantown, 5018 ... SD-3
Getaway, 100 ... SJ-12
Gettysburg, 525 ... NN-2
Gettysburg, 80 ... SB-1
Geyer, 50 ... NK-4
Ghent, 1600 ... NH-15
Gibisonville, 130 ... SD-11
Gibsonburg, 2435 ... NF-7
Gilboa, 168 ... NH-5
Gillivan ... SA-8
Gilmore, 85 ... NM-16
Ginghamsburg, 250 ... SB-4
Girard, 10069 ... NH-19
Girton, 50 ... NG-7
Givens, 75 ... SG-10
Gladstone, 30 ... SC-6
Glandorf, 955 ... NH-4
Glasgow, 40 ... NJ-19
Glasgow ... NM-15
Glen Este, 350 ... SM-6
Glen Karn, 40 ... SA-1
Glen Roy, 100 ... SG-11
Glencoe, 500 ... SA-19
Glendale, 2214 ... SJ-3
Glenford, 195 ... SB-12
Glenmont, 294 ... NK-13
Glenmoor, 2192 ... NJ-20
Glenmore, 50 ... NJ-1
Glouster, 2023 ... SD-13
Gnadenhutten, 1283 ... NL-16
Goes, 100 ... SC-5
Golf Manor, 3765 ... SL-4
Gomer, 350 ... NI-4
Good Hope, 220 ... SE-8
Goodyear Hts. ... NB-6
Gordon, 183 ... SB-2
Gore, 240 ... SD-12
Goshen, 200 ... NL-16
Goshen, 730 ... SF-4
Gould Pk. ... SG-20
Grafton, 6716 ... NG-13
Grand Rapids, 1015 ... NF-5
Grand River, 366 ... ND-17
Grandview, 60 ... SD-18
Grandview Hts., 170 ... NM-4
Grandview Hts., 6306 ... SB-9
Grange Hall, 20 ... SD-8
Granger, 90 ... NH-15
Grant, 60 ... NJ-6
Granville, 5382 ... SA-11
Grape Grove, 35 ... SC-6
Gratiot, 190 ... SB-13
Gratis, 878 ... SC-2
Graysville, 109 ... SC-18
Graytown, 180 ... NE-8
Green, 23404 ... NI-16
Green Camp, 313 ... NK-8
Green Springs, 1200 ... NG-9
Greenbush ... SD-2
Greenbush ... SG-5
Greencastle ... SC-10
Greendale, 30 ... SD-13
**GREENE CO., 147886 ... SC-6**
Greenfield, 5063 ... SE-7
Greenford, 700 ... NI-19
Greenhills, 3837 ... SJ-3
Greenland ... SE-9
Greensburg, 600 ... NI-16
Greentown, 3154 ... NI-16
Greenview ... SK-7
Greenville, 12983 ... NN-2
Greenwich, 1503 ... NH-11
Greenwood, 40 ... SB-16
Greer, 50 ... NK-13
Grelton, 110 ... NG-4
Groesbeck, 7202 ... SK-2
Grove City, 33829 ... SB-9
Groveport, 5240 ... SB-10
Grover Hill, 368 ... NH-2
Guerne, 250 ... NJ-14
**GUERNSEY CO., 40792 ... SA-17**
Guernsey, 50 ... NM-15
Guilford, 320 ... NJ-18
Gurneyville ... SD-5
Gustavus ... NF-19
Gutman, 35 ... NK-4
Guysville, 250 ... SF-14
Gypsum, 200 ... NF-9
Hackney, 40 ... SD-15
Hallock, 100 ... NE-2
Hallsville, 230 ... SE-10
Hambden, 180 ... NE-17
Hamburg ... SC-11
Hamden, 929 ... SF-11
Hamersville, 488 ... SH-5
**HAMILTON CO., 845303 ... SF-1**
Hamilton, 62477 ... SE-2
Hamilton Meadows, 2100 ... SB-9
Hamler, 645 ... NG-4
Hamlet, 160 ... SG-4
Hammansburg, 100 ... NG-6
Hammondsville, 600 ... NK-19
**HANCOCK CO., 71295 ... NI-5**
Hanging Rock, 289 ... SJ-10
Hanley Vil. ... NJ-11
Hannibal, 500 ... SC-19
Hanover ... NL-18
Hanover, 986 ... SA-13
Hanoverton, 377 ... NJ-18
Harbor View, 99 ... ND-7
**HARDIN CO., 31945 ... NJ-6**
Hardin, 80 ... NM-3
Harlem, 150 ... NN-10
Harlem Springs, 300 ... NK-18
Harmon, 50 ... NJ-15
Harmony, 250 ... SB-6
Harper, 70 ... NL-6
Harpersfield ... ND-18
Harpster, 191 ... NJ-8
Harriett, 30 ... SG-7
Harriettsville, 60 ... SD-17
Harrisburg, 100 ... NI-17
Harrisburg, 316 ... SC-8
**HARRISON CO., 15856 ... NM-17**
Harrison, 9387 ... SF-1
Harrisonville, 150 ... SG-13
Harrisville, 256 ... NM-18
Harrod, 470 ... NJ-5
Harshasville ... SH-7
Hartford, 350 ... NG-20
Hartland ... NH-11
Hartsgrove, 300 ... NE-18
Hartshorn ... SC-17
Hartville, 2553 ... NI-16
Harveysburg, 631 ... SD-5
Haselton ... ND-14
Haskins, 641 ... NF-6
Hatton, 50 ... NG-7
Havana, 110 ... NH-10
Havensport, 60 ... SC-11
Haverhill, 120 ... SJ-10
Haviland, 165 ... NH-2
Haydenville, 450 ... SD-12
Hayesville, 348 ... NJ-12
Heath, 8903 ... SA-12
Hebbardsville, 50 ... SF-13
Hebron, 2147 ... SB-12
Hecla, 200 ... SJ-11
Helena, 230 ... NG-8
Hemlock, 145 ... SD-13
Hemlock Grove, 70 ... SG-14
Hendrysburg, 150 ... SA-17
**HENRY CO., 29210 ... NG-4**
Hepburn, 100 ... NK-7
Hessville, 125 ... NF-8
Hicksville, 3408 ... NG-1
Higginsport, 276 ... SI-5
**HIGHLAND CO., 40875 ... SF-6**
Highland, 290 ... SE-6
Highland Hts., 8533 ... SK-20
Highland Hills, 1389 ... SM-19
Highland Pk., 700 ... NJ-15
Highlandtown, 100 ... NK-19
Highpoint, 600 ... SJ-5
Hill Grove, 100 ... NM-1
Hilliard, 27901 ... SA-8
Hills and Dales, 263 ... NB-9
Hillsboro, 6661 ... SF-6
Hinckley, 500 ... NG-14
Hiram, 1192 ... NG-17
Hiram Rapids, 250 ... NF-17
Hiramsburg ... SC-16
Hitchcock ... SH-11
Hoagland, 50 ... SF-6
**HOCKING CO., 28241 ... SE-11**
Hockingport, 300 ... SF-15
Holgate, 1113 ... NG-4
Holiday City, 72 ... NE-2
Holland, 1327 ... NB-1
Hollansburg, 206 ... SA-1
Hollister, 300 ... SD-13
Holloway, 326 ... NM-17
**HOLMES CO., 38943 ... NK-13**
Holmesville, 409 ... NK-14
Homer, 250 ... NM-11
Homerville, 80 ... NH-13
Homeworth, 400 ... NI-18
Honeytown, 40 ... NJ-14
Hooker ... SC-11
Hooven, 500 ... SF-1
Hopedale, 962 ... NM-18
Hopetown, 50 ... SE-9
Hopewell, 250 ... SB-13
Hopkinsville ... SE-4
Hoskinsville ... SC-16
Houcktown, 180 ... NI-6
Houston, 220 ... NM-3
Howard, 300 ... NL-12
Howenstein, 160 ... NJ-16
Howland Center, 6481 ... NA-12
Hoytville, 296 ... NG-5
Hubbard, 7688 ... NH-20
Huber Hts., 37148 ... SB-4
Huber Ridge, 4883 ... SF-19
Hudson, 23037 ... NG-16
Hulington, 100 ... SG-4
Hume, 135 ... NK-3
Hunter ... SB-18
Hunting Valley, 701 ... NF-16
Huntington, 220 ... NH-12
Huntsburg, 300 ... NE-18
Huntsville, 421 ... NL-5
**HURON CO., 59487 ... NH-11**
Huron, 7348 ... NF-11
Hustead, 120 ... SB-5
Hyatts, 50 ... NM-8
Iberia, 350 ... NK-10
Idaho, 50 ... SG-9
Iler ... NG-8
Ilesboro ... SE-12
Independence, 6770 ... NF-15
Independence, 100 ... NG-3
Ingomar, 60 ... SC-2
Irondale, 396 ... NK-19
Ironspot, 150 ... SB-13
Ironton, 11307 ... SJ-11
Irvington ... SK-7
Irwin, 150 ... NN-7
Island View, 300 ... NK-5
Isle St. George, 50 ... ND-10
Isleta ... NM-15
Ithaca, 98 ... SB-2
**JACKSON CO., 32641 ... SH-11**
Jackson, 6138 ... SG-11
Jackson Center, 1443 ... NL-4
Jacksonburg, 69 ... SD-2
Jacksontown, 350 ... SB-12
Jacksonville, 560 ... SE-13
Jacksonville, 75 ... SH-7
Jacobsburg, 200 ... SB-19
Jamestown, 1861 ... SC-6
Jasper, 250 ... SG-9
Jasper Mills, 40 ... SD-7
Jaysville, 30 ... SA-2
**JEFFERSON CO., 73894 ... NK-19**
Jefferson, 3416 ... ND-19
Jefferson, 200 ... NJ-13
Jeffersonville, 1229 ... SC-7
Jelloway, 90 ... NK-12
Jenera, 214 ... NI-6
Jericho, 200 ... SE-3
Jerome, 200 ... NN-8
Jeromesville, 483 ... NJ-13
Jerry City, 469 ... NG-6
Jersey, 250 ... SA-10
Jerusalem, 146 ... SB-18
Jewell, 160 ... NG-3
Jewett, 767 ... NL-18
Johnston, 250 ... NF-19
Johnstown, 4016 ... NN-10
Jonestown, 80 ... NJ-2
Joy ... SE-14
Junction, 100 ... NG-2
Junction City, 841 ... SC-12
Justus, 100 ... NJ-15
Kalida, 1254 ... NI-4
Kanauga, 100 ... SH-13
Kansas, 400 ... NG-8
Keene, 100 ... NL-14
Kelleys Island, 386 ... NE-10
Kelloggsville, 200 ... NC-19
Kemp, 70 ... NJ-3
Kendall Hts., 850 ... NB-8
Kennard, 40 ... NM-6
Kennonsburg, 50 ... SB-17
Kensington, 320 ... NJ-18
Kent, 27983 ... NH-16
Kenton, 8052 ... NK-6
Kenwood, 7423 ... SK-4
Kerr, 90 ... SH-13
Kessler, 100 ... SB-3
Kettering, 53708 ... SC-4
Kettlersville, 166 ... NL-3
Key ... SA-19
Kidron, 550 ... NJ-15
Kieferville, 40 ... NH-4
Kilbourne, 200 ... NM-9
Kileville, 25 ... NN-8
Kilgore, 100 ... NL-18
Killbuck, 887 ... NL-13
Kimball, 110 ... NG-10
Kimberly, 40 ... SE-13
Kimbolton, 190 ... NN-15
Kingman ... SD-5
Kings Creek, 300 ... NN-6
Kings Mills, 500 ... SE-3
Kingston, 1037 ... SE-10
Kingsville, 650 ... NC-19
Kingsway ... NF-8
Kinnikinnick, 65 ... SE-9
Kinsman, 600 ... NF-20
Kiousville ... SC-8
Kipling, 180 ... SA-16
Kipton, 245 ... NG-12
Kirby, 123 ... NJ-7
Kirkersville, 544 ... SB-11
Kirkpatrick, 80 ... NJ-9
Kirkwood, 60 ... NM-4
Kirtland, 7362 ... NE-16
Kitts Hill, 200 ... SJ-11
Klondike, 250 ... NG-19
Knockemstiff ... SF-9
**KNOX CO., 54500 ... NM-12**
Knoxville, 180 ... NK-19
Kossuth, 80 ... NK-3
Kunkle, 300 ... NE-2
Kyger, 60 ... SG-13
La Croft, 1307 ... NK-20
La Grange, 70 ... SJ-11
La Rue, 703 ... NK-7
Lacarne, 200 ... NF-9
Lafayette, 120 ... NH-14
Lafayette, 292 ... NJ-5
Lafayette, 130 ... SB-7
Lafferty, 500 ... NN-18
Lagrange, 2012 ... NG-13
Laings, 30 ... SC-18
**LAKE CO., 227511 ... NE-17**
Lake Cable, 2500 ... NB-9
Lake Fork ... NJ-13
Lake Milton, 1100 ... NH-18
Lake O'Springs, 600 ... NB-8
Lakeline, 162 ... NE-16
Lakemore, 2857 ... NH-16
Lakeside, 950 ... NE-10
Lakeside, 200 ... SB-11
Lakeview, 1065 ... NL-5
Lakeville, 130 ... NK-13
Lakewood, 50704 ... NF-14
Lancaster, 36976 ... SC-11
Landeck, 100 ... NJ-3
Langsville, 80 ... SG-13
Lansing, 400 ... SA-19
LaPorte, 400 ... NG-13
Latham, 230 ... SG-8
Lattasburg, 100 ... NI-13
Lattaville, 60 ... SE-8
Latty, 184 ... NH-2
Laura, 497 ... SA-3
Laurel, 150 ... SH-4
Laurel Ridge, 2350 ... NI-16
Laurelville, 547 ... SE-10
**LAWRENCE CO., 62319 ... SI-11**
Lawrenceville, 302 ... SA-5
Lawshe, 60 ... SH-7
Layhigh, 60 ... SE-1
Layland, 50 ... NL-14
Layman ... SE-15
Leavittsburg, 2200 ... NG-18
Leavittsville, 100 ... NK-17
Lebanon, 50 ... SD-17
Lebanon, 20535 ... SE-4
Lees Creek, 160 ... SE-6
Leesburg, 1304 ... SE-7
Leesville, 200 ... NJ-10
Leesville, 178 ... NL-17
Leetonia, 1996 ... NI-19
Leipsic, 2173 ... NH-5
Leistville, 40 ... SD-10
Lemert, 30 ... NI-8
Lemoyne, 300 ... NF-7
Lena, 100 ... NN-4
Lenox Center, 60 ... ND-19
Leo ... SF-11
Leonardsburg ... NL-9
Letart Falls, 180 ... SH-14
Levanna, 80 ... SI-5
Lewis Center, 350 ... NM-9
Lewisburg, 1726 ... SB-2
Lewistown, 260 ... NL-5
Lewisville, 224 ... SC-17
Lexington, 4092 ... NJ-11
Liberty, 240 ... SC-3
Liberty Center, 1080 ... NF-4
**LICKING CO., 145491 ... SA-12**
Lightsville, 30 ... NM-1
Lilly Chapel, 200 ... SB-8
Lima, 37829 ... NJ-4
Limaville, 183 ... NI-17
Lime City, 120 ... ND-2
Limecrest, 400 ... SB-6
Limestone, 50 ... NE-8
Limestone City, 40 ... NB-15
Lincoln Hts., 3911 ... SK-3
Lincoln Vil., 9482 ... SI-17
Lindale, 200 ... SG-4
Lindsey, 473 ... NF-8
Linndale, 89 ... SM-16
Linnville, 60 ... SB-12
Linnville, 90 ... SJ-12
Linworth ... SF-18
Lisbon, 2985 ... NJ-19
Litchfield, 400 ... NH-13
Lithopolis, 1070 ... SC-10
Little Hocking, 600 ... SF-15
Little Sandusky, 50 ... NJ-8
Little Walnut, 80 ... SC-9
Little Washington ... NJ-11
Little York, 400 ... SK-7
Lloydsville, 150 ... SA-18
Lock ... NM-11
Lockbourne, 268 ... SC-9
Lockington, 198 ... NM-3
Lockland, 3473 ... SK-3
Lockville, 90 ... SC-10
Lockwood, 30 ... NF-18
Locust Grove, 220 ... SG-7
Locust Ridge ... SG-5
Lodi, 3375 ... NH-13
**LOGAN CO., 46005 ... NL-5**
Logan, 7436 ... SD-12
Logan Elm Vil., 1062 ... SD-9
Logansville, 55 ... NL-5
Lombardsville, 40 ... SH-9
London, 9587 ... SB-7
Londonderry, 40 ... NN-17
Londonderry, 400 ... SF-10
Long Bottom, 70 ... SG-15
**LORAIN CO., 284664 ... NG-12**
Lorain, 70239 ... NF-13
Lordstown, 3522 ... NH-18
Lore City, 295 ... SA-16
Lottridge ... SF-14
Louden ... SG-7
Loudonville, 3069 ... NK-12
Louisville, 9459 ... NI-17
Loveland, 11734 ... SF-3
Loveland Pk., 1799 ... SJ-6
Lowell, 593 ... SD-16
Lowellville, 1146 ... NH-20
Lower Salem, 102 ... SD-17
**LUCAS CO., 455054 ... NE-5**
Lucas, 589 ... NJ-12
Lucasburg ... SB-16
Lucasville, 1588 ... SH-9
Luckey, 989 ... NF-7
Ludlow Falls, 210 ... SA-3
Lumberton, 150 ... SD-5
Luray, 110 ... SB-11
Lykens, 70 ... NI-9
Lynchburg, 1393 ... SF-6
Lyndhurst, 13848 ... NE-16
Lyndon, 200 ... SE-8
Lynx, 220 ... SI-7
Lyons, 586 ... ND-4
Lyra, 40 ... SI-10
Lytle, 200 ... SD-4
Macedon, 70 ... NL-1
Macedonia, 10435 ... NG-16
Mack, 2000 ... SL-1
Macksburg, 200 ... SD-16
Macon, 140 ... SH-6
Madeira, 8558 ... SF-3
**MADISON CO., 40213 ... SC-7**
Madison, 3104 ... ND-18
Madison Mills, 60 ... SC-8
Madison-on-the-Lake, 950 ... ND-18
Magnetic Springs, 314 ... NL-8
Magnolia, 924 ... NK-17
**MAHONING CO., 257555 ... NI-18**
Maineville, 1033 ... SE-4
Mainsville, 30 ... SC-13
Malaga, 180 ... SB-18
Malinta, 271 ... NG-4
Mallet Creek, 300 ... NH-14
Malta, 651 ... SC-14
Malvern, 1193 ... NJ-17
Manchester, 4000 ... NI-15
Manchester, 2096 ... SI-6
Mandale, 50 ... NH-3
Mansfield, 49579 ... NJ-11
Mantua, 1017 ... NG-17
Mantua Corners, 80 ... NG-17
Maple Hts., 23667 ... NF-15
Maple Valley ... NB-5
Mapleton, 90 ... NJ-17
Maplewood, 230 ... NL-4
Marathon, 300 ... SF-5
Marble Cliff, 650 ... SH-18
Marblehead, 831 ... NE-10
Marchand ... NA-9
Marcy ... SC-10
Marengo, 304 ... NL-10
Maria Stein, 300 ... NL-2
Mariemont, 3183 ... SL-4
Marietta, 14307 ... SE-16
**MARION CO., 66217 ... NK-7**
Marion, 35841 ... NK-8
Mark Center, 200 ... NG-2
Marlboro, 500 ... NI-17
Marne, 400 ... SA-12
Marseilles, 117 ... NJ-7
Marshall, 110 ... SF-7
Marshallville, 810 ... NI-15
Martel, 250 ... NK-9
Martins Ferry, 6654 ... NN-19
Martinsburg, 186 ... NM-12
Martinsville, 434 ... SE-5
Marysville, 18212 ... NM-7
Mason, 29682 ... SE-3
Massieville, 400 ... SF-9
Massillon, 32613 ... NJ-16
Masury, 2618 ... NG-20
Matville ... SC-9
Maud, 500 ... SE-3
Maumee, 13856 ... NC-1
Maximo, 200 ... NI-17
Maxville, 100 ... SD-12
May Hill, 50 ... SG-7
Mayfield, 3121 ... SK-20
Mayfield Hts., 17737 ... NE-16
Mayflower Vil. ... NC-7
Maynard, 400 ... NN-19
Maysville, 100 ... NJ-14
Maysville, 50 ... NJ-5
McArthur, 2022 ... SF-12
McCartyville, 160 ... NL-3
McClimansville, 60 ... SC-8
McClure, 713 ... NF-5
McComb, 1599 ... NH-5
McConnelsville, 1562 ... SC-14
McCuneville, 90 ... SD-13
McCutchenville, 300 ... NI-8
McDermott, 600 ... SH-9
McDonald, 3214 ... NH-19
McDonaldsville, 180 ... NI-16
McGill, 20 ... NH-1
McGonigle, 130 ... SE-2
McGuffey, 517 ... NJ-5
McKinley Hts., 700 ... NB-13
McLuney, 75 ... SC-13
McZena, 40 ... NJ-13
Mecca, 60 ... NF-19
Mechanicsburg, 1699 ... SA-7
Mechanicstown, 280 ... NK-18
**MEDINA CO., 151095 ... NH-13**
Medina, 26011 ... NH-14
Medway, 2000 ... SB-5
Meeker, 150 ... NK-8
**MEIGS CO., 23072 ... SG-13**
Meigs, 40 ... SC-15
Melbern, 200 ... NF-2
Melmore, 220 ... NH-8
Melrose, 298 ... NH-3
Melvin, 90 ... SE-6
Memphis ... SE-6
Mendon, 683 ... NJ-2
Mentor, 51825 ... NE-16
Mentor-on-the-Lake, 8288 ... ND-16
**MERCER CO., 40924 ... NK-1**
Mercer, 50 ... NK-2
Mercerville, 100 ... SI-13
Mermill ... NG-6
Mesopotamia, 300 ... NF-18
Metamora, 591 ... ND-5
Metzger, 200 ... SE-9
Mexico, 60 ... NI-8
Meyers Lake, 556 ... NC-9
**MIAMI CO., 98868 ... SA-4**
Miami Shores ... SN-7
Miami Villa ... SK-8
Miamisburg, 19793 ... SD-3
Miamitown, 730 ... SF-1
Miamiville, 300 ... SK-5
Middle Bass, 100 ... NE-10
Middle Pt., 564 ... NI-2
Middleboro, 50 ... SE-4
Middlebourne, 80 ... SA-17
Middlebranch, 220 ... NI-16
Middleburg, 220 ... NM-6
Middleburg, 50 ... SC-17
Middleburg Hts., 14928 ... NF-14
Middlebury, 30 ... NI-1
Middlefield, 2378 ... NF-17
Middleport, 2469 ... SG-14
Middleton, 80 ... NI-19
Middletown, 25 ... NM-6
Middletown, 51422 ... SD-3
Midland, 262 ... SF-5
Midvale, 583 ... NL-16
Midway, 262 ... SC-7
Mifflin, 144 ... NJ-12
Milan, 1345 ... NG-11
Milford, 6332 ... SF-3
Milford Center, 696 ... NM-7
Millbrook, 60 ... NJ-13
Millbury, 1161 ... NE-7
Milledgeville, 115 ... SD-7
Miller, 200 ... SJ-12
Miller City, 126 ... NH-4
Millersburg, 3611 ... NK-14
Millersport, 927 ... SB-11
Millerstown, 90 ... NN-5
Millersville, 150 ... NG-8
Millfield, 500 ... SE-13
Millport, 110 ... NJ-18
Millville, 891 ... SE-2
Millwood, 200 ... NL-12
Milton Center, 195 ... NG-5
Miltonsburg, 28 ... SB-18
Miltonville, 170 ... SD-3
Mineral ... SE-13
Mineral City, 841 ... NK-16
Mineral Ridge, 3900 ... NH-19
Minersville, 70 ... SG-14
Minerva, 3906 ... NJ-17
Minerva Pk., 1205 ... SA-9
Minford, 500 ... SH-10
Mingo, 100 ... NM-6
Mingo Junction, 3320 ... NL-20
Minster, 2781 ... NL-3
Mitiwanga, 130 ... NF-11
Modest ... SF-4
Mogadore, 3912 ... NH-16
Mohicanville, 100 ... NJ-13
Moline, 470 ... NC-2
Monclova, 120 ... NE-6
Monfort Hts., 2400 ... SL-2
Monnett, 60 ... NJ-9
**MONROE CO., 15180 ... SC-18**
Monroe, 12178 ... SE-3
Monroe Center, 100 ... ND-20
Monroeville, 1345 ... NG-10
Monroeville, 50 ... NK-19
Monterey, 120 ... SG-4
Montezuma, 185 ... NL-2
**MONTGOMERY CO., 559062 ... SC-3**
Montgomery, 10458 ... SK-5
Monticello, 40 ... NJ-2
Montpelier, 4015 ... NE-2
Montra, 100 ... NL-4
Montrose, 350 ... NH-15
Montville, 330 ... NE-18
Moorefield, 100 ... NM-17
Mooresville ... SE-10
Moraine, 6467 ... SC-4
Moreland, 100 ... NJ-14
Moreland Hills, 3064 ... SM-20
**MORGAN CO., 14897 ... SC-15**
Morgansville, 25 ... SD-14
Morgantown, 50 ... SG-8
Morning Sun, 200 ... SD-1
Morral, 368 ... NJ-8
Morristown, 292 ... SA-18
Morrisville, 70 ... SE-6
**MORROW CO., 31628 ... NK-9**
Morrow, 1544 ... SE-4
Moscow, 244 ... SH-4
Moss Run ... SD-17
Moulton, 130 ... NK-3
Mt. Air, 300 ... NN-9
Mt. Blanchard, 425 ... NI-6
Mt. Carmel, 4308 ... SM-5
Mt. Carmel Hts., 1200 ... SM-5
Mt. Cory, 187 ... NI-5
Mt. Eaton, 240 ... NJ-15
Mt. Ephraim, 80 ... SB-16
Mt. Forest Trails, 450 ... SM-5
Mt. Gilead, 3581 ... NK-10
Mt. Healthy, 6070 ... SK-2
Mt. Holly ... SD-5
Mt. Holly, 250 ... SG-4
Mt. Hope, 200 ... NK-14
Mt. Jefferson, 100 ... NM-3
Mt. Liberty, 250 ... NL-11
Mt. Olive ... SH-4
Mt. Orab, 2742 ... SG-5
Mt. Perry, 220 ... SB-13
Mt. Pisgah, 100 ... SG-4
Mt. Pleasant ... NA-9
Mt. Pleasant, 496 ... NM-19
Mt. Pleasant, 70 ... SE-12
Mt. Repose, 4102 ... SF-4
Mt. Sterling, 1835 ... SC-8
Mt. Vernon, 16037 ... NL-11
Mt. Victory, 592 ... NK-7
Mowrystown, 380 ... SG-6
Moxahala, 300 ... SC-13
Moxahala Pk., 200 ... SB-13
Mudsock, 25 ... SA-8
Mulberry, 3139 ... SF-4
Munroe Falls, 5165 ... NA-6
Murdock ... SF-4
Murlin Hts., 550 ... SB-4
Murray City, 462 ... SD-13
**MUSKINGUM CO., 84585 ... SB-14**
Mutual, 129 ... SA-6
Nankin, 400 ... NI-12
Napoleon, 8817 ... NF-4
Nashport, 260 ... SA-13
Nashville, 183 ... NK-13
Nashville, 100 ... SB-3
Navarre, 1896 ... NJ-16
Neapolis, 260 ... NF-5
Neelysville, 40 ... SD-15
Neffs, 1138 ... SA-19
Negley, 900 ... NJ-20
Nellie, 131 ... NL-13
Nelson, 80 ... NG-18
Nelsonville, 5375 ... SE-13
Neptune, 130 ... NK-2
Nettle Lake, 200 ... ND-1
Nevada, 763 ... NJ-8
Neville, 140 ... SH-4
New Albany, 75 ... NI-19
New Albany, 7088 ... SA-10
New Alexander, 50 ... NJ-18
New Alexandria, 210 ... NM-19
New Antioch, 220 ... SE-6
New Athens, 339 ... NM-18
New Baltimore, 250 ... NI-17
New Baltimore, 280 ... SF-2
New Bavaria, 76 ... NG-4
New Bedford, 220 ... NL-14
New Bloomington, 531 ... NK-7
New Boston, 2169 ... SI-9
New Bremen, 3071 ... NL-3
New Burlington, 400 ... SJ-2
New California, 35 ... NN-8
New Carlisle, 5543 ... SB-5
New Concord, 2589 ... SA-15
New Cumberland, 150 ... NK-17
New Dover, 130 ... NM-8
New England, 50 ... SE-14
New Franklin, 230 ... NJ-17
New Garden, 100 ... NJ-18
New Germany ... SL-9
New Guilford, 30 ... NM-13
New Hagerstown, 40 ... NL-17
New Hampshire, 200 ... NK-5
New Harrisburg, 100 ... NK-17
New Harrison, 50 ... NN-2
New Haven, 450 ... NH-10
New Haven, 320 ... SF-1
New Holland, 786 ... SD-8
New Hope, 150 ... SC-1
New Jasper, 100 ... SC-5
New Knoxville, 890 ... NL-3
New Lebanon, 4100 ... SC-3
New Lexington, 4576 ... SC-13
New Lexington, 100 ... SC-2
New London, 2597 ... NH-12
New Madison, 755 ... SB-1
New Market, 50 ... SG-6
New Marshfield, 400 ... SE-13
New Martinsburg, 150 ... SE-7
New Matamoras, 907 ... SD-18
New Miami, 2651 ... SE-2
New Middletown, 1553 ... NI-20
New Moorefield, 300 ... SA-6
New Moscow, 35 ... NM-14
New Paris, 1490 ... SB-1
New Petersburg, 150 ... SF-7
New Philadelphia, 17319 ... NL-16
New Pittsburg, 270 ... NJ-13
New Plymouth, 120 ... SE-12
New Richland, 100 ... NL-5
New Richmond, 2510 ... SH-3
New Riegel, 212 ... NH-7
New Rochester, 75 ... NF-7
New Rumley, 140 ... NL-18
New Salem, 130 ... SB-12
New Somerset, 150 ... NK-19
New Springfield, 600 ... NI-20
New Stark, 70 ... NJ-6
New Straitsville, 793 ... SD-13
New Vienna, 1373 ... SE-6
New Washington, 911 ... NI-10
New Waterford, 1345 ... NI-20
New Weston, 131 ... NL-1
New Westville, 90 ... SB-1
New Winchester, 40 ... NJ-9
Newark, 47236 ... SA-12
Newburgh Hts., 2131 ... SM-18
Newcastle, 100 ... NL-13
Newcomerstown, 3895 ... NM-15
Newell Run ... SE-17
Newhope, 150 ... SH-5
Newman, 300 ... NJ-15
Newport, 200 ... NL-16
Newport, 220 ... NM-3
Newport, 180 ... SB-7
Newport, 700 ... SE-17
Newton Falls, 4641 ... NG-18
Newtonsville, 537 ... SF-4
Newtown, 3889 ... SM-5
Ney, 336 ... NF-2
Nicholsville, 200 ... SH-4
Niles, 19268 ... NG-19
Nipgen, 100 ... SF-9
**NOBLE CO., 14058 ... SC-16**
N. Auburn, 40 ... NI-10
N. Baltimore, 3341 ... NH-6
N. Bend, 638 ... SF-1
N. Benton, 280 ... NI-18
N. Bloomfield, 500 ... NF-18
N. Bristol, 160 ... NF-18
N. Canton, 16925 ... NI-16
N. College Hill, 9549 ... SF-2
N. Creek, 80 ... NH-3
N. Fairfield, 559 ... NH-11
N. Fork Vil., 1726 ... SE-9
N. Georgetown, 150 ... NI-18
N. Hampton, 352 ... SA-5
N. Industry, 2700 ... NJ-16
N. Jackson, 900 ... NH-18
N. Kingsville, 2565 ... NC-19
N. Lawrence, 800 ... NI-15
N. Lewisburg, 1585 ... NM-6
N. Liberty, 80 ... NK-12
N. Lima, 900 ... NI-19
N. Madison, 8451 ... ND-18
N. Monroeville, 40 ... NG-10
N. Olmsted, 31319 ... NF-14
N. Perry, 964 ... ND-17
N. Randall, 830 ... SM-19
N. Ridgeville, 28153 ... NF-13
N. Robinson, 192 ... NJ-10
N. Royalton, 29331 ... NG-15
N. Salem, 100 ... NN-15
N. Star, 203 ... NM-2
N. Warren ... NA-12
N. Woodbury, 90 ... NK-10
N. Zanesville, 3013 ... SB-14
Northbrook, 11076 ... SK-2
Northfield, 3654 ... NG-15
Northridge, 6853 ... SA-6
Northridge, 8487 ... SB-4
Northup, 100 ... SI-13
Northwood, 5520 ... NE-7
Northwood, 80 ... NL-6
Norton, 11472 ... NH-15
Norton, 50 ... NL-9
Norwalk, 16637 ... NG-11
Norwich, 117 ... SA-15
Norwood, 20341 ... SF-3
Nova, 430 ... NH-12
Novelty, 500 ... NF-16
Oak Harbor, 2796 ... NF-8
Oak Hill, 1613 ... SH-11
Oakdale, 570 ... NB-8
Oakdale ... SD-13
Oakfield, 40 ... NF-18
Oakfield, 100 ... SC-13
Oakland, 120 ... SD-10
Oakland ... SD-5
Oakshade, 100 ... NE-4
Oakwood, 557 ... NH-3
Oakwood, 8439 ... SC-4
Oakwood, 3719 ... SN-20
Oberlin, 8381 ... NG-12
Obetz, 4219 ... SB-9
Oceola, 280 ... NJ-8
Octa, 79 ... SD-6
Ogden, 100 ... SE-5
Ohio City, 759 ... NJ-2
Ohio Furnace, 150 ... SJ-10
Okeana, 240 ... SE-1
Okolona, 140 ... NF-4
Old Fort, 350 ... NG-8
Old Washington, 258 ... SA-16
Oldtown, 80 ... SC-5
Olena, 350 ... NH-11
Olive Branch, 200 ... SG-4
Olive Green, 65 ... NM-10
Olivesburg, 80 ... NI-11
Olivett, 80 ... SA-18
Olmsted Falls, 8168 ... SN-14
Omega, 130 ... SF-10
Oneida, 150 ... NJ-17
Ontario, 5204 ... NJ-11
Opperman, 30 ... SB-16
Oran, 50 ... NM-3
Orange, 35 ... NM-15
Orange, 3258 ... SM-20
Orangeville, 178 ... NF-20
Orbiston, 40 ... SD-13
Orchard Island, 300 ... NL-5
Oregon, 18921 ... NE-7
Oregonia, 300 ... SE-4
Oreville, 150 ... SD-13
Orient, 269 ... SC-9
Orland, 50 ... SE-12
Orrville, 8387 ... NI-14
Orwell, 1473 ... NE-18
Osgood, 248 ... NL-2
Ostrander, 562 ... NM-8
Otsego, 50 ... NN-14
**OTTAWA CO., 40985 ... NE-7**
Ottawa, 4403 ... NH-4
Ottawa Hills, 4594 ... NB-1
Otterbien Home, 700 ... SE-3
Ottokee, 35 ... NE-4
Ottoville, 860 ... NI-3
Otway, 83 ... SH-8
Outville, 110 ... SA-11
Overlook ... SL-8
Overpeck, 300 ... SE-2
Overton, 140 ... NI-13
Owensville, 850 ... SG-4
Oxford, 22887 ... SD-1
Ozark, 40 ... SB-18
Padua, 70 ... NK-1
Page Manor ... SL-9
Pagetown, 40 ... NL-10
Painesville, 18474 ... ND-17
Painesville on the Lake, 850 ... ND-17
Painters Creek, 150 ... SA-3
Paintersville, 70 ... SD-5
Palestine, 163 ... SA-1
Palmyra, 240 ... NH-18
Pancoastburg, 150 ... SD-8
Pandora, 1184 ... NI-5
Pansy ... SE-5
Paris, 220 ... NJ-17
Park Ridge Acres, 300 ... NB-15
Parkertown, 25 ... NG-10
Parkman, 600 ... NF-18
Parma, 77947 ... NF-15
Parma Hts., 19757 ... NF-14
Parral, 233 ... NK-16
Parrott ... SD-7
Pasco, 50 ... NM-4
Pataskala, 12683 ... SA-11
Patmos ... NI-18
Patriot, 140 ... SI-12
Patterson, 136 ... NJ-7
Pattersonville, 30 ... NJ-18
**PAULDING CO., 20293 ... NH-1**
Paulding, 3355 ... NH-2
Pavonia, 100 ... NJ-11
Payne, 1152 ... NH-1
Pedro, 110 ... SJ-11
Peebles, 1828 ... SH-7
Pekin, 100 ... NJ-17
Pekin, 250 ... SD-4
Pemberton, 350 ... NM-4
Pemberville, 1358 ... NF-7
Penfield, 100 ... NH-13
Peninsula, 689 ... NG-15
Pennsville, 150 ... SD-14
Peoli, 30 ... NM-16
Peoria, 230 ... NM-7
Pepper Pike, 5715 ... NF-16
Perintown, 150 ... SF-4
**PERRY CO., 34078 ... SC-12**
Perry, 1282 ... ND-17
Perry Hts., 8900 ... NC-8
Perrysburg, 17049 ... NC-1
Perrysville, 816 ... NK-12
Perrysville, 60 ... NL-17
Perryton, 100 ... NN-12
Peru, 70 ... NH-10
Petersburg, 950 ... NI-20
Pettisville, 500 ... NE-3
Pfeiffer Station, 30 ... NK-7
Phalanx ... NG-18
Pharisburg, 50 ... NL-8
Phillipsburg, 588 ... SB-3
Philo, 761 ... SB-14
Philothea, 120 ... NL-1
Phoneton, 220 ... SB-4
**PICKAWAY CO., 52727 ... SD-9**
Pickerington, 17531 ... SB-10
Pickrelltown, 60 ... NM-6
Piedmont, 200 ... NM-17
Pierpont, 420 ... ND-20
Pigeon Run, 250 ... NJ-15
**PIKE CO., 27695 ... SG-9**
Piketon, 1971 ... SG-9
Pikeville, 40 ... NM-2
Pine Grove, 50 ... SJ-11
Piney Fork, 350 ... NM-19
Pioneer, 1383 ... ND-2
Piqua, 20582 ... NN-3
Pisgah, 1000 ... SE-3
Pitchin, 150 ... SB-6
Pitsburg, 380 ... SA-2
Pittsfield, 200 ... NG-12
Plain City, 3543 ... NN-8
Plainfield, 156 ... NM-15
Plants, 150 ... SH-14
Plantsville, 25 ... SE-14
Plattsburg, 160 ... SB-6
Plattsville, 100 ... NM-4
Pleasant Bend, 50 ... NG-4
Pleasant City, 425 ... SB-16
Pleasant Corners, 80 ... SB-9
Pleasant Grove, 130 ... NM-19
Pleasant Grove, 2016 ... SB-14
Pleasant Hill, 1134 ... SA-3
Pleasant Home, 200 ... NI-13
Pleasant Plain, 169 ... SF-4
Pleasant Run, 5267 ... SJ-2
Pleasant Run Farm, 4731 ... SJ-2
Pleasant Valley, 500 ... SE-9
Pleasant View, 35 ... SC-6
Pleasantville, 853 ... SC-11
Plumwood, 280 ... SA-7
Plymouth, 1803 ... NI-10
Plymouth Center, 50 ... ND-19
Poast Town, 150 ... SD-3
Pt. Isabel, 45 ... SH-4
Pt. Place ... NA-3
Pt. Pleasant, 35 ... SH-4
Poland, 2662 ... NH-20
Poland Center, 250 ... NE-14
Polk, 350 ... NI-12
Pomeroy, 1952 ... SG-14
Port Clinton, 6135 ... NF-9
Port Jefferson, 317 ... NM-4
Port Union, 100 ... SE-3
Port Washington, 565 ... NM-16
Port William, 253 ... SD-6
**PORTAGE CO., 152061 ... NH-17**
Portage, 430 ... NG-6
Portage Lakes, 9870 ... NI-15
Porter, 150 ... SH-13
Porterfield, 130 ... SF-15
Portersville, 85 ... SC-14
Portland, 120 ... SG-15
Portsmouth, 20297 ... SI-9
Possum Woods, 200 ... NC-16
Potsdam, 207 ... SB-3
Powell, 12814 ... NN-9
Powellsville, 50 ... SI-10
Powhatan Pt., 1655 ... SB-19
Pratts Fork ... SF-14
Prattsville, 30 ... SF-12
**PREBLE CO., 42337 ... SC-1**
Pricetown, 60 ... SF-6
Princeton, 120 ... SE-3
Proctorville, 622 ... SK-12
Prospect, 1092 ... NL-8
Pulaski, 220 ... NF-2
Pulaskiville, 50 ... NK-10
Put-in-Bay, 131 ... NE-10
**PUTNAM CO., 34726 ... NI-4**
Pyrmont, 120 ... SB-3
Pyro, 150 ... SH-11
Quaker City, 546 ... SB-17
Queen Acres, 650 ... SE-2
Quincy, 692 ... NM-5
Raccoon Island ... SI-13
Racine, 741 ... SG-14
Radcliff, 150 ... SF-12
Radnor, 250 ... NL-8
Ragersville, 150 ... NL-15
Rainsboro, 280 ... SF-7
Randolph, 750 ... NH-17
Range, 40 ... SC-7
Rarden, 171 ... SH-8
Ravenna, 11381 ... NH-17
Rawson, 433 ... NI-5
Ray, 100 ... SF-11
Rayland, 398 ... NM-19
Raymond, 300 ... NM-7
Reading, 10523 ... SF-3
Red Lion, 200 ... SD-3
Redbird, 2000 ... ND-17
Redfield, 50 ... SC-13
Redhaw, 50 ... NI-13
Redoak ... SH-6
Redtown ... SE-13
Reedsburg, 150 ... NJ-13
Reedsville, 300 ... SG-15
Reedtown, 50 ... NH-10
Reedurban, 4400 ... NJ-16
Reese, 400 ... SJ-19
Reesville, 250 ... SD-6
Rehoboth, 100 ... SC-13
Reily, 200 ... SE-1
Reinersville, 100 ... SC-15
Reminderville, 2741 ... NG-16
Remington, 300 ... SK-5
Remsen Corners, 75 ... NG-15
Rendville, 47 ... SD-13
Reno, 850 ... SE-17
Reno Beach, 430 ... NE-8
Rensselaer Pk., 850 ... SK-3
Republic, 577 ... NH-9
Resaca, 40 ... SA-7
Residence Pk. ... SL-7
Reynoldsburg, 33657 ... SB-10
Rice ... NH-3
Riceland, 200 ... NJ-14
Rich Hill, 75 ... NM-10
Richfield, 3576 ... NG-15
Richfield Center, 110 ... NE-5
**RICHLAND CO., 128852 ... NI-11**
Richmond, 450 ... NL-19
Richmond Center, 100 ... ND-20
Richmond Dale, 700 ... SF-10
Richmond Hts., 10174 ... NE-16
Richville, 1600 ... NJ-16
Richwood, 2175 ... NL-8
Ridgeland, 50 ... SG-11
Ridgeton, 20 ... NI-9
Ridgeville, 200 ... SD-4
Ridgeville Corners, 500 ... NF-3
Ridgeway, 344 ... NK-6
Ridgewood Hts. ... SM-7
Rimer, 50 ... NI-3
Rinard Mills, 30 ... SD-18
Ringgold, 200 ... SC-10
Ringgold, 25 ... SD-14
Rio Grande, 873 ... SH-12
Ripley, 1668 ... SI-5
Risingsun, 607 ... NG-7
Rittman, 6274 ... NI-14
River Corners, 60 ... NH-13
Riverlea, 477 ... SF-18
Riverside, 25327 ... SC-4
Rix Mills, 40 ... SB-15
Roachester, 500 ... SE-4
Roads, 160 ... SG-11
Roaming Shores, 1195 ... NE-19
Robertsville, 600 ... NJ-17
Robins, 90 ... SB-16
Robtown ... SC-9
Rochester, 181 ... NH-12
Rock Camp, 80 ... SJ-11
Rock Creek, 554 ... NE-18
Rock Mills, 30 ... SE-7
Rock Way, 160 ... SB-5
Rockbridge, 450 ... SD-11
Rockford, 1107 ... NJ-2
Rockport, 90 ... NI-4
Rocky Hill ... SG-11
Rocky Ridge, 384 ... NE-8
Rocky River, 18900 ... NF-14
Rodney, 250 ... SH-12
Rogers, 251 ... NJ-20
Rollersville, 40 ... NG-7
Rome, 200 ... NE-18
Rome, 90 ... NI-11
Rome, 1450 ... SK-12
Romohr Acres, 300 ... SL-6
Rootstown, 650 ... NH-17
Rose Farm, 200 ... SC-13
Rosedale, 180 ... SA-7
Roseland, 2150 ... NJ-11
Rosemont, 90 ... ND-11
Rosemount, 2043 ... SI-9
Roseville, 1897 ... SC-13
Rosewood, 250 ... NM-5
**ROSS CO., 73345 ... SE-9**
Ross, 1971 ... SE-2
Rossburg, 216 ... NM-2
Rossford, 6404 ... NE-6
Roswell, 279 ... NL-16
Roundhead, 300 ... NK-5
Rowsburg, 90 ... NI-13
Roxabell, 200 ... SE-8
Royalton, 260 ... SC-10
Rubyville ... SH-10
Rudolph, 500 ... NG-6
Ruggles, 40 ... NH-12
Ruggles Beach, 50 ... NF-11
Ruraldale, 60 ... SC-14
Rush Run, 250 ... NM-19
Rushmore, 50 ... NI-3
Rushsylvania, 516 ... NL-6
Rushtown ... SH-9
Rushville, 262 ... SC-12
Russell, 50 ... SF-6
Russell Center, 200 ... NF-16
Russells Pt., 1519 ... NL-5
Russellville, 425 ... SH-6
Russia, 615 ... NM-3
Rutland, 414 ... SG-13
Sabina, 2783 ... SD-6
Sagamore Hills, 1930 ... NG-15
Sahara Sands, 650 ... NB-8
St. Bernard, 4626 ... SL-3
St. Clairsville, 5053 ... SA-19
St. Henry, 2408 ... NL-2
St. Johns, 200 ... NK-4
St. Louisville, 336 ... NM-12
St. Martin, 109 ... SF-5
St. Marys, 8138 ... NK-3
St. Paris, 1965 ... NN-5
Salem, 11781 ... NI-18
Salem Center, 80 ... SG-13
Salesville, 149 ... SB-17
Salineville, 1318 ... NK-19
Saltillo, 110 ... SC-13
Samantha, 130 ... SF-6
Sampleville, 200 ... SC-2
San Margherita, 300 ... SH-17
**SANDUSKY CO., 61792 ... NF-8**
Sandusky, 25688 ... NF-10
Sandy Springs ... SJ-8
Sandyville, 420 ... NK-16
Santa Fe, 100 ... NL-4
Sarahsville, 192 ... SB-16
Sardinia, 828 ... SG-5
Sardis, 500 ... SD-19
Savannah, 369 ... NI-12
Savona, 50 ... SA-2
Sawyerwood, 1730 ... NC-6
Saybrook ... ND-18
Sayre ... SC-13
Schoenbrunn, 700 ... NL-16
Schumm, 25 ... NJ-1
Scienceville ... NC-14
Scio, 759 ... NL-18
**SCIOTO CO., 79195 ... SH-9**
Scioto Furnace, 250 ... SI-10
Sciotodale, 982 ... SI-10
Scotch Ridge, 90 ... NF-7
Scott, 310 ... NI-2
Scottown, 100 ... SJ-12
Seal, 25 ... NI-8
Seaman, 1074 ... SH-7
Sebring, 4534 ... NI-18
Secedar Corners, 300 ... NC-14
Selma, 130 ... SC-6
**SENECA CO., 58683 ... NH-8**
Senecaville, 439 ... SB-16
Senior ... SE-4
Seven Hills, 11616 ... NF-15
Seven Hills, 2000 ... SJ-2
Seven Mile, 770 ... SD-2
Seventeen, 120 ... NL-16
Seville, 2409 ... NI-14
Sewellsville, 35 ... NN-17
Shade, 120 ... SF-14
Shademore, 40 ... SL-5
Shadeville, 100 ... SB-9
Shadyside, 3530 ... SA-19
Shaker Hts., 26460 ... NF-15
Shalersville, 220 ... NG-17
Shandon, 400 ... SE-1
Sharon, 90 ... SC-16
Sharon Center, 350 ... NH-15
Sharonville, 13377 ... SF-3
Sharpsburg, 50 ... SE-14
Shauck, 200 ... NK-10
Shawnee, 631 ... SD-13
Shawnee Hills, 577 ... NN-8
Shawnee Hills, 2355 ... SC-6
Sheffield, 3515 ... NF-13
Sheffield Lake, 8905 ... NF-13
**SHELBY CO., 47910 ... NM-3**
Shelby, 9311 ... NI-10
Shenandoah, 100 ... NI-11
Sheridan, 500 ... SJ-11
Sherrodsville, 306 ... NK-17
Sherwood, 806 ... NG-2
Shiloh, 687 ... NI-11
Shiloh, 11272 ... SB-4
Shinrock ... NG-11
Shreve, 1496 ... NJ-13
Sidney, 19944 ... NM-4
Signal, 40 ... NJ-19
Silver Lake, 3100 ... NH-16
Silverton, 5155 ... SK-4
Simons, 70 ... NE-20
Sinking Spring, 162 ... SG-7
Skyline Acres, 500 ... SK-2
Sligo, 300 ... SE-5
Slocum ... SI-10
Smith Corners ... ND-12
Smithfield, 798 ... NM-19
Smithville, 1298 ... NI-14
Smyrna, 40 ... NM-17
Snyderville, 200 ... SB-5
Socialville ... SE-3
Solon, 21866 ... NF-16
Somerdale, 320 ... NK-16
Somerset, 1556 ... SC-12
Somerton, 180 ... SB-18
Somerville, 321 ... SD-2
Sonora, 250 ... SA-14
S. Amherst, 1743 ... NF-12
S. Bloomfield, 1642 ... SC-9
S. Bloomingville, 40 ... SE-11
S. Charleston, 1774 ... SB-6
S. Condit, 110 ... NM-10
S. Euclid, 21209 ... NE-15
S. Hill Pk., 1040 ... NB-1
S. Lebanon, 3691 ... SE-4
S. New Lyme, 120 ... NE-19
S. Newbury, 70 ... NF-17
S. Olive, 40 ... SC-16
S. Perry, 140 ... SD-11
S. Plymouth, 20 ... SD-7
S. Point, 3986 ... *B-2
S. Russell, 3913 ... NF-16
S. Salem, 219 ... SE-8
S. Solon, 387 ... SC-6
S. Vienna, 450 ... SB-6
S. Webster, 742 ... SH-10
S. West Hubbard ... NC-14
S. Woodbury, 30 ... NL-10
S. Zanesville, 2021 ... SB-14
Southington, 400 ... NG-18
Sparta, 195 ... NL-10
Spencer, 820 ... NH-13
Spencerville, 2160 ... NJ-3
Spring Mill, 150 ... NJ-11
Spring Mtn., 80 ... NL-13
Spring Valley, 495 ... SD-5
Springboro, 17551 ... SD-4
Springdale, 10384 ... SJ-3
Springfield, 62269 ... SB-6
Springhills, 150 ... NM-5
Springville, 40 ... NJ-13
Squirrel Town ... SI-7
Stafford, 82 ... SC-17
Standley ... NG-3
Stanleyville, 40 ... SD-16
Stanwood, 30 ... NJ-15
**STARK CO., 378098 ... NJ-17**
Starr, 25 ... SE-12
Staunton, 120 ... SA-4
Staunton, 100 ... SE-7
Steam Corners ... NJ-10
Steinersville, 200 ... SB-19
Stelvideo, 90 ... NN-2
Sterling, 600 ... NI-14
Steuben, 60 ... NH-10
Steubenville, 18820 ... NL-20
Stewart, 250 ... SE-14
Stewartsville, 200 ... SA-19
Stillwater, 150 ... NM-17
Stockdale, 200 ... SH-10
Stockport, 524 ... SD-15
Stone ... SC-14
Stone Creek, 180 ... NL-15
Stonelick, 100 ... SG-4
Stony Ridge, 400 ... NF-7
Stout, 123 ... SI-7
Stoutsville, 572 ... SD-10
Stovertown, 35 ... SB-14
Stow, 33899 ... NH-16
Strasburg, 2676 ... NK-16
Stratford, 80 ... NM-9
Stratton, 273 ... NK-20
Streetsboro, 14488 ... NG-16
Stringtown, 50 ... SH-4
Strongsville, 42617 ... NG-14
Struthers, 10734 ... NH-20
Stryker, 1354 ... NF-3
Suffield, 500 ... NH-16
Sugar Bush Knolls, 223 ... NG-16
Sugar Grove, 160 ... NK-19
Sugar Grove, 444 ... SD-11
Sugar Grove, 150 ... SI-9
Sugar Grove Hill, 300 ... NB-15
Sugar Ridge, 130 ... NF-6
Sugar Tree Ridge, 110 ... SG-6
Sugar Valley, 25 ... SC-1
Sugarcreek, 2193 ... NK-15
Sullivan, 450 ... NH-12
Sulphur Springs, 300 ... NI-9
Sulphurgrove ... SK-9
Summerfield, 287 ... SC-17
Summerford, 250 ... SB-7
Summerside, 5523 ... SG-3
Summerside Estates, 1700 ... SM-6
Summersville, 80 ... NL-7
**SUMMIT CO., 542899 ... NH-15**
Summit, 700 ... NC-13
Summithill, 20 ... SF-9
Summitville, 103 ... NJ-18
Sumner ... SF-14
Sunbury, 3375 ... NM-10
Sunbury, 110 ... SD-3
Sunnyland, 250 ... NB-15
Surrey Hill, 700 ... NB-12
Swanders, 80 ... NL-4
Swanton, 3651 ... NE-5
Sybene, 200 ... SK-12
Sycamore, 867 ... NI-8
Sycamore Valley, 25 ... SC-17
Sylvania, 19185 ... ND-6
Syracuse, 856 ... SG-14
Tacoma, 100 ... SA-18
Tallmadge, 17293 ... NH-16
Tarlton, 308 ... SD-10
Tawawa, 100 ... NM-4
Taylors Creek, 660 ... SK-1
Taylorsburg ... SK-6
Taylorsville ... SG-6
Tedrow, 160 ... NE-4
Temperanceville, 80 ... SB-17
Terrace Pk., 2220 ... SL-5
Terre Haute, 180 ... SA-5
Terry Acres, 165 ... SL-10
Texas, 130 ... NF-5
Thackery, 120 ... SA-5
The Bend, 100 ... NG-2
The Plains, 2931 ... SE-13
The Vil. of Indian Hill, 6030 ... SK-5
Thomastown ... NC-6
Thompson, 370 ... ND-18
Thornville, 1138 ... SB-12
Thrifton, 40 ... SE-8
Thurman, 142 ... SH-12
Thurston, 602 ... SB-11
Tiffin, 17347 ... NH-8
Tiltonsville, 1201 ... NM-19
Timberlake, 735 ... NE-16
Tipp City, 9254 ... SB-4
Tippecanoe, 200 ... NM-17
Tipton, 25 ... NH-1
Tiro, 251 ... NI-10
Tiverton, 80 ... NL-13
Toboso, 200 ... SA-13
Toledo, 293201 ... NE-7
Tontogany, 371 ... NF-6
Torch, 300 ... SF-15
Toronto, 5221 ... NL-20
Townwood, 25 ... NH-5
Trail ... NK-15
Tranquility, 30 ... SH-7
Tremont City, 342 ... SA-5
Trenton, 10866 ... SD-3
Triadelphia, 40 ... SC-14
Trimble, 480 ... SD-13
Trinway, 380 ... NN-13
Trotwood, 26039 ... SC-3
Trowbridge, 50 ... NE-8
Troy, 21975 ... SA-4
**TRUMBULL CO., 225116 ... NG-18**
Trumbull, 200 ... ND-18
Tucson ... SE-10
Tuppers Plains, 350 ... SF-14
Turpin Hills, 4960 ... SM-4
**TUSCARAWAS CO., 90914 ... NM-16**
Tuscarawas, 981 ... NL-16
Twenty Mile Stand, 300 ... SJ-6
Twightwee ... SJ-6
Twin Lakes, 300 ... NG-16
Twin Valley, 300 ... SI-9
Twinsburg, 17394 ... NG-16
Tymochtee, 25 ... NI-8
Tyndall, 150 ... NM-14
Uhrichsville, 5516 ... NL-16
**UNION CO., 40909 ... NL-7**
Union, 6293 ... SB-3
Union City, 1649 ... NM-1
Union Furnace, 430 ... SE-12
Union Station ... SA-11
Unionport, 150 ... NL-19
Uniontown, 2802 ... NI-16
Uniontown, 230 ... NN-18
Unionvale, 90 ... NM-18
Unionville Center, 286 ... NN-7
Uniopolis, 247 ... NK-4
Unity, 100 ... NI-20
Unity, 30 ... SH-7
University Hts., 12519 ... NF-15
University View, 500 ... SH-18
Upper Arlington, 31878 ... SA-9

## Oklahoma

Page locator

| Map keys | Atlas pages |
|---|---|
| 1 - 10 | 166 - 167 |
| 11 - 20 | 168 - 169 |



## Oregon

Page locator

| Map keys | Atlas pages |
|---|---|
| 1 - 10 | 170 - 171 |
| 11 - 20 | 172 - 173 |



*, †, ‡, §, ◊ See explanation under state title in this index.

County names are listed in capital letters and in boldface type.

## Pennsylvania

Page locator

| Map keys | Atlas pages |
|---|---|
| WA1 - WJ14 | 174 - 175 |
| WK1 - WT14 | 176 - 177 |
| EA1 - EJ14 | 178 - 179 |
| EK1 - ET14 | 180 - 181 |

* City keyed to p. 50
† City keyed to p. 137
‡ City keyed to pp. 182 - 183



*, †, ‡, §, ◊ See explanation under state title in this index.

County names are listed in capital letters and in boldface type.

Lawrence, 1100 WN-3
**LAWRENCE CO., 94643 WI-2**
Lawrence Pk., 4048 WC-4
Lawrenceville, 606 ED-3
Lawsonham, 50 WJ-6
Lawsville Center, 75 EE-9
Lawton, 85 EF-8
Le Raysville, 298 EE-7
Leaders Hts., 1200 WT-13
Leamersville, 80 WN-11
Lebanon, 24097 EN-6
**LEBANON CO., 120327 EL-7**
Leck Kill, 200 EL-5
Leckrone, 250 WQ-4
Lecontes Mills, 125 WJ-11
Lee Pk., 3600 EI-8
Leechburg, 2212 WL-5
Leeper, 200 WH-7
Lees Cross Roads, 200 EP-2
Leesburg, 100 WI-3
Leesport, 1908 EM-8
Leetsdale, 1117 WL-3
Lehigh, 100 EH-10
**LEHIGH CO., 312090 EK-12**
Lehighton, 5428 EK-9
Lehman, 250 EH-8
Leinbachs, 150 ES-11
Leisenring, 600 WP-5
Leith, 900 WP-5
Leithsville, 300 EM-11
Lemasters, 250 WQ-13
Lemon, EF-8
Lemont, 2116 WK-13
Lemont Furnace, 800 WP-5
Lemoyne, 3949 ET-3
Lenape, 200 EP-10
Lenhartsville, 169 EM-9
Lenoxville, 130 EF-10
Leola, 1100 EO-7
Leroy, 90 EF-5
Lester *G-2
Level Green, 2500 WN-5
Levittown, 53966 EO-13
Lewis Run, 566 WE-10
Lewisberry, 385 EO-4
Lewisburg, 5454 EJ-4
Lewistown, 8528 EL-1
Lewisville, 225 EQ-9
Liberty, 216 EG-3
Liberty, 2436 *M-8
Liberty Corners, 100 EF-6
Library, 200 *N-5
Lickingville, 60 WH-7
Lightstreet, 881 EJ-6
Ligonier, 1599 WN-7
Lilly, 861 WM-10
Lima, 3225 EP-11
Lime Ridge, 951 EJ-7
Limehill EF-7
Limeport, 250 EM-11
Limerick, 800 EN-10
Limestone, 150 WI-7
Limestoneville, 70 EJ-5
Lincoln, 1120 *N-8
Lincoln Hts., 300 WN-5
Lincoln Hill, 500 WO-2
Lincoln Pk., 1800 ET-11
Lincolnville, 112 WE-5
Linden, 270 EI-3
Linden, 100 WN-3
Linesville, 1091 WF-2
Linfield, 650 EO-10
Linglestown, 6414 EN-5
Linntown, 1542 EJ-4
Linwood, 3374 EQ-11
Lionville, 400 EP-10
Lippincott, 40 WP-3
Lisburn, 400 EO-4
Listie, 700 WP-8
Listonburg, WQ-7
Litchfield, 80 ED-6
Lithia Springs, 90 EK-5
Lititz, 9034 EO-7
Little Cooley, 50 WF-4
Little Corners, 45 WF-3
Little Gap, 300 EK-10
Little Hope, 60 WD-5
Little Marsh, 80 EE-2
Little Meadows, 282 ED-7
Littlestown, 4115 EQ-3
Liverpool, 884 EM-4
Llewellyn, 800 EL-7
Llewellyn Corners EC-8
Loag, 200 EO-9
Lobachsville EM-9
Lock Haven, 8541 EI-2
Locke Mills EL-1
Locust WK-8
Locust Gap, 450 EK-6
Locust Run, 100 EM-3
Locustdale, 70 EK-7
Loganton, 462 EJ-2
Loganville, 1023 EQ-5
London WI-3
London Grove, 250 EQ-9
Lone Pine, 50 WO-3
Long Pond, 900 EJ-10
Longfellow, 70 WM-14
Longstown, 100 WS-14
Longswamp, 300 EM-10
Loop, 60 WC-13
Lopez, 180 EG-7
Lorain, 677 WS-2
Lorane, 2994 EN-9
Lords Valley, 70 EH-12
Loretto, 1338 WM-10
Lost Creek, 500 EK-7
Lottsville, 100 WD-6
Lovejoy, 50 WL-8
Lovelton, 100 EG-7
Lover, WO-4
Lower Burrell, 12087 WM-5
Lowville, 130 WD-5
Loyalsockville, 400 EH-4
Loyalton, 450 EM-5
Loysburg, 330 WO-11
Loysville, 200 EN-2
Lucerne Mines, 951 WM-7
Lucinda, 300 WH-7
Lucisboro, 100 WM-8
Ludlow, 500 WF-8
Lumber City, 81 WJ-10
Lumberville, 200 EN-12
Lundys Lane, 180 WE-2
Lurgan, 90 WO-14
Luthersburg, 325 WJ-9
Luxor, 750 WN-6
Luzerne, 2751 EH-9
**LUZERNE CO., 319250 EF-10**
Lycippus, 90 WN-6
**LYCOMING CO., 120044 EE-4**
Lykens, 1848 EM-5
Lyleville, 135 WL-10
Lynch WG-8
Lyndell, 400 EP-9
Lyndon, 750 EP-7
Lyndora, 1300 WK-4
Lynn, 130 EF-8
Lynnport, 200 EL-9
Lynnville, 55 EL-9
Lyons, 510 EM-9
Lynnwood, 1300 EB-7
Macdonaldton, 130 WP-8
Mackeyville, 250 EJ-2
Macungie, 3123 EM-10
Maddensville, WO-13
Madera, 1000 WK-11
Madison, 478 WN-5
Madisonburg, 135 EK-1
Madisonville, 150 EH-10
Madley, WP-10
Mahaffey, 372 WK-9
Mahanoy City, 4375 EK-8
Maiden Creek, 200 EK-8
Mainesburg, 200 EF-4
Mainland, 2000 EN-11
Mainsville, 300 EP-1
Mainville, 83 EJ-6
Maitland, 200 EL-2
Malta EL-4
Malvern, 3098 EP-10
Mammoth, 450 WO-6
Mamont WM-6
Manatawny, 200 EN-9
Manchester, 2456 EP-5
Mandata, 60 EL-5
Manheim, 4646 EO-6
Manns Choice, 277 WP-10
Mannsville, 25 EM-3
Manor, 2905 WN-5
Manor Ridge, 1000 ES-8
Manorville, 369 WK-6
Mansfield, 3177 EE-3
Mantz, 300 EL-9
Maple Grove, 135 WJ-6
Maple Lake EH-10
Mapledale, 200 WH-4
Mapleton, 437 WN-13
Mapleton, 150 WQ-4
Maplewood EH-11
Marble, 130 WH-6
Marchand, 90 WK-8
Marcus Hook, 2228 EQ-11
Marianna, 579 WP-3
Mariasville WI-5
Marienville, 1325 WG-8
Marietta, 2593 EP-6
Marion, 800 WQ-14
Marion Center, 416 WK-8
Marion Hts., 684 EK-6
Markelsville, 25 EM-3
Markes, 100 WQ-13
Marklesburg, 211 WN-12
Markleton, 40 WQ-7
Markleysburg, 268 WQ-6
Markton WJ-8
Marlin, 640 EL-7
Mars, 1653 WL-4
Marsh EO-9
Marsh Hill EG-4
Marshallton, 1437 EK-6
Marshallton, 600 EP-10
Marshburg, 50 WE-9
Marshlands, 50 EF-1
Marstown, 150 EM-6
Martha Furnace, 25 WK-13
Marticville, 400 EP-7
Martin, 110 WQ-4
Martindale, 180 EO-8
Martindale, 140 WN-10
Martins Creek, 1200 EK-12
Martinsburg, 2121 WN-11
Martinsville, 100 EP-6
Marwood, 60 WK-5
Marysville, 2425 EN-4
Masontown, 3394 WQ-4
Mastersonville, 185 EO-6
Masthope, 50 EG-12
Matamoras, 2312 EH-14
Mather, 1000 WP-3
Mattawana, 245 WM-14
Mausdale, 75 EJ-5
Maxatawny, 250 EM-10
Mayburg WG-7
Mayfield, 1703 EG-10
Mayport, 80 WJ-7
Maysville WH-2
Maysville, 60 WM-6
Maytown EO-4
Maytown, 2604 EO-6
Maze EL-3
Mazeppa, 225 EJ-4
McAdoo, 2088 EK-8
McAlevys Fort, 70 WL-13
McAlisterville, 765 EL-3
McCartney, 70 WK-10
McClure, 925 EL-2
McConnellsburg, 1040 WP-13
McConnellstown, 440 WM-12
McCoysville, EM-1
McCracken, 60 WQ-1
McCrea, EN-2
McDonald, 2120 WN-3
McElhattan, 500 EI-2
McEwensville, 298 EJ-5
McGees Mills, 100 WK-9
McGovern, 2538 WN-3
McGrann, 700 WK-6
McIntyre, 220 WL-7
**MCKEAN CO., 45936 WF-10**
McKeansburg, 155 EL-8
McKee, 200 WN-11
McKee Half Falls, 130 EL-4
McKees Rocks, 6018 WM-3
McKeesport, 22130 WN-4
McKinney, EO-1
McKnightstown, 350 EQ-2
McLane, 75 WD-3
McMichael, 50 EJ-11
McMurray, 4726 WN-3
McSherrystown, 2791 EQ-4
McVeytown, 390 WM-14
McVille, WL-5
Meadow Lands, 900 WN-3
Meadowbrook, 1000 ‡A-5
Meadville, 13233 WF-3
Mechanics Grove, 120 EQ-8
Mechanicsburg, 8721 EO-4
Mechanicsville, 120 EA-2
Mechanicsville, 3099 EJ-6
Mechanicsville, 500 EN-12
Mechanicsville, 35 WI-7
Meckesville, 25 EJ-10
Mecks Corner, EN-3
Media, 5396 EP-11
Mehoopany, 350 EG-8
Meiserville, 100 EL-4
Melcroft, 250 WO-6
Melrose Pk., 6500 ‡B-5
Mendenhall, 600 EQ-10
Menges Mills, 200 EQ-4
Menno, 80 WM-14
Mentcle, 100 WL-9
Mercer, 2207 WI-3
**MERCER CO., 120293 WH-2**
Mercersburg, 1584 WQ-13
Meridian, 3794 WK-4
Merion Station, 700 ‡D-3
Mertztown, 300 EM-10
Meshoppen, 427 EG-8
Metal, 100 WP-13
Mexico, 279 EM-2
Meyersdale, 2273 WQ-8
Middle Creek, EK-3
Middle Lancaster, 100 WK-3
Middle Spring, 80 EO-1
Middleboro, 359 WD-3
Middleburg, 90 EJ-9
Middleburg, 1327 EK-4
Middlebury Center, 200 EE-3
Middleport, 433 EL-8
Middlesex, 250 EO-3
Middletown, 3411 EA-5
Middletown, 8841 EO-5
Midland, 2846 WL-2
Midvale EB-8
Midway, 924 WN-2
Mifflin, 605 EM-2
**MIFFLIN CO., 46486 WM-14**
Mifflinburg, 3517 EK-4
Mifflintown, 822 EM-2
Mifflinville, 1213 EJ-7
Milan, 300 EE-6
Milanville, 175 EF-12
Mildred, 200 EG-6
Milesburg, 1144 WJ-13
Milford, 1204 EH-13
Milford Square, 700 EM-11
Mill City, 450 EG-9
Mill Creek, 327 WM-13
Mill Grove, 40 EK-7
Mill Hall, 1478 EI-2
Mill Run, 250 WM-11
Mill Run, 400 WP-6
Mill Vil., 386 WE-4
Millardsville, 100 EN-7
Millbach EN-7
Millbourne, 905 ‡E-2
Milledgeville, 60 WG-4
Millersburg, 2460 EM-4
Millerstown, 683 EM-3
Millerstown WH-6
Millersville, 7341 EP-7
Millerton, 400 ED-4
Millertown, 200 WP-6
Millheim, 747 EK-2
Millmont, 150 EK-3
Millport, 80 WE-12
Millrift, 170 EH-14
Mills, 125 WE-14
Millsboro, 300 WP-4
Milltown, 170 EP-10
Millvale, 3656 ‡I-6
Millville, 944 EI-6
Millwood, 300 WN-7
Milmont ‡G-1
Milroy, 1386 EL-1
Milton, 6359 EJ-4
Milton Grove, 100 EO-6
Milwaukee EA-11
Mineral Springs, 300 WJ-11
Minersville, 4221 EL-7
Mingoville, 300 WJ-14
Minisink Hills, 350 EJ-12
Mocanaqua, 1000 EI-8
Modena, 598 EP-9
Mohnton, 3070 EN-8
Mohrsville, 700 EM-8
Monaca, 5749 WL-3
Monarch, 200 WP-5
Monessen, 8024 WO-4
Mongul, 60 EO-1
Monongahela, 4422 WO-4
Monroe, 90 WI-6
**MONROE CO., 138687 EH-13**
Monroeton, 480 EF-6
Monroeville, 27639 WM-5
Mont Alto, 1815 EQ-1
Mont Clare, 1800 EO-10
Montandon, 600 EJ-4
Montdale, 400 EG-10
Montello, 50 ET-11
Monterey, 100 EO-7
Monterey EQ-2
Montgomery, 1570 EI-4
**MONTGOMERY CO., 750097 EM-13**
Montgomery Ferry, 50 EM-4
Montgomeryville, 12031 EN-12
**MONTOUR CO., 18236 EH-6**
Montoursville, 4558 EI-4
Montrose, 1532 EE-9
Monument, 133 WI-14
Moon, 1000 ‡I-1
Moon Run, 400 ‡J-3
Mooresburg, 80 EJ-5
Moorestown, 250 EK-11
Moorheadville, WC-4
Moosic, 5758 EH-9
Morann, 400 WK-11
Moravia WJ-2
Morgan, 800 WN-3
Morgantown, 700 EO-9
Morris, 250 EG-2
Morris Run, 400 EF-4
Morrisdale, 600 WJ-12
Morrisville, 9594 EO-14
Morrisville, 1443 WP-3
Morton, 2634 ‡F-1
Moscow, 1938 EH-10
Moselem, 150 EM-9
Moselem Springs, 70 EM-9
Mosgrove, WK-6
Moshannon, 350 WJ-13
Mosherville, 200 ED-4
Mosiertown, 35 WE-3
Mostoller, 30 WO-8
Mt. Aetna, 350 EM-7
Mt. Allen, 2000 EO-4
Mt. Alton WE-10
Mt. Bethel, 650 EK-12
Mt. Braddock, 650 WP-5
Mt. Carmel, 5888 EK-6
Mt. Chestnut, 100 WK-4
Mt. Cobb, 2140 EH-10
Mt. Eagle, 80 WJ-14
Mt. Gretna, 234 EN-6
Mt. Holly Springs, 1909 EO-3
Mt. Jackson, 300 WJ-2
Mt. Jewett, 995 WF-10
Mt. Joy, 7249 EO-6
Mt. Lebanon, 33017 ‡L-4
Mt. Morris, 1300 WQ-3
Mt. Nebo, 300 EQ-7
Mt. Oliver, 3653 ‡K-6
Mt. Penn, 3063 EN-9
Mt. Pleasant, 200 EN-8
Mt. Pleasant, 65 EQ-3
Mt. Pleasant, 4380 WO-6
Mt. Pleasant Mills, 342 EL-4
Mt. Pocono, 2934 EI-11
Mt. Royal, 150 EP-4
Mt. Union, 100 EP-2
Mt. Union, 2344 WN-13
Mt. Vernon, EQ-8
Mt. Vernon, 2200 ‡N-9
Mt. Wilson, 200 EN-6
Mt. Wolf, 1331 EP-5
Mt. Zion EA-10
Mt. Zion, 350 EN-6
Mt. Zion, 100 ES-1
Mountain Top, 15269 EI-9
Mountainhome, 1169 EI-11
Mountville, 2820 EP-6
Mowersville, 40 EO-1
Muddy Creek Forks, 20 EQ-6
Muhlenberg, 200 EI-7
Muhlenberg Pk., 1300 ES-12
Muir, 450 EL-6
Mummasburg, 200 EQ-2
Muncy, 2459 EI-5
Muncy Valley, 270 EH-6
Munderf, 30 WH-8
Mundys Corner, 400 WM-9
Munhall, 11199 WM-4
Munson, 275 WJ-12
Munster, 300 WM-10
Murdocksville, 100 WM-2
Murrell, EO-7
Murrysville, 19488 WM-5
Muse, 1000 WN-3
Myerstown, 3120 EN-7
Myobeach, 65 EG-8
Myoma, 100 WK-4
Myrtle WD-12
Naginey, 120 EL-1
Nanticoke, 10199 EI-8
Nantmeal Vil., 250 EO-10
Nanty Glo, 2793 WM-9
Narberth, 4040 EP-12
Nashville, 200 EP-4
Natrona Hts., 2800 WL-5
Nauvoo, 40 EG-3
Nazareth, 6020 EL-11
Nectarine, WI-4
Needmore, 300 WQ-12
Neelyton, 80 WO-13
Neffs, 650 EL-10
Neffsville, 1000 EO-7
Neiltown, 25 WF-6
Nelson, 300 ED-3
Nemacolin, 1034 WP-4
Nescopeck, 1427 EJ-7
Nesquehoning, 3303 EK-9
New Albany, 286 EG-6
New Alexandria, 567 WM-6
New Baltimore, 153 WP-9
New Beaver, 1579 WK-2
New Bedford, 800 WI-2
New Berlin, 818 EK-4
New Berlinville, 800 EN-10
New Bethlehem, 981 WJ-7
New Bloomfield, 1090 EM-3
New Boston, 230 EK-7
New Bridgeville, 320 EP-6
New Brighton, 6068 WK-2
New Britain, 2265 EN-12
New Buena Vista, 80 WP-9
New Buffalo, 120 EM-4
New Castle, 24212 WJ-2
New Centerville, 178 WP-7
New Columbia, 435 EJ-4
New Columbus, 204 EI-7
New Cumberland, 7048 EO-4
New Danville, 500 ET-8
New Derry, 800 WN-7
New Eagle, 2201 WO-4
New Enterprise, 300 WO-11
New Era, 35 EF-7
New Florence, 726 WN-8
New Franklin, 275 WQ-14
New Freedom, 4205 EQ-5
New Freeport, 100 WQ-2
New Galilee, 385 WK-2
New Garden, 300 *B-6
New Geneva, 150 WQ-4
New Germantown, 150 EN-1
New Grenada, 90 WO-12
New Hamburg, 75 WH-2
New Holland, 5152 EO-8
New Hope, 2278 EN-13
New Jerusalem, 100 EM-9
New Kensington, 13667 ‡H-10
New Kingstown, 539 EO-3
New Lebanon, 195 WH-4
New Lexington, 50 WP-7
New London, 350 EQ-9
New Mahoning, 170 EK-9
New Milford, 806 EE-9
New Millport, 130 WK-10
New Oxford, 1764 EQ-4
New Paris, 201 WO-10
New Pk., 300 EQ-6
New Philadelphia, 1089 EL-8
New Providence, 300 EP-7
New Richmond, WF-4
New Ringgold, 278 EL-8
New Salem, 641 EP-5
New Salem, 850 WP-4
New Schaefferstown, 80 EM-7
New Sheffield, 2000 WL-2
New Stanton, 2099 WN-5
New Tripoli, 300 EL-9
New Vernon, 25 WH-3
New Washington, 82 WK-9
New Wilmington, 2253 WI-2
Newberrytown, 150 EO-5
Newburg, 250 EL-11
Newburg, 361 EO-1
Newburg, 76 WK-10
Newburg, 350 WM-11
Newchester, 225 EP-3
Newcomer, WP-5
Newell, 517 WO-4
Newfoundland, 650 EH-11
Newkirk, EK-8
Newmanstown, 1536 EN-7
Newmansville, 40 WH-7
Newport, 1463 EM-3
Newportville, 750 ‡A-9
Newry, 231 WN-11
Newton Hamilton, 257 WN-13
Newtown, 1400 EB-7
Newtown, 244 EL-7
Newtown, 2376 EN-13
Newtown Square, 11300 EP-11
Newville, 1303 EO-2
Newville, 350 EO-5
Niantic, 100 EN-10
Nicholson, 658 EF-9
Nickleville, WH-5
Nicktown, 200 WL-9
Niles EE-2
Niles Valley, 50 EE-2
Ninepoints, 120 EP-8
Nineveh, 100 WH-6
Nineveh, 125 WP-2
Nisbet, 550 EI-3
Nittany EJ-1
Nolo, 40 WM-8
Nordmont, 35 EH-6
Normal Square, 600 EK-9
Normalville, 700 WP-6
Norristown, 31843 EO-11
Norrisville, 50 WF-3
N. Apollo, 1324 WL-6
N. Belle Vernon, 1949 WO-4
N. Bend, 800 WH-14
N. Bingham, 50 WD-14
N. Braddock, 5821 ‡K-8
N. Catasauqua, 2834 EA-3
N. Charleroi, 1314 WO-4
N. East, 4184 WC-5
N. Fork, 30 WD-14
N. Hills, 3000 ‡A-4
N. Irwin, 838 WN-5
N. Liberty, 100 WI-3
N. Mountain, 60 EH-6
N. Oakland, WJ-5
N. Orwell, 100 EE-7
N. Point, 75 WK-8
N. Rome, 65 EE-6
N. Springfield, 130 WD-2
N. Towanda, 400 EF-6
N. Wales, 3226 EO-11
N. Warren, 2400 WE-8
N. Washington, 200 WJ-5
N. Washington, 500 WM-5
N. York, 1742 EP-5
Northampton, 9841 EL-10
**NORTHAMPTON CO., 267066 EI-14**
Northern Cambria, 3907 WL-9
Northumberland, 3512 EK-5
**NORTHUMBERLAND CO., 94556 EI-6**
Northvue, 700 WJ-4
Northwood, 340 WL-12
Norvelt, 1100 WO-6
Norwood, 5774 ‡F-2
Nottingham, 3000 ‡A-8
Nottingham, 200 EQ-8
Nowrytown, 150 WM-6
Noxen, 500 EH-8
Nuangola, 650 EI-8
Numidia, 254 EK-6
NuMine, 300 WK-7
Nuremberg, 231 EJ-7
Nyesville, 75 WP-14
Oak Forest, 60 WQ-3
Oak Grove, 200 WN-7
Oak Hall, WT-7
Oak Pk., 700 EN-11
Oakdale, 1435 WM-3
Oakland, 572 EE-10
Oakland, 1516 WJ-2
Oakland, 1600 WS-2
Oakland Mills, 110 EL-2
Oakleigh, 1000 ES-4
Oaklyn, 4030 *F-5
Oakmont, 6423 WM-4
Oaks, 850 EO-11
Oakville, 150 EO-2
Oberlin, 850 ET-4
Ogdensburg, 75 EG-4
Ogletown, 150 WO-9
Ohiopyle, 74 WQ-6
Ohioville, 3614 WL-2
Ohl, 75 WJ-7
Oil City, 10598 WG-5
Oklahoma, 600 WI-9
Oklahoma, 841 WL-6
Olanta, 100 WK-10
Old Boston, 200 EC-9
Old Concord, 30 WP-2
Old Forge, 8524 EH-9
Old Port, 90 EM-2
Oley, 900 EN-9
Oliphant Furnace, 250 WQ-5
Olivebug, 150 WJ-8
Oliver, 2925 WP-5
Olney, *C-5
Olyphant, 4956 EG-10
Oneida, 219 EK-8
Oneida, 100 WJ-4
Ono, 300 EN-6
Opp EI-5
Orange, 300 EH-9
Orangeville, 471 EJ-6
Orbisonia, 402 WN-13
Orchard Beach, 125 WC-5
Orchard Hills, 2152 WL-6
Ore Hill, 95 WN-11
Orefield, 750 EB-1
Oregon Hill, 100 EG-2
Oreland, 5509 *A-4
Oriental, 100 EL-4
Oriole, 100 EI-3
Ormrod, 300 EA-2
Ormsby, 100 WE-10
Orrstown, 230 EP-1
Orrtanna, 169 EQ-2
Orson, 125 EE-10
Orviston, 200 WI-14
Orwell, 100 EE-7
Orwigsburg, 2979 EL-8
Orwin, 300 EM-6
Osborne, 522 *H-2
Osceola, 400 ED-2
Osceola Mills, 1152 WK-11
Osgood, 40 WG-2
Osterburg, 250 WO-10
Osterhout, 30 EG-9
Oswayo, 140 WE-13
Ottsville, 250 EM-12
Oval, 180 EI-3
Overbrook, *D-3
Overlook, 200 ES-9
Overlook Hts., 1200 WS-5
Overton, 85 EG-6
Overview, 50 ER-2
Oxford, 4658 EQ-8
Packerton, 200 EK-9
Paint, 1034 WN-9
Paintersville EL-2
Paisley, 100 WQ-4
Palm, 370 EM-10
Palmer Hts., 3612 EL-11
Palmerton, 5209 EK-10
Palmyra, 6997 EN-6
Palo Alto, 1105 EL-7
Panic, 80 WJ-8
Paoli, 5425 EP-11
Paradise, 250 EM-6
Paradise, 1028 EP-8
Paradise Valley, EI-11
Pardoe, 50 WI-3
Paris, 200 WM-1
Park Forest Vil., 8830 WS-5
Park Hills, 2500 WT-5
Parker, 747 WI-5
Parker Ford, 800 EO-10
Parkesburg, 3430 EP-9
Parkhill, 300 WP-2
Parkside, 2176 EQ-11
Parkville, 6593 EC-4
Parryville, 504 EK-10
Patchinville, 30 WK-10
Patterson Hts., 614 WK-2
Patton, 1848 WL-10
Paupack, 600 EH-11
Pavia, 90 WN-10
Paxinos, 500 EK-6
Paxtang, 1486 ES-4
Paxton, 100 EL-4
Paxtonia, 5254 EA-4
Paxtonville, 221 EK-3
Peach Glen, 100 EP-3
Pearl, 50 WH-4
Pecks Pond EH-12
Pen Argyl, 3627 EK-11
Pen Mar, 400 ER-2
Penbrook, 2919 ES-3
Penfield, 650 WI-10
Penn, 422 WN-5
Penn Hall, 60 EK-1
Penn Hills, 46809 WM-4
Penn Lake Pk., 283 EI-9
Penn Run, 250 WL-8
Penn Valley, 3000 *C-2
Penn Wynne, 5382 *D-2
Penndel, 2470 EO-13
Pennline, 40 WF-2
Penns Creek 668 EK-3
Penns Park, 500 EN-13
Pennsburg, 3002 EN-10
Pennsbury Vil., 685 *K-4
Pennsdale, 110 EI-5
Pennside, 3000 ET-13
Pennside, 35 WE-2
Pennsville, 350 EK-10
Pennsville, 200 WO-6
Pennsylvania Furnace, 150 WL-13
Pennwyn, 650 ET-12
Penowa, 20 WN-1
Penryn, 700 EO-7
Pequea, 300 EQ-7
Perdix, 90 EN-4
Perkasie, 8636 EN-11
Perkiomen Hts., 700 EN-10
Perrine Corners, WH-4
**PERRY CO., 43602 EM-3**
Perryopolis, 1712 WO-5
Perrysville, *H-5
Perryville, 25 EH-3
Perryville, 180 WI-5
Perryville, 70 WM-6
Petersburg, 423 WM-12
Petrolia, 201 WJ-5
**PHILADELPHIA CO., 1447395 EO-12**
Philadelphia, 1447395 EP-12
Philipsburg, 2915 WK-12
Phillips, 225 WP-5
Phillipston, 45 WJ-6
Phillipsville, WD-4
Phoenixville, 15382 EO-10
Picture Rocks, 670 EH-5
Pigeon, 20 WG-8
**PIKE CO., 46302 EH-12**
Pilgrim Gardens, 1650 *E-1
Pillow, 288 EL-5
Pine, 80 EI-2
Pine Bank, 20 WQ-2
Pine Flats, 30 WL-8
Pine Glen, 210 WJ-12
Pine Grove, 2043 EM-6
Pine Grove Furnace, EP-3
Pine Grove Mills, 1141 WL-13
Pine Summit, 55 EI-6
Pinecroft, 100 WM-11
Pineville, 950 EN-13
Piney Fork, 400 *N-5
Pinola, 20 EP-1
Pipersville, 600 EM-12
Pitcairn, 3343 *K-9
Pitman, 300 EL-6
Pittsburgh, 310037 WM-4
Pittsfield, 500 WE-7
Pittston, 7547 EH-9
Pittsville, 35 WH-5
Plainfield, 376 EO-2
Plains EB-8
Plainsville, 350 EB-8
Platea, 438 WD-3
Pleasant Gap, 1611 WK-14
Pleasant Grove, 50 WO-2
Pleasant Hall, 150 WP-14
Pleasant Hills, 7751 *M-6
Pleasant Mt., 125 EF-11
Pleasant Union, WQ-9
Pleasant Unity, 600 WN-6
Pleasant Valley, 200 EM-11
Pleasantville, 808 WG-6
Pleasureville, 1200 WR-13
Plum, 26198 WM-5
Plumer, 150 WG-5
Plumsteadville, 1300 EN-12
Plumville, 317 WK-7
Plymouth, 6041 EI-8
Plymouth Meeting, 6034 *A-2
Pocahontas, 125 WQ-8
Pocono Lake, 750 EI-10
Pocono Manor, 300 EI-11
Pocono Pines, 1013 EI-11
Pocono Summit, 1500 EI-11
Point, WO-10
Pt. Marion, 1247 WQ-4
Pt. Pleasant, 800 EM-12
Poland, 75 WQ-4
Polk, 995 WH-4
Pomeroy, 750 EP-9
Pond Bank, 200 EQ-1
Pond Eddy, 50 EG-13
Pond Hill, 200 EI-8
Port Allegany, 2190 WE-11
Port Carbon, 1747 EL-7
Port Clinton, 277 EL-8
Port Griffith, 1350 EB-9
Port Matilda, 638 WK-12
Port Royal, 950 EM-2
Port Trevorton, 451 EL-4
Port Vue, 3843 *M-8
Portage, 2581 WN-10
Porter, 25 WJ-7
Porters Sideling, 100 EQ-4
Portersville, 250 WJ-3
Portland, 565 EJ-12
Portland Mills, 40 WH-9
Potter Brook, 200 EE-1
**POTTER CO., 18080 EE-1**
Potters Mills, 180 WK-14
Pottersdale, 20 WI-12
Potterville, 75 EE-7
Potts Grove, 140 EJ-5
Pottstown, 21288 EN-10
Pottsville, 14405 EL-7
Powell, 150 EE-6
Powells Valley, 600 EM-4
Powys, 100 EH-3
Prescottville, 100 WJ-9
President, 100 WG-6
Preston, 240 EB-6
Preston Pk., 200 EE-11
Pricetown, 180 EM-9
Primos, 900 *F-2
Primrose, 600 EL-7
Princeton, 150 WJ-3
Pringle, 937 EA-8
Prittstown WO-6
Proctor, 50 EH-5
Progress, 9647 ES-3
Prompton, 237 EG-11
Prospect, 1221 WJ-4
Prospect Pk., 6374 EP-11
Prospect Pk., 400 WG-12
Prospectville, 550 EO-12
Prosperity, 175 WO-2
Pulaski, 300 WI-2
Punxsutawney, 5944 WJ-8
Purcell, 25 WQ-11
Purchase Line, 40 WL-8
Putneyville, 140 WJ-7
Puzzletown, WN-10
Quakake, 300 EK-8
Quaker Hills, ET-8
Quaker Lake, 70 ED-8
Quakertown, 8643 EM-11
Quarryville, 2146 EP-7
Quecreek, 150 WO-8
Queen, 100 WN-10
Queens Run, 50 EI-2
Quentin, 529 EN-6
Quicktown, 50 EH-10
Quiggleville, 50 EH-3
Quincy, 400 EQ-1
Railroad, 302 EQ-5
Rainsburg, 139 WP-10
Ralston, 400 EG-4
Ramey, 500 WK-11
Ramsey, 45 EH-2
Rankin, 2108 *K-8
Ransom, 250 EH-9
Rathmel, 150 WI-9
Raubsville, 200 EL-12
Rauchtown, 300 EI-3
Ravine, 629 EM-6
Rawlinsville, 150 EQ-7
Raymilton WH-4
Raymond WE-13
Rea, 50 WN-2
Reading, 80506 EN-9
Reagantown, 70 WO-5
Reamstown, 3498 EO-8
Rebel Hill, 300 *B-1
Rebersburg, 492 EJ-2
Rebuck, 150 EL-5
Rector, 400 WO-7
Red Bridge, 200 WP-14
Red Cross, 70 EL-5
Red Hill, 2302 EN-10
Red Hill, 100 WA-12
Red Lion, 6041 EP-6
Red Rock, 30 EH-7
Redclyffe, 20 WH-8
Reeders, 350 EJ-11
Reeds Gap, 30 EM-1
Reedsville, 858 EL-1
Reels Corners, 25 WO-9
Reese, 80 WM-11
Reesedale, 50 WJ-6
Refton, 250 EP-7
Rehrersburg, 600 EM-7
Reidsburg, 40 WI-6
Reiffton, 2888 ET-13
Reinerton, 500 EL-6
Reinholds, 500 EN-8
Reistville, 100 EN-7
Renfrew, 350 WK-4
Renovo, 1217 WH-14
Republic, 1396 WP-4
Revere, 500 EM-12
Revloc, 650 WM-9
Rew, 300 WE-10
Rexford, 100 EF-2
Rexis, 45 WM-8
Reynoldsville, 2551 WI-9
Rheems, 1557 EO-6
Rices Landing, 408 WP-4
Riceville, 82 WE-5
Rich Hill, 50 EM-11
Richardsville, 40 WI-8
Richboro, 6678 EO-13
Richfield, 459 EL-3
Richland, 1492 EN-7
Richlandtown, 1327 EM-11
Richmond, 100 EK-12
Richmond Furnace, 50 WP-13
Richvale, 35 WN-14
Riddlesburg, 300 WO-11
Ridgebury, 170 ED-5
Ridgeview, 800 ER-4
Ridgway, 4096 WH-9
Ridgewood, 300 EC-9
Ridley Pk., 6979 *G-1
Riegelsville, 833 EL-12
Rillton, 900 WN-5
Rimer, 40 WJ-6
Rimersburg, 1013 WJ-6
Rinely, EQ-6
Ringgold, 225 WJ-7
Ringtown, 762 EK-7
River View Pk., 2000 ES-12
Riverside, 1853 EJ-5
Riverside, 700 WT-2
Rixford, 400 WE-10
Roaring Branch, 100 EG-4
Roaring Creek, 105 EK-6
Roaring Spring, 2252 WN-11
Robertsdale, 240 WO-12
Robesonia, 2049 EN-8
Robinson, 650 WM-7
Rochester, 3662 WL-2
Rochester Mills, 175 WK-8
Rock, 60 EM-7
Rock Glen, 300 EJ-7
Rockhill, 395 WO-13
Rockingham, 60 WO-9
Rockland, 30 WH-5
Rockledge, 2478 EO-12
Rockport, 85 EJ-9
Rockspring, WL-13
Rockton, 200 WJ-10
Rockville, 85 EN-4
Rockville WE-14
Rockville, 170 WR-4
Rockwood, 905 WP-7
Roedersville, 200 EL-7
Rogersville, 220 WP-2
Rogertown, 140 WE-8
Rohrerstown, 1200 ES-8
Rohrsburg, 164 EI-6
Rome, 361 EE-7
Ronco, 300 WP-4
Roscoe, 788 WO-4
Rose Pt., 90 WJ-3
Rose Valley, 924 EP-11
Rosebud, 450 WK-10
Rosecrans EJ-2
Roseglen, 110 EN-3
Rosemont *C-1
Roses, 140 WG-8
Roseto, 1644 EK-12
Roseville, 195 EE-4
Roseville, 100 EP-4
Roseville, 150 WI-8
Roslyn, 10000 *A-5
Rossiter, 790 WK-8
Rossland, 60 EK-11
Rosslyn Farms, 423 *K-4
Rossmoyne, 1100 ET-2
Rote, 250 EJ-2
Rothsville, 3017 EO-7
Roulette, 1000 WF-12
Round Top, 115 WT-9
Rouseville, 432 WG-5
Rouzerville, 862 EQ-1
Rowes Run, 400 WP-4
Rowland, 200 EG-12
Roxbury, 275 WO-14
Roxbury, 50 WP-8
Royal, 100 EF-10
Royalton, 946 EO-5
Royer, 45 WM-11
Royersford, 4356 EO-10
Ruff Creek WP-3
Ruffs Dale, 800 WO-5
Rummel, 600 WO-9
Rummerfield, 35 EF-7
Runville, 250 WJ-13
Rupert, 174 EJ-6
Rural Ridge, 1000 WL-4
Rural Valley, 852 WK-7
Rush, 100 EF-8
Rushland, 350 EN-13
Rushville, 45 EF-8
Russell, 1200 WE-8
Russell City, 60 WG-9
Russell Hill, EG-8
Russellton, 1530 WL-4
Russellville, 125 EQ-8
Russellville, WN-11
Rutherford, 3859 ES-4
Rutledge, 823 *F-1
Rydal, 1800 *A-5
Ryde, 70 WM-14
Rye EQ-5
Ryerson Station, 80 WP-2
Ryot, 30 WO-10
Sabinsville, 300 EE-1
Sabula, 200 WI-10
Sackett WG-9
Sacramento, 300 EL-6
Sadsburyville, 700 EP-9
Saegersville, 120 EL-10
Saegertown, 1036 WF-3
Safe Harbor, 150 EP-7
Sagamore, 450 WK-7
Saginaw, 200 EP-5
St. Augustine, 20 WL-10
St. Benedict, 450 WL-9
St. Boniface, 120 WL-10
St. Clair, 2987 EL-7
St. Clairsville, 81 WO-10
St. George, 50 WI-5
St. Johns, 300 EJ-8
St. Joseph, 150 EE-8
St. Lawrence, 1861 EN-9
St. Lawrence, 40 WL-10
St. Marys, 13409 WH-10
St. Michael, 850 WN-9
St. Paul, 75 WQ-8
St. Peters, 400 EO-9
St. Petersburg, 393 WI-5
St. Thomas, 900 WP-13
Salem, 150 EK-4
Salem, 30 WG-2
Salemville, 50 WO-11
Salina, 500 WM-6
Salisbury, 810 WQ-8
Salix, 900 WM-9
Salladasburg, 248 EI-3
Salona, 250 EJ-2
Saltillo, 325 WO-12
Saltsburg, 883 WM-6
Salunga, 950 EO-6
Saluvia, WP-12
Sanatoga, 7734 EN-10
Sand Beach, 300 EN-5
Sand Hill, 2345 EN-6
Sandts Eddy, EK-12
Sandy, 1687 WI-9
Sandy Hollow, 55 WJ-6
Sandy Lake, 691 WH-3
Sandy Ridge, 340 WK-12
Sandy Run, 60 EJ-9
Sandy Valley, 40 WJ-9
Sandyville, 25 EI-12
Sankertown, 655 WM-10
Santiago, 400 WM-2
Sarver, 300 WL-5
Sassamansville, 350 EN-10
Satterfield, EG-6
Saulsburg, WL-13
Saville, 20 EM-2
Saxonburg, 1593 WK-5
Saxton, 762 WO-11
Saybrook, WF-8
Saylorsburg, 500 EK-11
Sayre, 5440 ED-6
Scalp Level, 789 WN-9
Scandia, 25 WE-8
Scarlets Mill, 100 EO-9
Scenery Hill, 600 WO-3
Schaefferstown, 984 EN-7
Schellsburg, 306 WP-10
Schenley, 55 WL-5
Scherersville, 100 EB-2
Schnecksville, 1989 EL-10
Schoeneck, 650 EN-7
Schoenersville, 400 EA-4
School Lane Hills, 1840 ES-8
Schubert, 200 EM-7
Schuster Hts., 75 WL-5
**SCHUYLKILL CO., 150336 EK-8**
Schuylkill Haven, 5161 EL-7
Sciota, 700 EJ-11
Scotch Hill WH-7
Scotland, 650 EP-1
Scotrun, 650 EJ-11
Scottdale, 4409 WO-5
Scranton, 72233 EH-10
Scullton WP-7
Seanor, 130 WO-9
Searights, 250 WP-4
Sebring, 30 EG-3
Secane, 1700 *F-1
Seelyville, 300 EG-11
Seidersville, 250 EB-4
Seipstown, 30 EL-10
Seisholtzville, 100 EM-10
Seitzland, 70 EQ-5
Selinsgrove, 5344 EK-4
Sellersville, 4476 EN-11
Seminole, 250 WJ-7
Seneca, 966 WH-5
Sergeant, 40 WF-9
Seven Fields, 2550 WL-3
Seven Springs, 117 WP-7
Seven Stars, EL-3
Seven Stars, EQ-2
Seven Valleys, 507 EQ-5
Seward, 451 WM-8
Sewickley, 3562 WM-3
Sewickley Hts., 918 *H-3
Sewickley Hills, 700 WL-3
Shade Gap, 91 WO-13
Shado-wood Vil., 350 WL-7
Shady Grove, 280 WQ-14
Shaffer, 120 WI-9
Shamokin, 7367 EK-6
Shamokin Dam, 1438 EK-5
Shanesville, 70 EN-10
Shanksville, 222 WP-8
Shannondale, 75 WJ-7
Sharon, 14869 WI-2
Sharon Center, 30 WE-12
Sharon Hill, 5308 *F-2
Sharpsburg, 3263 WM-4
Sharpsville, 4100 WH-2
Shartlesville, 300 EM-8
Shavertown, 2000 EA-8
Shawmut WH-9
Shawnee on Delaware, 400 EJ-12
Shawville, 100 WJ-11
Shay WL-6
Sheakleyville, 151 WG-3
Sheffield, 1268 WF-8
Shehawken, 40 EE-11
Shellsville, 200 EN-5
Shelly, 125 EM-11
Shelocta, 119 WL-7
Shenandoah, 5155 EK-7
Shenandoah Hts., 1298 EK-7
Shenango, 80 WH-2
Shepherdstown, 200 EO-4
Sheppton, 239 EK-8
Shermans Dale, 120 EN-3
Shermansville, 50 WF-2
Sherrett, 125 WJ-6
Shesequin, 150 EE-6
Shickshinny, 892 EI-7
Shillington, 4997 EN-8
Shiloh, 10192 EP-5
Shimerville, 130 EM-10
Shimpstown, 80 WQ-13
Shindle, 70 EL-2
Shinglehouse, 1105 WD-12
Shingletown, 150 WK-13
Shintown, 75 WH-13
Shippensburg, 5578 EP-1
Shippenville, 470 WI-6
Shippingport, 219 WL-2
Shiremanstown, 1464 EO-4
Shirleysburg, 130 WN-13
Shoemakers, EJ-12
Shoemakersville, 1393 EM-8
Shohola, 160 EG-13
Shrader, 100 EL-1
Shrewsbury, 3573 EQ-5
Shumans, 140 EJ-7
Shunk, 45 EG-5
Siddonsburg, 110 EO-4
Sidman, 300 WN-9
Sigel, 250 WH-8
Siglerville, 100 EL-1
Silkworth, 100 EH-8
Silvara, 80 EF-7
Silver Lake, 150 EE-8
Silver Spring, 200 EP-6
Silverdale, 965 EN-11
Simpson, 1500 EG-10
Sinking Spring, 3605 EN-8
Sinnemahoning, 350 WH-12
Sipesville, 350 WO-8
Sizerville, 35 WG-12
Skelp, 60 WL-11
Skinners Eddy, 40 EF-7
Skippack, 2889 EN-11
Skytop, 200 EI-11
Slate Lick, 150 WK-5
Slate Run, 60 EG-2
Slatedale, 600 EL-10
Slateford, 250 EJ-12
Slatington, 4404 EL-10
Slickville, 372 WM-6
Sligo, 690 WI-6
Slippery Rock, 3068 WJ-4
Slocum Corners, 150 EI-8
Slovan, 1100 WN-2
Smethport, 1563 WE-11
Smicksburg, 45 WK-7
Smithfield, 4181 WM-12
Smithfield, 803 WQ-4
Smithport, 100 WK-9
Smithton, 407 WO-5
Smithville, EP-7
Smock, 600 WP-5
Smoke Run, 150 WK-11
Smoketown, 350 EP-7
Smullton, 125 EJ-2
Snow Shoe, 774 WJ-13
**SNYDER CO., 37546 EJ-5**
Snyders, 90 EL-9
Snydersburg, 100 WH-7
Snydersville, 150 EJ-11
Snydertown, 331 EK-5
Somers Lane, 300 ED-3
Somerset, 6359 WP-8
**SOMERSET CO., 80023 WP-8**
Somerton, *A-7
Sonestown, 200 EH-6
Soudersburg, 200 EP-8
Souderton, 6553 EN-11
S. Altoona, WB-13
S. Bend, 40 WL-6
S. Bethlehem, 412 WJ-7
S. Bradford, 150 WE-10
S. Canaan, 250 EG-11
S. Coatesville, 1060 EP-9
S. Connellsville, 2151 WP-5
S. Fork, 1028 WN-9
S. Gibson, 175 EF-10
S. Greensburg, 2217 WN-6
South Heights, 492 WL-3
S. Lakemont, 1100 WB-13
S. Montrose, 400 EE-9
S. Mountain, 600 EQ-2
S. New Castle, 737 WJ-2
S. Philipsburg, 438 WK-12
S. Pottstown, 2135 EN-10
S. Renovo, 517 WH-14
S. Sterling, 150 EH-11
S. Tamaqua, 200 EK-8
S. Temple, 1400 ER-13
S. Towanda, 250 EF-6
S. Uniontown, 3500 WP-5
S. Waverly, 964 ED-6
S. Williamsport, 6017 EI-4
Southampton, 11500 EO-12
Southmont, 2063 WN-8
Southview, 240 WN-3
Southwest, 600 WO-6
Southwest Greensburg, 2201 WN-6
Sparta WP-2
Spartansburg, 314 WE-5
Speers, 1165 WO-4
Spinnerstown, 350 EM-11
Sporting Hill, 300 EO-6
Sporting Hill, 200 ET-2
Spraggs, 30 WQ-3
Sprankle Mills, 60 WJ-8
Spring Church, 450 WL-6
Spring City, 3375 EO-10
Spring Creek, 100 WE-6
Spring Garden, 45 EI-4
Spring Glen, 300 EL-5
Spring Grove, 2206 EQ-4
Spring Hill, 970 WN-10
Spring House, 3290 EO-12
Spring Mills, 289 EK-1
Spring Mt., 2205 EN-11
Spring Mt., 75 WL-12
Spring Run, 350 WO-14
Springboro, 459 WE-2
Springdale, 3487 WM-5
Springfield, 23677 *E-1
Springfield, EE-5
Springhope, 50 WO-10
Springmont, 650 ET-11
Springs, 300 WQ-7
Springtown, 750 EM-11
Springvale, 120 EP-6
Springville, 400 EF-8
Sproul, 250 WN-10
Spruce Creek, 90 WL-12
Spruce Hill, 40 EM-2
Spry, 4903 WT-14
Stafore Estates, 1200 EA-4
Stahlstown, 300 WO-7
Standing Stone, 60 EF-7
Stanton, 70 WJ-8
Star Junction, 500 WO-5
Starbrick, 800 WE-7
Starford, 250 WL-8
Starkville, EG-9
Starlight, 135 EE-11
Starr WG-7
Starrucca, 224 EE-10
Starview, 100 EP-5
State College, 39419 WK-13
State Hill, 40 EN-8
State Line, 1100 WQ-14
Stateline, 20 WC-5
Steamburg, WF-2
Steel City, 650 EB-5
Steelstown, 100 EN-6
Steelton, 5598 EO-4
Steinsburg, 100 EM-11
Sterling, 150 EH-11
Sterling Run, 120 WH-12
Sterrettania, WD-3
Stevens, 400 EO-8
Stevensville, 250 EF-7
Stewartstown, 1997 EQ-6
Stewartsville, 700 *M-10
Stickney, WD-9
Stifflertown, 120 WL-9
Stiles, 800 EA-2
Stillwater, 203 EI-7
Stiltz, 50 EQ-5
Stockdale, 520 WO-4
Stockertown, 753 EK-11
Stoddartsville, 130 EI-10
Stone Church, 150 EK-12
Stoneboro, 1015 WH-3
Stoneham, 250 WE-8
Stonerstown, 250 WO-11
Stonington, 100 EK-5
Stony Creek Mills, 2500 ET-13
Stony Run, 225 EL-9
Stonybrook, 1500 EP-5
Stonyfork, 150 EF-2
Stormstown, 1602 WK-12
Stormville EJ-11
Stouffferstown, 400 WP-14
Stoughstown, 75 EO-2
Stoverstown, 250 EQ-5
Stowe, 3585 EN-10
Stowell, 60 EG-7
Stoystown, 391 WO-8
Strabane, 1100 WN-3
Strafford, 4500 EP-11
Strasburg, 2777 EP-7
Strattanville, 531 WI-7
Strausstown, 334 EM-7
Strawberry Ridge, 30 EJ-5
Strinestown, 250 EO-5
Strodes Mills, 350 WM-14
Strongstown, 400 WM-8
Stroudsburg, 6294 EJ-12
Studa, 100 WN-2
Stump Creek, 300 WJ-9
Sturgeon, 1000 WM-3
Suedberg, 200 EM-6
Sugar Grove, 555 WD-7
Sugar Hill, 60 WI-9
Sugar Notch, 956 EI-8
Sugar Run, 100 EF-7
Sugarcreek, 4969 WH-4
**SULLIVAN CO., 6556 EE-8**
Summer Hill, 350 EJ-7
Summerdale, 800 ES-2
Summerhill, 477 WN-9
Summerville, 493 WI-7
Summit, 250 WM-10
Summit Hill, 2956 EK-9
Summit Lawn, 420 EC-3
Summit Mills, 175 WQ-8
Summit Station, 208 EM-7
Sunbrook, 300 WC-12
Sunbury, 9819 EK-5
Sunderlinville, 50 WE-14
Sunnyburn, 80 EQ-7
Sunset Valley, 2700 *N-10
Sunville, 30 WG-5
Surveyor, 35 WJ-11
Suscon, 700 EH-9
Susquehanna, 1560 EE-10
**SUSQUEHANNA CO., 42238 EC-11**
Sutersville, 587 WN-5
Swanville, 150 WB-8
Swarthmore, 6090 *F-1
Swedeland, 500 *B-1
Sweden Valley, 100 WF-13
Swedesburg, *A-1
Sweet Valley, 500 EH-7
Swiftwater, 600 EI-11
Swissvale, 8765 *K-8
Swoyersville, 4836 EH-9
Sybertsville, 300 EJ-8
Sycamore, 60 WP-3
Sykesville, 1170 WJ-9
Sylvan, 20 WQ-12
Sylvan Hills, 1000 WB-13
Sylvania, 186 EE-4
Table Rock, 45 EP-3
Tacony *C-6
Tafton, 600 EH-12
Talmage, 325 EP-7
Tamaqua, 6581 EK-8
Tamiment, 125 EI-12
Tannersville, 1200 EJ-11
Tannery, 70 EJ-5
Tarentum, 4534 WL-5
Tarrs, 600 WO-5
Tarrtown, 150 WK-6
Tatamy, 1079 EK-11
Tatesville, 175 WP-11
Taylor, 6151 EH-9
Taylorstown, 250 WO-2
Teepleville, 20 WF-4
Telford, 4599 EN-11
Temple EM-9
Templeton, 400 WJ-6
Ten Mile, 60 WP-3
Terre Hill, 1261 EO-8
Terrytown, 100 EF-7
Thomas, WN-3
Thomas Mills, 50 WO-8
Thomasville, 450 EP-4
Thompson, 275 EE-10
Thompsontown, 688 EM-3
Thompsontown, 30 WK-10
Thornburg, 429 *J-4
Thorndale, 3561 EP-9
Thornhurst, 250 EI-10
Three Springs, 424 WO-13
Throop, 4051 EC-13
Tidal, 75 WJ-6
Tidioute, 718 WF-6
Tillotson, 20 WE-5
Timberly Hts., 400 WJ-4
Timblin, 142 WJ-7
Tingley, 80 EE-9
Tinicum, 100 EM-12
Tioga, 582 EE-3
**TIOGA CO., 41373 EC-3**
Tioga Junction, 100 EE-3
Tiona, 400 WF-8
Tionesta, 567 WG-6
Tipton, 1225 WL-11
Tire Hill, 900 WN-8
Titusville, 5774 WF-5
Tivoli, 25 EH-5
Toby, 100 WH-10
Tobyhanna, 1200 EI-11
Todd, 70 WN-12
Tolna, 40 EQ-5
Tomhicken, 160 EJ-8
Tompkinsville, 200 EG-10
Tomstown, 180 EQ-1
Topton, 1965 EM-9
Torpedo, 30 WE-6
Torrance, 250 WM-7
Torresdale, *C-7
Toughkenamon, 1375 EQ-9
Towanda, 2826 EF-6
Tower City, 1318 EL-6
Townville, 286 WF-4
Trade City, 60 WK-8
Trafford, 3011 WN-5
Trainer, 1831 *B-10
Transfer, 500 WH-2
Trappe, 3347 EO-11
Treichlers, 600 EL-10
Tremont, 1696 EL-6
Trent WP-7
Tresckow, 964 EK-8
Treveskyn, 1200 WN-3
Trevorton, 2010 EK-5
Trexler EF-2
Trexlertown, 1000 EM-10
Trooper, 6061 EO-11
Trout Run, 350 EH-3
Troutville, 239 WJ-9
Troxelville, 192 EK-3
Troy, 1457 EF-5
Troy Center WF-5
Trucksville, 1650 EA-8
Truittsburg, 15 WJ-7
Truman, 75 WG-11
Trumbauersville, 1071 EM-11
Tryonville, 50 WF-5
Tuckerton, 300 ER-12
Tullytown, 1956 *T-12
Tunkhannock, 1773 EG-8
Tunnelhill, 378 WM-10
Tunnelton, 110 WM-6
Turbotville, 645 EJ-5
Turkey City, 100 WI-5
Turnersville, 35 WG-2
Turnip Hole, 20 WI-6
Turtle Creek, 5547 WM-4
Turtlepoint, 90 WE-11
Tuscarora, 939 EK-8
Tuscarora, 60 EM-2
Tusseyville, 30 WK-14
Twin Lakes, 125 EH-13
Twin Rocks, 600 WM-9
Two Taverns, 50 EQ-3
Tyler, 40 WI-10
Tyler Hill, 450 EF-12
Tylersburg, 210 WH-7
Tylersport, 320 EN-11
Tylersville, 225 EJ-2
Tyrone, 5197 WL-11
Uhlerstown, 100 EM-12
Uledi, 250 WP-5
Ulster, 300 EE-6
Ulysses, 631 WE-14
Union City, 3276 WE-5
**UNION CO., 41624 EI-4**
Union Dale, 341 EF-10
Union Deposit, 700 ES-6
Uniontown, 100 WM-9
Uniontown, 11682 WP-5
Unionville, 900 EP-9
Unionville, 100 WJ-4
Unionville, 287 WK-13
Unionville, 350 WK-3
Unityville, 60 EI-6
Universal *J-9
Upland, 2864 EQ-11
Upper Black Eddy, 550 EM-12
Upper Darby, *F-2
Upper Exeter, 150 EH-9
Upper Saint Clair, 20053 WN-3
Upper Strasburg, 225 WP-14
Upton, 100 WQ-13
Urban, 45 EL-5
Urey WK-9
Ursina, 233 WQ-7
Utahville, 160 WL-10
Utica, 193 WG-4
Vail WL-12
Valencia, 537 WL-4
Valier, 275 WK-8
Valley Forge, 1500 EO-11
Valley View, 1677 EL-6
Valley View, 600 EO-7
Valley View, 150 WK-13
Van, 60 WH-5
Van Ormer, 180 WL-10
Van Voorhis, 400 WO-4
Vanderbilt, 517 WP-5
Vandergrift, 4997 WL-6
Vandling, 700 EF-10
Vandyke, 80 EM-2
Vankirk WO-3
Varden, 150 EG-11
Venango, 267 WE-3
**VENANGO CO., 57565 WG-5**
Venetia, 450 WN-3
Venus, 150 WH-6
Vera Cruz, 250 EM-10
Verona, 2842 WM-4
Versailles, 1566 WN-4
Vestaburg, 975 WP-4
Vicksburg, 200 EJ-4
Victory Hts., 250 WH-5
Villa Maria, 150 WI-2
Villanova *C-1
Vinco, 1429 WM-9
Vinemont, 100 EN-8
Vintage, 200 EP-8
Vintondale, 477 WM-8
Vira, 150 EL-1
Virginville, 300 EM-9
Volant, 110 WI-3
Vosburg EG-8
Vowinckel, 50 WH-7
Wagner, 65 EL-2
Wagnerville, EA-6
Wagontown, 500 EQ-9
Wakefield, 160 EQ-7
Walbert, 200 EB-1
Wall, 664 *L-9
Wallaceton, 327 WJ-11
Waller, 55 EI-6
Wallingford *F-1
Walls Corners, 30 EG-9
Wallsville, 50 EG-9
Walmo, 900 WJ-2
Walnut, 115 EM-2
Walnut Bottom, 300 EO-2
Walnutport, 2185 EK-10
Walnuttown, 110 EM-9
Walston, 275 WJ-8
Waltersburg, 275 WP-5
Waltonville, 125 EE-6
Wampum, 623 WK-2
Wanamakers, 55 EL-9
Wanamie, 600 EI-8
Wapwallopen, 150 EJ-8
Warfordsburg, 135 WQ-12
Warminster, 31400 EO-12
Warren, 9403 WE-8
Warren Center, 200 EE-7
**WARREN CO., 43863 WE-7**
Warrendale, 500 WL-3
Warrensville, 300 EH-4
Warrington, 7000 EN-12
Warrior Run, 581 EI-8
Warriors Mark, 350 WL-12
Warwick, 200 EO-9
Washington, 14754 WO-3
Washington Boro, 430 EP-6
**WASHINGTON CO., 202897 WN-2**
Washington Crossing, 850 EN-13
Washingtonville, 187 EJ-5
Wassergass, 200 EL-11
Water Street, 15 WM-12
Waterfall, 100 WO-12
Waterford, 1413 WD-4
Waterford, 300 WN-7
Waterloo, 50 WN-14
Waterman, 140 WM-8
Waterside, 50 WO-11
Waterville, 150 EH-2
Watrous, 150 EF-1
Watsontown, 2102 EI-4
Wattersonville, 150 WJ-6
Wattsburg, 344 WD-5
Waverly, 1000 EG-9
Waymart, 3135 EG-11
**WAYNE CO., 47722 EC-14**
Wayne Hts., 1805 EQ-1
Waynesboro, 9947 EQ-1
Waynesburg, 4147 WP-3
Weatherly, 2588 EJ-9
Webster, 400 WO-4
Webster Mills, 25 WQ-12
Weedville, 950 WH-10
Weigelstown, 10117 EP-5
Weikert, 215 EK-2
Weishample, 100 EL-6
Weissport, 424 EK-9
Weldbank, 100 WF-8
Wellersburg, 162 WQ-9
Wells Tannery, 175 WO-12
Wellsboro, 3242 EF-2
Wellsville, 274 EP-4
Welsh Run, WQ-13
Wenksville, 120 EP-2
Wernersville, 2444 EN-8
Wertzville, 30 EN-4
Wescosville, 1250 EC-2
Wesley, 50 WI-4
Wesleyville, 3299 WC-4
W. Alexander, 301 WO-1
W. Auburn, 100 EF-8
W. Bangor, 30 EQ-7
W. Bingham, 50 WE-13
W. Bristol, 5000 *A-9
W. Brownsville, 1021 WP-4
W. Burlington, 90 EF-5
W. Cameron, 165 EL-5
W. Catasauqua, 700 EA-3
W. Chester, 18315 EP-10
W. Clifford, 50 EF-10
W. Conshohocken, 1495 *B-1
W. Decatur, 600 WJ-11
W. Elizabeth, 516 WN-4
W. Enola, 800 ES-2
W. Fairfield, 60 WN-8
W. Fairview ES-2
W. Finley, 100 WP-2
W. Franklin, 40 EF-5
W. Freedom, 240 WI-5
W. Grove, 2761 EQ-9
W. Hazleton, 3302 EJ-8
W. Hickory, 300 WG-6
W. Homestead, 1997 *K-7
W. Kittanning, 1138 WK-6
W. Lancaster, 800 ES-8
W. Lawn, 1597 ET-11
W. Lebanon, 125 WL-7
W. Leechburg, 1256 WL-5
W. Leisenring, 600 WP-5
W. Liberty, 334 WJ-4
W. Liberty, 70 WJ-9
W. Mayfield, 1079 WK-2
W. Middlesex, 848 WI-2
W. Middletown, 133 WN-2
W. Mifflin, 20669 WN-4
W. Milton, 850 EJ-4
W. Monterey, 75 WJ-5
W. Nanticoke, 1200 EI-8
W. Newton, 2848 WO-5

*, †, ‡, §, ◊ See explanation under state title in this index.

County names are listed in capital letters and in boldface type.

## Rhode Island

Page locator
Map keys A - N | Atlas pages 184 - 185
* City keyed to p. 100



## South Carolina

Page locator
Map keys A - J | Atlas pages 186 - 187
* City keyed to p. 59



## South Dakota

Page locator
Map keys A - J | Atlas pages 188 - 189



## Tennessee

Page locator
Map keys 1 - 10 | Atlas pages 190 - 191
Map keys 11 - 20 | Atlas pages 192 - 193
* City keyed to p. 155
† City keyed to p. 195



*, †, ‡, §, ◊ See explanation under state title in this index.

County names are listed in capital letters and in boldface type.

## Texas

Page locator

| Map keys | Atlas pages |
|---|---|
| WA1 - WJ14 | 198 - 199 |
| WK1 - WT14 | 200 - 201 |
| EA1 - EJ14 | 202 - 203 |
| EK1 - ET14 | 204 - 205 |

* City keyed to pp. 194 - 195
† City keyed to pp. 196 - 197



*, †, ‡, §, ◊ See explanation under state title in this index.

County names are listed in capital letters and in boldface type.

## Vermont

Page locator

| Map keys | Atlas pages |
|---|---|
| A - N | 210 - 211 |



## Virginia

Page locator

| Map keys | Atlas pages |
|---|---|
| 1 - 10 | 212 - 213 |
| 11 - 20 | 214 - 215 |

* City keyed to p. 195
† City keyed to pp. 216 - 217
‡ City keyed to pp. 224 - 225
~ Independent city not included in any county



*, †, ‡, §, ◊ See explanation under state title in this index.

County names are listed in capital letters and in boldface type.

## Washington

Page locator

| Map keys | Atlas pages |
|---|---|
| 1 - 10 | 218 - 219 |
| 11 - 20 | 220 - 221 |

* City keyed to pp. 222 - 223



## West Virginia

Page locator

| Map keys | Atlas pages |
|---|---|
| A - J | 226 - 227 |

* City keyed to p. 95



*, †, ‡, §, ◊ See explanation under state title in this index.

County names are listed in capital letters and in boldface type.

## Wisconsin

Page locator

| Map keys | Atlas pages |
|---|---|
| A - J | 228 - 229 |
| K - T | 230 - 231 |

* City Keyed to p. 109
† City keyed to pp. 232 - 233



## Wyoming

Page locator

| Map keys | Atlas pages |
|---|---|
| A - J | 234 - 235 |



*, †, ‡, §, ◊ See explanation under state title in this index.

County names are listed in capital letters and in boldface type.